Ocean Circulation in Three Dimensions

Notable advances of the last quarter-century have deepened our appreciation of the three-dimensional nature of the ocean's large-scale circulation. This circulation has important implications for ocean chemistry and biology, atmospheric science, and climate. *Ocean Circulation in Three Dimensions* surveys both observations and theories of the time-mean circulation, enabling readers to see the relevance and limitations of leading theories, as well as the patterns linking the behavior of different oceans. The book covers "classical" topics of horizontal circulation, and expands them to include shallow wind-driven overturning, the deep global "conveyor belt", high latitudes, the role of eddies, and the ocean's role in heat transport. Solutions to exercises are available online for instructor use. This textbook is ideal for students of physical oceanography, chemical oceanography and climate. It is also suitable for readers from related fields as it includes a summary of introductory topics.

Barry A. Klinger is Associate Professor in the Department of Atmospheric, Oceanic and Earth Sciences at George Mason University, Virginia. His primary research interests are large-scale ocean circulation and its effects on climate, which he investigates with numerical models and theory. Since the early 2000s he has been affiliated with the Climate Dynamics Doctoral Program at George Mason University.

Thomas W. N. Haine is Professor in the Department of Earth and Planetary Sciences at The Johns Hopkins University. He studies and teaches ocean circulation and the ocean's role in climate and has participated in many oceanographic expeditions. He has pioneered understanding of how the ocean stores and transports trace substances, and is currently investigating how the Arctic and sub-Arctic oceans are responding to, and influencing, climate change.

"Klinger and Haine have created a *tour-de-force* discussion of the quasi-steady ocean circulation. They manage to combine a clear discussion of both basic and advanced theoretical principles, with the complexities of real data, and major points of controversy. At the same time, the reader comes away with an appreciation for the numerous and important regional variations of large-scale oceanic flows."

– Carl Wunsch, Harvard University/
Massachusetts Institute of Technology

"*Ocean Circulation in Three Dimensions* is a new oceanography textbook which combines classical physical oceanography with a modern view of the ocean. The textbook is cleverly constructed to provide lucid physical descriptions of ocean current systems, supported by detailed theoretical derivations – to the benefit of all readers, irrespective of background and experience. This is an expansive and thorough textbook which deserves a place on the bookshelves of all oceanographers."

– Andy Hogg, Australian National University

"Klinger and Haine's writing and thinking is clear, and their take on oceanography is thoroughly modern – featuring dynamics in three dimensions plus time, color images in beautiful map projections and revealing 3D schematics, challenging problems, and a keen take on modern observations by ship, autonomous platform, or satellite. This book will guide young scientists entering oceanography and climate modeling and update those who learned about the oceans the old way, too."

– Baylor Fox-Kemper, Brown University

"*Ocean Circulation in Three Dimensions* offers a refreshingly innovative approach to introductory physical oceanography. The book explores the ocean not just in the sense of three dimensions of motion, but also in the sense of three dimensions of understanding: observations, concepts, and theory. This presentation is well-honed for stand-alone physical oceanography courses or for independent study. Readers should gain from this book a descriptive understanding of how the ocean moves, as well as the conceptual and theoretical framework needed to explain why the ocean moves as it does. The approach has been field-tested over several years in undergraduate and graduate classes at George Mason University, where Barry Klinger teaches, and at Johns Hopkins University, where Tom Haine teaches. The book is imbued with a certain wisdom about what students really need to learn. Open-ended exercises at the end of each chapter eschew the "plug-and-chug" approach and encourage thoughtful reflection on the material. The scope of the book is sufficient for a year-long course sequence, and the authors suggest strategies for subdividing the material between semesters or for selecting a coherent set of chapters for a quarter or semester course."

– Sarah Gille, Scripps Institution of Oceanography,
University of California, San Diego

OCEAN CIRCULATION IN THREE DIMENSIONS

BARRY A. KLINGER

George Mason University,
USA

and

THOMAS W. N. HAINE

Johns Hopkins University,
USA

CAMBRIDGE
UNIVERSITY PRESS

CAMBRIDGE
UNIVERSITY PRESS

University Printing House, Cambridge CB2 8BS, United Kingdom

One Liberty Plaza, 20th Floor, New York, NY 10006, USA

477 Williamstown Road, Port Melbourne, VIC 3207, Australia

314–321, 3rd Floor, Plot 3, Splendor Forum, Jasola District Centre, New Delhi – 110025, India

79 Anson Road, #06–04/06, Singapore 079906

Cambridge University Press is part of the University of Cambridge.

It furthers the University's mission by disseminating knowledge in the pursuit of education, learning, and research at the highest international levels of excellence.

www.cambridge.org
Information on this title: www.cambridge.org/9780521768436
DOI: 10.1017/9781139015721

First published 2019

Printed in the United Kingdom by TJ International Ltd. Padstow Cornwall

A catalogue record for this publication is available from the British Library.

Library of Congress Cataloging-in-Publication Data
Names: Klinger, Barry A., 1964– author. | Haine, Thomas W. N., 1967– author.
Title: Ocean circulation in three dimensions / Barry A. Klinger, George Mason University, and Thomas W. N. Haine, Johns Hopkins University.
Description: New York : Cambridge University Press, [2019]
Identifiers: LCCN 2018037698 | ISBN 9780521768436
Subjects: LCSH: Ocean circulation.
Classification: LCC GC228.5 .K55 2019 | DDC 551.46/2–dc23
LC record available at https://lccn.loc.gov/2018037698

ISBN 978-0-521-76843-6 Hardback

Additional resources for this publication at www.cambridge.org/oc3d.

To Elise Berliner, and to Carrie, Bradley, Sullie, Brooks, and Robert.

Contents

Color plates section can be found between pages 196 and 197

Preface

Scope and Organization of the Book

The general circulation of the ocean is an important and beautiful subject, one in which significant progress has been made in the last few decades. Some of these advances have deepened (appropriately enough) our appreciation of the three-dimensional nature of the circulation. In addition to the traditional description of horizontal flow, the vertical flow, and vertical variations in the horizontal flow, are important aspects of the circulation.

Nonetheless, students who want to go beyond a two-dimensional picture face significant barriers. This is a shame, because the three-dimensional circulation has important implications for sibling fields, including climate, atmospheric science, and ocean chemistry and biology. Scientists in these disciplines are often most interested in how properties are transported and transformed by the ocean. Such properties include heat, carbon, and nutrients. For transport and transformation, slow vertical travel through hundreds of meters of ocean depth can be as important as faster horizontal motion over thousands of kilometers of ocean width.

Many students in the above-mentioned fields take a single course in physical oceanography that introduces observational and dynamical methods and results. Such courses sketch the observed circulation and may include some theoretical topics such as the western intensification of the Gulf Stream. More advanced texts on ocean circulation tend to be rather mathematical. This may deter some students, and often even physical oceanography students work hard on the equations but lose sight of the underlying concepts. Although an appropriate response to difficult material, this habit obscures the deeper links between theory and the observations. Accordingly, we emphasize conceptual explanations for the general circulation here.

The reference to three dimensions in the book's title emphasizes the three-dimensional nature of the ocean circulation, meaning its variations with longitude, latitude, and depth. Equally, it signals our attempt to present three dimensions of knowledge: observations of the ocean, concepts to explain the observations (or try to!), and theories to quantify and justify the conceptual explanations. These dimensions guide the organization of the book, so that individual chapters contain observational, conceptual, and theoretical sections (Figure 0.1). We confine challenging math to the Theory sections, which makes the Concepts and Observations sections accessible to a broader audience. Separating concepts from theory highlights the *ideas* and then shows how they are illuminated by mathematical derivations. Separating observations distinguishes between real ocean phenomena and our

CHAPTER 1 Dynamics & Methods of Physical Oceanography

1 General Circulation	Time & Space Scales	Governing Eqs.
2 Sensors	Models	Solving Governing Eqs
3 Platforms	Equation of State	Conserved Quantities
4 Data Synthesis		
5 Global T & S		
6 Topography		

CHAPTER 2 Rotating & Shallow–Water Flow

1	Coriolis	Geostrophy & Ekman
2	Hydrostasy, Geostrophy	Geostrophic Example
3	Ekman	Vort, SF, Geostrophic Vort
4	Div, Vort, SF	Shallow–Water Eqs
5	Shallow Water & PV	Shallow Water Pot Vort

CHAPTER 3 Two–Dimensional Horizontal Circulation

1 Wind Stress	Uniform–Density	Scale Analysis
2 Gyres	Vorticity of Gyres	Sverdrup Failure
3 WBCs	Vorticity of WBCs	WBCs
4 Vertically Integrated	Sverdrup	Recirculation
5	Sverdrup Limitations	Integral Balances

CHAPTER 4 Surface & Mixed Layer Properties

1 surface properties	Surface Fluxes	Restoring SST & S Flux
2 ML Structure	ML evolution	Freshwater Flux
3		1D ML Model

CHAPTER 5 Depth–Dependent Geostrophic Gyre Circulation

1 Pycnocline Topography	Wind Paradox	Neglecting Mixing
2 Depth Distribution	Ventilated	Elements of Models
3 Density Distribution	Unventilated	Ventilated
4	ML Variations	Unventilated
5	Open Questions	

CHAPTER 6 Upwelling, Overturning, and Equatorial Circulation

1 Equator & STC	EUC	Incomplete Theories
2 Upwell, Downwell	3D Uniform Density	EUC Scaling
3 Eastern Boundary	STC in Models	Meridional SF
4	Gyre + Cell	Layer Models
5		Gyre with STC

CHAPTER 7 Eddies & Small Scale Mixing

1 Mesoscale Eddies	Eddies & Large Scale	Eddy Correlation Flux
2 Small–Scale Mixing	Transport	Diffusion of Tracers
3	Parameterization	Thickness Advection

CHAPTER 8 Deep Meridional Overturning

1 Evidence	Thermohaline Flow	Energetics
2 Horizontal Flow	Equilibrium Constraint	Overturning Scaling
3 Overturning	Overturning Strength	Horizontal Flow Patterns
4 Sinking	DWBCs	
5	Dynamics of Sinking	

Chapter 9 The Southern Ocean Nexus

1 Global Overturning	Global Overturning	Wind/Buoy Scaling
2 ACC	Remote Wind–Driven	Models Eddying ACC
3	Elements of ACC Theory	
4	Sensitivity to Forcing	

Chapter 10 The Arctic

1 Arctic Circulation	Arctic T–S	Eurasian Surface S
2 Arctic T,S, Freshwater	Beaufort Gyre	Transformation
3		Beaufort Gyre

Chapter 11 Heat Flux, Freshwater Flux, and Climate

1 Energy Balance	Heat Transport	Surface Heat Flux
2 Energy Transport	Interative Surface	Two–Box Model
3 Surface Heat Flux	Multiple States	
4 Freshwater Forcing	Model States	
5 Paleo Water Masses		

Appendix A Data Sources

Appendix B Vector Calculus and Spherical Coordinates

Appendix C Notation and Useful Values

Figure 0.1   Outline of the book showing **Observations** (left column), **Concepts** (central column), and **Theory** (right column) for each chapter, along with an illustration of each chapter's topic (far right column).

explanations of them. Throughout, we try to indicate how well-established each concept is. We recognize that the separation into three dimensions of knowledge is sometimes ambiguous and sometimes subjective. Therefore, we attempt to organize and present the material in a coherent way with abundant cross-references.

Another important component of physical oceanography is numerical modeling, which we integrate into the other three categories. Diagnostic models improve our observational understanding, model experiments illustrate conceptual arguments, and realistic simulations test the relevance of mathematical theories to the real ocean.

Though aimed at readers who have already taken a course in oceanic or atmospheric dynamics (who may want to skip or skim the first two chapters), the book can be the core of a two-semester graduate course. In this way, an introductory course might cover all of Chapters 1–3, most of Chapter 5, plus some topics from, say, Chapters 4, 8, and 11. An advanced second semester course might cover Chapters 6, 7, 9, and 10. Many sections, particularly in Observations, are written to be accessible to undergraduates.

The book is based on notes from the General Circulation course that Klinger teaches to students in the Climate Dynamics PhD Program at George Mason University. Similarly, Haine teaches Ocean General Circulation to graduate (and some advanced undergraduate) students at Johns Hopkins University using material in the book. Our philosophy may help broaden enrollments in such classes to include students who might otherwise consider an ocean circulation theory class esoteric or inaccessible. While the book's fulcrum lies in concepts, it includes serious treatments of theory and presents modern observations of the general circulation throughout.

Ocean Circulation in Three Pages

It is convenient to group the chapters into a few **themes** (Figure 0.2). Here we take a tour of both the subject and the textbook.

Foundations

The Foundations chapters set the scene with material that is used extensively throughout the book. They review topics that will be familiar to students who have already studied physical oceanography.

Chapter 1 gives an overview of observational techniques, the mathematical framework for understanding the circulation, and some physical characteristics of the oceans. The ocean has variability on a broad range of time and space scales; we loosely define the large-scale steady circulation to be a multiyear average of features that are at least about 100 km wide. Ship-based, satellite-based, and autonomous instruments observe ocean properties. These observations are then compiled into estimates of the steady state. Circulation is influenced by ocean density, which is calculated from measurements of temperature and salinity. Ocean dynamics is based on a handful of equations representing physical laws. Solutions to simplified versions of the equations give insight into the ocean and for realistic cases the solutions can be approximated with computers (with some caveats).

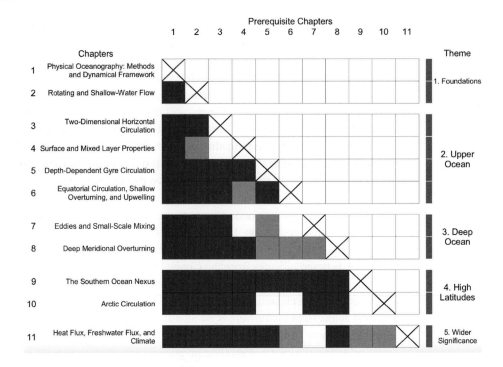

Figure 0.2 Paths through the book, prerequisites, and themes. For each chapter read from left to right to see which other chapters are required (dark gray) or recommended (light gray).

The rotation of the Earth (Chapter 2) influences every aspect of the large-scale circulation. Frictional stress from the wind drives **Ekman transport** at the sea surface. The circulation is generally the sum of the Ekman velocity and **geostrophic velocity**. We characterize two-dimensional slices of three-dimensional fluid motion with the mathematical tools **divergence**, **vorticity**, and **streamfunction**. The effects of density stratification can be approximated by idealizing the ocean as a stack of uniform-density layers governed by an approximation called the **shallow water equations**.

Upper-Ocean Circulation

A primary flow structure in the ocean is the **gyre**: a horizontal circulation loop which spans the ocean in a roughly 2000 km latitude band (Chapter 3). The gyre circulation consists of a slow (order 1 cm/s), broad flow which concentrates into a fast (1 m/s), 100 km wide **western boundary current** such as the Gulf Stream. A theory of wind stress on the surface of an idealized, two-dimensional, uniform-density ocean on a rotating sphere gives reasonable predictions of the location, strength, and other characteristics of the gyres. Several topics in Chapter 3 are also discussed in introductory classes.

Geostrophic gyre flow is strongest in the top kilometer of the ocean, and to understand why, we need to also understand the distribution of small ($<$ 1%) variations in ocean **density**. Chapter 4 begins this analysis by describing surface temperature, salinity, and other properties. Surface properties are strongly influenced by atmospheric interactions with the **mixed layer**, a vertically homogeneous region near the surface that is typically a few tens of meters thick.

The book returns several times to the question of how circulation and subsurface density variations influence each other, and how both are constrained by surface density. The **pycnocline**, a strong vertical density gradient, separates the mixed layer from denser **abyssal water** below a kilometer or so. Chapter 5 shows how linking the undersea "hills" and "bowls" of the pycnocline to the gyre circulations allows us to model the vertical variation in the horizontal circulation. In the **subtropical gyres**, Ekman pumping makes water slide downward along uniform-density (**isopycnic**) surfaces as it flows around the gyre.

Chapter 6 highlights shallow **overturning cells** in which **meridional** (northward or southward) surface Ekman transport is linked via upwelling and downwelling to geostrophic flows in the opposite direction. The cells link neighboring gyres and are strongest in the tropics. The cells are also linked to strong pairs of currents in the upwelling zones of the equator and ocean eastern boundaries. Chapter 6 demonstrates that even the idealization of a purely wind-driven, uniform density ocean can have complex **three-dimensional flow** involving both gyre and overturning patterns. In the real ocean, the geostrophic flow is primarily in the pycnocline.

Deep Ocean Circulation

Density variations create buoyancy forces which can drive circulation even in the absence of wind. An essential feature of a buoyancy-driven steady circulation is cross-isopycnal flow, in which a water parcel's density changes as the parcel travels. Chapter 7 describes two processes which can produce these density changes by mixing. **Mesoscale eddies** are horizontal current loops which typically are about 100 km wide and evolve over several weeks. **Turbulent** diapycnal mixing is caused by three-dimensional time-varying currents with length scales of millimeters to tens of meters and timescales of seconds to hours. The chapter explains how these processes may affect density and other properties of the steady circulation.

Chapter 8 discusses the **deep meridional overturning circulation**, a series of overturning cells linking the abyss to the pycnocline and mixed layer. The deep overturning is typically weaker than flow in the upper ocean, but it has important effects on climate, dominates deep water, and is especially prominent in the Atlantic Ocean. Chapter 8 discusses the processes controlling **deep water formation**, the strength of the meridional cells, and gyre-like horizontal circulations associated with the cells. While turbulence is often a means of dissipating energy, the deep overturning gains its strength from diapycnal mixing.

High Latitudes

Although the entire ocean is influenced by both wind stress and buoyancy forces, the circulation concepts in the prior themes mainly concern wind-driven upper-ocean flow and buoyancy-driven deep-ocean flow. At high latitudes, this division breaks down, so we discuss the Southern Ocean (Chapter 9) and Arctic (Chapter 10) separately.

Chapter 9 discusses the **Antarctic Circumpolar Current**, the meridional **Deacon Cell**, and the deep overturning links between the Southern Ocean and the Atlantic, Indian, and Pacific Oceans. Wind-driven geostrophic flow (Chapter 3), wind-driven meridional overturning (Chapter 6), turbulence and mesoscale eddies (Chapter 7), and the deep overturning circulation (Chapter 8) all play key roles. Moreover, northward Ekman transport from wind stress in the Southern Ocean may help drive deep meridional overturning extending far into the northern hemisphere.

Similarly, a mix of topics is difficult to disentangle in the circulation within the Arctic and in the Arctic's exchange of water with the Atlantic. The **cryosphere**, particularly sea ice, also plays an especially big role in the Arctic. Chapters 9 and 10 are less definitive than previous parts of the book, reflecting oceanographers' less complete understanding of high latitude dynamics.

Wider Significance

The ocean circulation is one component of the **global climate system**. The circulation is especially important for transporting heat around the globe, cooling the atmosphere in some places and warming it in others. Chapter 11 discusses how the circulation affects, and is affected by, **heat and freshwater transport** (flux). It describes the mutual influence of ocean surface properties and the atmospheric state. Geological data indicate that the Earth's climate state, including Atlantic overturning, often changed abruptly (on geological timescales), and theory shows how the circulation may fall into one of several states for given external constraints. The physical climate itself is part of the Earth System, which includes ocean and land biogeochemistry, land surface characteristics, the cryosphere, and even human intervention, but that is a subject for another book.

Exercises

To deepen their understanding, students need to work through problems. Exercises at the end of each chapter attempt to guide the learning process. They are progressive and graduated in terms of difficulty. As with the layout of each chapter, they address Observations, Concepts (including quantitative estimates), and Theory. Many of them are open-ended and provide material for class discussion and term papers.

To complete some Exercises, use of high-quality data analysis, mathematical, and plotting software is essential. This reflects the nature of the field, which uses analytically unsolvable equations to model structures which must be characterized by data sets that are too large to process by hand. Many programming languages are adequate for the task, but MATLAB and Python are two that are widely used by physical oceanographers and that can be easily integrated with the TEOS-10 software package used to compute thermodynamic properties of seawater. Calculations for the exercises do not overwhelm the computing resources of a 2018 laptop, however, and none of them involve solving the equations of motion with numerical circulation models. To help students get started, several problems include oceanographic datasets, and links to sites containing them, plus simple template MATLAB scripts.

Advice applicable to all Exercises is as follows. First, each problem is associated with specific sections of the text, which are important to identify. Second, Appendix A summarizes useful datasets while Appendix C contains nomenclature and useful characteristic values of important quantities. Finally, students should make, reflect on, and clearly state, suitable approximations where necessary. The wisdom of knowing what to include, and what to neglect, shines through the conceptual explanations in our field. Cultivating that wisdom, sometimes bewildering to neophytes, is an important aim of the Exercises.

The Exercises have been field-tested by many students at George Mason, Johns Hopkins, and Oxford Universities. Comments on the Exercises and suggestions for more should be addressed to the authors. Model answers are available to instructors on the book's companion website.

Acknowledgments

Barry Klinger thanks his George Mason colleagues, especially for discussions with and encouragement from Tim DelSole, Jagadish Shukla, Jim Kinter, Jennifer Adams, and Laurie Trenary, and Climate Dynamics students Gary Bucher, Hua Chen, Ioana Colfescu, Oluwayemi Garuba, Michael Gelfeld, Olivia Gozdz, Abheera Hazra, Liwei Jia, Yan Jin, Keri Kodama, Emerson LaJoie, Nick Lybarger, Jyothi Nattala, Xiaoqin Yan, and Liang Yu; and finally for discussions with Joe LaCasce and Alex Yankovsky. Benjamin Klinger helped with graphics for the outline figure. Barry Klinger also thanks his parents, Linda and Rudy Klinger, sister Elise Klinger, children Benjamin and David Klinger, and his wife, Elise Berliner, the star to his wand'ring bark.

Tom Haine thanks students and colleagues including: Mattia Almansi, Julia Chavarray, Alex Fuller, Anand Gnanadesikan, Chris Holder, Stephen Jeffress, Michael Karcher, Richard Kelson, Grace Kim, Torge Martin, Georgy Manucharyan, Sarah Ragan, Eshwan Ramudu, Bert Rudels, Atousa Saberi, Michael Spall, Molly Syme, Jordan Thomas, David Trossman, Takamasa Tsubouchi, Olga Tweedy, Amandeep Vashisht, and Darryn Waugh. Tom was Morton K. Blaustein Chair and Professor of Earth and Planetary Sciences while writing this book. In addition, he was supported by Johns Hopkins University and the National Science Foundation. It would have been impossible without the love and support of his family, and in particular Carrie.

Matt Lloyd at Cambridge University Press encouraged us to write. Zoë Pruce skillfully guided the book through production. For insightful reviews and feedback, both authors thank Tim DelSole, Baylor Fox-Kemper, Sarah Gille, Andy Hogg, Mark Inall, Helen Johnson, David Marshall, Jay McCreary, Kelvin Richards, Bert Rudels, Geoff Vallis, Ric Williams, and Carl Wunsch. The authors are solely responsible for errors.

Physical Oceanography: Methods and Dynamical Framework

Physical oceanography extends well beyond the study of the general ocean circulation. Physical oceanographers investigate tides, wind-driven surface waves, air/sea interaction, sound in the ocean, light in the ocean, and physical processes at the sea bed, on beaches and on man-made structures like breakwaters and ships. Physical oceanography is also key to chemical and biological oceanography, and marine geophysics. Through its coupled interaction with the atmosphere, cryosphere, biosphere, and land, the physical state of the ocean also plays a critical part in climate dynamics and Earth system science.

Nevertheless, study of the ocean general circulation is central to physical oceanography and is playing a central role in the development of the field. This development involves an interplay between physical observations of the ocean, theories of the fluid dynamics, and numerical models of the circulation. The focus in this book is on the time-averaged, global ocean currents. In order to approach this topic we must set the stage by reviewing some preliminary ideas about physical oceanography. We are selective in the coverage, only discussing topics that recur later.

For comprehensive discussions of empirical methods in physical oceanography, consult the books by Emery and Thompson (2001) or Wunsch (2015). For more details on dynamical oceanography consult Vallis (2006), Marshall and Plumb (2008), or Pedlosky (1996), among others. For reflections on the field's history, Jochem and Murtugudde (2006) and Mills (2011) are good starting points .

1.1 Observations

1.1.1 What Is the General Circulation?

The **ocean general circulation** is a description of the three-dimensional velocity field of the ocean, $\mathbf{u}$. It is difficult to define the concept more exactly, or to precisely distinguish it from the rest of physical oceanography. Nevertheless, it essentially means *the slowly changing, large-scale velocity field*. By "slowly changing" and "large-scale" we usually mean timescales of seasons and longer, and length scales ranging from about 100 to 10,000 kilometers.

In discussing the general circulation, we include other key physical properties such as temperature, salinity, density, sea-surface height, and pressure. These properties are intimately related to the dynamics and thermodynamics which determine the velocity field itself. Physical oceanographers study how all these physical properties vary from place to place, and analyze physical principles which underlie the ocean behavior of different regions of the ocean. Application of these principles leads us to try to understand the ocean circulation on global scales.

Most of this book discusses an abstraction known as the **steady** circulation. We know that the circulation is not steady and actually varies on a wide range of timescales. For instance, observations of the eastward component of velocity in the equatorial Pacific show variations from year to year, (Figure 1.1a), month to month (Figure 1.1b), day to day (Figure 1.1c), and even within an hour (Figure 1.1d). Despite this temporal variability, Figure 1.1 shows that the current (the equatorial undercurrent) at that particular location is most often eastward with strength of around 0.6 to 0.9 m/s (Subsection 6.1.1).

How does this current compare to flow at other locations? The most intuitive way to compare is simply to take a **time average** over a long time. How long that time should be is not well defined, because, as we discuss in Subsection 1.2.1, ocean characteristics vary from seconds to geological timescales. As with the short time variations shown in Figure 1.1, sporadic measurements have indicated that many physical ocean features are stable over the last century or so. In that case, averaging over any time interval from a few years to a few centuries will give a similar answer, so the precise averaging interval is not so important. However, the more precisely we try to define the circulation, the more sensitive our answer is to the averaging interval.

Such a long-term average often produces what is known as a **climatology**. It is analogous to the original meaning of **climate** as a description of long-term averages of atmospheric characteristics. The atmospheric meaning of climate often includes physical ocean characteristics because these play an intimate role in determining atmospheric conditions.

Sometimes the long-term average circulation is considered a synonym of the general circulation, but a broader definition usually includes some measure of time variability. Like the atmosphere, the ocean has many features that strongly depend on the time of the year, so a climatology is often defined as the average (over many years) of the evolution of a variable over the course of the year. For other timescales, it is the *statistics* of the variability rather than the detailed evolution over time that we usually consider part of the general circulation. For instance, we think of the mean value and the standard deviation of the time series of the equatorial undercurrent data (see the histogram in Figure 1.1).

Similar issues arise when we consider spatial variations in the general circulation. Ocean parameters such as current speed vary on all length scales from millimeters to the width of the basin (Subsection 1.2.1). Taking averages over multiple years tends to eliminate features with horizontal scales less than a few hundred kilometers. Larger scales are affected by smaller-scale steady structures in some regions and by the statistics of time-varying small-scale structures in many areas (Chapter 7).

Like all physical science, physical oceanography is an empirical discipline, but one in which performing controlled experiments is generally impossible because of the great scale

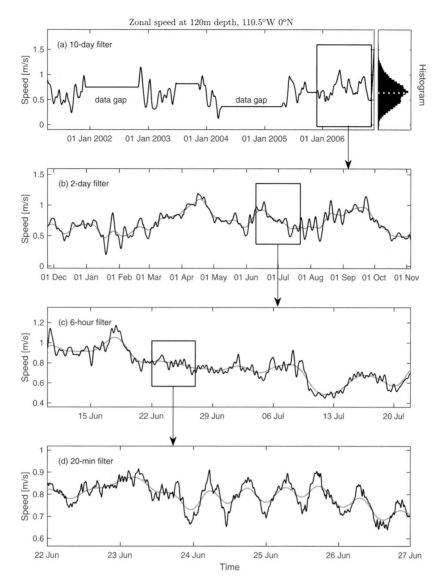

Figure 1.1 Time series of zonal current at 120 m depth in the equatorial Pacific ocean. Several different periods are shown, ranging from days (bottom) to years (top). At top right is a histogram of the speed data with the mean shown by the dotted line. Data are from the OceanSITES Tropical Atmosphere Ocean/Triangle Trans-Ocean Buoy Network moored buoy array (Appendix A).

of the ocean and ethical considerations. Also, we do not have additional oceans to tinker with! These limitations are shared with other fields in Earth and space sciences. Our knowledge comes from careful observations of the natural system evolving in its own way. However, laboratory experiments in rotating tanks of water provide analogues to ocean phenomena upon which we can conduct experiments.

Table 1.1 Comparison of sensor technologies for observing physical properties of the ocean.

	Property	Unit	Accuracy	Sensor
In-situ properties:				
p	pressure	Pa, db	0.001–0.1 db	strain-gauge sensor
T	temperature	°C	0.001 °C	thermistor
S	salinity	(none)	0.005	conductivity sensor
ρ	density	$\mathrm{kg\,m^{-3}}$	$0.01\ \mathrm{kg\,m^{-3}}$	(from T, S, p)
u	velocity	$\mathrm{m\,s^{-1}}$	$O(0.01)\ \mathrm{ms^{-1}}$	current meter
χ	tracer	$\mathrm{Mol\,kg^{-3}}$	varies	sensor or lab. analysis
Remote-sensed sea-surface properties:				
η	height	m	3 cm	radar altimeter
T	temperature	°C	0.7°C	IR & microwave radiometry
S	salinity	g/kg	0.1–0.2	microwave radiometry
Navigational position:				
H	water depth	m	$O(1)$ m	Echo sounder, satellite
p	pressure	Pa, db	0.001–0.1 db	strain-gauge sensor
ϕ	longitude	degrees	5 m	Global Positioning System
θ	latitude	degrees	5 m	Global Positioning System

1.1.2 Sensors to Observe the Ocean

Observations are a key part of physical oceanography. Here we summarize how to make physical observations of the ocean. First, we discuss sensor technologies, then we discuss instrument platforms. Table 1.1 summarizes this information. The notation we introduce here is used throughout the book, and is also summarized in the tables in Appendix C, which contains a useful summary of numbers relevant to physical oceanography. Appendix A is a reference for the datasets we show throughout the book.

Sensors to observe the ocean are diverse and constantly being improved. Our brief discussion splits sensors into two categories: in-situ sensors, which are physically in contact with the water, and *remote* sensors, which observe the ocean from above or by acoustical methods.

In-Situ Sensors

In-situ instruments measure physical properties of seawater by direct immersion in the sea. The most important properties measured this way are:

Pressure (p, see Table 1.1) is the force per unit area exerted by seawater at a particular place in the ocean and is a key quantity in understanding currents in a fluid. The force is measured perpendicular to a specific flat surface, but it does not depend on the orientation of that surface and is thus a scalar. (More generally, pressure arises from the normal components of the stress tensor in continuum mechanics. Detailed knowledge of tensors is not required for this book, although they are used in Chapter 7; Griffies, 2004 provides details.) We can think of pressure as the force of molecules banging against each other. Pressure does not have a direction because pushing in one direction on a volume of water causes the force to be transmitted in all directions. This is illustrated by a vertical tube of water, in which the pressure is due to the downward force of gravity but the water exerts a pressure force on the inside of the tube, perhaps even breaking the tube if the force is strong enough.

Various units of pressure are commonly used. The SI unit is $N\,m^{-2}$ or Pascals (Pa; $1\,N\,m^{-2} = 1$ Pa), but oceanographers often use decibars (db, dbar; 1 tenth of a bar, where 1 bar $= 10^5\,N\,m^{-2}$). A pressure sensor is typically made with a silicon strain gauge or an oscillating quartz crystal. Modern instruments have accuracies of around 0.02–0.1% of their full design range of pressures, and full-ocean depth coverage is possible. Long-term sensor stability is about 0.02–0.1% of full-range pressure per year. Although this is a small fraction, it corresponds to a water column uncertainty of a few meters each year, which is a large dynamical signal (we explain why in Subsection 2.2.2).

The value of the pressure is given accurately by the **hydrostatic balance**, which equates the seawater pressure to the weight of water overhead (see Subsection 2.1.2). The hydrostatic balance is an excellent approximation for the large scales we focus on in this book (Section 1.1), though it breaks down for phenomena characterized by length scales of less than a few kilometers and timescales of less than a day. Observational oceanographers usually use pressure – which is easily measured – as a surrogate for depth – which is not. Because density is nearly constant in the ocean (Subsection 1.2.3), converting pressure to depth is straightforward: roughly speaking, 1 db corresponds to a vertical water column of 1 meter. Sea-level atmospheric pressure is typically close to one "standard atmosphere," which is $101,325\,N\,m^{-2}$, so 1 db$\approx$1 m$\approx$10% of a standard atmosphere. In other words, the weight of 10 m of seawater causes a pressure equal to about 1 atmosphere (a fact familiar to all scuba divers). Ignoring the small variations in density, and some other subtleties concerning hydrostatic pressure, can cause errors of up to a few meters in equating depth and pressure. These errors are small compared to the depths of most features we will discuss, which are typically hundreds of meters or more.

In-situ temperature (T, Table 1.1) is measured using a thermometer in contact with the water. Typically, a thermistor (a thermal resistor) is used in oceanography. Mercury-in-glass thermometers were widely used before the era of micro-electronics began in the 1960s, but are now mainly museum pieces.

Moving a seawater sample deeper without exchanging any heat with the surroundings (an **adiabatic** movement) increases its pressure, and hence slightly compresses the sample. This squeezing does work on the sample and therefore increases its temperature. Similarly, lifting a sample to the surface decreases the pressure, causing the sample to expand and therefore cool. The atmosphere behaves similarly but with greater temperature change because air is much more compressible. **Potential temperature** (θ) is the temperature a sample would have if it were adiabatically moved to a **reference pressure**, typically zero pressure, taken to be at the sea surface. Potential temperature allows easy comparison of temperatures for water parcels at different pressures. Most often we want to know how the temperatures would differ if the parcels were brought to the same pressure as each other, so in this book we use potential, rather than in situ, temperature. Potential temperature θ is calculated from T using an empirically derived formula that mainly depends on T and p, and is a few tenths of a degree Celsius cooler for full ocean depth, as seen in Figure 1.6a. Where the context removes ambiguity, we use T to mean potential temperature rather than θ. Potential temperature is now being superceded by **conservative temperature** Θ (Subsection 1.2.3), which is essentially similar for our purposes.

Salinity (S, see Table 1.1 and Subsection 1.2.3) measures the concentration of dissolved salts in seawater. Salinity varies in the ocean due to concentration (from evaporation) and dilution (from precipitation and river runoff), but the abundance of different salts (sodium chloride, potassium chloride, etc.) relative to each other is nearly constant throughout the ocean. The most recent definition of salinity, called the **absolute salinity**, S_A, is the mass of dissolved salts per mass of seawater (Equation 1.7). Thus, $S_A = 35$ g/kg implies a concentration of 35 g salts in 1 kg of seawater. Much of the oceanographic literature uses the 1978 practical salinity scale (pss-78), which is based on seawater electrical conductivity and measured in practical salinity units (psu) or parts per thousand ("ppt," "per mille," or "‰"). Here we ignore the numerically small differences between absolute salinity and practical salinity.

Salinity is measured by oceanographers using an immersed conductivity cell. A thermistor and a pressure gauge for simultaneous temperature and pressure measurement are also often needed. Such observations of temperature and salinity at different pressures are called **hydrographic** observations (literally, mapping of water properties). Laboratory calibration of a sensor using standard seawater solutions is required for best results. A key challenge for long-term deployment of conductivity cells is to avoid fouling of the cell by biological material. The long-term drift of modern salinity sensors is 0.05 g/kg or better over several years, however.

Seawater **density** (ρ, see Table 1.1, Subsection 1.2.3) is the mass of a seawater sample divided by its volume, and is a critical variable in physical oceanography. Density is not measured directly at sea. Instead, density is inferred from the temperature, salinity, and pressure of a seawater sample using an empirical formula called the **equation of state**. The formula is estimated experimentally using laboratory samples of seawater and highly accurate analytical instruments. The accuracy of density estimates using in-situ temperature, salinity, and pressure measurements is around 0.01 kg m^{-3}, the main source of error being uncertainty in the equation of state itself.

Velocity (**u**, see Table 1.1) is the current vector of the fluid flow, and is measured directly with a **current meter**. The many types of current meter can be broadly categorized as mechanical, acoustic, or electromagnetic devices. Mechanical current meters often consist of a rotor whose spin is proportional to current speed, and a vane to turn the meter to the current direction which is measured by a compass. Mechanical current meters have been used for many decades (see Figure 1.2) and are often deployed on moorings (Subsection 1.1.3) which are designed to hold the current meter at a fixed position in the water. A modern mechanical current meter (for example, the Aanderaa RCM-8) can measure currents from 0.02 m/s (below which the rotor stalls) to about 3 m/s, with an accuracy of the greater of 0.01 ms^{-1} and 4% of the measured current.

Figure 1.2   A mechanical current meter constructed by Vagn Walfrid Ekman in the early twentieth century. The essential design, such as the rotor and vane, is still in use today. From the Frammuseet, Oslo.

Many modern current velocity measurements are made with **Acoustic Doppler Current Profilers** (ADCPs), which exploit the Doppler effect on high-frequency sound waves reflecting off plankton and other suspended material drifting in the current. The shift in frequency of the reflected sound yields the speed of the flow. The ADCP uses a range of return times from sound radiated in different directions to estimate the three-dimensional velocity vector over a range of depths. Ranges of several hundred meters, resolution of a few meters, and accuracies of a few cm^{-1} are common. Figure 1.1 shows data from an ADCP attached to a mooring (Subsection 1.1.3) which was recovered and redeployed several times over the decade.

Some current meters are electromagnetic and work on Faraday's principle of induction for a moving conductor (seawater) in the Earth's magnetic field. This type of instrument is well adapted to turbulent near-shore environments, in the presence of surface waves, for example, and regimes with very low or very high currents. They are not often used for current measurements in deep water environments, however.

Finally, we refer to a whole class of in-situ methods as **serendipitous current measurements**. These include mariner's reports, tracking the location of drifting objects, either at the surface or at depth, and interpreting the spread of natural, or deliberately released, chemical **tracers**, such as dissolved oxygen gas. The surface ocean circulation was described in a basic way using these methods long before systematic scientific instruments were deployed. Indeed, chemical tracers are still used to give information about deep currents. Some kind of "clock" is needed so that changes in tracer concentration from place to place, or time to time, can be associated with the ocean current. For example, radioactive decay of carbon-14 shows that the deep Atlantic is much "younger" than the deep North Pacific. Man-made compounds like chlorofluorocarbons (CFCs) with known transient atmospheric time histories are another example, and have been used since the 1980s to trace deep and abyssal flows. A large database of such tracer measurements now exists for the global ocean (see Figure 1.4). Interpreting the tracer data in an unambiguous way is not always easy, but they provide information on ocean circulation that complements that from hydrography and current meters.

Remote Sensors

Remote sensors operate by probing the ocean with some sort of wave. Acoustic waves are a good example because the ocean is transparent to sound. An ADCP (see above in this subsection) exploits this fact to measure currents at a distance from the instrument, and an echo sounder measures water depth this way. Acoustic networks have also been deployed to monitor average ocean temperature over basin and even antipodal scales. These networks have not yet been operated in an ongoing mode, however. See the book by Munk et al. (1995) for more details (and Wunsch, 2015; Dushaw and Menemenlis, 2014); we do not discuss acoustic methods further here.

The ocean is opaque to electromagnetic waves, which means that remote sensors observing the ocean from satellites (or aircraft) are generally limited to measuring surface properties. Nevertheless, observing the ocean from space using satellites has revolutionized physical oceanography in the last 30 or 40 years. Satellites provide much more comprehensive coverage in space and time than in-situ instruments. The most important satellite *remotely sensed* properties are as follows:

Sea-surface temperature (SST) has been measured over the globe from space since 1970. The technique involves observations of infrared radiation from the Earth's surface in multiple wavebands. There are some drawbacks, however. First, the infrared radiation observed by the satellite is emitted from a thin surface layer 10–20×10^{-6} m thick. This layer is colder than the bulk liquid by $O(0.1)°C$, but the difference varies with environmental conditions and can be hard to estimate (see Subsection 4.1.1). More importantly, clouds block infrared radiation from the sea surface, so many SST measurements are obscured. Some satellites avoid this problem by looking through cloud with microwave radiation rather than infrared, but the spatial resolution and data accuracy are not as good.

Nowadays, data centers provide daily global SST maps based on merged observations from multiple satellites. The coverage is at a resolution of up to about 5 km with an accuracy of about $0.7°C$. These data are invaluable to oceanographers and climate scientists, and SST is the best-measured parameter in marine science. An example of a satellite SST map is in Figure 4.1.

Recently, remote sensing of **sea-surface salinity** (SSS) has become practical through the measurement of microwaves emitted by the sea surface combined with SST and surface roughness measurements. One of the first major missions, Aquarius/SAC-D, was operated from 2011 to 2015 and achieved 0.1–0.2 g/kg accuracy with a spatial resolution of 100–300 km and a repeat cycle of 7–30 days. Aquarius collected as much salinity data as the entire historical data base many times over and is becoming an important

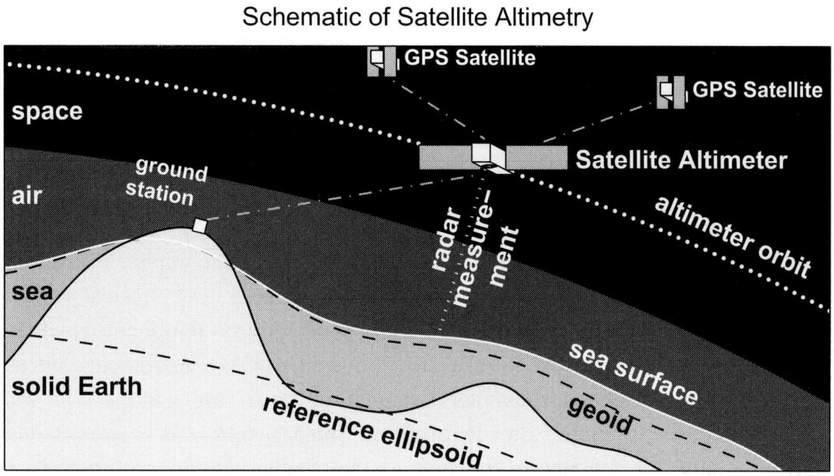

Schematic of Satellite Altimetry

Figure 1.3 Schematic of SSH measurement with TOPEX/Poseidon satellite altimeter. Image from NASA Jet Propulsion Laboratory.

new data source to oceanographers. An example of a sea-surface salinity map is in Figure 4.4.

Sea-surface height (SSH, see Table 1.1 and Figure 1.3), sometimes called ocean surface topography, is measured by an altimeter, either flying on an aircraft or a satellite. The instrument uses radar to measure the distance between itself and the sea surface. Given accurate information about the instrument position, the height of the sea-surface can be estimated to within a few cm. Every ten days, altimeters TOPEX/Poseidon (1992–2005) and Jason 1 and 2 (2001 to present) have been covering most of the global SSH field with 7 km resolution along tracks that are 315 km apart at the equator.

Sea-surface height is a critical quantity to physical oceanographers because it is directly connected to the surface ocean current via the simple **geostrophic relation** (see Subsection 2.1.2). In many places, the geostrophic current is a good approximation to the time-average ocean velocity (see Subsection 2.2.1). Over the globe, sea surface height varies over 3–4 m due to ocean currents.

Sea surface height is defined with respect to the **geoid**, which is the surface of equal gravitational potential, meaning that Earth's gravitational force is perpendicular to the geoid. Sea level would assume the shape of the geoid in the absence of ocean currents and other forces, such as those that generate the tide. The geoid itself is a complex surface that has peaks and troughs of $O(10–100)$ m because Earth's gravity field varies from place to place. SSH variability is relatively easy to measure to an accuracy of a few cm, because the geoid is generally constant on the decadal timescales of satellite missions (though tides must be eliminated, among other corrections). Uncertainties in the exact shape of the geoid have limited the ability of altimetry to estimate long-term averages of SSH which are most relevant to this book, however. Uncertainties are particularly large in determining geoid features on a scale of 100 km or less. Only in the past decade have such uncertainties also been reduced to a few cm.

1.1.3 Platforms to Observe the Ocean

Understanding how oceanographers glean information about the general circulation does not rely simply on understanding oceanographic sensors. We must also appreciate the different ways that sensors are deployed in practice, and there is a tremendous diversity of ocean observing *platforms* on which to deploy sensors. The most important platforms are as follows.

The **ship** is the traditional platform for oceanographic measurements. The ship lowers instruments on a cable in order to measure how physical variables depend on depth at a given latitude and longitude. A **conductivity–temperature–depth (CTD)** instrument is commonly used to measure T, S, and p. CTD instruments are usually mounted on a "rosette" of water samplers arranged in a circle on a frame. The water sampler is called a "Niskin bottle" after its inventor, Shale Niskin, and is an open plastic tube that floods with water as the rosette sinks. A mechanism closes the tubes in order to capture water samples at separate depths. Chemical concentrations, including salinity (to calibrate the CTD conductivity measurements), and trace chemicals, such as oxygen and CFCs, can then be measured from the water sample in a laboratory, often onboard. Other instruments such as ADCPs can also be mounted on the rosette frame or attached to the ship's hull.

Ships can also take measurements while underway. Expendable instruments, such as an **expendable bathythermograph**, and more recently an expendable CTD, can be dropped off a ship and measure a vertical profile of temperature and salinity. The data are transmitted up a thin wire, typically several hundred meters long. The wire eventually breaks and the instrument is abandoned to fall to the seafloor. Many more T observations are taken with expendable bathythermographs than with CTDs, though the expendable bathythermograph has an inferior accuracy of only about 0.1°C and correcting for temperature biases can be challenging. Ships also sample meteorological variables such as sea-level pressure and wind speed, SST (often estimated from the ship's intake of engine-cooling water), and, less frequently, surface salinity. Until the mid-twentieth century, surface temperature measurements were made with a bucket lowered over the side, which is less accurate and has biases compared to more modern techniques.

Autonomous vehicles are devices that are released into the sea. They carry in-situ instruments to measure the properties of the water and deliver their observations to satellites from the sea surface, or acoustically to a data logger while submerged. Autonomous vehicles also provide important information about currents via successive fixes on their positions to yield their displacements over a known period of time.

Three important types of autonomous vehicle are drifters, floats, and gliders. There exist many other types of robotic underwater and surface vehicles with interesting characteristics, but they have been (so far) less relevant for studying the general circulation.

Drifters float at the sea surface and drift with the currents. They measure surface currents and SST, and sometimes salinity, atmospheric pressure, and wind. The Global Drifter Program coordinates drifter deployment and assembles the data. It maintains a fleet of 1250 drifters distributed throughout the global ocean.

Floats are devices that float below the surface at a specified depth, typically 1000 m. They carry CTDs and periodically rise to the surface (for example, every ten days) to make a profile of temperature and salinity. At the surface the float position is fixed to determine its displacement since last contact and the data are telemetered to satellites. The **Argo** program is a multinational collaboration to operate such floats and process their data. It began in the early 2000s and now maintains a fleet of around 3000 floats distributed around the world.

Gliders are autonomous devices that do not passively drift with the current. Instead they glide from place to place, at about $0.5\ \mathrm{ms^{-1}}$, exploiting the hydrodynamic forces on the glider's hull as it rises or sinks through the water column. Gliders carry CTDs to measure temperature and salinity. Position fixes and knowledge of the hydrodynamic characteristics of the device allow currents to be estimated too. Gliders are controlled via internet and satellite links and can be programmed on a daily basis. This flexibility allows researchers to actively explore the ocean from their offices. This new innovative technology is ideally suited to studies of regional processes, and will likely continue to grow.

In recent years, small autonomous temperature and pressure sensors have been attached to pinnipeds (usually seals). In high latitudes these data provide very useful observations in sea ice and throughout the year. With traditional instrument platforms (ships, floats) these areas have been inaccessible.

Moorings are devices to suspend instruments in the water at a fixed position. They consist of a heavy anchor attached to a long wire that is held taut by buoyant spheres at the top. A release mechanism can be acoustically triggered, allowing the anchor to separate from the wire for recovery by a ship. Instruments are clamped to the wire at various depths. Current meters are often deployed this way (Figure 1.2). Devices to periodically crawl up and down the wire, called **profilers**, are also now common. Profilers often carry CTDs and sometimes ADCPs too. Deep-ocean moorings are deployed for one or two years, usually as part of an array of several moorings designed to intercept a strong current system. Much of what we know about deep ocean circulation comes from current meters on mooring arrays, and the technique has been used for many decades in various forms. Nevertheless, the extant current-meter mooring database is not large, and current meters vastly under-sample the global ocean circulation. Moreover, current-meter moorings sample the ocean with a heavy bias towards narrow, fast currents, rather than a random sampling of the ocean circulation.

Navigation – that is, determining one's position and orientation in space – is an important task for all instrument platforms. For physical oceanography this means knowledge of longitude ϕ, latitude θ, and depth (or height z, which increases upward and is often referenced to zero at or near the sea surface).

In the past, latitude and longitude far from shore were determined by astronomical observations and accurate timekeeping. Since 1994 the satellite-based Global Positioning System has provided latitude and longitude to about 5 m accuracy. Depth of the sea floor can be measured acoustically using an echo sounder, or by altimeters from space by analyzing variations in the geoid, which are caused by changes in ocean depth (Subsection 1.1.2). The depth of submerged instruments is measured using pressure as a surrogate for depth, which gives the position relative to the sea surface within O(0.001–0.1) m when the density profile above the instrument is known. This determination is still not accurate enough for all purposes, however, because sea-level variations as small as a few centimeters with respect to the geoid are dynamically significant (see Subsection 1.1.2). Measuring depth with this accuracy is the unsolved navigation problem of the modern era.

1.1.4 Synthesis of Ocean Data

Collecting ocean data is technically demanding, but our main observational uncertainty is not the instrumental accuracy. Instead, it concerns the challenge of constructing a representative picture of the ocean from sparse individual measurements, given variability in time (as in Figure 1.1) and space. The averaging discussed in Subsection 1.1.1 is only accurate if enough data are taken at each point so that the average is not biased by a short-term fluctuation. For some variables, such as velocity in a strong permanent current or vertical profiles of temperature in the tropics, the fluctuations are small enough that a few measurements give an estimate with error bars much smaller than the average. For many variables, however, such as broad mid-ocean currents, fluctuations can reach an order of magnitude larger than the long-term time mean, and so an accurate average requires many individual measurements.

Vertical T and S profiles are called **stations** or sometimes **casts**. Profiles taken from ships are usually arranged in linear **sections** (measurements along a single transect). Floats and drifters gather data along Lagrangian tracks (moving with the current). In 2017, the US National Centers for Environmental Information (NCEI, see Appendix A) possessed data extending to depths greater than 1000 m at about 2×10^6 locations distributed as shown in

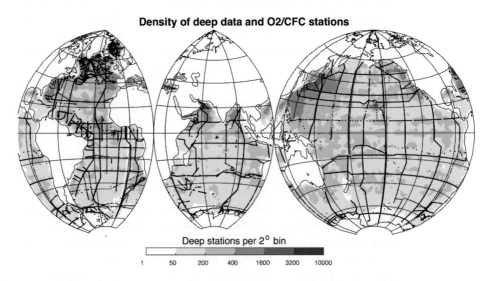

Density of deep data and O2/CFC stations

Deep stations per 2° bin

| 1 | 50 | 200 | 400 | 1600 | 3200 | 10000 |

Figure 1.4   Coverage of deep (> 1000 m) temperature and salinity data in 2017, as found in the US National Oceanographic Data Center (Appendix A). The shading shows the total number of stations in each $2 \times 2^\circ$ bin. The dots show the stations where, in addition to T and S, dissolved oxygen and chlorofluorocarbon data are available.

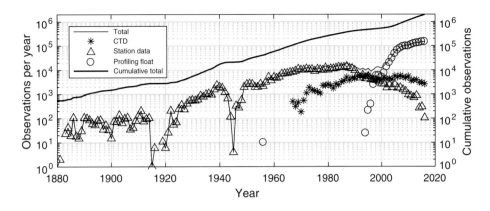

Coverage of deep (> 1000 m) temperature and salinity data over time using the database maintained by the US National Oceanographic Data Center.

Figure 1.4. Dissolved oxygen and chlorofluorocarbon observations are more scarce, with fewer than 2×10^4 stations (Figure 1.4).

Many of the sections were collected during global observing programs like the Meteor expedition of the 1920s, the International Geophysical Year (1957), and the World Ocean Circulation Experiment (WOCE) of the 1990s. Ship-based measurement density is relatively high in the North Atlantic and Pacific and much lower in the Southern Ocean. Coverage from the Argo program (Subsection 1.1.3) is more uniform, although Argo profiles typically extend to only 2000 m, not the sea bed. These initiatives have been augmented by regional observing programs, and by a large number of expendable bathythermograph and surface measurements from naval and merchant vessels.

Figure 1.5 shows how the database has grown over time (see Davis, 2006 for historical perspectives). Notice how the observing technology has changed. Ship station data extend back to the 1800s and grew from about 100 to 10,000 observations per year during the twentieth century, with two gaps for the world wars. By the end of the 1900s, high-resolution ship-borne CTD data superseded station data. CTD data has a resolution of around 1 m in the vertical, whereas station data typically has one or two dozen measurements in total over the water column. Profiling floats, especially as part of the Argo program, now dominate, however. They contribute 100,000 observations per year.

Moored data have similarly increased in coverage, with early moorings mostly limited to research projects concentrating on a limited area and running for a few years or less. Starting in the 1990s, a large number of moorings were placed across the equatorial Pacific to monitor El Niño Southern Oscillation and continue today as the Tropical Atmosphere Ocean/Triangle Trans-Ocean Buoy Network array. Later a tropical Atlantic mooring array (the Prediction and Research Moored Array in the Atlantic) and an Indian Ocean array (called the Research Moored Array for African-Asian-Australian Monsoon Analysis and Prediction) were added. These moorings, Argo, and other programs such as the Rapid Climate Change program (to measure the deep circulation in the North Atlantic) are based on the idea that routine and systematic measurements are needed in order to build a quantitative picture of the ocean and monitor its effect on climate variability.

Satellite-based measurements have greatly increased the coverage of sea-surface temperature, height, and salinity, starting in the 1970s, 1990s, and 2010s, respectively. Satellite data approaches the sampling frequency necessary to allow signals to exceed noise over large areas.

There are many ways of turning the collection of ocean measurements into a systematic picture of ocean circulation. For example, since the 1980s, the U.S. National Center for Environmental Information has periodically updated World Ocean Atlas, a gridded climatology of ocean temperature and salinity (see also Subsection 1.1.1 and Appendix A). Originally, this product was built for comparison with global circulation model output (Subsection 1.2.2). Another approach, **geophysical inverse modelling**, combines observations with physical constraints (such as conservation of mass) to improve estimates of properties (Wunsch, 2006). A more elaborate approach is to use a **forward numerical model** (Subsection 1.2.2) in which ocean data are "assimilated" by adjusting various uncertain model parameters to make the model consistent with (or close to) the observations. This technique is especially useful for reconstructing the interannual variability of the ocean.

1.1.5 Global Temperature and Salinity

Vertical variations in temperature, salinity, and density are qualitatively similar in much of the world ocean. When a ship lowers an instrument at a given location, it creates an oceanographic profile that typically looks like Figure 1.6, which is calculated from a climatology rather than an individual measurement.

Nearly everywhere in the ocean, the deeper one goes, the colder the water gets. Temperature is vertically uniform in the surface **mixed layer** (see Chapter 4), has large gradients in the **permanent thermocline** (see Chapter 5), and has a weak stratification below the thermocline in a region sometimes known as **deep water** or the **abyss** (see Chapter 8).

In mid-latitudes, the boundary between the mixed layer and the thermocline shows a strong annual cycle (Figure 1.6a), with a cold, deep mixed layer erasing the upper part of the thermocline during the winter (Chapter 4). The part of the thermocline that is only present during the summer is known as the **seasonal thermocline**. There is no universally agreed precise definition of the borders between the temperature layers. This is especially true of the bottom of the thermocline because the temperature gradient smoothly decreases with depth. Sometimes the deepest parts of the thermocline are considered a separate region known as **intermediate water**, which may be distinguished by a different salinity from the thermocline in addition to colder water and smaller vertical temperature gradient. The thermocline is confined to about the top kilometer of the ocean, so most of the ocean's volume consists of abyssal water. The abyssal water is sometimes itself subdivided between **bottom water**, adjacent to the sea floor, and deep water between the bottom water and the pycnocline.

The vertical salinity profile is harder to characterize than temperature. Like temperature, it also has relatively small vertical gradients in the mixed layer and the abyss, and the biggest gradients in between (Figure 1.6b). Unlike temperature, the sign of the gradient (that is, whether S increases or decreases with depth) differs in different geographical

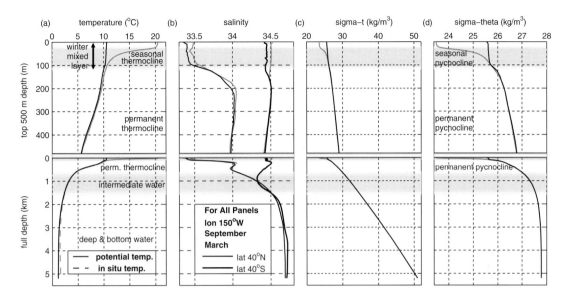

Figure 1.6 Vertical structure of (a) temperature, (b) salinity, (c) σ_t (density minus 1000 kg/m^3), and (d) σ_θ, at 40°N (thin curves) and 40°S (thick curves, salinity only), March (black) and September (gray), top 500 m (top panels) and whole water column (bottom panels). All profiles are in the Pacific at 150°W. Data from World Ocean Atlas 2013 (see Appendix A).

regions, and in some places does not change monotonically with depth. Figure 1.6b shows examples of quite distinct profiles at 40°N and 40°S.

The largest variation in ocean density is caused by pressure, so that vertical profiles of density are dominated by a nearly constant vertical gradient (Figure 1.6c). For that reason, oceanographers define **potential density** (Subsection 1.2.3) to only capture variations due to temperature and salinity. Ocean potential density profiles are qualitatively similar to temperature profiles in most places in the ocean (Figure 1.6d). Although the density of seawater depends on both temperature and salinity, the temperature makes a bigger contribution than salinity to density stratification. Such temperature-dominated places are referred to as **alpha oceans** (α is the thermal expansion coefficient; see (1.4) in Subsection 1.2.3, Carmack, 2007, and Exercise 5.3). The exceptions are shallow coastal waters, estuaries, and the polar oceans. Those places are called **beta oceans** because salinity controls their stratification (β is the haline contraction coefficient, (1.5)).

A more global view is provided by a **volumetric census**, which displays the distribution (histogram) of different temperatures and salinities. It divides the θ–S plane into bins, each of which consists of a small range of temperature and salinity. For example, Figure 1.7 uses bins of 0.05 g/kg in S and 0.5°C in θ. The relative uniformity of the abyssal ocean means that most of the ocean's volume has a relatively small range of θ and S values (Figure 1.7). Nearly 80% of seawater has $\theta < 6$°C and $34 \leq S \leq 35$ g/kg (Figure 1.7a). This largely fits in the density range of σ_θ between 27 and 28 and σ_2 between 35 and 37.5. (See Subsection 1.2.3 for definitions of σ_θ and σ_2). Ninety-six percent of the ocean

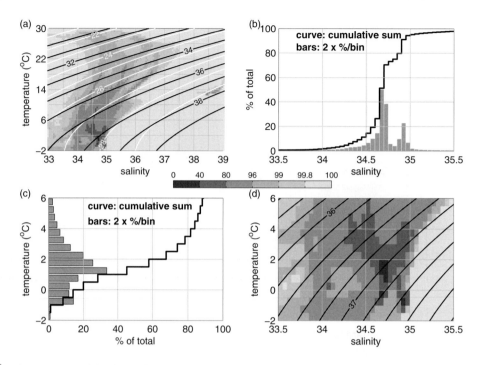

Figure 1.7 Volumetric census of global ocean from CARS Data (see Appendix A) as a function of salinity and potential temperature ((a) and (d)), salinity only (b), and temperature only (c). In (a) and (d), the ocean is subdivided into 0.5°C-by-0.05 g/kg bins in θ-S space which are shaded according to how large a percentage of ocean volume each bin occupies. The larger the volume, the darker the shade (see color bar in figure center), with the darkest shade representing the 40% of the ocean with θ-S bins occupying the largest volume and the lightest shade representing the 0.2% occupying the smallest volume. Contours show σ_θ (white; (a) only) and σ_2 (black) as a function of θ and S. In (b) and (c), bars represent percentage of ocean within a given range of one variable (S in (b) and θ in (c)) independent of the other variable. Stairstep graphs in (b) and (c) represent the cumulative percentage volume from lowest (33 g/kg in (b) and -2°C in (c)) to given value. Bar lengths are doubled for better visibility.

is included in a somewhat wider range below about 20°C and above 33.5 g/kg. The most common seven bins in Figure 1.7d account for 40% of the ocean's volume.

Including water of all temperatures (Figure 1.7b), we see that the salinity distribution has two maxima: at about 34.7 g/kg and a smaller one at about 34.9 g/kg; over half the ocean has salinity between 34.6 and 34.75 g/kg. The temperature distribution has a broad maximum centered at 1–1.5°C and a much weaker one above -1°C (Figure 1.7c). Over half the ocean volume has $\theta \leq 2$°C and nearly 90% has $\theta \leq 6$°C.

Geographical complexities of the ocean are reflected in the θ–S distribution. In Figure 1.7a, the isolated patch of water with $S > 38$ g/kg is due to the relatively salty, isolated Mediterranean. At the other extreme, the relatively fresh, cold water in the lower left-hand area of the panel is due to surface water at high latitudes (Chapter 10). The subsurface water of the relatively cold, salty Arctic is represented by the volumetric maximum at about -1°C and 34.9 g/kg (Figure 1.7d).

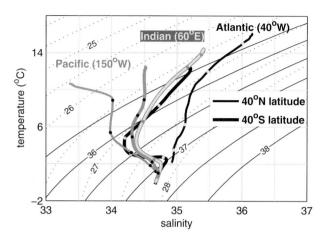

Figure 1.8 Temperature versus salinity for vertical profiles at 40°W (Atlantic, black), 60°E (Indian, light gray), and 150°W (Pacific, dark gray) for 40°N (thin) and 40°S (thick). Dots are values at depths of 250 m, 500 m, 1000 m, 2000 m, 4000 m. Data is for local winter, March/September in northern/southern hemispheres. Contours are σ_2 (solid) and σ_θ (dotted), with contour interval of 0.5 kg/m^3. Data from CARS climatology (see Appendix A).

The salinity differences between different basins (Figure 4.4) is reflected by the three branches visible in Figure 1.7a (for instance, at 14°C, salinities of 34.5, 35.25, and 35.9 g/kg). This can be seen more clearly by plotting θ and S for single vertical profiles at 40°N and 40°S in individual basins (Figure 1.8). Seawater volumes with particular θ–S characteristics are referred to as **water masses**. Water mass analysis often reveals much about the path, amount of mixing, and last contact with the mixed layer via water mass analysis. In Chapter 8 we trace the deep flow using salinity patterns. From Figure 1.8, we see that there is a sequence from saltier to fresher water as we travel from North Atlantic to South Atlantic to South Indian to South Pacific to North Pacific. This is the same sequence as the inter-basin trend in surface salinity visible in Figure 4.4.

1.1.6 Bottom Topography of the World Ocean

Surface elevations of the solid Earth are largely divided between continental masses and ocean **abyssal plains**, which reflects the two types of crustal rocks on Earth; continental and oceanic crust. Ocean topography is usually called **bathymetry**. As Figure 1.9 shows, most of the ocean is 3–6 km deep. A smaller fraction consists of **continental shelves** which are typically shallower than 250 m depth. The **continental slopes** leading from shelf to abyss are relatively narrow and steep, often less than 100 km wide, so that the 0.25–3 km depth range also comprises a relatively small fraction of the sea floor.

The narrowness of the continental slopes compared to the ocean basin widths has encouraged physical oceanographers to idealize the ocean as a basin with an approximately flat bottom (about 5 km deep) and straight sidewalls (the continental slopes) like a really big swimming pool. This view simplifies mathematical models and also eliminates problems

Global Topography (km), 0.5° Resolution

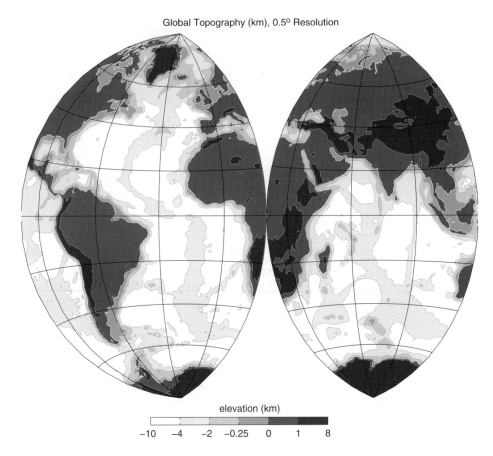

elevation (km)

−10   −4   −2   −0.25   0   1   8

Figure 1.9   Global topography (color scale on figure), from ETOPO5 averaged to 0.5° resolution (see Appendix A) and with some additional smoothing.

numerical models have in resolving steep topographic slopes. To the extent that basin-scale structures of velocity and density have horizontal scales that are large compared to the continental slope width, this is not a bad approximation. It is well to remember that in reality the ocean has no walls, only floors that rise to sea level. Boundary currents and continental slopes have similar widths, so much theoretical work done with the "straight sidewall" approximation has to be revisited with "no sidewall" geometry.

For clarity, graphs of ocean properties as a function of latitude and depth (as in Chapter 5) exaggerate the vertical axis relative to the horizontal, which makes bottom slopes look much steeper than they really are. The **aspect ratio** (As, height/width) for the real ocean is quite small: about 2.5×10^{-4} for the Pacific, so a scale model would be an Olympic swimming pool filled with about 1 cm of water. A continental shelf that descends 5 km over a horizontal distance of 100 km is relatively steep by ocean standards, but only has a slope of 5% and would be considered a mild incline to walk up on land.

The deep ocean floor has important topographic features, most notably the series of **ridges** that typically rise a few kilometers above the abyssal plains. These ridges include

Global Topography (km), 0.5° Resolution

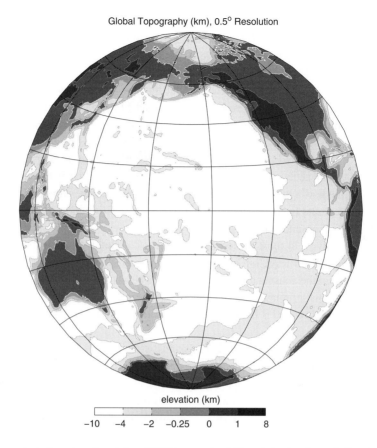

elevation (km)

| −10 | −4 | −2 | −0.25 | 0 | 1 | 8 |

Figure 1.9 Global topography (color scale on figure), from ETOPO5 averaged to 0.5° resolution (see Appendix A) and with some additional smoothing.

the **Mid-Atlantic Ridge** running north–south along the entire length of the Atlantic, the **East Pacific Rise** in the eastern South Pacific, the **Central Indian Ridge** and the rather narrow **Ninety East Ridge** (named for its longitude). Mid-ocean ridges break the sea surface at some locations, such as at Iceland on the Mid-Atlantic Ridge. New Zealand is the center of a series of ridges including the **Kermadec Ridge** to its north, and Kerguelen Island in the Indian sector of the Southern Ocean is the summit of another large rise.

Deep **trenches** are typically found at tectonic subduction zones next to a continental slope or ridge in such places as the eastern South Pacific, the northern Caribbean, east of the Mariana Islands in the western North Pacific, and east of the Kermadec Ridge. These trenches are generally less than 100 km wide, but can reach 10 km depth and 1000 km length. They are not known to have a big effect on the circulation above.

Examining bathymetry gives us some insight into the connection between oceans. Continental shelves extending between New Guinea and Australia and southward from Indochina form barriers between the Indian and Pacific for all but the shallowest 50 m or so of the water column (Figure 1.9). The two oceans connect through an archipelago consisting of Indonesia and other countries (Figure 3.8 shows a more detailed view). The narrow **Bering Strait** between the Pacific and the Arctic (Figure 10.1) is also blocked below 50 m (see also Figure 10.2). The Atlantic connection to the Arctic is partly blocked by the **Greenland–Scotland Ridge** (Figure 8.6), but with three passages ranging in depth from a few hundred meters to 850 m. The Mediterranean Sea, Black Sea, and Red Sea (examples of **marginal seas**) all extend to depths greater than a kilometer, but connect to the rest of the ocean through narrow straits no more than a few hundred meters deep (Figure 8.9).

Compared to the other oceans, a large proportion of the Arctic consists of broad continental shelves (Figures 1.9, 10.1). The region bounded by Greenland, Norway, Iceland, and the northern archipelago of Svalbard is known as the **Nordic Seas** and is often considered separate from the Arctic proper. Ice caps, covering most of Greenland and Antarctica, can be thousands of meters thick and redefine the "coast" by extending beyond the land as **ice shelves**. In most places, the coast is taken simply to be where the land elevation is 0 m above sea level, but in parts of Antarctica the ice shelves extend tens of kilometers away from the coast and hundreds of meters below sea level. The Greenland ice shelves consist of smaller tongues of ice reaching the ocean.

Although this book emphasizes large-scale circulation patterns, we should also remember the smaller-scale features that are not visible at the 0.5° resolution used in Figure 1.9. The ocean has thousands of islands with widths less than 50 km, with especially dense concentrations of islands in the tropical western Pacific. Generally speaking, the ridge systems have rougher topography than the abyssal plains, with many thousands of sea mounts and sea hills with a variety of horizontal and vertical scales. Valleys through the ridges, such as the **Charlie Gibbs Fracture Zone** (52°N) and **Vema Fracture Zone** (11°N) in the Mid-Atlantic Ridge, can be important connections between abyssal basins separated by ridges. Indeed, the bathymetry of the ocean has **fractal** characteristics (Gagnon et al., 2006). This means that the wiggles in bathymetry contours show similar statistics at many different scales.

1.2 Concepts

Many basic concepts underpin our understanding of the ocean general circulation. They include an appreciation of the range of time- and space-scales relevant to the ocean currents, and numerical computer models of the currents. These topics are covered first in the subsections below. Then we discuss the seawater equation of state, which is used repeatedly in this book.

1.2.1 Oceanic Time and Space Scales

The plethora of temporal and spatial scales discussed in Section 1.1 are caused by many processes in the ocean. We summarize these scales and processes in a schematic diagram

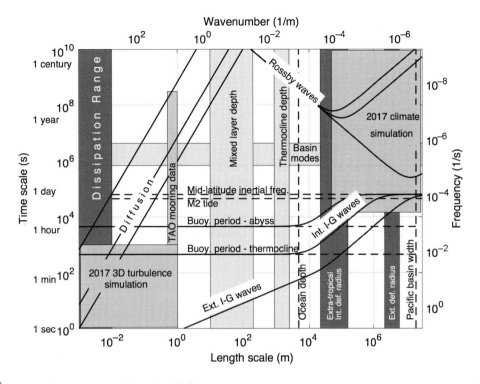

Figure 1.10 Diagram of space/time variability in the global ocean.

(Figure 1.10) which provides a framework to think about the ocean circulation. The diagram shows information about the energy of the ocean circulation as a function of length scale and timescale (equivalently, wavenumber k and wavelength λ where $k = 2\pi/\lambda$). The scales depicted stretch across the full range relevant to physical oceanography. There are about ten orders of magnitude in space and time, a vast span.

The diagram is useful for several reasons. First, it shows important characteristic scales. Some of them are supplied by factors external to oceanography (**extrinsic factors**). These include the widths of the ocean basins themselves and the depth of the ocean (see Subsection 1.1.6). Extrinsic timescales include the characteristic periods of the tides which are set by celestial mechanics (the name M2 in Figure 1.10 refers to the twice-daily tidal frequency due to the Moon). Another periodic motion called an **inertial oscillation** arises from the Earth's rotation (the frequency of inertial oscillation is set by the rotation rate of the Earth; Subsection 2.1.1). In mid-latitudes the period of the inertial oscillation lies between the M2 tidal period and one day, as shown.

An important extrinsic length scale arises because of Earth's rotation: the **external deformation radius**. This is a key length-scale in geophysical fluid dynamics (Vallis, 2006). It depends on the ocean depth, gravitational acceleration, and the vertical component of Earth's rotation. The external deformation radius is shown with a range in Figure 1.10 because it depends on latitude through the vertical component of Earth's rotation: it is several thousand kilometers in mid-latitudes.

The three **diffusion** lines are another guide: As a function of length-scale, they show the time taken for molecular diffusion to smooth out gradients in speed, temperature, and salinity (which takes the longest time). These are predictions for diffusive relaxation timescales in the absence of any circulation: they underestimate the actual timescale in the presence of circulation (see Chapter 7).

Most scales are determined internally, however, by **intrinsic factors**. Explaining what determines these intrinsic scales is a basic issue in physical oceanography. In many cases, we can measure the intrinsic scales without fully understanding what sets them. Knowing the scales allows us to understand other aspects of the ocean. A good example is the density stratification (see Figure 1.6). We cannot yet explain completely why the ocean is stratified in the way we observe, but we can exploit the stratification measurements to understand other things. Specifically, once we know the stratification we know the characteristic depths of the thermocline and the mixed layer (see Chapter 4). In Figure 1.10 the stratification is quantified by the **buoyancy frequency**, N (sometimes called the Brunt–Väisälä frequency):

$$\text{\underline{\textbf{buoyancy frequency, } N, :}} \qquad N^2 = -\frac{g}{\rho}\left(\frac{\partial \rho}{\partial z} - \frac{1}{C_s^2}\frac{\partial p}{\partial z}\right), \qquad (1.1)$$

where C_s is the sound speed (see Subsection 1.3.1 and Exercise 1.5). Two limiting cases are shown: the thermocline has a buoyancy frequency of about 0.014 Hz (a period of about 500 s: Hz means Hertz, namely cycles per second) and the abyss has a buoyancy frequency about ten times smaller.

Another example of an important intrinsic scale is the **internal deformation radius**. The internal deformation radius depends on the ocean depth, the vertical component of Earth's rotation rate, and the stratification (see (8.2)). These quantities vary from place to place so typical ranges are shown in the schematic diagram. Deformation radii are important for dynamical oceanography in several ways. For example, the internal deformation radius determines the typical width of mesoscale eddies (Subsection 7.1.1), which is tens to hundreds of kilometers.

Geophysical fluid dynamical theory provides information on length and timescales too. In particular, it tells us how the frequency depends on the wavenumber for various waves. For example, **inertia–gravity waves** are marked with three "I–G" lines on Figure 1.10 (we plot the so-called wave dispersion relation; see Vallis, 2006 for details). The line with highest frequency for a given length scale (at the bottom) depicts external inertia–gravity waves. Over length scales of 1–100 m these waves are driven by the wind, for instance, and break in surf zones at coasts. The two other lines depict internal inertia–gravity waves. These waves have a restoring force that depends on the strength of the density stratification. For this reason, two examples are shown corresponding to the relatively strong stratification of the thermocline and the relatively weak stratification of the abyssal ocean. In practice the gap between these limiting cases is filled by inertia–gravity waves riding on stratification of intermediate strength.

Another important class of waves that depend on Earth's rotation and the ocean stratification are **Rossby waves**. Examples of Rossby waves in the tropics are plotted in Figure 1.10. The high-frequency branch is the external Rossby wave and the other two are internal Rossby waves corresponding to the strong stratification (thermocline, middle branch) and low stratification (abyss, low-frequency branch). In the real ocean, Rossby waves can exist with other, similar scales. Extra-tropical Rossby waves have lower frequencies for a given wave length than those in the tropics, for example. Be aware that the dispersion relations do not show how strongly the waves are forced in practice. Rossby waves are forced by changes in the wind and are most common with large length scales.

A related type of oscillation is called a **basin mode**. A basin mode is a type of standing Rossby wave that is akin to water sloshing round in a bucket. The periods of basin modes can be deduced from knowledge of the shape of the ocean basin, gravity, and the Earth's rotation rate, which gives periods of weeks to months. Evidence of their excitation in the ocean is elusive, however.

Knowing how strongly the ocean is forced by the atmosphere is also revealing. For instance, we can observe how the ocean exchanges momentum, heat, freshwater, and radiation with the atmosphere to deduce other information. Characteristics of the surface **mixed layer** can be computed, for example, such as its depth (tens to hundreds of meters, Figure 1.10), temperature, and salinity (see Chapter 4). And knowledge from observations of turbulence intensity provides information on the **dissipation range** at the smallest scales shown in Figure 1.10 (on the left-hand side) where viscosity dominates and removes kinetic energy from the circulation (see Subsection 1.3.1 for more on viscosity). For typical turbulence levels in the ocean, the scale at which viscosity dominates is between about 0.1 and 10 mm.

These various length- and timescales are shown in Figure 1.10 to indicate, roughly, which processes are important at which scale. The diagram is schematic and the numbers are approximate because many details are currently unknown. To see this, think of the sampling technologies described in Section 1.1. For example, recall Figure 1.1 which shows Equatorial Pacific TAO current meter data at a single point over several years. This dataset is one of the best available by oceanographic standards because it is long. It consists of current meter observations at a fixed place, however, and therefore contains very little information about variations in space. (The TAO project maintains an array of current meter moorings in the Equatorial Pacific, but each instrument measures over a small volume.) For this reason, it spans a wide range of frequencies in Figure 1.10 but a narrow range of wavenumbers. The observing technologies described in Section 1.1 only measure small regions of the space time diagram.

Numerical computer models of the ocean currents are another important source of information about ocean circulation (see Subsection 1.2.2). They help fill out parts of Figure 1.10, but they do not solve the problem. Figure 1.10 shows the range of scales accessible by climate models circa 2017 at long space- and timescales. A different class of numerical model simulates fluid turbulence at the other end of the spectrum. There is a large gap between them! Computer resources suffice to simultaneously represent two or three decades in scale, but no more. Computer modelers pick which range of scales they will resolve when they design their models. Even with continued increases in computer power, numerical models will give access to only a modest part of the diagram.

Filling in the details with theoretical arguments appears unlikely too. The equations of ocean circulation are well known (Section 1.3), but they are nonlinear and very hard to solve. The main nonlinearity is related to the fact that the currents transport properties, like heat and momentum, that influence the currents themselves. This also means that all scales are connected: The ocean circulation has energy distributed across all the scales in Figure 1.10 and energy is exchanged across all the scales too. Another way to say this is that no gap exists in the spectrum of ocean circulation to break these connections. Like numerical models, theories of the ocean circulation focus on a small range of space- and timescales. They make simplifying assumptions about how unresolved scales affect the scales of interest. Many examples of this type of theoretical explanation appear in this book.

Figure 1.10 simplifies the issue in some ways. It ignores anisotropy for instance and lumps all length-scales into a single variable. In other words the diagram makes no distinction between horizontal or vertical length-scales. In practice horizontal length-scales are distinctly different to vertical length-scales, however. One might even argue that ten decades in time is not enough. Some ocean processes have periods shorter than 1 s (like acoustics); others last longer than 100 years. For example, climate fluctuations are known to exist on all periods up to the $O(1–10)$ million years it takes for continental drift to rearrange the shape of the ocean basins.

The time-varying global ocean circulation inhabits the gap between the viscous diffusion line (which is the lowest one) and the external gravity wave dispersion relation in Figure 1.10. This book is about steady large-scale ocean general circulation, which is a subset of this gap. That is, about the motions towards the upper right corner of the diagram (hundreds to thousands of kilometers and seasons to centuries). For this reason, several of the scales and processes in Figure 1.10 are ignored or treated very lightly. The preceding discussion suggests that all scales are ultimately connected and interact. It may sometimes be hard, therefore, to understand one specific range of scales on its own. Also, there are no direct observations of the ocean circulation on the scales of interest. None of the measuring technologies discussed in Section 1.1 measure the steady large-scale currents. For this reason, we must use observations in creative ways to infer information about the general circulation. Finally, be aware that the very notion of the "steady" circulation is an approximation. This apparently innocent restriction may sometimes be misleading.

Despite these caveats, a tremendous amount can be understood about the real ocean circulation. Many of the phenomena seen at large scales and low frequencies can be understood reasonably well without referring to other scales. Many kinds of data can be used judiciously to infer knowledge about the large-scale flow. Numerical models offer a vital chance to perform experiments on isolated pieces of the system to elucidate their nature in ways that are impossible for the real ocean. And approximate theories can be constructed that yield deep and surprising insights. From this perspective, the achievements described in this book are great triumphs. They inspire us to continue studying and trying to understand the ocean circulation at all its scales.

1.2.2 Models of the Circulation

Mathematical Challenges of Ocean Theory

Observations are the foundation of physical science but not the entire edifice. We also want to understand *why* the general circulation looks the way it does. Such understanding is pleasing in itself, forms a basis for further advancing our knowledge of the ocean, and may help us predict useful things about the circulation such as how it will influence climate in the future.

Physical oceanographers are confident that a handful of governing equations give an accurate description of the physical behavior of the ocean (see Subsection 1.3.1). These laws take the form of partial differential equations representing seawater, and boundary conditions representing the sea floor shape and interaction with the atmosphere at the ocean's surface. The surface boundary conditions are often referred to as the **forcing** of the ocean, because they represent the exchange of momentum, heat, and water that is directly responsible for driving the ocean circulation (tidal forces and geothermal heating also play a role). The governing equations (or **equations of motion**) are solved by functions of space and time representing fields of velocity, temperature, and salinity throughout the domain.

In **linear** differential equations, each term in the equation is proportional to a single dependent variable. For a linear system of equations, a given set of boundary conditions produces a single solution, steady for steady forcing and periodic for periodic forcing. However, the equations governing the ocean are **nonlinear**. For such equations, it is not clear how many solutions exist. As we will discuss in Chapter 11, circulation we observe in the ocean may not be the only one possible for given atmospheric conditions. Nonlinearity allows a system with simple forcing to generate flow structures that change in a complex way in time and space (Chapter 7). For example, flow in a cylindrical pipe can generate turbulent motions much smaller than the radius of the pipe. The forcing of the Earth's atmosphere consists primarily of steady sunshine modulated by daily and annual periodicities of the Earth's motion, but instability in the atmosphere produces weather which can change by the day or by the hour. Instabilities in the ocean produce $O(100 \text{ km})$ wide eddies which are analogous to mid-latitude weather systems.

The nonlinearity of the governing equations accounts for many intriguing features in the ocean, but it also makes the equations difficult to solve. The complexity of the basin shape and atmospheric forcing also adds to the difficulty of solving the equations for the real ocean. Historically, physical oceanographers have dealt with this problem by solving simplified or "idealized" problems. Such idealized problems usually isolate and approximate one aspect of the circulation in one geographical region of the ocean. Often we consider an imaginary ocean in which the geometry or forcing is simpler than the real world. Similarly, we may assume parameters are such that the nonlinear terms in the equation are small compared to the linear terms and hence can be neglected. In some studies, the equations themselves are altered to a more easily solvable form, with the hope that solving the altered equations will reveal behavior analogous to the solution of the correct equations.

Physical oceanographers have had remarkable success with solving idealized problems. They have shown the cause of ubiquitous circulation features (like western boundary currents, Chapter 3), explained the behavior of other structures (like the equatorial undercurrents, Chapter 6), and even predicted some phenomena before they were clearly revealed by observations (like the deep western boundary currents, Chapter 8). While the mathematics involved in solving such simplified systems may be demanding, the solutions themselves are usually presented in formulae that show how the resulting circulation feature depends on parameters such as basin size or wind strength. On the other hand, it is usually difficult for such simplified solutions to give detailed, quantitative predictions about the more complicated behavior of the real ocean.

The Use of Ocean Models

The limits of mathematical techniques to solve differential equations, and the growing power of computers, have led to the widespread use of computer models of the circulation. Such models find numerical solutions to the governing equations, that is to say, they represent the approximate solution to the equations as a set of numbers rather than an algebraic formula. The most common solution technique is to **discretize** the equations in time and space. Though ocean variables such as velocity are continuous functions of time and space, the equations can be approximated by solving for each variable at a finite set of points, usually arranged in some kind of regular grid in latitude, longitude, and depth (Figure 1.11). Given the values of all the variables on all the grid points at a certain time, the governing equations can tell us, to a good approximation, what the values are a short time (or time step) later. The process can be repeated over many time steps to show how

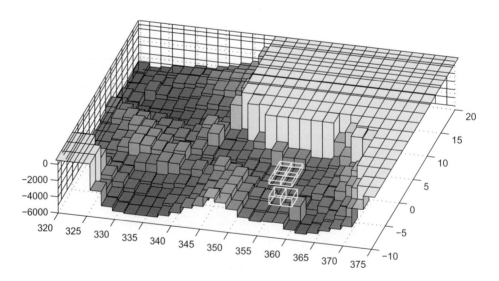

Figure 1.11 Illustration of longitude–latitude–depth numerical grids used for coarse resolution general circulation models. Gray planes represent topography (Africa is at top right). The ocean is divided into grid cells illustrated by representative cells shown at lower right of the figure (four near-surface cells and two deeper cells).

the ocean circulation evolves in time in response to the initial state or under the influence of changing forcing. It is analogous to the use of atmospheric models in making weather forecasts. If the forcing is steady or periodic, integrating the model long enough will allow for time-averaging as an approximation of the steady circulation as discussed in Section 1.1.

Since the 1990s, the most commonly used computer models have solved the primitive equations (Subsection 1.3.1), which are a relatively fundamental statement of the governing equations. Such a model, when it includes the ability to solve the equations with realistic topography and forcing, is called a **general circulation model (GCM)**. GCMs consist of hundreds of subroutines and tens of thousands of lines of code, most commonly in modern versions of the Fortran computer language, and are generally supported by a team at a university or national laboratory and used by the scientific community at large. Modern GCMs are designed to take advantage of massively parallel computers in which hundreds or thousands of computer processors work simultaneously to calculate a simulation more quickly. Computationally intensive simulations can take days or weeks to run.

One of the first GCMs was the Geophysical Fluid Dynamics Laboratory Model (GFDL Model) developed in the 1960s and 1970s (Bryan and Cox, 1967; Bryan and Lewis, 1979). This model gave rise to the Modular Ocean Model (MOM) and the Parallel Ocean Program (POP). Other GCMs in use around the world include the Massachusetts Institute of Technology GCM (MITgcm), Hybrid-coordinate Ocean Model (HYCOM), and Nucleus for European Modelling of the Ocean (NEMO). The books by Griffies (2004) and Miller (2007) provide entry points to the field of ocean circulation modeling.

To the extent that ocean models are performing mathematical calculations based on the equations of motion, a numerical model can be considered an analytic or theoretical study of the ocean. However, because the availability of a model frees a scientist from thinking about techniques for solving the equation, numerical studies also resemble laboratory experiments. As in the laboratory, the numerical experimenter creates an experimental configuration (determining geometry, boundary conditions, and other parameters to be used by the model), conducts an experiment (by running the model), and analyzes the output of the experiment. GCMs can reproduce some ocean characteristics, such as the sphericity of the Earth, that are difficult to achieve in a physical laboratory.

GCMs are used in a variety of ways. Simplified configurations give insight into ocean processes. More realistic models test whether the physical laws represented by the GCM do in fact give an accurate representation of the circulation in the real ocean. Models combined with ocean observations via techniques such as **data assimilation** reveal circulation details that are difficult to resolve IRL[1] due to sampling limitations. Finally, GCMs can be used to predict the future state of the ocean, much as atmospheric models make weather forecasts.

While GCMs offer a giant step towards realism in ocean models, their behavior is not identical to the real world. One of the most important simplification is a consequence of model resolution, or the distance between neighboring grid points. Features in the real ocean must be considerably larger than the grid spacing to be resolved by the model.

---

[1] In Real Life

A key aspect of resolution is the representation of mesoscale eddies, which generally range in width from tens to hundreds of kilometers (Chapter 7). Models are usually **eddy-suppressing** (no eddies present) for horizontal grid spacing 50–400 km, **eddy-resolving** (realistic eddies reproduced) at 10 km or less, and **eddy-permitting** (eddies present but unrealistically wide and slow) at intermediate grid spacing. Choosing a model resolution involves a trade-off between more realistic behavior at higher resolution and less computing power needed at lower resolution.

Even eddy-resolving models leave out smaller structures such as internal waves and turbulence (Figure 1.10). Thus all models must **parameterize** the effect of unresolved motion on the resolved scales. This is most often done with diffusion of momentum, temperature, salinity, and other quantities (which we discuss in Subsection 1.3.1 and Chapter 7). The division of flow into resolved and unresolved components is a central idea in physical oceanography and important not only to numerical modeling but to theory and observation as well.

Coupled Models

Much interest in ocean modeling occurs in the context of the ocean as one component of a larger system. Atmospheric GCMs (**AGCMs**) represent atmospheric circulation and related properties, while land models represent soil moisture, water runoff, and even some aspects of surface vegetation. Climate models combine AGCMs, OGCMs, models of ice behavior, and land models. If we know atmospheric conditions and river runoff, then we can drive an ocean model, but those atmospheric conditions themselves depend on the behavior of the ocean. For instance, atmospheric circulation is influenced by ocean surface temperature, which in turn is modified by ocean currents, so that specifying atmospheric conditions to drive an ocean model is an incomplete answer to the question of what determines the ocean general circulation. A more complete answer considers how the atmosphere, ocean, cryosphere (ice and snow), and land surface all adjust to produce the observed state of the entire system.

Climate models have become steadily more ambitious in attempting to predict the monthly average atmosphere and ocean behavior several months in advance. For longer timescales, the question of human-induced climate change through greenhouse gas emissions and other influences has been a tremendous impetus for developing climate models (Randall et al., 2007). Climate models have been developed and run by several groups around the world, including the National Center for Atmospheric Research (NCAR), the National Aeronautics and Space Administration (NASA), and the Geophysical Fluid Dynamics Laboratory in the United states, the Met Office in the United Kingdom, and the Max Planck Institute (MPI) for Meteorology in Germany.

Modeling ocean biogeochemistry is also an important task for many ocean GCMs and climate models. One good reason is that the ocean is a key player in Earth's carbon cycle. In order to model carbon in the ocean many biogeochemical processes must be considered. Models simulate the chemistry of biologically active elements such as oxygen, nitrogen, phosphorus, and silicon. Ecological components of models simulate populations of marine plants (phytoplankton), animals (zooplankton), and their interactions with the environment and with each other. Models explore the feedbacks between marine life, ocean circulation, atmospheric chemistry, and climate processes.

Finally, ocean GCMs often include packages to model biologically inactive chemicals. Geochemical tracers of this type are revealing about processes in the real ocean, especially on long timescales where direct measurements are impossible. For example, stable isotopes of oxygen tell us about processes that influence the hydrosphere, like freezing or evaporation at the sea surface. Similarly, natural and man-made tracers, like carbon-14 and CFCs, are useful for showing rates and pathways of ocean currents over the global ocean and decades. When a GCM simulates these tracers in a realistic way, confidence in our knowledge of the large-scale circulation increases.

1.2.3 Seawater Equation of State

Seawater is a complex solution whose density depends on temperature, salinity, and pressure (Section 1.1). No adequate theoretical model exists to predict density as a function of these variables (unlike the simple ideal gas law for air). Therefore the density of seawater is measured in a laboratory under varying conditions and these data are used to estimate an interpolating function called the seawater **equation of state**, $\rho = \rho(\theta, S, p)$.

Seawater density only varies by a few percent in the ocean, so ρ can often be treated as a constant, ρ_0. Much of the density range in the ocean is due to the (slight) compressibility of seawater, so that density mainly increases with pressure. For $\theta = 4°C$ and $S = 34$ g/kg (typical oceanic values), ρ increases from about 1027 kg m^{-3} at the surface (0 db) to 1049 kg m^{-3} at 5000 db. The remaining smaller variations, typically 8 kg m^{-3} or less, are due to temperature and salinity. They constitute tiny fractional changes in density but they play a large role in ocean dynamics. Small differences in density at different horizontal locations cause hydrostatic pressure differences which are responsible for ocean currents (Subsection 2.1.2).

To eliminate the relatively uninteresting dependence on pressure, it is convenient to define the **potential density anomaly** σ_0. This variable is called "potential" density for the same reasons θ is called potential temperature: it is the density a sample of seawater would have if it were moved adiabatically (no exchange of heat) to a reference pressure. In the case of σ_0, the reference pressure is the surface (like potential temperature), hence the subscript zero. The "anomaly" part of the name means that 1000 kg m^{-3} is subtracted. Therefore, we have:

$$\sigma_0 = [\rho(\theta, S, 0 \text{ db}) - 1000] \text{ kg m}^{-3}. \tag{1.2}$$

Other reference pressures can be used, for instance, σ_4 is the density anomaly at 4000 db. Often, σ_0 is written as σ_θ or just σ and sometimes the units (kgm^{-3}) are omitted.

Seawater density *increases* for *decreasing* temperature and *increasing* salinity: warm, fresh water is lightest and cold, salty water is densest (Figure 1.12a). Density is an approximately linear function of S but depends roughly quadratically on θ (Figure 1.12b), so that density is much less sensitive to temperature near the freezing point (about $-1.9°C$ for surface water with $S = 35$g/kg, see Figure 10.5) than at higher temperatures. A revealing way to express the dependence of ρ on θ, S, and p is

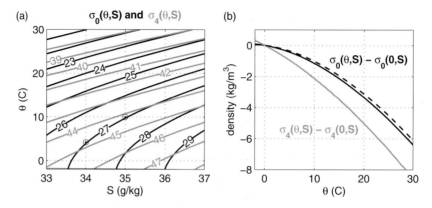

Figure 1.12 Equation of state represented by σ_0 (black) and σ_4 (gray) shown as (a) functions of S and θ in a contour graph, and (b) functions of θ, relative to a reference value at $\theta = 0°$C and $S = 35$g/kg (solid) and $S = 30$g/kg (dashed). In (a) the two dots represent water parcels with given (S, θ) values.

$$\rho = \rho_0 + \rho_1(p) + \rho_0\left[-\alpha(\theta, S, p)\theta + \beta(\theta, S, p)S\right], \qquad (1.3)$$

where ρ_0 is the constant **reference density** and ρ_1 expresses the increase of density with pressure. The parameters

Thermal expansion coefficient: $\qquad \alpha = -\dfrac{1}{\rho}\dfrac{\partial \rho}{\partial T}, \qquad (1.4)$

and

Haline contraction coefficient: $\qquad \beta = \dfrac{1}{\rho}\dfrac{\partial \rho}{\partial S}, \qquad (1.5)$

quantify how density depends on temperature and salinity (β here should not be confused with the gradient in planetary vorticity; Subsection 2.2.3). Typical values are: $\alpha \approx 2 \times 10^{-4}°$C^{-1} and $\beta \approx 8 \times 10^{-4}$, although they vary with temperature, salinity, and pressure. The choice of ρ_0, ρ_1, α and β are not unique for a given ρ. For typical oceanographic values, α and β are much less sensitive to S than to θ and p. We can approximate (1.3) with constant values of α and β, which is an increasingly accurate approximation for decreasing ranges of p and θ. That is,

Linear Equation of State: $\qquad \rho = \rho_0\left(1 - \alpha\Delta\theta + \beta\Delta S\right), \qquad (1.6)$

where $\Delta\theta$ and ΔS are anomalies of potential temperature and salinity from reference values. It is often convenient to use this linearized equation of state in theoretical studies (see Chapter 4 and Chapter 7).

The dependence of α on p, known as **thermobaricity** (IOC, SCOR, and IAPSO, 2010; Stewart and Haine, 2016), becomes significant over depth ranges of several kilometers in the ocean (Figure 1.12b), with density more sensitive to temperature changes in deeper water. This becomes troubling when we want to compare the density of water at different depths, which often occurs when we are investigating the three-dimensional flow in the ocean. Consider two water parcels at two locations: we represent each parcel by a single point in the (S, θ) plane (Figure 1.12a). At the surface, these two points would have the same density, since they both lie on the $\sigma_0 = 27$ contour (Figure 1.12a). However, at 4 km depth, Figure 1.12a shows that the colder one would be denser (σ_4 nearly 45 as compared to σ_4 slightly greater than 44). Thus if the two parcels move through the ocean so that they approach each other, which one is lighter (and hence should sit above the other parcel if they both go to the same horizontal location) depends on the depth at which they meet. Sometimes oceanographers use another measure of density called **neutral density** (IOC, SCOR, and IAPSO, 2010). Neutral density is usually numerically similar to σ_0 but is calculated in a way that facilitates comparison of density between different depths.

Thermodynamic Equation of State – 2010

The **Thermodynamic Equation of State – 2010 (TEOS-10)** is a new framework for describing the properties of seawater and ice. It incorporates the latest high-accuracy measurements of seawater properties in a thermodynamically consistent approach. It expresses the saltiness of seawater using **absolute salinity** instead of practical salinity (Subsection 1.1.2). Absolute salinity (S_A, units g/kg) is the mass fraction of salt in seawater and has the advantage of being an SI unit. Namely,

$$S_A = \frac{m}{M + m}, \tag{1.7}$$

where m is the mass of all the dissolved salts and M is the mass of the freshwater. Hence $m + M$ is the mass of seawater. In TEOS-10, practical salinity based on the 1978 Practical Salinity Scale (defined in terms of the conductivity of seawater) is referred to with the variable S_P (this quantity we call S in Subsection 1.1.2). Absolute salinity is greater than practical salinity, by 0.03 to 0.20 g/kg (0.17 g/kg on average) for global ocean values.

The temperature variable is also new in TEOS-10; **conservative temperature** (Θ, °C), which is a more accurate measure of heat content and is more accurately conserved by adiabatic motion. Conservative temperature is numerically similar to potential temperature for typical ocean salinities and temperatures (see Exercise 1.10).

For the purposes of this book, the differences between the older system and the new TEOS-10 variables are small, and we refer to practical salinity and potential temperature unless stated otherwise. Nevertheless, the TEOS-10 convention is the way of the future and is worth adopting, particularly if you are new to oceanography. The TEOS-10 manual is an excellent reference on the thermodynamic properties of seawater (Pawlowicz, 2010; IOC, SCOR, and IAPSO, 2010) and `www.teos-10.org` has details, clear advice, and well-documented software in multiple languages to compute all the properties of seawater.

1.3 Theory

The theory of ocean general circulation is a branch of **geophysical fluid dynamics**. It concerns the physics of natural fluids which rotate and are stratified by density differences, like planetary oceans and atmospheres. Geophysical fluid dynamics is a mature field and several textbooks are devoted to it. In this section we describe some important topics for use later in the book.

1.3.1 Governing Equations

Basis for Governing Equations

The circulation is largely determined by a small number of equations governing the behavior of the three-dimensional velocity $\mathbf{u}$, the pressure p, the density ρ, the (potential) temperature θ, and the salinity S. These governing equations, also known as the **equations of motion**, derive from fundamental laws about the **conservation** of extensive physical quantities (an extensive quantity has the property that it equals the sum of the quantity over all the constituent elements; mass and heat are examples). In this way, the velocity, temperature, and salinity equations come from fundamental laws that say momentum, heat, and freshwater can be exchanged, but not created or destroyed. The ocean circulation moves these quantities around and exchanges them with the atmosphere and the sea floor.

Generally if some physical quantity $\chi(\mathbf{x}, t)$ is transported with a **flux** Φ and is transformed to other forms at a rate $\mathcal{F}$, its evolution within a given region is

$$\frac{\partial}{\partial t} \int \chi \, dV = -\oint \Phi \cdot \hat{\mathbf{n}} \, dS + \int \mathcal{F} \, dV. \tag{1.8}$$

In this equation $\int dV$ is a volume integral over the region and $\oint dS$ is the surface integral over the surface bounding the region, which has unit outward normal $\hat{\mathbf{n}}$. The three terms are (from left to right) the evolution term, the **transport** term, and the **source** term. The source term represents the rate of creation or destruction of the quantity within the region, while the transport (flux) is the rate at which the quantity is leaving the region. The **heat transport** has attracted intense scrutiny as a key to interactions between the ocean and the atmosphere. By the divergence theorem, (1.8) can be rewritten to apply to individual points in the ocean as

$$\frac{\partial \chi}{\partial t} = -\nabla \cdot \Phi + \mathcal{F}. \tag{1.9}$$

Any quantity that obeys (1.9) with Φ is called a **conservative tracer** or simply a tracer.

In a fluid, the main way that properties are transported are by the resolved fluid flow $\mathbf{u}$ (called **advection** or **stirring**) and by **mixing** due to molecular motion. The flux of a property by advection is given by

$$\Phi_A = \mathbf{u}\chi \tag{1.10}$$

and we write the flux due to mixing as

$$\mathbf{D} = -\kappa \nabla \chi, \tag{1.11}$$

where κ is the diffusivity coefficient. There is always mixing due to molecular processes that irreversibly remove variance in χ via diffusion. Molecular mixing is associated with a (nearly) uniform value of κ that does not depend on $\mathbf{u}$. This removal of variance is sometimes called dissipation (see Subsection 1.2.1 and Subsection 7.1.1). For large-scale flows, this makes a small contribution to transport. In that case, unresolved motion is often also characterized with a diffusion law in which κ is much bigger than the molecular value and may depend on $\mathbf{u}$ and other factors.

Using these expressions for flux components, we can write (1.9) as

$$\frac{\partial \chi}{\partial t} = -\nabla \cdot (\chi \mathbf{u}) - \nabla \cdot \mathbf{D} + \mathcal{F}, \qquad (1.12)$$

and if we use the product rule on the advection term and approximate seawater as an incompressible fluid for which $\nabla \cdot \mathbf{u} = 0$ (Subsection 2.1.4), then standard vector identities tell us that

$$\nabla \cdot (\chi \mathbf{u}) = \mathbf{u} \cdot \nabla \chi + \chi \nabla \cdot \mathbf{u} = \mathbf{u} \cdot \nabla \chi. \qquad (1.13)$$

It is conceptually revealing to define the operator

$$\frac{D}{Dt} = \frac{\partial}{\partial t} + \mathbf{u} \cdot \nabla. \qquad (1.14)$$

which allows us to write (1.12) as

$$\frac{D\chi}{Dt} = -\nabla \cdot \mathbf{D} + \mathcal{F}. \qquad (1.15)$$

The operator D/Dt is called the **material derivative** and represents the rate of change of properties of a given parcel of fluid. For example, a steady flow across a salinity gradient will have $\partial S/\partial t = 0$ because S is not changing at any point, but $DS/dt \neq 0$ because the salinity of the water parcel must change as it flows (for example) to a region of higher S. As (1.15) shows, change in some property of a given water parcel must be caused by mixing at the boundary of the parcel and/or some source or sink within the parcel.

The two ways of writing the conservation equation represent two ways of conceptualizing the behavior of a fluid. The **Eulerian** view analyzes motion at a fixed location in space, and so (1.12) describes the evolution of a quantity at that location in terms of fluxes to/from the location as well as forcing there. The **Lagrangian** view analyzes the behavior of a parcel of water as it travels through the domain, and (1.15) describes the evolution of a water parcel due to forcing of the parcel and diffusive (but not advective) fluxes at the parcel boundary. Note that measurement platforms described in (Subsection 1.1.2) include Eulerian observations (for instance, a current meter), Lagrangian measurements (for instance, a drifter), and observations that do not correspond closely to either (for instance, a glider). Eulerian and Lagrangian views of the dynamics are equally and simultaneously valid, but often each gives unique insights.

Synopsis of Governing Equations

Some key conservation laws for seawater are conservation of mass, conservation of salt, and conservation of internal energy. Note that the concept of internal energy involves thermodynamic subtleties (IOC, SCOR, and IAPSO, 2010) which we neglect here. For an incompressible fluid, conservation of mass is equivalent to conservation of volume and can be simplified to the condition of zero divergence for the fluid. Salt mass per volume is $\rho_s = \rho S$, and internal energy per volume is $E = c_p \rho \theta$, where $c_p \approx 4000$ J m^{-3} K^{-1} is the specific heat capacity of seawater (the amount of energy it takes to raise the temperature of one kilogram of water 1 K). Thus conservation of salt and internal energy give us conservation of S and θ, respectively. Sometimes θ and S are referred to as tracers because they have the same conservation equation as chemical tracers. Because they affect density via the equation of state, and hence the current, they are called dynamically active (chemical tracers that have no effect on the flow are dynamically passive).

Conservation of momentum is expressed by Newton's second law, which on a per-volume basis can be expressed as $\mathbf{a} = \mathbf{F}/\rho$, where $\mathbf{a}$ is acceleration and $\mathbf{F}$ is force per volume. The connection of the momentum equation for a fluid and conservation is a bit more complicated than for the tracers. In addition, the governing equations are written for a coordinate system fixed to the Earth, which has a rotation characterized by an angular velocity vector $\mathbf{\Omega}$ (for more, see Subsection 2.1.1). As we will see, such a rotating frame of reference necessitates an additional term that further complicates the equations. Gathering all the governing equations together (the so-called primitive equations; see also Appendix B):

$$\textbf{Momentum:} \quad \frac{D\mathbf{u}}{Dt} = -\frac{1}{\rho_0}\nabla p - \frac{\rho}{\rho_0}g\hat{\mathbf{z}} - 2\mathbf{\Omega} \times \mathbf{u} - \nabla \cdot \mathbf{D_u}(\mathbf{u}),$$

$$\textbf{Continuity:} \quad \nabla \cdot \mathbf{u} = 0,$$

$$\textbf{Heat:} \quad \frac{D\theta}{Dt} = -\nabla \cdot \mathbf{D}_\theta(\theta) + \mathcal{F}_\theta,$$

$$\textbf{Salt:} \quad \frac{DS}{Dt} = -\nabla \cdot \mathbf{D}_S(S),$$

$$\textbf{Eq. of State:} \quad \rho = \rho\,(\theta, S, z)\,. \tag{1.16}$$

For more information on the governing equations, see Vallis (2006) or McWilliams (2006). The source term $\mathcal{F}_\theta$ for heat is mainly due to radiation from sunlight near the sea surface and latent heat from evaporation (Subsection 4.2.1, Subsection 11.1.3). The radiative heating can be thought of as the divergence of an energy flux because the radiation is traveling into and out of any given water parcel. Friction can also generate heat, but in the ocean this is several orders of magnitude smaller than other terms in the equation. Because salt is not created or destroyed in the ocean, the salinity equation has no corresponding source term.

The right-hand side of the momentum equation represents the forces on a parcel of water, including (from left to right) the pressure gradient, gravity (with gravitational acceleration g), the Coriolis force (see Subsection 2.1.1), and mixing. Note that the material derivative is defined for a scalar χ rather than a vector such as $\mathbf{u}$, but we can think of the left-hand side of the momentum equation as giving material derivatives of each scalar component of $\mathbf{u}$. For density variations of magnitude $\delta\rho$, we assume that $\delta\rho/\rho_0 \ll 1$ (the **Boussinesq approximation**; this ratio is a few percent for the ocean; see Subsection 1.2.3 and Exercise 1.13).

The assumption of incompressibility eliminates acoustic waves and applies as the **Mach number** vanishes. The Mach number is $Ma = |\mathbf{u}|/C_s$ where the sound speed $C_s = \sqrt{\partial p/\partial \rho}$ is near 1500 ms^{-1}. Hence typical values of the Mach number are $O(10^{-3})$ and the ocean general circulation is, to good accuracy, non-divergent.

The mixing terms include molecular diffusion with values of momentum diffusivity (also called **kinematic viscosity**) $\nu \approx 10^{-6}$ m^{2}s^{-1}, **thermal diffusivity** $\kappa_\theta \approx 10^{-7}$ m^{2}s^{-1}, and **freshwater diffusivity** $\kappa_S \approx 10^{-9}$ m^{2}s^{-1}. As mentioned above, the more relevant source of mixing is turbulent mixing, which is discussed in Chapter 7.

For accurate representation of circulation on the Earth, it is convenient to solve the primitive equations in a spherical coordinate system using longitude, latitude, and height as coordinates. The $\hat{\mathbf{z}}$ in (1.16) is then the unit vertical vector pointing up, antiparallel to gravity. Writing the full equations in a spherical coordinate system introduces some extra **metric terms** which arise from the gradient operator ∇ in curvilinear coordinates (Appendix B). For idealized studies, the equations are often written in a Cartesian coordinate system; one that is tangent to the spherical Earth is a good approximation for a small domain. In any event, (1.16) is a set of vector equations that applies regardless of the specific coordinates one picks.

1.3.2 Solving the Governing Equations

Boundary Conditions

For a mathematically well-posed problem we need to supplement (1.16) with **boundary conditions** on velocity (components normal to and parallel to any boundary), temperature and salinity. Typically the boundary condition imposes a value (as a function of location and perhaps time) of the quantity itself or its derivative in the direction normal to the boundary. These conditions are referred to as **Dirichlet** and **Neumann** conditions, respectively. Specifying the Neumann condition is equivalent to specifying the flux of the quantity, so the Neumann condition is also referred to as a flux boundary condition.

We distinguish between solid boundaries such as the sea floor or (in models or theories) a vertical side wall, and the "free-surface" at the air/sea interface. We generally assume solid bodies are **impermeable** (no normal flow; $\mathbf{u} \cdot \hat{\mathbf{n}} = 0$) and **insulating** ($\nabla(\kappa\theta) \cdot \hat{\mathbf{n}} = 0$ and a similar expression for S). The tangential component of velocity at the boundaries can be set to zero, which is called the **no-slip** condition and applies to real seawater. However, the region next to the boundary where the velocity vanishes is usually too narrow to be represented realistically by the resolved flow, so the effect of unresolved processes can be represented by permitting the resolved circulation to freely slide along the boundary. This is called a **free-slip** boundary condition. The choice of velocity boundary condition affects the momentum flux (stress) at the boundary: no-slip walls support a nonzero stress, but free-slip walls do not.

The location of the free surface is not fixed but occurs at the height $z = \eta(x, y, t)$. The equivalent of the no-normal-flow condition is the **kinematic free-surface** boundary condition in this case. It states that water at the surface cannot flow through the surface, that is, $w = D\eta/Dt$. In reality, water *can* flow through the surface as precipitation P, evaporation E, and runoff R from rivers (collectively given by freshwater flux $E - P - R$). There is also interaction with ice (see Chapter 10). These fluxes (with units of volume per unit area per unit time, or m s^{-1}) represent the speed at which the sea surface would rise or fall in the imaginary case with no lateral exchange of water. Combining the kinematic free surface condition and these terms, we get the free-surface boundary condition

$$w = \frac{D\eta}{Dt} + P - E + R. \tag{1.17}$$

Variations in sea-surface height in the real ocean are usually no more than about a meter, much less than the vertical extent of most circulation features, which are tens to thousands of meters deep. Therefore, the free-surface boundary condition can be applied at $z = 0$, rather than at $z = \eta$, to good accuracy. Similarly, w at the sea surface (averaged over distances large compared to the surface gravity waves that make a small boat bob up and down) is very small compared to subsurface values associated with the general circulation, and so the free-surface boundary condition is often simplified as $w = 0$.

Boundary conditions at the sea surface are crucial for driving the circulation. The stress at the surface (a momentum flux) is associated with the near-surface wind and is usually specified from observations or from an atmospheric model. For temperature, in some cases the value itself is specified (a Dirichlet condition), in others the heat flux is specified (a Neumann condition), and sometimes a combination of the two (a **Robin** boundary condition; see Subsection 4.3.1 for an example). Choices for salinity are similar (see subsection 4.3.2), except that physically it is the freshwater flux, not the salt flux, that is specified: freshwater flux affects salinity through concentration and dilution.

We also must state **initial conditions** in order to integrate the primitive equations. That usually requires specifying the currents, temperature, salinity, and free surface height at the starting time.

Time-Stepping Solutions

Given these boundary and initial conditions one can, in principle, solve (1.16) for future currents, temperature, and salinity. The equations are nonlinear, however, due to the advective term $\mathbf{u} \cdot \nabla$ from the material derivative in the momentum, temperature, and salinity equations, and the nonlinear equation of state $\rho(\theta, S, z)$. Nonlinearity makes solving the governing equations very challenging. It is not even clear that the mathematical problem outlined above is well posed. At the time of writing, the Clay Mathematics Institute offers a $1M prize for proving the existence of solutions to the (related) incompressible Navier–Stokes equations describing fluid motion!

In practice, the solution to the equations governing ocean general circulation are often approximated numerically (Subsection 1.2.2). To get an impression of this process, think as follows: The evolution of the ocean state (represented by u, v, w, θ, S, p, and ρ) occurs because some of the equations in (1.16) are of the form

$$\frac{\partial \Psi}{\partial t} = F, \tag{1.18}$$

where Ψ is a variable and F represents all the other terms in the equation. If we know the state of the system at time t, we can estimate Ψ a short time Δt later with

$$\Psi(t + \Delta t) \approx \Psi(t) + F\Delta t \qquad (1.19)$$

(usually, more accurate but less simple rules are used). We can estimate u, v, θ and S at time $t + \Delta t$ in this way. These then allow us to calculate $\rho(\theta, S, z)$ at time $t + \Delta t$. Estimating the pressure p involves integrating the hydrostatic relation (2.4), but is tricky because the height of the sea-surface $\eta(t + \Delta t)$ is unknown. Usually, a separate equation for η is written in the form of (1.18), which assumes that the ocean has uniform density. Stepping the η equation, as in (1.19), gives an estimate of $\eta(t + \Delta t)$, and then the pressure p can be computed. Finally, the vertical velocity w is found from the incompressibility relation and the horizontal velocity components u and v. This scheme can then be iterated to estimate the variables at the next time step, $t + 2\Delta t$, and so on.

1.3.3 Conserved and Derived Quantities

We refer to (1.8) as a *global* conservation law because it applies to the physical system as a whole. If mixing and source terms are sufficiently small for some variable χ, than (1.15) can be written

$$\frac{D\chi}{Dt} = 0. \qquad (1.20)$$

This is a conservation law that applies *locally* to every parcel moving with the flow, and we say that the quantity in question is **materially conserved**.

In the momentum equation, the magnitude of the advection terms relative to the diffusion terms is given by the nondimensional **Reynolds number** $Re = |\mathbf{u}|L/\nu$. Large-scale flows, for which length scale L ranges from tens of meters in the vertical to thousands of kilometers in the horizontal, have molecular Reynolds numbers ranging from about 10^6 to greater than 10^{12}. Analogous nondimensional numbers exist for θ and S. Using eddy values instead of molecular values for the diffusivities still gives large Reynolds numbers (and its analogues) for much of the ocean.

In some theoretical models of the ocean, these diffusion coefficients are set to zero for convenience. Such models are called **inviscid** ($\nu = 0$) or **adiabatic** ($\kappa_\theta = \kappa_S = 0$). The properties of the inviscid, adiabatic equations are important and provide dynamical insight into the real, weakly-dissipative (high Reynolds number) ocean. Much of the dissipation occurs in boundary layers (for example, Ekman layers, Subsection 2.1.3), so there are large regions where it makes some sense to neglect dissipation. Temperature, salinity, and other tracers are unchanging as one follows the flow in the interior (away from boundary fluxes) in these idealized, adiabatic models of the ocean. As we see later (in Chapter 5, for example), these are powerful constraints on the circulation itself. Nevertheless, viscosity and diffusion play critical roles in practice. For example, a steady-state ocean circulation driven by the input of momentum by the wind must have a mechanism to drain the momentum (see Chapter 3, for example) .

Not all globally conserved quantities are materially conserved, even if dissipation vanishes. Notably, energy and momentum are not materially conserved for adiabatic, inviscid conditions. Transport of these quantities is not limited to advection and mixing because pressure, gravity, and Coriolis forces constitute source terms. These quantities are influenced by waves of various kinds which can carry energy and momentum from place to place and disrupt the material conservation law. Temperature, salinity, and tracer concentration are not affected by waves in this way.

The ocean is not a closed system because it interacts with the atmosphere, the sea floor, and outer space. Integrating (1.8) over the entire ocean, shows that the net exchange with the ocean of the following quantities is relatively small compared to the total that the ocean stores:

- The total mass and volume of seawater. At the ocean surface (and sea floor) the important freshwater fluxes are precipitation, runoff from land, evaporation, and in polar regions freezing or melting of ice. Exchange with terrestrial ice can slowly change mass of the ocean a small amount (for example, sea level changed by around 100 m over a few thousand years during the last de-glaciation). The combination of incompressibility and Boussinesq approximations eliminate volume change (though not mass change) from the system governed by (1.16).
- The total mass of dissolved salts. Runoff from land provides a source of dissolved salts and the sediments at the sea floor provide a net sink. For example, the total amount of silicon, an important biogeochemical, does not change over the O(1000)-year timescale that the ocean circulates.
- The total amount of momentum. At the surface the wind adds momentum by exerting stresses on the water. At the sea floor viscous drag removes momentum as do pressure differences across topography (see Figure 9.16).
- The total amount of mechanical energy. Mechanical energy consists of kinetic energy, due to the currents, and potential energy, due to the presence of dense fluid above light fluid. Mechanical stress at boundaries is a source or sink of kinetic energy and freshwater and heat fluxes at boundaries affect the potential energy. Constructing budgets of energy and the conversion of different types of energy between energy reservoirs is an important and challenging task in physical oceanography (see Subsection 4.3.3 and Subsection 8.3.1).

1.4 Excursions: Paths to Oceanography

For many, the path to oceanography begins with wanting to go to sea. That desire can be fed by early experiences fishing, surfing, sailing, or scuba diving, for example. For others, inspiration flows from celebrated accounts of pioneers who endured great hardships to explore. For them, the call of the wild is irresistible.

Losing sight of land below the horizon for the first time is memorable. Indeed, there are many memorable sights at sea: smoking volcanic islands, soaring albatross, majestic bad weather, to name a few. As the outside world recedes, literally and figuratively, oceanographers immerse themselves in their field. You devote yourself to data collection, quality control, and analysis; sometimes in discomfort, often with camaraderie. The occasional moments of discovery, of elucidation, are sublime.

When the authors of this book entered oceanography in the late 1980s, the field was on the brink of a technological revolution. Oceanographic satellites were experimental, GPS quality questionable, and seagoing email and internet access unavailable. On research expeditions we remember annotating the paper depth sounder record, transferring data between computers by walking across the room holding a floppy disk, and, occasionally, resorting to heavy "messengers" to slide down the wire and mechanically trigger water bottles to close. The digital revolution has made all these tasks obsolete.

The digital revolution has changed oceanography in other important ways too. For example, ocean circulation models have co-evolved with supercomputers to (sometimes) become so realistic that pseudo-data from them are hard to distinguish from the real thing. Robotic exploration of the global ocean from orbit, or in situ, is now the norm and the future. Oceanographers now access and share data, and have tools to manipulate them, in ways and at rates that were unthinkable thirty years ago.

The lure of the sea is still tempting. But for many oceanographers, hybrid expertise in observations, theory, and modeling is now more attractive. Indeed, as one esteemed scholar of oceanography used to tell us: "you don't need to be an astronaut to be an astronomer."

Exercises

1.1 How much table salt (NaCl) must you add to 1 litre of freshwater to make a solution with the global average salinity of seawater? How does the solution differ from global average seawater?

1.2 What instruments are most useful for deducing the distribution of ocean velocity and why (name as many as you like)?

1.3 What is the justification for neglecting high-frequency variability (subseasonal and shorter timescales) in discussing the general circulation? What are the dangers of neglecting it? If ocean circulation is not constant in time, what do we mean when we discuss the "steady circulation?"

1.4 One might say that the real ocean is bounded only by a sea-floor and by the sea surface, but ocean models often have side walls too. Is there any physical justification for this? Why is this approximation made?

1.5 Consider the definition of buoyancy frequency (1.1). Show that the second term involving the speed of sound amounts to an N^2 correction of g^2/C_s^2, making a suitable assumption about the pressure field. Explain the origin of this term.

1.6 What kind of processes or phenomena can cause errors in numerical general circulation models at a given spatial resolution? How does one determine if a particular resolution is high enough? What are the tradeoffs as spatial resolution is increased?

1.7 Imagine a 4000m deep column of seawater that has uniform potential density (i.e., the same at all depths) and is resting. The (in-situ) temperature decreases by 2°C for every 1000m depth increase. The surface temperature and salinity are 10°C and 35.00 g/kg, respectively. What are the temperature, salinity, and pressure at the bottom? (Assume seawater is incompressible, and make a sensible assumption for the equation of state.)

1.8 (a) Explain the meaning of potential temperature and how it differs from in-situ temperature.

(b) What is the advantage of conservative temperature over potential temperature?

Hint: Consult Pawlowicz, (2010).

(c) Why is a "potential salinity" variable not necessary?

1.9 Download the MATLAB code `make_profile_plot.m` and datafile `deep_cast.csv` to plot hydrographic profiles for the subtropical Atlantic Ocean.

(a) Add code to plot potential temperature, conservative temperature Θ, absolute salinity S_A, potential density anomalies σ_0, σ_4, and in situ density ρ.

(b) Mark the mixed layer, thermocline, halocline, and pycnocline.

(c) Write code to make a (Θ, S_A) diagram showing the hydrographic profile and contours of σ_θ. Include a line marking the freezing point of seawater.

(d) Repeat (a)-(c) for full-depth hydrographic profiles for the equatorial Pacific, the subpolar north Atlantic, the subtropical south Pacific, and the Arctic oceans. Pick suitable locations and download the hydrographic profile data from the WODselect website. (Ship-based CTD data is the best option and you will need to spend some time selecting appropriate stations. State your assumptions and the reasons for your choices.).

1.10 Using a global gridded synthesis of temperature and salinity data (for instance the WOA13 climatology, see Appendix A), make figures of the following:

(a) Vertical profiles of T, Θ, S_A, σ_θ, and ρ for: the subtropical North Atlantic, the equatorial Pacific, the subpolar north Atlantic, the subtropical south Pacific, and the Arctic oceans. Compare and contrast to your corresponding profiles from Exercise 1.9.

(b) Surface maps of T, S_A, and σ_θ for the global ocean.

(c) Vertical, meridional (latitude, depth) sections of Θ, S_A, σ_θ, and σ_4 for 30°W and 170°W.

(d) Vertical, zonal (longitude, depth) sections of Θ, S_A, and σ_θ for 30°N and 30°S. In each case, label hydrographic and circulation features.

1.11 What is the main source of error in constructing climatological estimates of the circulation? (Consider your answers to Exercises 1.9 and 1.10.)

1.12 Using a global gridded synthesis of temperature and salinity data (for instance, from Exercise 1.10), and explaining your methods, estimate:

(a) The mean, mode, and median values of: T, Θ, S_A, σ_θ, and ρ for the global ocean.

(b) Repeat (a) for the 1 and 99 percentiles.

1.13 (a) How is the speed of sound C_s related to the equation of state $\rho(T, S, p)$?

(b) Using your answers from Exercise 1.12, estimate the "Boussinesq number" $= \Delta(\sigma_\theta + 1000)/\rho_0$ for the ocean, and explain its significance.

(c) Repeat (b), now defining the Boussinesq number as $= \Delta\rho/\rho_0$ where is ρ the in-situ seawater density. Comment on your result.

(d) Name a non-Boussinesq fluid.

1.14 Estimate how much global sea level is lowered by the compressibility of seawater, making suitable assumptions that you clearly state.

Rotating and Shallow-Water Flow

This chapter continues our review of introductory material. The rotation of the Earth has a large influence on ocean circulation through the Coriolis force. Paradoxically, the complication of the Earth's rotation allows for simplifications in the equations of motion. Much of the ocean circulation is in a balance between Coriolis forces, pressure gradients (the geostrophic flow), and wind stress (the Ekman flow). While ocean circulation has a rich vertical structure, much of the flow can be described by a simplified configuration in which the ocean is represented by a small number of uniform-density layers. Approximations associated with this representation comprise the shallow water equations, which arise because of the ocean's small aspect ratio (Subsection 1.1.6). They are used extensively in the rest of the book.

In this chapter we depart from the practice of first introducing observations to be explained by theory.

2.1 Concepts

2.1.1 Coriolis Force

The ocean general circulation is strongly influenced by the Earth's rotation, primarily through the influence of the **Coriolis force**.

The motion of an object is governed by Newton's Second Law $\mathbf{F} = m\mathbf{a}$, which is a statement that acceleration $\mathbf{a}$ (rate of change in magnitude or direction of velocity) is produced by the force $\mathbf{F}$ on the object (with mass m). This law is only true in a frame of reference which itself is not accelerating. (Accelerating relative to what? Apparently to some average of the mass of the universe: Rothman, 2017.) It is convenient to measure ocean currents in a reference frame which is stationary relative to the Earth, hence rotating relative to the universe. It can be shown that, in this rotating frame, $\mathbf{F} = m\mathbf{a}$ must be modified by including **fictitious forces**, which are terms that do not represent physical interactions such as gravity or magnetism. Fictitious forces account for the tendency of objects to *appear* to accelerate due to the *observer's* acceleration associated with the reference frame's rotation. A familiar fictitious force is the **centrifugal force**, which causes riders to stay in a roller coaster when it is upside down in a vertical loop, for example. The general circulations of the ocean and atmosphere are strongly influenced by another fictitious force known as the Coriolis force.

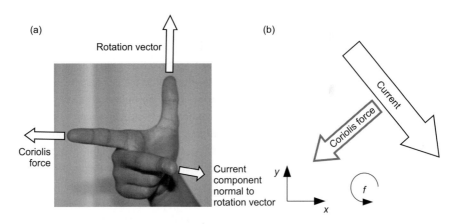

Figure 2.1 (a) Right hand rule for the Coriolis force. The vector Coriolis force equals $-2m\mathbf{\Omega} \times \mathbf{u}$ where $\mathbf{\Omega}$ is the angular velocity vector of the rotating frame, m is the mass of the moving object, and $\mathbf{u}$ is the vector flow. (b) Coriolis force resolved in the local horizontal (x, y) plane. The force is of magnitude $mf|\mathbf{u}|$ to the right of the horizontal current $\mathbf{u}$. Here f is the Coriolis parameter (see Equation 2.1). These rules also work for geostrophic flow (Subsection 2.1.2): in that case an equal pressure-gradient force opposes the Coriolis force (not shown in these diagrams).

The Coriolis force is a sideways push on objects moving over the Earth's surface. Specifically it is a fictitious deflective force in a rotating frame. For a rotating frame of reference with a rotation period of T and an angular velocity vector $\mathbf{\Omega}$ with magnitude $\Omega = 2\pi/T$, an object of mass m moving with velocity $\mathbf{u}$ relative to the rotating frame experiences a Coriolis force given by $-2m\mathbf{\Omega} \times \mathbf{u}$. The expression $-2\mathbf{\Omega} \times \mathbf{u}$ is therefore the Coriolis acceleration, or the force per unit mass. The Coriolis force is always perpendicular to the rotation axis and the flow direction. It can be constructed using a simple right hand rule as follows (Figure 2.1a): take the thumb of your right hand in the direction of the rotation vector $\mathbf{\Omega}$, your second finger in the direction of flow $\mathbf{u}$ (more precisely, the component of $\mathbf{u}$ normal to $\mathbf{\Omega}$), and then your first finger indicates the direction of the Coriolis force $-2m\mathbf{\Omega} \times \mathbf{u}$. This rule works in either hemisphere, just remember that the Coriolis force includes components in both the local vertical and the local horizontal directions in general (the local vertical direction on a sphere is in the direction of increasing distance from the center of the Earth).

On the rotating Earth, the Coriolis force always lies in a plane parallel to the Earth's equatorial plane (which is perpendicular to $\mathbf{\Omega}$), so it always has a vertical component (except at the poles). For a horizontal current, the horizontal component is to the right in the northern hemisphere and to the left in the southern. The vertical component of the Coriolis force is almost always neglected because it is dwarfed by other forces. Similarly, the Coriolis force due to vertical motion is almost always neglected. But the horizontal Coriolis force due to horizontal motion has profound importance for large-scale ocean circulation. Figure 2.1b illustrates this rule as an alternative to the vector rule with your right hand. The horizontal component of the Coriolis force depends on the vertical ($\hat{\mathbf{z}}$) component of the rotation vector. This quantity (times two) is called the **Coriolis parameter**. At latitude θ it is

Coriolis Parameter: $f = \hat{\mathbf{z}} \cdot (2\boldsymbol{\Omega}) = 2|\boldsymbol{\Omega}| \sin\theta.$ (2.1)

The Coriolis parameter has a maximum value of $1.46 \times 10^{-4} \text{ s}^{-1}$. The horizontal component of the Coriolis force is then $f|\mathbf{u}|$; on the right of the current in the northern hemisphere and on the left in the southern hemisphere where f is negative.

Physically, the Coriolis force arises from an unbalanced centrifugal force in a rotating system. Because the gravitational potential is conventionally defined to include the centripetal force associated with rotation at rate $\boldsymbol{\Omega}$, any relative motion ($\mathbf{u}$) with respect to the rotating body causes a change in the centripetal force. This change is conventionally called the Coriolis force. The Coriolis force can change only the direction of velocity but not the speed, hence it does no work.

For any moving object in a rotating reference frame, the Coriolis effect is important if the other competing forces are relatively small. For a steady circulation of speed $|\mathbf{u}|$ and length scale L, this condition is equivalent to the nondimensional **Rossby number** $Ro = u/fL$ being smaller than 1. Coriolis forces are unfamiliar to people because other forces (like frictional forces on the soles of our feet) are relatively large. However, Coriolis forces can influence rapidly moving human-sized objects over short periods, such as artillery rounds (see Exercises 2.1 and 2.2).

Coriolis forces are responsible for inertial oscillations, which are periodic movements of seawater in horizontal circles in response to an impulsive push, such as changing wind from a storm. The water is continuously deflected to the right (in the northern hemisphere) by the Coriolis force. The frequency of the inertial oscillation equals the Coriolis parameter f: in fact, sometimes f is called the inertial frequency for this reason.

2.1.2 Geostrophy, Hydrostasy, and Thermal Wind

As shown in Subsection 2.2.1, under some circumstances the momentum equation (1.16) can be approximated as a balance of a few terms. Three such balances govern much of large-scale physical oceanography.

When the Coriolis force (Subsection 2.1.1) is dominant (Rossby number $\ll 1$) and friction is sufficiently small, the flow satisfies the dynamical constraint of **geostrophy**. Geostrophy is a balance between the horizontal component of the Coriolis force per unit mass and the horizontal pressure gradient force per unit mass. For fluid density ρ_0, pressure p, and Coriolis parameter f (2.1), we have

Geostrophic Balance: $f\hat{\mathbf{z}} \times \mathbf{u}_G = -\dfrac{1}{\rho_0}\nabla_h p,$

$$f\mathbf{u}_G = \hat{\mathbf{z}} \times \frac{1}{\rho_0}\nabla_h p. \qquad (2.2)$$

The second equation follows from the first because of the vector triple product rule. Namely, for any three vectors $(\mathbf{a}, \mathbf{b}, \mathbf{c})$ we have $\mathbf{a} \times (\mathbf{b} \times \mathbf{c}) = (\mathbf{a} \cdot \mathbf{c})\mathbf{b} - (\mathbf{a} \cdot \mathbf{b})\mathbf{c}$. Note that $\hat{\mathbf{z}} \times \frac{1}{\rho_0}\nabla_h p$ (and similar expressions involving $\hat{\mathbf{z}}\times$) means the horizontal vector $\nabla_h p$ is treated as a three-dimensional vector with zero vertical projection.

Equation 2.2 shows that the geostrophic velocity has only horizontal components, which in Cartesian coordinates are

$$fv_G = \frac{1}{\rho_0} \frac{\partial p}{\partial x},$$

$$fu_G = -\frac{1}{\rho_0} \frac{\partial p}{\partial y} \tag{2.3}$$

(the derivatives are at constant height z; see (5.7) for spherical coordinates).

In large scale ocean circulation, the vertical component of the pressure gradient force is balanced not by the Coriolis force but by the much larger force of gravity. This is called the **hydrostatic balance** or **hydrostasy**. For gravitational acceleration $g = 9.8 \text{ ms}^{-2}$ (considered to be a constant for our purposes),

Hydrostatic Balance: $\dfrac{\partial p}{\partial z} + \rho g = 0.$ $\tag{2.4}$

In hydrostatic fluids, pressure at any point is simply the weight per unit area of all the fluid above that point (see also Subsection 1.1.2). This can be demonstrated by integrating (2.4) vertically between a depth z and the sea surface at $z = \eta(x, y)$, so that pressure is given by

Hydrostatic Pressure: $p = \displaystyle\int_z^{\eta} g\rho \; dz.$ $\tag{2.5}$

The weight of the atmosphere should also be included for an accurate calculation of p, but we will ignore the atmospheric pressure contribution in this book. Because the main influence of pressure on dynamics (such as in geostrophy) is through the horizontal pressure gradient, we are often interested only in the horizontal variation in p (explained in Subsection 2.2.2). In that case, it is useful to write the density in terms of constant reference density ρ_0 and varying part $\rho'(\mathbf{x}, t)$,

$$\rho = \rho_0 + \rho'. \tag{2.6}$$

Then (2.5) allows us to define

$$p = p_0(z) + p'(\mathbf{x}, t), \tag{2.7}$$

in which $p_0(z \leq 0) = -g\rho_0 z$ is a background pressure that always increases with depth (remember that z decreases as depth increases), and p' is the **dynamic pressure** given by

$$p'(\mathbf{x}) = \int_0^{\eta} g\rho_0 \; dz + \int_z^{\eta} g\rho' \; dz. \tag{2.8}$$

The first term in (2.8) represents pressure variations associated with sea-surface height, and the second term represents pressure variations associated with density. In other words, pressure at some level z is influenced by whether a taller or shorter column of water is piled above, and whether the overlying water is denser or lighter. Often neither term is negligible: the first term integrates a relatively large factor, $g\rho_0$, over a small distance η, and the second term integrates a relatively small factor, $g\rho' \ll g\rho_0$, over a big distance $\eta - z \approx |z|$. Note that the first term is independent of z: sea surface tilt has the same influence on pressure all the way down to the bottom of the ocean. All of the dependence on depth of dynamic pressure comes from the density variations. The example calculation in Subsection 2.2.2 makes these points clearly.

In observations, ρ' is usually better known than η, so oceanographers frequently have more complete knowledge of the density term in (2.8). The direct effect of density on geostrophic velocity is captured by the **thermal wind equation**.

We can eliminate pressure p' from geostrophy by taking the z derivative of (2.2) and combining with the hydrostatic equation (2.4). The result is (see Exercise 2.6)

$$\textbf{Thermal Wind Equation:} \qquad f\hat{\mathbf{z}} \times \frac{\partial \mathbf{u}}{\partial z} = \frac{g}{\rho}\nabla_p\rho',$$

$$f\frac{\partial \mathbf{u}}{\partial z} = -\frac{g}{\rho}\hat{\mathbf{z}} \times \nabla_p\rho', \tag{2.9}$$

or in terms of individual components,

$$f\frac{\partial v_G}{\partial z} = -\frac{g}{\rho}\frac{\partial \rho'}{\partial x},$$

$$f\frac{\partial u_G}{\partial z} = \frac{g}{\rho}\frac{\partial \rho'}{\partial y}. \tag{2.10}$$

Unlike ∇_h, which represents the gradient in the horizontal direction for constant height, ∇_p is calculated on surfaces of constant pressure. If the Boussinesq approximation is made, then the gradients at constant pressure in these equations equal gradients at constant height z.

The thermal wind equations link vertical variations in the horizontal geostrophic current to the horizontal gradient of the density. An alternative form for layers of constant density appears in Subsection 2.2.2 (Equation 2.46). These formulae are central to observational physical oceanography because the density derivatives ($\partial\rho/\partial x$, $\partial\rho/\partial y$) (at constant pressure) can be estimated from hydrographic data, whereas the pressure derivatives ($\partial p/\partial x$, $\partial p/\partial y$) (at constant depth) in the geostrophic formulae cannot. This problem arises because, in practice, oceanographers cannot measure pressure p as a function of depth z accurately enough. Instead oceanographers use pressure measurements as a surrogate for depth (Subsection 1.1.2). We can estimate $\partial\rho/\partial x$ at constant z using $\partial\rho/\partial x$ at constant p and the Boussinesq approximation. But the corresponding approximation for the horizontal pressure derivatives obviously does not work. We see how this issue plays out in detail in the worked numerical example below (Subsection 2.2.2).

We can see from thermal wind that a uniform-density geostrophic flow has zero vertical shear. **Shear** is the gradient in current speed perpendicular to the direction of the current (see Subsection 2.1.3). This can lead to the interesting phenomenon of the **Taylor column**. A current flowing towards a small obstacle such as a low sea-mount will go around, not over, the obstacle and continue downstream. An unstratified geostrophic current has no vertical shear, as explained above. Therefore, in such a case, the flow far above the seamount takes the same path as the deep flow. The water curves around as if the sea-mount reached all the way to the sea surface. The cylinder of isolated fluid above the seamount is the Taylor column. Taylor columns are difficult to observe in their pure state in the ocean because stratification permits vertical shear in the horizontal geostrophic current. But they are readily seen in the laboratory, they illustrate an important feature of geostrophic flow, and they point to the strong influence of bottom topography in the real ocean, especially where the stratification is weak.

The fluid is referred to as **barotropic** if the isobars (lines of constant pressure) coincide with the isopycnals (lines of constant density). The fluid is referred to as **baroclinic** if the isobars are inclined with respect to the isopycnals. In a baroclinic fluid, the mis-alignment of the isolines generates a current that varies with depth (through the thermal wind equations for a geostrophic circulation). The use of *barotropic* and *baroclinic* can be confusing in the literature, however. Sometimes *barotropic* is used to mean "depth-averaged" and *baroclinic* to mean "depth-dependent." Sometimes, the current is split into *barotropic and baroclinic components* this way (for example, the "barotropic streamfunction" in Subsection 3.2.2). Although the general sense is clear, be aware that different authors use the same words to mean different things.

Finally, note that a right hand grip rule exists for the thermal wind equation (Figure 2.2): In the northern hemisphere, take your right hand with your fingers curled. Point your thumb in the direction of the horizontal component of the density gradient (thumb points to denser water). The curl of the fingers then represents the sense of the thermal wind shear in the geostrophic current with depth.

The geostrophic (and hydrostatic) equations are approximations to the full momentum equations, but they have a fundamentally different character: They are *diagnostic* relations, not *prognostic* relations. This means that geostrophy and hydrostasy contain no information about how the flow will evolve in time: the acceleration has been neglected because it is small compared to the pressure gradient force. If we know $p'(x, y, z, t)$, we can calculate the geostrophic velocity, but we generally need some other information to calculate p'. To construct a theory of some circulation feature in the ocean, we usually need to calculate p' together with velocity in order to find both of them.

The geostrophic relations are so useful because they are diagnostic: We can glean information about the currents immediately by considering the pressure field, for instance through observations in the real ocean. However, geostrophy may seem strange. People familiar with classical Newtonian mechanics usually think of motion *changing* under the

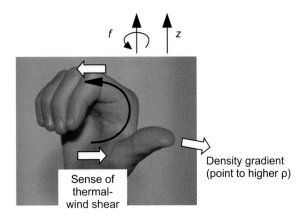

Density gradient (point to higher ρ)

Sense of thermal-wind shear

Figure 2.2 Right hand grip rule for thermal wind. In the northern hemisphere point your right thumb towards the dense (cold) water in the horizontal plane. The curl of the fingers then indicates the sense of the thermal wind shear. Use your left hand in the southern hemisphere.

action of a force (Newton's second law). In a geostrophic flow the motion (current speed) is *proportional* to the pressure gradient force! If the force stops, then so does the flow (the force would have to change sufficiently slowly so that geostrophic balance applied at all times). Moreover, the current moves at right angles to the applied force, as in Figure 2.1: again, this is counterintuitive. The precession of a rapidly spinning gyroscope is an analogous phenomenon (Haine and Cherian, 2013).

2.1.3 Ekman Boundary Layers

In Subsection 1.1.2 we defined pressure on a volume of fluid as the force per unit area perpendicular to the boundary of the volume. Another important quantity, usually called **stress**, is the force per unit area parallel to the boundary. In fluids, stress is generally proportional to shear. An example of shear is a flow $u(z)$ in the x direction that varies in the z direction (Figure 2.3). The shear in this case is du/dz (units of s^{-1}). Interactions between neighboring molecules generates the stress, sometimes referred to as **viscous** or **frictional** stress. We can think of stress as the flux of momentum in a direction perpendicular to the momentum.

The stress can be caused by random motion of individual molecules; the strength of this random motion is characterized by the kinematic **viscosity**, which is the proportionality constant between stress and shear. Turbulent flow is much more effective than molecular motion at transporting momentum and other properties. Two key sources of stress on the ocean are the momentum source due to wind blowing over the ocean surface and the momentum sink of water flowing over the sea floor.

In the ocean, the Earth's rotation confines most of the frictional stress to relatively thin zones (50–100 m thick) at the top and bottom of the ocean. Such confinement is in contrast to the non-rotating case, in which stress at the surface of a body of water would, after a long enough time, transmit stress through the entire water column. The stress regions are known as Ekman layers because Vagn Walfrid Ekman made the pioneering theoretical explanation of them in 1905 for his doctoral thesis.

The flow within the Ekman layer can be split into two conceptual pieces: First, there is flow due to the geostrophic current. This component usually does not vary much vertically over the Ekman layer because the layer is relatively thin. Second, there is a flow that is driven by stresses within the layer. This stress-driven flow fades away to zero moving out of the Ekman layer into the ocean interior (the Ekman layer is defined as the region with significant stress-driven flow).

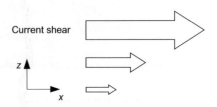

Current shear

z

x

Figure 2.3   Schematic of a horizontal current $u(z)$ in the x direction that is sheared in the vertical z direction. Here the shear, du/dz is positive.

For present purposes, the most important result of Ekman theory connects the stress-driven flow to $\boldsymbol{\tau}$, the stress at the boundary itself. For the Ekman layer at the sea surface the relation says that, in steady state,

$$\underline{\textbf{Ekman transport:}}\ f\hat{\mathbf{z}} \times \mathbf{U}_{Ek} = \mathbf{X},$$
$$f\mathbf{U}_{Ek} = -\hat{\mathbf{z}} \times \mathbf{X}, \qquad (2.11)$$

where $\mathbf{X} = (X, Y) = \boldsymbol{\tau}/\rho$; in this book we will often refer to $\mathbf{X}$ as the stress though it is actually the stress divided by density (with units of $m^2 s^{-2}$). In terms of individual components,

$$U_{Ek} = \frac{Y}{f},$$
$$V_{Ek} = -\frac{X}{f}. \qquad (2.12)$$

In these formulae $\mathbf{U}_{Ek} = (U_{Ek}, V_{Ek})$ is the stress-driven current integrated in the vertical over the Ekman layer (called the **Ekman transport** with units of $m^2 s^{-1}$; it is not the same as the Ekman velocity itself, $\mathbf{u}_{Ek}$, see (2.38)).

Figure 2.4a illustrates this relationship. Begin with the wind stress vector which is estimated from the sea-level wind speed or measured directly using a satellite scatterometer (Subsection 3.1.1 and see Figure 3.1 for an example). The Ekman transport lies at right angles to this stress on the right- (left-) hand side in the northern (southern) hemisphere (see Figure 6.4). A strong breeze of 10 ms^{-1} corresponds to a stress of about 0.1 $N m^{-2}$ (Subsection 3.1.1). The Ekman transport is therefore about 1 $m^2 s^{-1}$ in mid-latitudes giving an average speed for the stress-driven current of 1–2 cms^{-1} over the 50–100 m thick layer. The Ekman transport is at right angles to the wind. This surprising result is due to the Coriolis effect: over the whole Ekman layer the drag of the wind on the sea is balanced by an equal and opposite Coriolis force. Associated with the Coriolis force is a current at

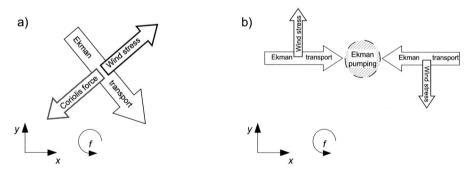

Figure 2.4 Schematic of Ekman transport in the surface wind-driven Ekman layer. (a) The wind stress on the sea $\boldsymbol{\tau}$ is balanced by a Coriolis force associated with the Ekman transport at right angles $\mathbf{U}_{Ek}$ (see Equation 2.11). (b) Where the wind stress direction changes in space, the Ekman transport can converge or diverge. This drives a weak vertical current, called Ekman pumping or suction, that is very important for the ocean circulation (see Equation 2.13). The case of the mid-latitude northern hemisphere is shown.

right angles: the Ekman transport. This idea provides the essential explanation of Nansen's observation that Arctic sea ice drifts to the right of the wind.

For ocean circulation there is a more profound effect. Where the sea-level wind changes direction from place to place, the Ekman transport can converge or diverge. Figure 2.4b shows a schematic example. Convergence drives water out of the Ekman layer into the interior (as shown) and divergence draws it in. Specifically, the speed of vertical flow into the Ekman layer from below is

$$\text{\underline{Ekman suction velocity:}} \; w_{Ek} = \nabla \cdot \mathbf{U}_{Ek},$$

$$= \hat{\mathbf{z}} \cdot \nabla \times \left(\frac{\mathbf{X}}{f} \right). \tag{2.13}$$

In words, the Ekman suction velocity is related to the *curl of the wind stress* (the right hand grip rule of Figure 2.7 applies). The vertical Ekman current is weak, typically 10^{-6} ms^{-1}, 10 cmd^{-1}, or 30 myr^{-1}. But it directly connects to the general circulation because the interior geostrophic current must diverge or converge to accommodate the flow leaving or entering the Ekman layer. How **Ekman suction** (w_{Ek} positive) or **pumping** (w_{Ek} negative) relates to the interior geostrophic flow is explained in Chapter 3.

Many models for ocean general circulation begin by specifying the Ekman pumping or suction caused by the wind. They assume that the Ekman layer is thin compared to the ocean depth and the processes in the Ekman layer do not interact with the interior flow. This is an example of a common argument: The effects of viscosity are confined to small isolated places called **boundary layers**. Relative to their small size, boundary layers play an important role in the circulation. Another important boundary layer is the mixed layer, which is explained in Chapter 4. Often oceanographers assume that damping by viscosity can be neglected outside these thin boundary layers. This inviscid assumption is a very useful approximation, but is not entirely accurate. The viscosity of seawater is small (the Reynolds number is large; see Subsection 1.3.1), but not zero.

2.1.4 Divergence, Vorticity, and Streamfunction

In describing and explaining the velocity field $\mathbf{u}(\mathbf{x}, t)$ of the ocean, it is often useful to consider three other fields that are derived from velocity. We introduce these fields here and discuss them with more complete mathematics in Subsection 2.2.3 (see also Appendix B for background on vector calculus).

The **divergence** of the velocity field $\mathbf{u}$,

$$D = \nabla \cdot \mathbf{u}, \tag{2.14}$$

represents the rate at which water is leaving a volume (units s^{-1}; convergence is minus the divergence). The small compressibility of water discussed in Subsection 1.2.3 only allows a negligible divergence of water volume, so oceanographers usually idealize water as incompressible and set the divergence to zero (see Subsection 1.3.1).

The divergence of the components of velocity along a two-dimensional surface may be nonzero if flow is entering or leaving the plane from the direction perpendicular to the

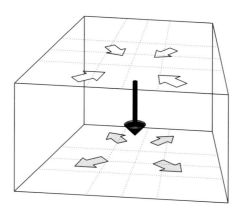

Figure 2.5   Convergence and divergence in the horizontal plane are associated with vertical motion in an incompressible fluid, like seawater. Here, the water converges near the surface, downwells, and then diverges at depth. For mathematical details, see Subsection 2.2.3. For the rotating version of this schematic see Figure 8.17.

plane. For example, as explained in Subsection 2.1.3, the horizontal flow at the sea surface is often **convergent** (water parcels move closer together), or **divergent** (parcels move away from each other). The horizontal divergence (referring here to either converging or diverging flow) $u_x + v_y$ is related to the vertical velocity; downward motion (for example) has converging water parcels above and diverging parcels below (Figure 2.5). If we neglect the relatively small flow through the surface of the ocean that is associated with evaporation, precipitation, and runoff, the vertical integral of the horizontal flow has zero divergence even if some levels have converging or diverging water.

Divergence is here applied to the divergence of the velocity vector, which is our most common usage. It can be applied to any vector, however (see Appendix B). For example, it applies to the flux of a quantity, as explained in Subsection 1.3.1 and Subsection 7.3.2.

The **vorticity**

$$\zeta = \nabla \times \mathbf{u} \tag{2.15}$$

represents the spin of the fluid (units s^{-1}). In its most general form vorticity is a vector that represents the magnitude of the spin and the orientation of the spin axis. Often we are only concerned with the vertical component of vorticity (although the horizontal component is usually much stronger; see Exercise 2.10),

$$\zeta = \hat{\mathbf{z}} \cdot \boldsymbol{\zeta} = v_x - u_y, \tag{2.16}$$

where $\hat{\mathbf{z}}$ is the unit vertical vector (see Appendix B) and the subscripts indicate differentiation (this notation is used in many places). The vorticity measures how fast a solid object such as a paddle wheel rotates as it drifts in a flow. A flow may have nonzero vorticity because of curving streamlines (Figure 2.6a), but flow with entirely straight streamlines can also have nonzero vorticity if there is shear in the circulation (Figure 2.6b), and it is possible for curving streamlines to have zero vorticity (Figure 2.6c).

Based on (2.16), counterclockwise spin has positive vorticity and is called **cyclonic** (clockwise is **anticyclonic**) in the northern hemisphere. Remember this convention using

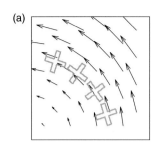 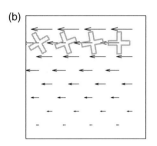 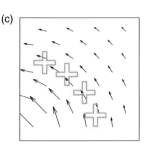

Figure 2.6   Successive positions of a paddle wheel (crosses) as it drifts and rotates in the flow (arrows) for (a) azimuthal velocity with speed $\sqrt{x^2 + y^2}$, (b) zonal velocity with speed y, and (c) azimuthal velocity with speed $1/\sqrt{x^2 + y^2}$. For mathematical details, see Subsection 2.2.3.

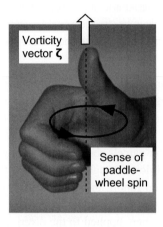

Figure 2.7   Convention for vorticity (Equation 2.15): imagine the local spin of the fluid turning a small paddle wheel, indicated by the curl of the fingers of your right hand. Then the vorticity vector ζ is in the direction of your thumb (paddle wheel axis), with size proportional to the speed of spin.

your right hand (Figure 2.7): Imagine the little crosses in Figure 2.6 spinning counter-clockwise in the plane of the page. The fingers of your right hand indicate this spin and your right thumb, aligned with the axis of spin, indicates the direction of ζ (upwards is positive).

Streamlines are curves that are tangent to **u**, and for a steady flow they show the paths of water parcels. All fluid flow has streamlines, but if the velocity tangent to a two-dimensional surface has zero divergence, the streamlines can be expressed as contours of a **streamfunction** $\Psi(\mathbf{x}_S)$, where $\mathbf{x}_S$ is the two-dimensional position vector defining locations on the surface. Some examples of surfaces relevant for oceanography are a plane, a surface along which density is constant (Chapter 5), and the spherical surface of the Earth.

It is difficult to visualize the full three-dimensional flow of a fluid such as the ocean, so oceanographers often resort to two-dimensional representations of the flow. We discuss flow in the horizontal plane in Chapter 3 and in the latitude–depth plane in Chapter 6 and

Chapter 8. For horizontal circulation, the depth integral from the sea floor at $z = -H(x, y)$ to the sea surface at $z = \eta(x, y)$,

$$\mathbf{U}(x, y) = \int_{-H}^{\eta} \mathbf{u} \, dz, \tag{2.17}$$

is often represented by the so-called **barotropic streamfunction** Ψ

$$\mathbf{U} = \hat{\mathbf{z}} \times \nabla \Psi, \tag{2.18}$$

or in Cartesian coordinates,

$$U = -\frac{\partial \Psi}{\partial y},$$

$$V = \frac{\partial \Psi}{\partial x} \tag{2.19}$$

(Subsection 2.1.2 discusses the meaning of *barotropic*). Here Ψ has units of $m^3 s^{-1}$, but sometimes the same name refers to a similar quantity based on the depth-average $\mathbf{u}$, with units of $m^2 s^{-1}$. Be aware that some authors use a different sign convention for streamfunction in (2.19) (for instance, Kundu et al., 2012).

Another common two-dimensional view shows vertical and meridional components of flow as a function of latitude and depth. Rather than a *vertical* integral, the appropriate quantity is the *zonal* integral from the western boundary at $x = W(y, z)$ to the eastern boundary $x = E(y, z)$:

$$V(y, z) = \int_{W}^{E} v \, dx. \tag{2.20}$$

If there is no flow through the eastern and western boundaries (for instance, at solid boundaries), then this flow can be represented by a **meridional streamfunction**. The meridional streamfunction is often called the **meridional overturning streamfunction** or the **(deep) meridional overturning circulation**. It is useful because ocean properties vary most dramatically with latitude and depth. Subsection 6.3.3 introduces several varieties of meridional streamfunction and Chapters 6 and 8–11 use them to describe ocean circulation. One could also consider vertical and zonal components, but so far (to our knowledge) this has not been done.

The difference in streamfunction value between any two streamlines is the volume flux flowing in the space between the streamlines. This volume flux is often called the volume transport or simply the **transport**. Figure 2.8 shows schematically how to interpret a streamfunction map. In physical oceanography, the volume transport unit of m^3/s is usually replaced with the **Sverdrup**: 1 Sv $= 10^6$ m^3/s.

2.1.5 Shallow Water and Potential Vorticity

In trying to understand the horizontal circulation, we often consider a simpler vertical structure than that represented by the primitive equations of motion. This approach can be applied to regions where the horizontal current does not vary much with depth, where the vertical variations are relatively simple (for example, upper ocean currents flowing in

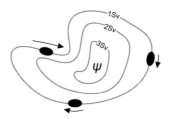

Figure 2.8   Streamfunction Ψ for hypothetical two-dimensional clockwise-circulation showing magnitude of Ψ. The water moves along contours of Ψ like beads on a wire.

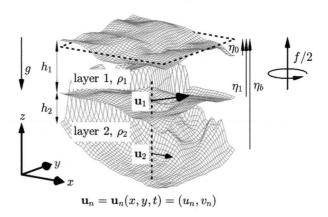

$$\mathbf{u}_n = \mathbf{u}_n(x, y, t) = (u_n, v_n)$$

Figure 2.9   Schematic of rotating shallow water flow for $N = 2$ layers. The two-dimensional layer currents, $\mathbf{u}_n$, and the layer thicknesses, h_n (or, equivalently, the layer interface heights, η_n), describe the flow.

the opposite direction to deep currents), or when our focus is on the vertical average, as in Subsection 2.1.4. The dynamical equations in question are called the **shallow water equations**.

In the shallow water equations the ocean is approximated as a hydrostatic stack of N layers (Figure 2.9). Each layer is homogeneous with a smaller density than the layers below it and has a layer thickness $h_n(x, y, t)$ (for the nth layer). As we increase N, the system more closely resembles a continuously-stratified fluid, but as N gets small, the equations of motion become easier to solve and understand. While the ocean generally has continuous stratification rather than distinct layers, the shallow water model is physically possible and can be well-approximated in the laboratory, at least with a small number of layers. In such a laboratory experiment, multiple layers are best represented by immiscible fluids (such as oil and water); otherwise the sharp density gradient between each layer grows more diffuse. There is no geostrophic shear within each layer, so that water in each layer moves as a Taylor column (Subsection 2.1.2), with a jump in horizontal velocity from layer to layer.

Shallow water models can help us visualize an important quantity called **potential vorticity**. Subsection 2.1.4 defines the vorticity of a fluid. In a rotating reference frame, ζ in (2.15) is the **relative vorticity** vector. The vorticity in the non-rotating frame is called the

absolute vorticity. If **u** is calculated in a rotating frame defined by a rotation vector $\boldsymbol{\Omega}$, then the vertical component (on the Earth, the direction given by $\hat{\mathbf{z}}$ at each point) is

$$\zeta_a = \hat{\mathbf{z}} \cdot \boldsymbol{\zeta}_a = \hat{\mathbf{z}} \cdot (2\boldsymbol{\Omega} + \boldsymbol{\zeta}) = f + \zeta. \tag{2.21}$$

In (2.21), the term consisting of the Coriolis parameter f is the **planetary vorticity**, (2.1), representing the vorticity due to spinning with the Earth (Section 1.2), and $\zeta = \hat{\mathbf{z}} \cdot \boldsymbol{\zeta}$ is the vertical component of the relative vorticity. Potential vorticity is defined for the shallow water model as

$$\text{\underline{\textbf{Shallow Water Potential Vorticity:}}} \ q = \frac{\zeta_a}{h} = \frac{f + \zeta}{h}, \tag{2.22}$$

where we have dropped the n index from all variables.

Potential vorticity is important in physical oceanography for several reasons. Arguably, the most important is that it is governed by a conservation law,

$$\frac{Dq}{Dt} = \frac{q\Delta w_*}{h} + \frac{1}{h}\left[\delta_T \hat{\mathbf{z}} \cdot \nabla \times (\mathbf{X}/h) - \delta_B r\zeta + v_h \nabla_h^2 \zeta\right], \tag{2.23}$$

where Δw_* is the net transport out of the layer to layers above and below. We derive (2.23) in Subsection 2.2.5. If all the terms on the right are zero, then (2.23) shows that q of a column of water in a layer stays constant as the column moves around.

Potential vorticity can remain constant while the water column is traveling throughout the domain if individual terms in (2.22) change together. The factor of h in the denominator of (2.22) can be understood in terms of angular momentum. If a water column is stretched (h increased), for instance, then for constant f and conserved q, $\zeta = hq - f$ must also increase. As the volume of the water column is conserved (by the continuity law in the governing equations), its horizontal area must decrease when h increases. The angular momentum is the integral of $\rho v_a r$, where v_a is the azimuthal speed of the column of water at radius r from the center of the column. The increase in ζ of the stretching column corresponds to the increase in v_a as r decreases. The opposite happens when the water column is squashed.

The right-hand terms in (2.23) show what processes can change the potential vorticity. Perhaps the most intuitive one is the wind stress curl (only applicable to the top layer of a shallow water model), which applies a torque to the water column, changing its angular momentum and hence its potential vorticity. The $-r\zeta$ term (only applicable to the bottom layer of a shallow water model) represents the loss of angular momentum through a torque with the solid bottom of the domain. The bigger the relative vorticity, the greater the torque on the column. This is based on an assumption that frictional force is proportional to the horizontal velocity of the water column; correct for laminar flow but not for turbulent flow (although a torque still exists). The $v_h \nabla_h^2 \zeta$ term represents horizontal diffusion of vorticity and ultimately torque with the lateral boundaries of the domain. This term also tends to remove vorticity of the same sign as ζ and assumes that horizontal mixing of momentum is given by a diffusion law.

The $q\Delta w_*$ term changes q by stretching or squashing. For instance, if water is leaving the layer ($\Delta w_* > 0$, corresponding to horizontal divergence in the layer), vorticity of the same sign as q is added to the column, thus increasing the column's spin as we expect from angular momentum considerations. Note that, unlike the friction terms which are proportional to ζ because they depend on the motion of the water relative to the boundaries, the stretching/squashing forcing term is proportional to q.

The shallow water potential vorticity is a special case of the more general **Ertel potential vorticity**:

$$\text{Ertel potential vorticity: } Q = \frac{\boldsymbol{\zeta}_a \cdot \nabla \rho}{\rho_0},$$

$$= (\nabla \times \mathbf{u} + 2\boldsymbol{\Omega}) \cdot \nabla \rho / \rho_0. \tag{2.24}$$

In the factor $\nabla \rho$, $\partial \rho / \partial z$ is much larger than horizontal derivatives. Thus we can write $|\nabla \rho| \approx \Delta \rho / \Delta z$, so for two given isopycnals (hence a fixed $\Delta \rho$), $|\nabla \rho|$ is proportional to $1/\Delta z$. Therefore the $\nabla \rho / \rho_0$ factor in (2.24) is analogous to the $1/h$ factor in (2.22).

Ertel PV obeys a generalized material conservation law, like (2.23). It acts like (the concentration of) a chemical tracer that is dynamically active. It is also useful because it diagnoses the hydrodynamic stability of a particular flow. This topic is only treated lightly in this book: see Vallis (2006) or Haine and Marshall (1998) for more information.

Another useful form of the Ertel PV is

$$Q = (f + \zeta) \frac{N^2}{g} + \frac{(\boldsymbol{\zeta}_h + e \hat{\mathbf{y}}) \cdot \nabla_h \rho}{\rho_0}, \tag{2.25}$$

where the vorticity vector is split into vertical, ζ, and horizontal, $\boldsymbol{\zeta}_h$, parts, and N is the buoyancy frequency (1.1). In (2.25) the Coriolis term $e = \hat{\mathbf{y}} \cdot (2\boldsymbol{\Omega}) = 2|\boldsymbol{\Omega}| \cos \theta$ is the horizontal component of the Earth's rotation vector at latitude θ, analogous to (2.1). This form of the Ertel PV is written so that the size and importance of the four different terms on the right-hand side diminishes from left to right. The horizontal planetary vorticity component is usually much smaller than the horizontal vorticity (see Subsection 2.1.1). Often, the Ertel PV simplifies further because the whole second term is relatively small. Often, ζ is small compared to f, and the Ertel PV is simply $Q \approx fN^2/g$. See Exercise 2.10 for an example where the individual terms are computed and compared.

2.2 Theory

2.2.1 Geostrophic Flow and Ekman Boundary Layers

Momentum Equations

The geostrophic and Ekman balances discussed in Subsection 2.1.2 and Subsection 2.1.3 can be recovered from the momentum equation (Equation 1.16 in Subsection 1.3.1). We restate it here for horizontal components alone, and neglect the small vertical Coriolis acceleration:

$$\frac{\partial \mathbf{u}_h}{\partial t} + \mathbf{u}_h \cdot \nabla \mathbf{u}_h = -\frac{1}{\rho_0} \nabla p - f\hat{\mathbf{z}} \times \mathbf{u}_h + \frac{\partial}{\partial z}\left(v_v \frac{\partial \mathbf{u}_h}{\partial z}\right) + \nabla \cdot \left(v_h \nabla \mathbf{u}_h\right) - w\frac{\partial \mathbf{u}_h}{\partial z}. \tag{2.26}$$

We represent the dissipation using eddy viscosity coefficients, ν_v and ν_h, which may take different values because horizontal mixing is thought to be caused by different processes than vertical mixing (Chapter 7).

It is useful to decompose the velocity into different components,

$$\mathbf{u}_h = \mathbf{u}_G + \mathbf{u}_{Ek} + \mathbf{u}_I + \mathbf{u}_a. \tag{2.27}$$

In this expression, the geostrophic velocity $\mathbf{u}_G$ is defined by (2.2), when all terms in (2.26) are eliminated except the Coriolis and pressure gradient terms. Ekman velocity $\mathbf{u}_{Ek}$ (units ms^{-1}) is defined by

$$f\hat{\mathbf{z}} \times \mathbf{u}_{Ek} = \frac{\partial}{\partial z}\left(\nu_v \frac{\partial \mathbf{u}_h}{\partial z}\right), \tag{2.28}$$

and occurs when the only terms are the Coriolis and vertical stress terms. Inertial velocity $\mathbf{u}_I$ is given by the balance between Coriolis and acceleration terms:

$$\frac{\partial \mathbf{u}_I}{\partial t} = -f\hat{\mathbf{z}} \times \mathbf{u}_I. \tag{2.29}$$

This balance occurs in periodic **inertial oscillations** (Subsection 1.2.1) in which the current rotates anticyclonically (clockwise in the northern hemisphere) with frequency f. Inertial oscillations are ubiquitous because they are generated by daily changes in wind stress, which occur over much of the ocean. As f is of order the Earth's rotation frequency (at least in the extra-tropics) these inertial oscillations are fast compared to the timescales we consider in this book.

The remaining velocity component $\mathbf{u}_a$, is merely the residual from the other terms. The division of $\mathbf{u}$ into $\mathbf{u}_G$ etc. is true by definition, but it is only useful if the individual components tell us something interesting about the velocity. For large-scale flows with timescales of weeks or more, the interesting fact is that we often have

$$\mathbf{u}_h \approx \mathbf{u}_G + \mathbf{u}_{Ek}. \tag{2.30}$$

The vertical momentum equation differs from the horizontal in the unimportance of the Coriolis force and the presence of a term representing gravity:

$$\frac{\partial w}{\partial t} + \mathbf{u} \cdot \nabla w = -g - \frac{1}{\rho_0}\frac{\partial p}{\partial z} + \frac{\partial}{\partial z}\left(\nu_v \frac{\partial w}{\partial z}\right) + \nabla \cdot \left(\nu_h \nabla w\right). \tag{2.31}$$

The hydrostatic approximation (Subsection 2.1.2) follows from the vertical momentum equation when all terms are removed except the gravity term $-g$ and the pressure term.

Scale Analysis

What is the basis for determining if and when we can ignore terms in the governing equations (including terms already missing from (2.26) and (2.31))? The assumption is that under certain circumstances, some terms are small enough that the solution of the equation without the terms is approximately equal to the solution to the complete equation. Mathematically, this can be thought of as a **regular perturbation** (Bender and Orszag, 1987), in contrast to a **singular perturbation** (Subsection 3.3.2) for which this is not true.

A **scale analysis** shows us under what conditions we expect a given part of the equation to be small. The results of a scale analysis do not have the finality of a mathematical proof. A scale analysis generally uses a single number to represent the magnitude of each variable (including variables which may change their magnitude and sign across the domain of interest). It oversimplifies the equation to be analyzed and gives results that strongly depend on the appropriateness of assumptions about the solution to the equation. Despite these risks, scale analysis has proved very useful in many branches of fluid dynamics, including physical oceanography. It provides hints about the solution to an equation. When a scale analysis indicates that some terms in an equation may be neglected, we can solve the resulting simplified equation and check that the assumptions used for the scale analysis are consistent with the solution. We can also check the scale analysis against nature, laboratory experiment, or a numerical solution of the equation.

As both $\mathbf{u}_G$ and $\mathbf{u}_{Ek}$ involve the Coriolis force, we compare the magnitude of other terms to the Coriolis term. In our scale analysis, for velocity scale U (not the depth-integrated zonal current here) and horizontal length scale L, we assume that $(u, v) \sim U$ and $(x, y) \sim L$, where the sign $\sim$ represents "approximately equal to." In that case, the relative size of the nonlinear advection terms is given by

$$\frac{|\mathbf{u}_h \cdot \nabla \mathbf{u}_h|}{|f\hat{\mathbf{z}} \times \mathbf{u}_h|} \sim \frac{U^2/L}{fU} \sim \frac{U}{fL} = Ro. \tag{2.32}$$

Thus the advective terms are small if the Rossby number Ro is small, which it is for large-scale circulation except near the equator (Subsection 2.1.1). Similarly, the vertical advection term scales as:

$$\frac{|w\frac{\partial \mathbf{u}_h}{\partial z}|}{|f\hat{\mathbf{z}} \times \mathbf{u}_h|} \sim \frac{U^2/L}{fU} \sim \frac{U}{fL} = Ro, \tag{2.33}$$

because the vertical velocity $w \sim UH/L$ from continuity (with vertical length scale H). Next, note that $\partial \mathbf{u}_h/\partial t$ is small for the flows of interest that evolve on timescales long compared to an inertial period $1/f$.

The relative magnitude of vertical friction is

$$\frac{\frac{\partial}{\partial z}\left(\nu_v \frac{\partial \mathbf{u}_h}{\partial z}\right)}{|f\hat{\mathbf{z}} \times \mathbf{u}_h|} \sim \frac{\nu_v U/H^2}{fU} \sim \frac{\nu_v}{fH^2} \equiv Ek \tag{2.34}$$

where Ek is the **Ekman number**. If we take typical ocean values of $\nu_v \sim 10^{-5}$ m²s⁻¹ and $H \sim 1000$ m, then $Ek \sim 10^{-7}$ and is very small. However, within the Ekman layer the appropriate numbers are more like $H \sim 10$ m and (because of turbulence generated by air–sea interaction) $\nu_v \sim 10^{-2}$ m²s⁻¹, so that $Ek \sim 1$ and within the Ekman layer vertical friction is important.

A similar analysis for horizontal friction is

$$\frac{|\nabla \cdot (\nu_h \nabla \mathbf{u}_h)|}{|f\hat{\mathbf{z}} \times \mathbf{u}_h|} \sim \frac{\nu_h U/L^2}{fU} \sim \frac{\nu_h}{fL^2}. \tag{2.35}$$

The nondimensional number v_h/fL^2 is sometimes called the horizontal Ekman number. The proper eddy value for v_h is unclear, but values of at least 100 m^{2}s^{-1} are plausible. For $f \sim 10^{-4}$ s^{-1} and $L \geq 100$ km, $v_h/fL^2 < 10^{-4}$, hence horizontal friction is small.

In the vertical momentum equation, a scale analysis (see Vallis, 2006) shows that for a uniform-density layer of thickness H, a circulation is hydrostatic if $(H/L)^2 \ll 1$. For a continuously stratified current, it is $(H/L)^2(U/NH)^2 = (U/NL)^2 \ll 1$, where N is the buoyancy frequency (1.1). N is the maximum frequency of internal gravity waves in a stratified ocean, typically around 10^{-2} s^{-1}, and NH is approximately their maximum propagation speed. Typically H/L, the aspect ratio, is small for the general circulation, and the current speed is slower than the gravity wave speed, so the hydrostatic approximation is quite accurate for scales discussed in this book. Hydrostatic balance applies when the stratification is strong, the current speed is weak, and/or the horizontal scale of interest is large, as one intuitively expects. Nonhydrostatic effects become important as the horizontal length scale shrinks and stratification weakens, for example in high latitude boundary layer flow driven by strong surface buoyancy losses.

While $\mathbf{u}_G$ can be diagnosed if we are given p (from the stratification and hydrostatic balance; see Subsection 2.1.2 and Subsection 2.2.2), $\mathbf{u}_{Ek}$ is defined by a differential equation in z which must be solved to find the Ekman velocity. Ekman considered the dynamics of the surface ocean driven by the wind, dissipated by viscous damping, and influenced by Earth's rotation. For viscous stress given by

$$\boldsymbol{\tau}(z) = \rho_0 v_v \frac{\partial \mathbf{u}}{\partial z}, \tag{2.36}$$

the Ekman velocity $\mathbf{u}_{Ek}$ can be written

$$f\hat{\mathbf{z}} \times \mathbf{u}_{Ek} = \frac{1}{\rho_0} \frac{\partial \boldsymbol{\tau}}{\partial z}, \tag{2.37}$$

where $\mathbf{u}_{Ek}$ and $\boldsymbol{\tau}$ here depend on z. At the sea surface, $\boldsymbol{\tau}$ is the wind stress, which is often considered known (either stipulated in a theoretical problem or taken from observations). Notice that the sign convention says that a positive stress component $\boldsymbol{\tau} \cdot \mathbf{x}_h$ means a downward flux of momentum in the $\mathbf{x}_h$ direction. A convergence in the vertical in the stress ($\partial \boldsymbol{\tau} \cdot \mathbf{x}_h/\partial z$ positive) therefore means the stress is depositing $\mathbf{x}_h$ momentum into the flow and $\mathbf{u}_h \cdot \mathbf{x}_h$ is being accelerated by the stress ($\partial \mathbf{u}_h \cdot \mathbf{x}_h/\partial t$ is positive; Figure 2.3 shows an example where $\mathbf{x}_h$ is the x direction).

Ekman Spirals and Ekman Transport

Ekman theory describes how the wind stress penetrates into the ocean. There are two logical parts: One describes how the variables in (2.37) change with z. It predicts that the currents change direction and amplitude in a helix approaching the sea surface (if the current vectors are plotted on a plane figure, they inscribe a spiral, which was one of Ekman's main discoveries). Textbooks on geophysical fluid dynamics cover the details, but we skip them because they depend on how the stress changes with depth, which is poorly known. Also, the current spiral is hard to observe in practice. And the details of the depth-varying currents within the Ekman layer are not very important for the large-scale ocean circulation.

The second part of Ekman theory concerns the depth-integrated currents which are described by the **Ekman Transport** $\mathbf{U}_{Ek}$ with components (U_{Ek}, V_{Ek}). We obtain an expression for the depth-integrated Ekman velocity $\mathbf{U}_{Ek}$ with components (U_{Ek}, V_{Ek})

$$\mathbf{U}_{Ek} = \int_{z_B}^{z_T} \mathbf{u}_{Ek}\, dz. \tag{2.38}$$

For the surface Ekman layer, the appropriate integration limits are the sea surface for z_T and a z_B deep enough for $\mathbf{X}(z_B)$ to be negligible compared to $\boldsymbol{\tau}$. We integrate (2.37) in z to get

$$f\hat{\mathbf{z}} \times \mathbf{U}_{Ek} = \mathbf{X}. \tag{2.39}$$

Thus we recover the expression (2.11) which relates Ekman transport to the wind stress. Ekman transport is insensitive to the details of how stress changes with depth and is very important for the ocean circulation.

In the argument above we integrated far enough so that the stress vanished (z_B is deep enough so that the stress $\boldsymbol{\tau}(z_B) \approx 0$). How deep is that? Observations suggest that $z_B \approx -100$m (e.g., Lenn and Chereskin, 2009; remember that z is measured positive upwards from the sea surface). This depth defines the **thickness of the Ekman layer**,

$$H_{Ek} = \sqrt{\frac{2\nu_v}{f}} \tag{2.40}$$

(about 30 m), which is thin compared to the full depth of the ocean. In the layer itself, both the pressure-driven part $\mathbf{u}_G$ and the stress-driven part $\mathbf{u}_{Ek}$ contribute to the current. Below the layer, only the pressure-driven part remains, which is called the **geostrophic interior**.

Similar arguments apply to the **bottom Ekman layer** at the sea bed (see (2.73)). There, frictional stresses drive Ekman transports in similar ways. Where these bottom Ekman transports converge, vertical currents pump water out of the Ekman layer into the geostrophic interior having profound effects on the ocean circulation.

2.2.2 Example: Hydrostatic, Geostrophic, Thermal Wind, and Dynamic Height Calculation

Hydrostatic and geostrophic balance (Subsection 2.1.2), and their corollary, the thermal wind relation, are fundamental to understanding the large-scale ocean circulation. The calculation presented in this section builds expertise and intuition in analyzing hydrographic (namely, temperature and salinity) field observations and understanding the links between the ocean's mass distribution and its geostrophic flow.

Consider the idealized vertical section between two oceanographic stations, 1 and 2, shown in Figure 2.10a. The ocean consists of three homogeneous slabs, A, B, and C, with densities ρ_A, ρ_B, ρ_C, as indicated. We assume that the pressure field is in hydrostatic and geostrophic balance. The task is to compute the currents, u_A, u_B, u_C, perpendicular to the section in slabs A, B, and C.

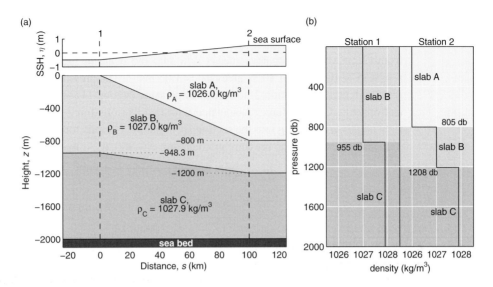

Figure 2.10   (a) Geostrophic current calculation involving three slabs of homogeneous fluid, A, B, and C, and two hydrographic stations, 1 and 2. The latitude is 40°N. (b) Density profiles from the two hydrographic stations in (a).

Method 1: Hydrostasy and Geostrophy

First, notice that the interfaces between the slabs, and the sea surface, have constant slope between the stations. This fact may, or may not, accurately represent the real ocean in that region. Often, we assume for simplicity that the slopes are constant in this way. If the assumption is inaccurate, the argument below can be modified to consider the gap between two stations that are close enough that the slopes are constant. An important result follows from constant interface slopes: the horizontal pressure gradient does not vary with distance s between the stations or with height z in any of the slabs. Think of slab A, for example. The horizontal pressure gradient is dp_A/ds. From hydrostatic balance, the pressure at z (measured positive upwards as shown) in slab A is (from (2.4), (2.8))

$$p_A(z) = \int_z^\eta \rho_A g\, dz = (\eta - z)\rho_A g, \tag{2.41}$$

where, for convenience, the air pressure pushing on the sea surface is neglected. Hence, $dp_A/ds = \rho_A g d\eta/ds$ (at fixed z) is constant because the sea surface slope $d\eta/ds$ is constant. The argument for the deeper slabs follows similarly. Because the horizontal pressure gradient is constant between the stations, the geostrophic current u_A is also uniform between the stations.

To compute the currents, we therefore just need to pick three convenient depths, calculate the pressure gradient, and use geostrophy. In slab A, consider the pressure gradient at z:

$$\frac{dp_A}{ds} = \frac{p_A^2(z) - p_A^1(z)}{\Delta s} = \frac{\left[(\eta^2 - z) - (\eta^1 - z)\right]\rho_A g}{\Delta s} = \frac{(\eta^2 - \eta^1)\rho_A g}{\Delta s}, \tag{2.42}$$

where the superscripts indicate the station in question. Inserting $\eta^1 = -0.5$ m, $\eta^2 = 0.5$ m, $\rho_A = 1026$ kg m^{-3}, $g = 9.81$ ms^{-2}, and $\Delta s = 100$ km gives $dp_A/ds = 0.101$ N m^{-3}. The geostrophic current u_A is therefore

$$u_A = \frac{1}{f\rho_A}\frac{dp_A}{ds} \quad (2.43)$$

from (2.2), where positive u_A means flow *into* the plane of the vertical section. Using $f = 9.35 \times 10^{-5}$ s^{-1} at 40°N gives $u_A = 1.05$ ms^{-1}. The pressure gradient force pushes from high pressure (station 2) towards low pressure (station 1) and is balanced by an equal and opposite Coriolis force. Associated with that Coriolis force is the horizontal flow at right angles, u_A, flowing into the section such that the Coriolis force is pushing to its right.

In slab B, consider the pressure gradient at $z = -900$ m. At that depth,

$$\frac{dp_B}{ds} = \frac{p_B^2(z=-900) - p_B^1(z=-900)}{\Delta s}$$
$$= \frac{\left(800.5\rho_A + 100\rho_B - 899.5\rho_B\right)g}{\Delta s}. \quad (2.44)$$

In this formula notice how the hydrostatic law has been applied to both slabs A and B at station 2 (800.5 m of slab A and 100 m of slab B). The pressures at $z = -900$ m are: $p_B^1 = 9.06235 \times 10^6$ and $p_B^2 = 9.06457 \times 10^6$ N m^{-2}, only different from each other by 0.02%. Compared to the absolute pressures, the horizontal pressure difference driving the geostrophic flow, $p_B^2 - p_B^1 = 2.22 \times 10^3$ N m^{-2}, is tiny! Very high accuracy is required in computing the absolute hydrostatic pressures in order to obtain an accurate value for the pressure difference. From the analog of (2.43) for slab B, and taking $\rho_B = 1027$ kg m^{-3}, we find that the geostrophic current in slab B is $u_B = 0.23$ ms^{-1} into the section. The current is weaker with depth because the horizontal pressure gradient in slab B is smaller than in slab A. Although the water column is shorter at station 1, it is denser than at station 2.

The argument for slab C is similar. Taking $z = -1500$ m,

$$\frac{dp_C}{ds} = \frac{p_C^2(z=-1500) - p_C^1(z=-1500)}{\Delta s}$$
$$= \frac{\left(800.5\rho_A + 400\rho_B + 300\rho_C - 947.8\rho_B - 551.7\rho_C\right)g}{\Delta s}. \quad (2.45)$$

Make sure you understand how the hydrostatic pressure is computed for each station. We find that $p_C^1 = 1.51121 \times 10^7$, $p_C^2 = 1.51121 \times 10^7$ N m^{-2}, hence $dp_C/ds = 0$, and there is zero current in slab C. The slab speeds have been determined.

Method 2: Thermal Wind

The preceding argument (Method 1) is straightforward to generalize and adapt to other configurations as long as the hydrostatic pressures are computed accurately. But its practical utility is limited because we assume knowledge of all the information in Figure 2.10a, including the sea-surface heights at the two stations. Often that information is unavailable, for example, if we only possess hydrographic data from a ship or profiling float. What can be said then?

To answer that question, begin by plotting the density profiles as a function of pressure for the stations 1 and 2 (Figure 2.10b). This density information is calculated from temperature, salinity, and pressure measurements in practice (Subsection 1.1.2 and Subsection 1.2.3). From it, one can construct a diagram very similar to the large lower panel in

Figure 2.10a (without the small panel of sea-surface height). The vertical scale is pressure and exploiting the 1 db $\approx$ 1 m rule of thumb (Subsection 1.1.2) gives a moderately accurate depth estimate. This data cannot be used to compute the horizontal pressure gradients because η is missing.

The thermal wind relation (2.9) can be applied, however. A useful alternate version of the thermal wind formula for two layers of constant density is:

$$\Delta u_{A/B} = \frac{g(\rho_A - \rho_B)}{f\rho_0}\frac{dz_{A/B}}{ds}, \tag{2.46}$$

(see Exercise 2.7). In this formula, positive $\Delta u_{A/B}$ means that the current flowing into the plane of the section *increases* with z. The slab interface height is $z_{A/B}$ and ρ_0 is the reference density. Apply (2.46) to the interface between slabs A and B in Figure 2.10a. Take the interface slope to be $-805/100$ dbkm^{-1}, or -8.05×10^{-3} dbm^{-1} using the 1 dbm^{-1} rule of thumb above (a more accurate conversion is $|dp/dz| = \rho_A g = 1.0065$ dbm^{-1} for this slab density, which is insignificantly different). Thus, $\Delta u_{A/B} = 0.82$ ms^{-1} (using $\rho_0 = 1026$ kg m^{-3}; other reasonable choices make little difference). As expected, this value is consistent with the geostrophic calculation above. That is, $\Delta u_{A/B} = u_A - u_B = 1.05 - 0.23 = 0.82$ ms^{-1}. Positive $\Delta u_{A/B}$ means that the geostrophic current increases speed into the plane of the section as height increases (pressure decreases). This sense is consistent with the right hand grip rule (Figure 2.2): take your right hand and point your thumb horizontally towards the denser water (from station 2 to 1). The curl of your fingers shows the current speed u_A is greater than u_B.

Similarly, (2.46) applied to the interface between slabs B and C gives

$$\begin{aligned}\Delta u_{B/C} &= \frac{g(\rho_B - \rho_C)}{f\rho_0}\frac{dz_{B/C}}{ds}\\ &= \frac{9.81 \times 0.9}{9.35 \times 10^{-4} \times 1026} \times \frac{1208 - 955}{100 \times 10^3} = 0.23\text{ms}^{-1}.\end{aligned} \tag{2.47}$$

Again, we see that $\Delta u_{B/C} = u_B - u_C = 0.23 - 0 = 0.23$ ms^{-1}, as expected. The thermal wind relation applied to hydrographic (density) data yields the change in geostrophic speed perpendicular to the section between the slabs, but it cannot give the absolute geostrophic speed in any single slab.

Method 3: Dynamic Height

In practice, hydrographic station data have been used to compute the thermal wind using a different approach. For completeness, and because of its historical importance, we include this method here. It involves computation of a quantity called **dynamic height**, which appears later in the book (Subsection 5.1.2 and Subsection 5.1.3, Figure 5.6).

Dynamic height (units of m^{2}s^{-2} = Jkg^{-1}), Ψ, is defined by:

$$\Psi(p, p_0) = -\int_{p_0}^{p} \frac{1}{\rho}\, dp \tag{2.48}$$

(sometimes dynamic height is reported in meters, meaning Ψ/g). Note that dynamic height is a function of pressure-of-interest p and reference pressure p_0. In (2.48) the reciprocal of the density is called the **specific volume**. Traditionally, the specific volume in (2.48) was replaced with the specific volume anomaly, $\rho^{-1} - \rho_{\text{ref}}^{-1}$, to compute the dynamic height anomaly, where $\rho_{\text{ref}}(p)$ is the density at a reference salinity and temperature (still a function of pressure). The reason is that using the dynamic height anomaly to estimate the geostrophic current from (2.49) below is more accurate than using the dynamic height itself with low precision arithmetic. This issue has now been overcome with fast double-precision computers (IOC, SCOR, and IAPSO, 2010).

Dynamic height can be computed for any hydrographic station where density (hence specific volume) is measured as a function of pressure. It is useful because the geostrophic current $u(p)$ at pressure p with respect to the current $u(p_0)$ at p_0 (namely, the difference in geostrophic currents between them) is

$$u(p) = \frac{1}{f} \frac{\partial \Psi}{\partial s} + u(p_0). \tag{2.49}$$

This formula follows from the thermal wind relation (see Exercise 2.8). So, given the difference in dynamic heights between two hydrographic stations the geostrophic current can be computed (relative to the current at p_0).

Dynamic height is closely related to the **geopotential** Φ, which is the gravitational potential energy per unit mass (units of $\text{m}^2\text{s}^{-2} = \text{Jkg}^{-1}$). The geopotential is commonly used in atmospheric sciences to depict the geostrophic flow: the gradient of Φ on a pressure surface gives the geostrophic wind, similar to (2.49). Apart from a constant offset that has no dynamical significance, dynamic height and geopotential are identical, given hydrostatic balance. Dynamic height is also closely related to the **steric height** (units of m); specifically, the steric height equals the dynamic height divided by the gravitational acceleration (which is considered constant for this purpose). Steric height is therefore akin to the sea-surface height, and can be directly compared to it. The geostrophic current (relative to the pressure p_0) can be found from the steric height gradient: an example is shown in Chapter 5.

In the present problem (Figure 2.10a), compute the dynamic heights at stations 1 and 2 at $p = 900$ db, which is in slab B. At station 1, we have, for $p_0 = 2000$ db and referring to Figure 2.10b,

$$\Psi^1(900, 2000) = 10^4 \times \left(\frac{2000 - 955}{1027.9} + \frac{955 - 900}{1027.0} \right) = 10701.899 \text{ m}^2\text{s}^{-2}, \tag{2.50}$$

and for station 2,

$$\Psi^2(900, 2000) = 10^4 \times \left(\frac{2000 - 1208}{1027.9} + \frac{1208 - 900}{1027.0} \right) = 10704.056 \text{ m}^2\text{s}^{-2} \tag{2.51}$$

(the factor of 10^4 is the conversion between pressure in db and Pa; see Subsection 1.1.2). Thus, the gradient of the dynamic height is

$$\frac{\partial \Psi}{\partial s} \approx \frac{\Psi^2 - \Psi^1}{\Delta s} = 2.16 \times 10^{-5} \text{ ms}^{-2}, \tag{2.52}$$

and hence the geostrophic current at $p = 900$ db with respect to that at $p_0 = 2000$ db equals

$$u(p) - u(p_0) = \frac{1}{f} \frac{\partial \Psi}{\partial s} \approx \frac{2.16 \times 10^{-5}}{9.35 \times 10^{-5}} = 0.23 \text{ ms}^{-1}, \tag{2.53}$$

which is what we found before. The calculation to compute the difference in speed between slabs A and B ($(p, p_0) = (0, 900)$ db), or A and C ($(p, p_0) = (0, 2000)$ db) follows similarly.

The advantage of the dynamic height method is that it is well-adapted to hydrographic field data (temperature, salinity – hence density and specific volume – profiles as a function of pressure; no Boussinesq assumption). It also yields a streamfunction for the geostrophic flow – the dynamic height – that can be mapped (see Chapter 5). As described here, Methods 1 and 2 cannot. The TEOS-10 software (Subsection 1.2.3) contains functions to compute the dynamic height from hydrographic field data.

As for Method 2, the dynamic height calculation cannot give the absolute geostrophic current, only the relative current with respect to the current at the reference pressure p_0. Sometimes the current at p_0 is assumed to be negligible; in which case p_0 is called a **level of no motion** (see Chapters 3, 5–8, 11). This is a dubious assumption, however, because direct current measurements suggest that, in general, there is no depth at which the geostrophic current vanishes. It is safer to remember that geostrophic currents computed from the thermal wind equations (Methods 2 and 3) give relative currents with an unknown constant offset. (To see an interesting idea to avoid this issue, consult Exercise 5.14.)

Finally, notice that this example calculation focuses on computing the geostrophic speed into the plane of the section in Figure 2.10a. Does that mean that the geostrophic current is guaranteed to be perpendicular to the section? No, it does not. There may also be a pressure gradient at right angles to the section and therefore a geostrophic current in the plane of the section. We have no information about that component of the geostrophic flow.

2.2.3 Vorticity, Streamfunction, and Geostrophic Divergence

Decomposition of Two-Dimensional Velocity Fields

Subsection 2.1.4 asserts that under some circumstances, flow can be described with a **streamfunction**; here we describe more fully the relationship between streamfunction, vorticity, and related quantities.

A two-dimensional vector field, such as the vertically integrated velocity $\mathbf{U}$, can always be expressed in terms of two scalar fields, a (velocity) **potential** Φ and a streamfunction Ψ. Potential fields are used in physics, for example to quantify gravitational or electrostatic forces, but streamfunction fields may be less familiar. The decomposition works as follows:

$$\mathbf{U} = \hat{\mathbf{z}} \times \nabla\Psi + \nabla\Phi \tag{2.54}$$

($\hat{\mathbf{z}}$ is here the unit vector perpendicular to the surface defined by $\mathbf{U}$). As we will see below, dividing the flow field into components represented by a potential and streamfunction is useful because each component has distinct characteristics. In particular, the part of the flow described by the streamfunction Ψ is non-divergent, and the part of the flow described by the potential Φ is irrotational. Note that adding a constant to either Ψ or Φ does not change $\mathbf{U}$.

Taking the divergence of (2.54) gives

$$\nabla \cdot \mathbf{U} = \nabla^2\Phi, \tag{2.55}$$

where the divergence of the first term in (2.54) is identically zero from vector identities. As discussed in Subsection 2.1.4, the full three-dimensional velocity of water is approximately nondivergent, but the two-dimensional flow in a plane can be divergent. For a two-dimensional velocity field, divergence represents the net flow into or out of an area A, which follows from the **divergence theorem** from vector calculus (see also Appendix B):

$$\int_A \nabla \cdot \mathbf{U} \, dA = \oint_S \mathbf{U} \cdot \hat{\mathbf{n}} \, ds, \qquad (2.56)$$

where the right-hand side is the line integral around the perimeter S of the area and $\hat{\mathbf{n}}$ is the unit normal vector on the perimeter pointing outward (Figure 3.21). If there is no divergence, as is approximately true for $\mathbf{U}$ in the ocean, then $\nabla^2 \Phi = 0$. For a closed basin with no exchange of fluid from outside, Φ is uniform over the boundaries and hence must be uniform over the entire basin. Therefore, we can describe the two components of depth-integrated flow, $\mathbf{U} = (U, V)$, with a *single* scalar function, the streamfunction Ψ.

The vorticity, or curl of $\mathbf{U}$, is given by

$$\zeta = \hat{\mathbf{z}} \cdot (\nabla \times \mathbf{U}) = \nabla^2 \Psi, \qquad (2.57)$$

(the curl of the second term in (2.54) is identically zero from vector identities: to be precise, ζ in (2.57) is the relative vorticity of the vertical integral of the flow, not the relative vorticity itself). The curl represents the flow around the perimeter of an area, which is shown by Stokes' theorem from vector calculus (see also Appendix B):

$$\int_A \hat{\mathbf{z}} \cdot (\nabla \times \mathbf{U}) \, dA = \oint_S \mathbf{U} \cdot \hat{\mathbf{t}} \, ds. \qquad (2.58)$$

Equation 2.58 looks very similar to (2.56), but $\hat{\mathbf{t}}$ represents a unit vector *tangential* to the perimeter (pointing in the counterclockwise direction, Figure 3.21) rather than perpendicular to it like $\hat{\mathbf{n}}$. The right-hand side of (2.58) is called the fluid **circulation** (not to be confused with the ocean circulation).

In a closed ocean basin the divergence vanishes and (2.57) shows that there is an intimate relationship between Ψ and ζ. Given ζ and boundary conditions on Ψ, it is relatively straightforward to solve (2.57) for Ψ, analytically for simple cases, or numerically for more complex basin geometry and $\zeta(x, y)$.

Alternatively, a great deal of classical fluid mechanics in a non-rotating reference frame concerns the analogous case of **irrotational** flow, in which the vorticity vanishes and the two-dimensional velocity field is determined by the potential Φ. This case is not very relevant to the large-scale ocean currents in which relative vorticity (Subsection 2.1.4) is easily created by the conversion of planetary vorticity.

Illustrative Examples of Vorticity

Consider again how the simple flows in Figure 2.6 have vorticity, or not. Look at Figure 2.6b first because it can be analyzed using Cartesian coordinates. The velocity components, divergence, and vorticity are:

$$U = -\Psi_y + \Phi_x,$$
$$V = \Psi_x + \Phi_y,$$
$$D = U_x + V_y = \Phi_{xx} + \Phi_{yy},$$
$$\zeta = V_x - U_y = \Psi_{xx} + \Psi_{yy}. \tag{2.59}$$

In Figure 2.6b, $V = 0$ everywhere so $\zeta = -U_y = \Psi_{yy}$ which is a positive number, indicating cyclonic vorticity (counterclockwise spin). This example shows that even if the velocity is all in the same direction, the vorticity can be nonzero provided there is nonzero shear. Notice also how the relative vorticity depends only on the streamfunction, not on the potential.

The other examples in Figure 2.6 are understood more easily in polar coordinates (r, λ), which represent radial distance from the origin and azimuthal angle, respectively. We choose (U, V) to represent velocity components in the (r, λ) directions. In polar coordinates,

$$\zeta = \frac{1}{r}(rV)_r - \frac{1}{r}U_\lambda \tag{2.60}$$

(see Batchelor, 1967, Appendix 2). For the axisymmetric examples in Figure 2.6a, c, $U = 0$ and $V(r, \lambda) = V(r)$. Figure 2.6a has $V = C\sqrt{x^2 + y^2} = Cr$, for some positive constant C, which means that the flow moves as if it were a solid-body with angular velocity C. The vorticity is therefore $\zeta = 2C$, twice the angular velocity, and it is the same everywhere.

In Figure 2.6c, we have $V = C/r$ and therefore, apparently, the vorticity vanishes from (2.60). The cross orbits the center of the flow in this case, but it does not spin. Notice, however, that the speed V gets arbitrarily large as one approaches $r = 0$. This flow has a nonzero circulation that is independent of the patch radius. Computation of the right-hand side of (2.58) shows the circulation equals $2\pi C$. Thus the flow must have a nonzero vorticity integrated over a circular area A around the origin. Therefore we conclude that the flow in Figure 2.6c actually has a singular vorticity distribution at the origin (a Dirac delta function). This flow is called a point vortex.

Consideration of (2.58) for the point vortex shows that if we take a region A of fluid which has $\zeta = 0$ except for a patch which is small compared to A, then the circulation around the perimeter of the region is determined by the patch. This is a mathematical property of the vorticity and circulation that arises from the definition of the vector field $\mathbf{U}$. It is tempting – but misleading – to think of vorticity as some kind of force that influences the fluid at remote locations. It is better to think of ζ and $\mathbf{U}$ (or Ψ) as two descriptions of the same flow which are mathematically linked. If we know one of these fields, we can mathematically deduce the other.

Geostrophic Horizontal Divergence and the β-effect

The second formula (2.2) between geostrophic velocity and pressure looks similar to the relationship between velocity and the streamfunction Ψ in (2.18). For a rotating system with uniform f, pressure p is indeed a streamfunction. Thus not only does depth-integrated flow have zero horizontal divergence, but the geostrophic flow at each depth also has zero horizontal divergence. This is a powerful constraint on the circulation. For example, incompressibility (1.16) gives us

$$w_z = -\nabla_h \cdot \mathbf{u}, \tag{2.61}$$

and if $w \approx 0$ for some z, such as the surface or the sea floor, $\nabla_h \cdot \mathbf{u}_g = 0$ gives us $w = 0$ through the entire water column. In such a system the vertical flow is associated with the weak ageostrophic components of flow only.

For the ocean, $f = 2\Omega \sin\theta$ depends on latitude θ, and so p is a streamfunction for the vector field $f\mathbf{u}$ but not for $\mathbf{u}$ itself. When f varies, inserting the geostrophic velocity (2.3) into the divergence formula (2.61) gives (using Cartesian coordinates):

$$\textbf{Geostrophic Divergence:} \qquad \frac{\partial u_g}{\partial x} + \frac{\partial v_g}{\partial y} = -\frac{\partial w}{\partial z} = -\frac{\beta}{f}v_g \tag{2.62}$$

(also called the geostrophic Sverdrup relation, see (3.24)). Here,

$$\textbf{Planetary Vorticity Gradient:} \qquad \beta = \frac{df}{dy} = \frac{2\Omega}{R_E}\cos\theta, \tag{2.63}$$

for Earth radius R_E, is the rate of change of the Coriolis parameter with latitude (not to be confused with the haline contraction coefficient). Equation 2.62 shows that the geostrophic divergence is not zero as long as $v_g \neq 0$, which we expect to be true in general. Therefore a nonzero vertical current must exist for incompressible flow. Connecting this vertical current with the horizontal geostrophic current is essential to understanding the large-scale circulation. Note that the size of the vertical speed is much smaller than the size of the horizontal velocity. Chapter 3 explains these ideas in detail and Subsection 3.3.1 shows that (2.62) implies that the geostrophic divergence becomes small as the length scale of circulation structures become small compared to the Earth's radius. In that case using p as a geostrophic streamfunction may be a reasonable approximation. Neglecting variations in f this way is called the f-**plane** approximation. Similarly, when we consider relatively small meridional scales, the variation of the Coriolis parameter with latitude is taken to be linear:

$$f = f_0 + \beta_0 y \tag{2.64}$$

for constant β_0. This is called the β-**plane** approximation; the β-**effect** refers to variations in f with latitude (see also Subsection 3.2.4).

2.2.4 Shallow-Water Equations

The shallow water equations comprise an approximation to the equations of motion for the stack of N homogeneous layers introduced in Subsection 2.1.5. Each layer has density $\rho_n(\mathbf{x}_h, t) > \rho_{n-1}$, where the larger the layer index n, the deeper the layer; $\mathbf{x}_h$ is the horizontal position (latitude and longitude, or x and y, as the case may be). In representing an ocean with shallow water equations, we must choose the values of all the ρ_n as external parameters.

For each layer, we derive equations governing the depth-average horizontal velocity, and the layer thickness h_n. The layer thickness is related to η_n, the height of the interface at the bottom of the layer, by

$$h_n = \eta_{n-1} - \eta_n. \tag{2.65}$$

Note that for N layers, there are N different h_n and $N + 1$ different η_n (because η_0 represents the sea surface height); η_N is the depth of the basin, however, which is supplied.

Hydrostatic Pressure

A key assumption of the shallow water model is that the pressure is hydrostatic, so pressure in each layer can also be written in terms of the interface heights and hence in terms of the h_n. To show this, we start by restating (2.5),

$$p = \int_z^{\eta_0} \rho g \, dz'. \tag{2.66}$$

We can use this integral to write an expression for the increase in pressure to a given depth z_n in layer n from a depth z_{n-1} in the layer just above (see Figures 2.9 and 2.11). For the case of uniform-density layers, the pressure difference is:

$$p_n = p_{n-1} + g\rho_{n-1}(z_{n-1} - \eta_{n-1}) + g\rho_n(\eta_{n-1} - z_n), \tag{2.67}$$

where $p_{n-1} = p(z_{n-1})$. It is convenient to define a quantity known as **reduced gravity** (or buoyancy jump, (8.1)),

$$\gamma_n \equiv \frac{g}{\rho_0}(\rho_{n+1} - \rho_n), \tag{2.68}$$

so that we can write the pressure relation as

$$p_n = p_{n-1} + \rho_0\gamma_{n-1}\eta_{n-1} + g(\rho_{n-1}z_{n-1} - \rho_n z_n). \tag{2.69}$$

Ultimately we are interested in the horizontal pressure gradient, and the last two terms do not contribute to it (because within each layer they only depend on z_n and z_{n-1}, which are horizontal surfaces). Therefore we define **dynamic pressure** as

$$p'_n = p'_{n-1} + \rho_0\gamma_{n-1}\eta_{n-1} \tag{2.70}$$

(see Exercise 2.9 and (5.8)). Iterating this recursion relation to eliminate dynamic pressure from the right-hand side, where p'_0 is assumed to be zero (neglecting variations in surface atmospheric pressure), we get

$$p'_n = \rho_0 \sum_{m=0}^{n-1} \gamma_m\eta_m. \tag{2.71}$$

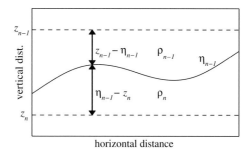

Figure 2.11   Schematic for calculating hydrostatic pressure for vertical density integral across a density interface.

Vertical Shear Within Layers

From (2.71), the horizontal pressure gradient is independent of z within the layer (see also Subsection 2.2.2). Equation 2.2 then implies that the geostrophic current is also independent of z within a layer. The depth-independence of horizontal velocity is used in deriving the shallow water equations. However, Ekman velocity near the sea surface and sea floor is not depth-independent. One way to avoid the contradiction is to limit the shallow water equations to describing the water outside the Ekman layers. In that case, Ekman transport affects the geostrophic layers via an exchange of water across the top interface of the top geostrophic layer and the bottom interface of the bottom geostrophic layer.

As the geostrophic flow excludes the Ekman flow, the depth-integrated horizontal geostrophic flow may be divergent and hence not completely described by a horizontal streamfunction. Indeed, as discussed in Chapter 6, in some regions of the ocean the surface Ekman transport is not negligible. Then, it may be useful to include the Ekman transport *within* the surface density layer. In that case the shallow water equations correspond to the (usually unrealistic) physical situation of the Ekman velocity being made vertically uniform within the top layer by strong mixing. Even for non-geostrophic flow, the assumption of no vertical shear within the layer becomes less restrictive as one invokes more layers to resolve any vertical $\mathbf{u}_h$ variations of interest. Moreover, in some circumstances, one can argue that no process exists to change any nonzero vertical variation in $\mathbf{u}_h$ over time, and hence the vertical variation in $\mathbf{u}_h$ decouples from its vertical average (see Vallis, 2006 and Pedlosky, 1987a for details).

Momentum Equations

Henceforth when we describe individual layers, we drop the subscript n and use the subscript T for the top of the layer (for instance $\eta_T = \eta_{n-1}$) and B for the bottom of the layer (for instance $\eta_B = \eta_n$). The momentum equations look similar to the governing equations stated in Subsection 1.3.1. The horizontal components are

$$\frac{D\mathbf{u}_h}{Dt} + f\hat{\mathbf{z}} \times \mathbf{u}_h = -\frac{1}{\rho_0}\nabla p + \frac{\partial}{\partial z}\left(\nu_v \frac{\partial \mathbf{u}_h}{\partial z}\right) + \nabla \cdot (\nu_h \nabla \mathbf{u}_h), \qquad (2.72)$$

where we have assumed a diffusive form for friction with the expectation that vertical and horizontal fluxes have different eddy viscosities (ν_v and ν_h). Away from the Ekman layers, effects of vertical friction are usually assumed to be negligible, and the lack of vertical shear eliminates the $\partial/\partial z$ term in the material derivative. If we include Ekman flow within the model layer(s), then we want to know the stress on the sea surface and sea floor. Stress on these surfaces represents the vertical transport of horizontal momentum, and is given by (2.36). The wind stress $\mathbf{X}_T$ at the sea surface is generally specified as a variable that is external to the ocean. (Surface stress is actually proportional to the difference between air speed and water speed at the atmosphere-ocean interface, but generally the wind contribution dominates.) At the sea floor, the stress depends on the ocean circulation itself. A commonly used model of stress is the laminar Ekman layer, which leaves out important features like turbulence and bottom boundary roughness but gives some insight into the relationship between bottom stress and the geostrophic velocity. A laminar Ekman layer has (Vallis, 2006, Exercise 3.6)

$$\mathbf{X}_B = \frac{\nu_v}{H_{Ek}}(\mathbf{u}_G + \hat{\mathbf{z}} \times \mathbf{u}_G), \tag{2.73}$$

where H_{Ek} is the Ekman layer thickness (see Subsection 2.2.1) and $\mathbf{u}_G$ is the value of the horizontal velocity just outside the Ekman layer. We denote the vertical average, over the bottom Ekman layer, of Ekman velocity as $\mathbf{u}_{Ek}$. The average velocity of a layer of thickness h is given by

$$\mathbf{u} = \mathbf{u}_G + \delta_{Ek}\mathbf{u}_{Ek}, \tag{2.74}$$

where $\delta_{Ek} = H_{Ek}/h$ is the fractional Ekman layer depth. As the Ekman velocity is known (Vallis, 2006), $\mathbf{u}_G$ can be determined from $\mathbf{u}$. As the bottom Ekman layer is determined by the need for Ekman velocity to cancel the depth-independent component of velocity in order to fulfill the bottom no-slip condition, $\mathbf{u}_{Ek}$ is of the same order of magnitude as $\mathbf{u}_G$. In a shallow water model, the bottom (or in some cases only) layer often has $\delta_{Ek} \ll 1$, in which case $\mathbf{u}_G \approx \mathbf{u}$.

We can rewrite the momentum equation to include these stresses by integrating vertically over the layer and dividing by h in order to find the vertical average. This does not affect the pressure term in (2.72), as $\nabla_h p$ is independent of z, or $\nabla \cdot (\nu_h \nabla \mathbf{u}_h)$, as we can interpret $\mathbf{u}_h$ as a depth-averaged quantity. The problem with violating the assumption of no vertical shear occurs in the material derivative term, because in general,

$$\overline{\mathbf{u} \cdot \nabla \mathbf{u}} \neq \overline{\mathbf{u}} \cdot \nabla \overline{\mathbf{u}}, \tag{2.75}$$

where the bar denotes the vertical average $(1/h) \int dz$. Therefore, it is not clear how realistic the shallow water equations are when (1) some individual layer(s) have significant vertical velocity shear due to inclusion of the Ekman layer and (2) the advection terms are important to the dynamics. Setting these caveats aside, we state the resulting version of the horizontal momentum equations integrated vertically over a layer as

$$\frac{D\mathbf{u}_h}{Dt} + f\hat{\mathbf{z}} \times \mathbf{u}_h = -\frac{1}{\rho_0}\nabla p + \nabla \cdot (\nu_h \nabla \mathbf{u}_h) + \delta_T \frac{\mathbf{X}_T}{h} - \delta_B \frac{\nu_v}{h^2 \delta_{Ek}}(\mathbf{u}_G + \hat{\mathbf{z}} \times \mathbf{u}_G), \tag{2.76}$$

where $\boldsymbol{\tau}_T/\rho_0 = \mathbf{X}_T$ is the wind stress and (δ_T, δ_B) are each zero unless we are in a layer that includes the Ekman layer at the top or bottom, respectively, in which case they are one.

Continuity Equation

We now transform the continuity equation, $\nabla \cdot \mathbf{u} = 0$ (1.16), into an equation for h. Integrating in the vertical between the lower and upper layer interfaces of an individual layer, we find

$$\int_{\eta_B}^{\eta_T} \left(\nabla \cdot \mathbf{u}_h + \frac{\partial w}{\partial z} \right) dz = 0, \tag{2.77}$$

which can be written

$$w(\mathbf{x}_h, t, \eta_T) - w(\mathbf{x}_h, t, \eta_B) + h\nabla \cdot \mathbf{u}_h = 0, \tag{2.78}$$

assuming that either horizontal velocity is independent of z within the layer or h is constant (in which case $\mathbf{u}_h$ represents a depth-average for the layer). If there is no flow through a layer interface, then the vertical speed of that interface is $w = D\eta/Dt$ (as at the sea surface; see Subsection 1.3.1). Alternatively, we may want to allow water to flow *through* the interface with an **interfacial vertical velocity** w_* which is positive upwards. Including this flow across the interface, the continuity relation becomes

$$\frac{D}{Dt}\left(\eta_T - \eta_B\right) + h\nabla \cdot \mathbf{u}_h + w_{*T} - w_{*B} = 0, \tag{2.79}$$

or substituting for the interface heights,

$$\frac{Dh}{Dt} + h\nabla \cdot \mathbf{u}_h = -(w_{*T} - w_{*B}). \tag{2.80}$$

Equation 2.80 can also be written in flux form (Subsection 1.3.1). As $\mathbf{u}_h \cdot \nabla h + h\nabla \cdot \mathbf{u}_h = \nabla \cdot (\mathbf{u}_h h)$, (2.80) becomes

$$\frac{\partial h}{\partial t} + \nabla \cdot (\mathbf{u}_h h) = -(w_{*T} - w_{*B}). \tag{2.81}$$

The flux form can be interpreted as telling us that h at some location grows due to transport of volume in from the sides and through the layer interface from above and below.

What is the physical significance of the interfacial velocity? If the shallow water layers exclude Ekman layers, w_* on the upper boundary of the top layer or the lower boundary of the bottom layer represents pumping/suction (2.13) from the Ekman layers. Thus frictional forcing by the wind and damping by the sea floor can be either represented explicitly in the momentum equation (2.76) or implicitly through the continuity equation (2.80).

Interfacial vertical velocity between layers represents water changing its density as it moves from one layer to a neighboring one. Because of the discrete densities, the density of a parcel must instantly jump to that of the layer it enters. A layer model is a crude representation of this process, though as the number of layers representing a particular stratification increases, the density jump between layers decreases and the density change of a parcel becomes smoother. If $w_* \neq 0$, we need a rule relating w_* at each interface to other variables such as layer thickness in order to represent density-changing processes such as mixing and air–sea exchanges of heat and freshwater. Volumes of any individual layers remain constant in time if the area averages of $w_{*T} = w_{*B} = 0$, otherwise they may evolve as water moves between layers.

To model any given circulation with a shallow water system, we need to specify the basin geometry, number of layers, density of each layer, and wind stress. For each layer, equations (2.76) and (2.80) constitute three scalar equations in unknowns u_h, v_h, h, p, w_* for each layer. Pressure for all the layers can be calculated from interface heights for all the layers with (2.71), and the interface heights can be related to h with (2.65). If we have a rule for w_* (either $w_* = 0$, Ekman pumping, or a parameterization of w_* in terms of other variables, as appropriate), then the system is complete except for lateral boundary conditions such as impermeable side walls. If we are looking for a solution that evolves over time rather than an equilibrium solution, we need initial conditions on the variables as well.

2.2.5 Shallow-Water Potential Vorticity Budget

Horizontal large-scale velocity in the ocean is closely related to the vertical component of vorticity, ζ. One can derive an evolution equation for ζ from the shallow water equations; however, we get more insight into the fluid behavior by considering the potential vorticity instead. Therefore we will derive the budget equation for potential vorticity (2.23). As mentioned in Subsection 2.1.5, potential vorticity can be defined in several ways. Here, we consider the shallow water version (2.22),

$$q = \frac{f + \zeta}{h}. \tag{2.82}$$

In this section we drop the subscripts representing layer number, except as noted; results here apply to any layer in a shallow water model. For generality, we consider a layer with cross-isopycnal flow, forcing by wind stress, bottom drag, and horizontal viscosity. Thus the shallow water equations are:

$$\mathbf{u}_t + \mathbf{u} \cdot \nabla \mathbf{u} + f\hat{\mathbf{z}} \times \mathbf{u} = -\frac{1}{\rho}\nabla p + \delta_T(\mathbf{X}/h) - \delta_B r \mathbf{u} + \nu_h \nabla^2 \mathbf{u}, \tag{2.83}$$

$$h_t + \nabla \cdot (h\mathbf{u}) = w_{*_B} - w_{*_T}, \tag{2.84}$$

where w_{*_T} and w_{*_B} are interfacial vertical velocities at top and bottom, respectively, of the layer. Similarly, we include terms that are only non-zero in a single layer using δ_T and δ_B, which equal one for top and bottom layers, respectively, and zero otherwise. Pressure is given by (2.71) which relates it to interface height and ultimately to layer thickness h.

To derive the potential vorticity equation, we first form the vorticity equation by taking the curl of the momentum equations, eliminating the pressure gradient terms in (2.83). We want to transform the advective and Coriolis terms into a more useful form. For this we need the following identity which applies to an arbitrary vector function $\mathbf{A}$,

$$(\mathbf{A} \cdot \nabla)\mathbf{A} = \frac{1}{2}\nabla(\mathbf{A} \cdot \mathbf{A}) - \mathbf{A} \times (\nabla \times \mathbf{A}), \tag{2.85}$$

which gives us

$$(\mathbf{u} \cdot \nabla)\mathbf{u} = \frac{1}{2}\nabla(\mathbf{u} \cdot \mathbf{u}) - \mathbf{u} \times (\hat{\mathbf{z}}\zeta). \tag{2.86}$$

Now take the curl of this equation, noting that the gradient term is zero, so that

$$\nabla \times \left[(\mathbf{u} \cdot \nabla)\mathbf{u} + (\hat{\mathbf{z}}f \times \mathbf{u})\right] = \nabla \times \left[(f + \zeta)\hat{\mathbf{z}} \times \mathbf{u}\right]. \tag{2.87}$$

It is convenient to use another vector identity to deal with the curl of a cross product (for vector functions $\mathbf{A}$ and $\mathbf{B}$):

$$\nabla \times (\mathbf{A} \times \mathbf{B}) = (\mathbf{B} \cdot \nabla + \nabla \cdot \mathbf{B})\mathbf{A} - (\mathbf{A} \cdot \nabla + \nabla \cdot \mathbf{A})\mathbf{B}. \tag{2.88}$$

When applied to the right-hand side of (2.87), the final group of terms from this vector identity vanish, as $\nabla \cdot [(f + \zeta)\hat{\mathbf{z}}] = 0$ and $\mathbf{u}_z = 0$, leaving

$$\nabla \times \left[(\mathbf{u} \cdot \nabla)\mathbf{u} + (\hat{\mathbf{z}}f \times \mathbf{u})\right] = \hat{\mathbf{z}}\left[\mathbf{u} \cdot \nabla(f + \zeta) + (\nabla \cdot \mathbf{u})(f + \zeta)\right]. \tag{2.89}$$

This allows us to write the curl (i.e., $\hat{\mathbf{z}} \cdot \nabla \times$) of the momentum equation (2.83) as

$$\frac{D}{Dt}(f + \zeta) + (f + \zeta)\nabla \cdot \mathbf{u} = \delta_T \hat{\mathbf{z}} \cdot \nabla \times (\mathbf{X}/h) - \delta_B r\zeta + \nu_h \nabla^2 \zeta. \tag{2.90}$$

This form of the vorticity equation shows that the absolute vorticity of the layer, $f + \zeta$, is advected by the flow (first term on the left), changes by vortex stretching/squashing (second term on the left), and frictional forcing and damping (right-hand terms).

Next, we use the continuity equation to eliminate the horizontal divergence term $\nabla \cdot \mathbf{u}$. Using the product rule for derivatives to rearrange the height equation (2.84) gives:

$$\frac{Dh}{Dt} + h\nabla \cdot \mathbf{u} = w_{*_B} - w_{*_T}. \tag{2.91}$$

Substitution into (2.90) yields

$$\frac{D}{Dt}(f + \zeta) + \left(\frac{f + \zeta}{h}\right)\left(w_{*_B} - w_{*_T} - \frac{Dh}{Dt}\right) = \delta_T \hat{\mathbf{z}} \cdot \nabla \times (\mathbf{X}/h) - \delta_B r\zeta + \nu_h \nabla_h^2 \zeta. \tag{2.92}$$

Finally, rearrange to write the shallow water potential vorticity on the left-hand side. Noting that

$$\frac{D}{Dt}\left(\frac{f + \zeta}{h}\right) = \frac{1}{h}\frac{D}{Dt}(f + \zeta) - \left(\frac{f + \zeta}{h^2}\right)\frac{Dh}{Dt}, \tag{2.93}$$

dividing (2.92) by h gives

$$\frac{D}{Dt}\left(\frac{f + \zeta}{h}\right) = \left(\frac{f + \zeta}{h^2}\right)(w_{*_T} - w_{*_B}) + \frac{1}{h}\left[\delta_T \hat{\mathbf{z}} \cdot \nabla \times (\mathbf{X}/h) - \delta_B r\zeta + \nu_h \nabla_h^2 \zeta\right]. \tag{2.94}$$

This equation shows that if there is no cross-isopycnal flow into or out of a layer, and if forcing and damping are negligible (namely, the right-hand side vanishes), then potential vorticity q is materially conserved. It is the basis of (2.23) in Subsection 2.1.5. This is an important constraint on the circulation and it appears in various guises in many places (see Chapters 3, 5, and 6).

Notice that the right-hand side forcing and dissipation terms are all quite uncertain. They all represent the parametrized effects of unresolved processes. For this reason, one has scope in how to specify them exactly. For instance, one may linearize the terms in the cross-isopycnal flow and wind stress curl (replace h with its average value in the first two terms), or replace the dissipation of vorticity with dissipation of potential vorticity (replace ζ_n with q_n in the final two terms). The exact forms of these terms are not particularly important; they are placeholders for the unresolved forcing and dissipation processes. Indeed, they are rarely all present: the (common) adiabatic assumption removes the cross-isopycnal flow term, wind forcing only applies to the surface layer, bottom drag only applies to the bottom layer, and the (common) inviscid assumption removes the final viscous term.

2.3 Excursions: Walfrid Ekman, Fridtjof Nansen, and the Fram Expedition

Frictional effects are mainly important in thin zones at the sea surface and sea floor. The zones are known as Ekman layers because Swedish oceanographer Vagn Walfrid Ekman (1874–1954) explained them with a pioneering theory in 1905 for his doctoral thesis. Ekman was interested in understanding the observation that Arctic sea ice drifts at an angle of 20–40° to the right of the wind. Ekman's key insight was to balance the stress on the ice with turbulent friction and the Coriolis force. The dynamics of the Ekman layer, and its interaction with the rest of the ocean, are profoundly important for ocean circulation, and rotating fluids in general (see Jenkins and Bye, 2006).

The observation that ice drifts to the right of the wind was made by Fridtjof Nansen (1891–1930), during his attempt to reach the North Pole. Nansen believed that a current crosses the pole from Siberia to Greenland. He therefore procured the *Fram*, a purpose-built ship designed to withstand being crushed by sea ice, and mounted an expedition. The *Fram* became caught in the ice in September 1893 and drifted across the Arctic Ocean for nearly three years. Although Nansen failed to reach the pole, the expedition made abundant discoveries. In addition to the sideways drift of sea ice, which led to Ekman's theory, they include the great depth of the Arctic Ocean, the transpolar drift of ice and water, and the layer of warm salty water below the surface, now called Atlantic water.

See also Figure 1.2 for a picture of one of Ekman's current meters in the Norwegian museum where the *Fram* is displayed.

Exercises

2.1   A baseball is thrown due South 40 m at Camden Yards, Baltimore in 1 second.

   (a) What is the Coriolis parameter?
   (b) What is the deflection due to the Coriolis force?
   (c) Estimate the Rossby number in this case.

2.2   A 60 kg sprinter runs 100 m in 10 s at the London Olympics. If she runs due East estimate her apparent reduction in weight due to the vertical Coriolis force.

2.3   Reflect on your answer to Exercise 1.2. Update your answer based on what you learned in Chapter 2.

2.4   The Straits of Dover which separate England and France are aligned east to west with the French coast to the south. The Straits are at 51°N and are 35 km wide. The water is well mixed by tidal currents and strong winds. The mean geostrophic current is toward the east.

   (a) If the mean geostrophic current is 0.2 ms^{-1}, what is the slope of the sea surface?

(b) Is the sea surface higher on the French or English side?

(c) How much higher?

2.5 (a) Estimate the angle and speed that icebergs drift with respect to the direction of the surface wind over the sea at $60°$S. Ignore the direct effect of the wind on the ice and make reasonable simplifying assumptions that you clearly state.

(b) Fridtjof Nansen observed that Arctic sea ice moves at 20–$40°$ to the right of the wind during the Arctic *Fram* expedition. What conclusions can you draw about the direct effect of the wind on sea ice?

Hint: Be aware of the difference between icebergs and sea ice.

2.6 (a) A useful expression relating derivatives of field χ at constant height z to derivatives at constant pressure p is

$$\nabla_p \chi = \nabla_z \chi + \frac{\partial \chi}{\partial z} \nabla_p z. \tag{2.95}$$

Derive this formula, for example, by sketching the χ field as a function of (x, z) and (x, p).

(b) Using (2.95), or otherwise, derive the thermal-wind equations.

2.7 (a) A useful version of the hydrostatic law for two layers of constant-density seawater is:

$$\nabla p_2 - \nabla p_1 = (\rho_1 - \rho_2) g \nabla h. \tag{2.96}$$

In this formula the gradients are in the horizontal plane, the pressures are p_1 and p_2 in layers 1 and 2, the densities are ρ_1 and ρ_2 ($\rho_2 > \rho_1$; layer 1 is above layer 2), g is acceleration due to gravity, and h is the depth of the interface between the layers. Derive the formula assuming that the pressure field is continuous across the interface.

(b) Derive the alternate thermal-wind formula (2.46) using your result from (a), or otherwise.

2.8 Derive formula (2.49) relating the dynamic height gradient to the geostrophic current.

2.9 Derive the following useful formula (2.70) for the hydrostatic pressure in layer n of a shallow water model with reduced gravity $\gamma_n = g(\rho_{n+1} - \rho_n)/\rho_0$, interface heights η_n, and gravitational acceleration g:

$$p'_n = p'_{n-1} + \rho_0 \gamma_{n-1} \eta_{n-1}. \tag{2.97}$$

2.10 Consider the idealized vertical hydrographic section of a mid-latitude eddy in Figure 2.12. The text datafile `station_data.txt` and the MATLAB file `station_data.mat` contain the (pressure, conservative temperature) data from the stations. Using the methods described in Subsection 2.2.2 (or otherwise), and in each case stating the maximum and minimum, compute and plot:

(a) The geostrophic current relative to the deepest common depth.

(b) The vertical and horizontal components of relative vorticity of the current from (a). Comment on their relative sizes.

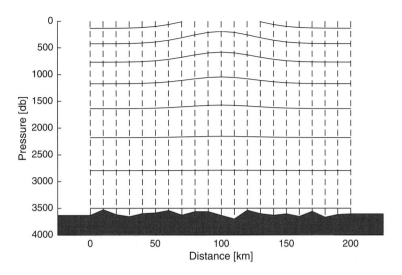

Figure 2.12 Geostrophic current calculation for Exercise 2.10. Temperature contours are shown for several hydrographic stations (dashed lines). The latitude is 40°N and the station spacing is 10 km. The salinity of the water is uniform at 35 g/kg. See also Figure 7.5.

(c) The Ertel potential vorticity of the flow from your answers to (a) and (b). Comment on the relative sizes of the terms from the planetary vorticity, the vertical relative vorticity, and the horizontal vorticity.

(d) The streamfunction for the depth-integrated flow from (a).

(e) Assume that the section is zonal (at constant latitude). Compute and plot the vertical velocity using the geostrophic divergence of the meridional current from (a). Make a sensible assumption about the vertical velocity at the deepest common depth and state it.

(f) Review your answer to (a) and comment on how well the geostrophic current relative to the deepest common depth estimates the true geostrophic current if: (i) the velocity at the deepest common depth is not zero, and (ii) the true flow contains a nonzero component in the plane of the section. Repeat this process by considering how well your results in (b)–(e) estimate the vorticity components, Ertel potential vorticity, streamfunction, and vertical velocity of the true flow.

Hint: Several of the tasks in this problem can be tackled with the TEOS-10 software. You will need to approximate derivatives with finite-difference formulae and integrals with sums.

2.11 This problem involves estimating the geostrophic current along a hydrographic section of CTD stations. The hydrographic data are used to compute the thermal wind and hence to estimate the geostrophic current. Work on Exercise 2.10 first to build experience and talk to the instructor about how to do each step, as necessary.

(a) Pick a long zonal CTD section in subtropical latitudes, for instance from the database at http://cchdo.ucsd.edu. For example, A5 along 24°N in the

North Atlantic is a good choice (there are several years available; Figure 11.11 is also useful). These data consist of profiles of CTD in-situ temperature and salinity (on the practical salinity scale) as functions of seawater pressure (excluding atmospheric pressure), longitude, and latitude. Load the CTD profiles from the archive. Plot the station locations on a map of bathymetry (see Appendix A).

(b) Make a Θ-S_A diagram from the cruise data and contour σ_θ on it.

(c) Make vertical sections of Θ, S_A, and σ_0 along the cruise track (i.e., make contour plots with section distance on the abscissa and pressure (or depth) on the ordinate). To facilitate plotting, interpolate each CTD profile onto a standard pressure grid.

(d) Use the thermal wind relationship and the density data to compute the geostrophic velocity with respect to a deep common pressure, p_0, as a function of along-section distance and pressure. Contour your results and state the maximum and minimum current speeds.

(e) Compute the streamfunction for the pressure-integrated flow from (d). Hence estimate the volume transport of the subtropical gyre.

Hints: You may need to remove a few CTD stations when they do not fall along the main section. Experiment with different deep reference pressures p_0 to try and capture the subtropical gyre circulation as clearly as you can.

2.12 Why might rainfall or evaporation interfere with our ability to describe the circulation in the ocean with a horizontal streamfunction? Is this a major problem?

2.13 The governing equations for ocean velocity, temperature, and salinity, are given by (1.16). These equations are statements of how each quantity changes over time. Thus if there is a given surface forcing (wind stress and conditions on temperature and salinity) and a given initial condition, the equations can be integrated in time to find how each quantity evolves. In some theoretical or numerical models, $\partial/\partial t$ terms are removed from the momentum equations, however (for instance, in the hydrostatic and geostrophic equations; Subsection 2.1.2). How can velocity change over time in that case?

Harder question (optional): What difference does removing those terms from the momentum equations make in the evolution of the flow?

3 Two-Dimensional Horizontal Circulation

The horizontal circulation of the ocean is dominated by the transfer of momentum from the atmosphere, which is measured by the wind stress $\boldsymbol{\tau}$. For that reason, in this chapter we examine observations of wind stress and horizontal ocean circulation (especially near-surface), and then use conceptual and theoretical models to relate the horizontal ocean circulation to the wind stress. Specifically, we present observational evidence of the major wind-driven current systems, we explain the existence of gyres, the presence of western boundary currents and their recirculations. We also derive theories for gyre strength (Sverdrup balance), western boundary currents (β plane versus f plane, boundary layer theory), and zonal current strength.

3.1 Observations

3.1.1 Wind Stress Forcing

Different strategies have been used to produce a climatology of wind stress over the global ocean. One method is to compile observations from ships and buoys of near-surface wind velocity (typically at 10 m height) and other meteorological variables. Such observations typically do not include direct measurement of wind stress, but analysis of near-surface atmospheric behavior has led to **bulk formulae**, which convert a set of measurements at a given time and place into an estimate of the stress of the wind on the ocean surface $\boldsymbol{\tau} = (\tau, \sigma)$ (see for instance Fairall et al., 2003; Edson et al., 2013). Observations over many years can then be interpolated and averaged on to regular grids, as was done by Hellermann and Rosenstein (1983), and Woodruff et al. (1987) and Josey et al. (2002). Another approach is to incorporate observations into numerical weather models to produce an estimate of the wind stress and other variables, such as the NCEP-NCAR Reanalysis (Kalnay et al., 1996) and the ERA-Interim Reanalysis (Dee et al., 2011); see Appendix A. This method potentially improves on the observations by allowing the model dynamics to estimate data values at times and places where data is missing, but may also introduce additional errors because of imperfections in the models. Finally, remote sensing from satellite-borne instruments promises greater amounts of data (especially in regions not often visited by ships, such as large sectors of the southern hemisphere), though such data is only available for recent years. Since the 1990s, scatterometers such as QuikSCAT have used radar signals reflected from small-scale surface waves to estimate both the direction and magnitude of wind stress (Risien and Chelton, 2008).

The general pattern of winds was known to mariners before any of the climatologies mentioned above were produced. It has been more difficult to create a quantitatively accurate picture of the wind stress. Recent attempts such as Risien and Chelton (2008) are probably accurate to 20% or better.

The annual average wind stress has a broadly similar pattern in the Atlantic, Pacific, and South Indian oceans (Figure 3.1a). The tropics are dominated by the **Trade Winds**, bands of **easterly** wind (that is, wind blowing from the east). The Trades also have a strong equatorward component, especially on the eastern side of basins. They represent the surface

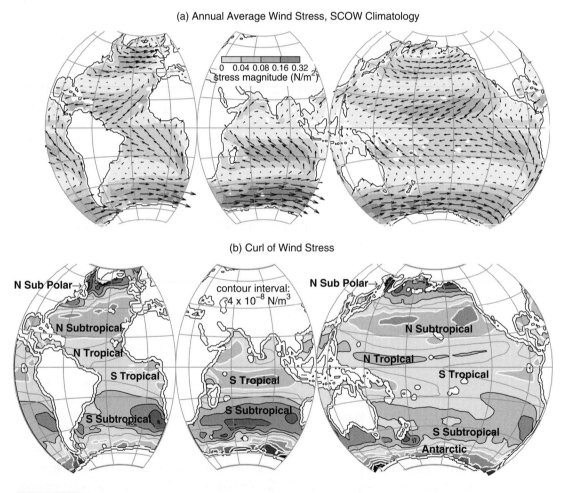

Figure 3.1 Annual average wind stress data from the SCOW product averaged on 1° cells. See Appendix A and Risien and Chelton (2008). (a) Wind stress, (b) wind stress curl. For (b), contour interval $= 4 \times 10^{-8}$ N/m^3, black and white indicating positive (counterclockwise) and negative (clockwise) values, respectively. A black-and-white version of this figure (Figure 3.1b) appears in some formats. For the color version, please refer to the plate section.

component of the **Hadley Circulation**, an overturning circulation in the tropical atmosphere in which temperature contrasts between the equator and the subtropics causes air to rise near the equator, blow poleward in the upper troposphere, sink in the subtropics, and return near the surface. The strong zonal component of the Hadley circulation is due to the Coriolis force (Subsection 2.1.1). The surface wind near the equator is relatively weak and is known as the **doldrums**. This marks the upward-flowing part of the Hadley Circulation. One might expect this to occur right on the equator, but it is actually a bit to the north, where it is called the **Intertropical Convergence Zone**. In the Indian Ocean, the equatorial wind is especially weak (and westerly rather than easterly), and the Trade Winds are absent in the northern hemisphere annual mean.

Poleward of about 30° latitude in all oceans, there is, on average, a strong, nearly zonal wind known simply as the **westerlies** due to its direction. The westerlies grow weaker as we go further poleward in the northern North Atlantic and Southern Ocean. The full explanation of why the Hadley circulation does not extend all the way to the poles is beyond our scope (see Marshall and Plumb, 2008, Holton and Hakim, 2013, for instance), but note that at higher latitudes, synoptic atmospheric eddies (roughly 500 km wide weather systems of clockwise and counterclockwise flows that last for a few days) form and these time-dependent eddies create the time-average westerlies. The region of relatively weak wind between the trades and the westerlies is known to sailors (for obscure reasons) as the **Horse Latitudes**. Much of the large-scale pattern can be seen in the zonal average of zonal wind stress (Figure 3.2).

The global wind field has a strong seasonal cycle, with the wind stress stronger in the winter hemisphere (Figure 3.2). The midlatitude southern hemisphere has less annual variation than the rest of the ocean. There is strong daily variability (**weather**) in the wind field, especially in the extra-tropics due to the synoptic eddies mentioned above. Tropical cyclones (typical diameter of a few hundred kilometers) and sporadic development

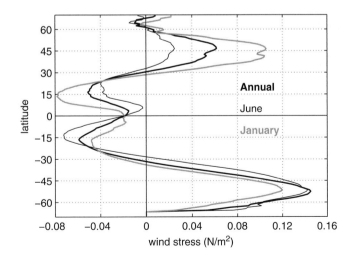

Figure 3.2   Zonal average wind stress from SCOW (see Appendix A) climatology for annual (thick black), June (thin black), and January (gray) averages.

of thunderstorms and other small-scale atmospheric convection also make large changes in the local wind. For a linear system, the time-average response equals the response of the time-average. Many aspects of large-scale ocean circulation depend linearly on τ, so it is appropriate to consider the time-average wind stress when trying to understand the time-average ocean circulation. Nevertheless, anticipating the discussion in Chapter 7 and Chapter 9, we must remain open to the possibility that nonlinear dynamics are important.

As we see below (see also Subsection 2.1.4 and Subsection 2.1.3) an important quantity in relating circulation to wind is the wind stress curl,

$$\mathrm{curl}\ \tau = \hat{\mathbf{z}} \cdot (\nabla \times \tau), \tag{3.1}$$

where $\hat{\mathbf{z}}$ is the unit vector in the vertical direction. Just as wind stress represents the transfer of momentum from the air to the sea, wind stress curl represents the transfer of "spin," as will be discussed more precisely in Subsection 3.2.1 and Subsection 3.2.2 (see also Subsection 2.2.5). Each basin has alternating bands of positive (counterclockwise) and negative (clockwise) wind stress curl (Figure 3.1b). We refer to rotation in the same direction as the Earth's rotation as **cyclonic** (counterclockwise in the northern hemisphere and clockwise in the southern), and in the opposite direction as **anticyclonic**. Thus the wind stress curl is anticyclonic in the subtropics of both hemispheres, and cyclonic in the subpolar regions and (to a lesser extent) in the tropics.

If it seems absurd that the northern hemisphere and southern hemisphere should be spinning in opposite directions, do the following experiment: Look down on the north pole of a globe and rotate it counterclockwise. While continuing to rotate it in the same direction, raise it above your head so that you are looking up at the south pole. It should appear to be moving clockwise.

3.1.2 Ocean Gyres

As discussed in Chapter 2 (Subsection 2.1.1 and Subsection 2.1.2), large scale flows are geostrophic with an Ekman component near the surface (Subsection 2.1.3). The surface geostrophic flow is indicative of the flow in the top kilometer of the ocean, whereas the Ekman flow is confined to the upper 30–100 m. In much of the ocean, especially poleward of about $10°$ latitude, the surface flow is also quite similar to the geostrophic flow. We discuss the Ekman flow and the total surface flow in more detail in Chapter 6. Here we concentrate on the geostrophic flow.

Like the time-mean wind stress, the ocean's surface circulation is dominated by relatively simple patterns repeated among the basins and hemispheres. Sea surface height, derived from satellite altimetry (Subsection 1.1.2), traces the approximate streamlines of the geostrophic flow (Figure 3.3). This chart shows that most of the ocean surface is covered by a series of **gyres**, each of which consists of either cyclonic or anticyclonic flow around closed streamlines. Each gyre spans the basin in the east–west direction and has northern and southern boundaries a few tens of degrees apart. Just as the smooth time-mean wind stress patterns (shown in Figure 3.1a) hide strong daily weather variability, the time-mean flow in Figure 3.4 hides intense transient currents from eddies and tides that are discussed more fully in Chapter 7.

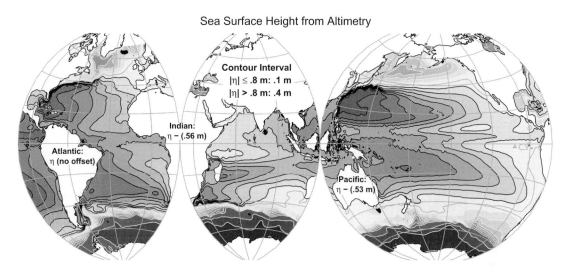

Sea Surface Height from Altimetry

Figure 3.3 Sea surface height (SSH), 1993–2012 average, from satellite-derived altimetry (data source described in Appendix A). Contour intervals are described in figure. Darker shading represents higher (regions with black contours) or lower (white contours) sea surface. Height is offset by a different amount in each basin in order to highlight the gyres. In low-speed regions, each point is the average of neighboring grid points in a 25×25 array.

The largest and most obvious gyres are the anticyclonic **subtropical gyres**, which are centered around 30° latitude and occur in the North and South Pacific, North and South Atlantic, and South Indian oceans. North of the subtropical gyres in the North Atlantic and North Pacific are the cyclonic **subpolar gyres**. Zonal boundaries defining the separate basins are largely absent in the subpolar southern hemisphere, and so there is no obvious subpolar gyre there. Instead, the Southern Ocean is dominated by an eastward flow called the **Antarctic Circumpolar Current (ACC)**. Whereas the gyres are confined by coasts and circulate water around an oceanic center, the outer perimeter of the ACC is not confined by coasts and instead a continent (Antarctica) is in the *center* of its circulation. In the Atlantic sector of the Antarctic, the region just to the east of the **Antarctic Peninsula** is the **Weddell Sea**, which has an additional cyclonic gyre between the ACC and Antarctica. Chapter 9 discusses the circulation of the Southern Ocean and Chapter 10 discusses the Arctic Ocean.

The geostrophic flow in low latitudes shows some sign of zonally elongated cyclonic **tropical gyres**. They consist of westward currents equatorward of the subtropical gyres and eastward flow just north of the equator in the Atlantic and Pacific and just south of the equator in the Indian Ocean. The eastward flow is known as the **Equatorial Countercurrent** and the westward tropical flows to either side of the Equatorial Countercurrent are known as the **North Equatorial Current** and the **South Equatorial Current** (see also Subsection 6.1.1). Because, in the Atlantic and Pacific, the South Equatorial Current straddles the equator, it is partially in the Northern Hemisphere despite its name.

There is some resemblance between the wind stress pattern (Figure 3.1a) and the gyres (Figure 3.4), especially in the Southern Ocean where a strong eastward current flows

Surface Geostrophic Velocity From Altimetry

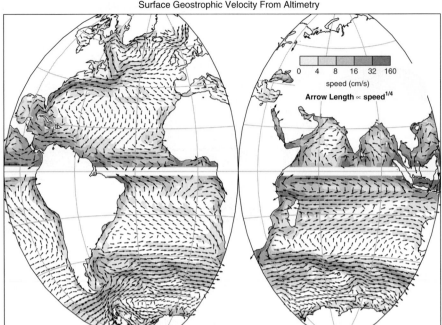

Figure 3.4   Surface geostrophic velocity, as in Figure 3.3 (arrows) with shading representing speed (contours at 2, 8, 32, 128 cm/s). Close to the equator, where geostrophic balance fails, the speeds are masked.

directly below a strong eastward (westerly) wind. However, the poleward flow on the western side of the subtropical gyres does not generally correspond to the poleward wind stress, and the strong eastward Equatorial Countercurrent flows against a fairly strong westward wind stress. The correspondence between the gyres and the wind stress curl (Figure 3.1b) is much closer, with the gyre locations and circulation direction corresponding to the location and sign of wind stress curl. As we will see in Subsection 3.2.1, it is wrong to explain the surface ocean currents as being simply proportional to the wind stress.

3.1.3 Western Boundary Currents

A striking feature of the gyres is the asymmetry between flow on their eastern and western sides. This can be seen in a map of surface speeds (Figure 3.4). Much of the gyres have speeds less than 2 cm/s, while speeds of greater than 8 cm/s can be found near the equator, in the ACC, and along the western edges of the basins. The dynamic center of each gyre is where the current switches from northward to southward flow and can be seen most easily by the local maximum or minimum in sea surface height (SSH, Figure 3.3). This point occurs close to the western boundary of the basin in each gyre, rather than in the middle. Thus in the subtropical gyres, the poleward flow in the west is much narrower and stronger than the equatorward flow in the rest of the basin. Similarly, the equatorward flow in the west of the subpolar gyres is strong and narrow. This phenomenon is called **western intensification** and its causes are discussed in Subsections 3.2.2, 3.3.2 and 3.3.3.

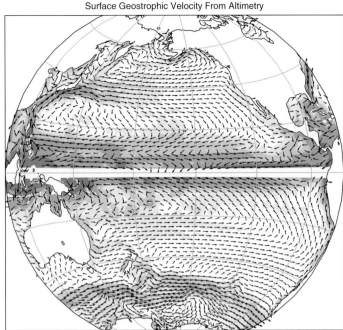

Surface Geostrophic Velocity From Altimetry

Figure 3.4 Surface geostrophic velocity, as in Figure 3.3 (arrows) with shading representing speed (contours at 2, 8, 32, 128 cm/s). Close to the equator, where geostrophic balance fails, the speeds are masked.

The details of these **western boundary currents** are hard to see in a global chart, so we look more closely at some examples as illustrated by sea surface height (Figure 3.5). The western boundary currents are typically about 100 km wide. Speeds in the western boundary current jet-cores often exceed 1 m/s. In comparison, typical annual-average current speeds outside the western boundary currents are a few cm/s in the extra-tropics. In the tropics, the zonal surface currents can be tens of cm/s. Figure 5.7 shows a detailed Gulf Stream velocity profile as a function of depth and distance across the jet.

The Gulf Stream is a continuation of a western boundary current that flows along the north coast of South America (as the **north Brazil current**) into the Caribbean, near Central America, into the Gulf of Mexico (as the **Loop Current**), and through the Florida Strait (as the **Florida Current**). The subtropical gyre western boundary current is named the **Kuroshio Current** in the North Pacific (Figure 3.5b, Kuroshio means "black stream" in Japanese), the **East Australian Current** in the South Pacific, the **Brazil Current** in the South Atlantic (Figure 3.5c), and the **Agulhas Current** in the South Indian (Figure 3.5d). The subpolar gyre western boundary current is known as the **Oyashio Current** in the North Pacific (Figure 3.5b) and (in part) the **Labrador Current** in the Labrador Sea (Figure 3.5a). Though a South Atlantic subpolar gyre is not generally recognized north of the Antarctic Peninsula, the South Atlantic possesses what looks to be a typical equatorward-flowing subpolar western boundary current known as the **Falklands Current** (sometimes known as the Malvinas Current, after the Spanish name for the Falklands Islands) to the south of the Brazil Current (Figure 3.5c).

(a) North Atlantic   (b) North Pacific

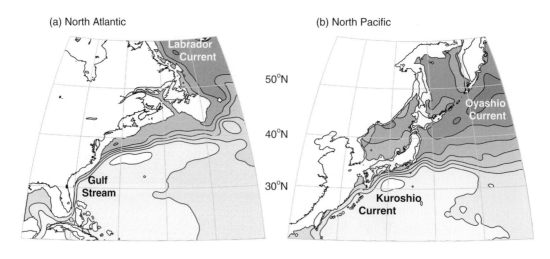

(c) South Atlantic   (d) South Indian

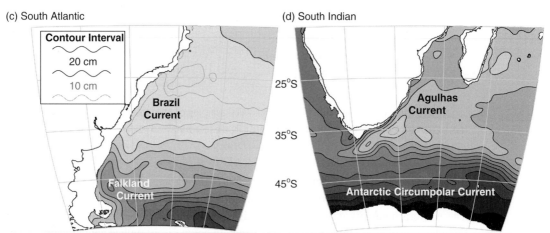

Figure 3.5   Sea surface height from same data set as in Figure 3.3 (but with no horizontal smoothing), showing western boundary currents for (a) North Atlantic, (b) North Pacific, (c) South Atlantic, and (d) Indian oceans. Darker shading represents lower SSH. Coasts are indicated, and all ocean areas with water depth less than 200 m are whited out. See also Figure 4.2.

Theory typically portrays western boundary currents as flowing along a "wall" at the edge of the ocean, and indeed, over much of their path western boundary currents do hug the continental slope on the western side of the basin. In some places, such as off the southeast United States (Figure 3.5a) or eastern South Africa (Figure 3.5d), the continental shelf is narrow and the western boundary current comes very close to the coast. In other cases, such as between Taiwan and Japan (Figure 3.5b) and off Argentina (Figure 3.5c), a broad continental shelf separates the western boundary current from the coast. More complicated geometries also exist. In Figure 3.5b, the Oyashio Current seems to leave the western boundary as it travels across the mouth of the Sea of Okhotsk, though the Kuril Islands (barely visible in the figure) form a chain that acts something like a western boundary. The Alaska Peninsula, which juts into the central subpolar gyre of the North Pacific and continues as a shallow ridge on which the Aleutian Islands sit (Figure 1.9), forms a western boundary along which the **Alaska Stream** flows. In the subpolar Atlantic, the western boundary current starts as the **East Greenland Current**, flows around the southern tip of Greenland and continues on the *eastern* boundary of the Labrador Sea as the **West Greenland Current**, and crosses the Labrador Sea near Davis Strait to feed the Labrador Current mentioned above. The Gulf Stream separates from the continental slope near **Cape Hatteras**, North Carolina. The flow downstream of the Grand Banks of Nova Scotia is often called the **Gulf Stream Extension** (or the **North Atlantic Current**). The Agulhas Current (Figure 3.4d) is fed by two western boundary currents, one along the African coast and one along the east coast of Madagascar.

There is a transition region where the western boundary current leaves the coast and rejoins the slow part of the gyre. The subtropical western boundary currents, and to a lesser extent the subpolar western boundary currents, retain their jet-like structure for thousands of kilometers downstream of the separation point (Figure 3.5), though eventually the separated current becomes broader and slower (Figure 3.4). The Agulhas Current "runs out of coast" by extending further south than the tip of Africa, where it makes a sharp turn to the east known as the **Agulhas Retroflection** (Figure 3.5d).

3.1.4 Vertically Integrated Velocity

The geostrophic velocity at the surface or other given levels is difficult to predict theoretically. As we see in the Concepts and Theory sections of this chapter, theories are better at addressing the depth-integrated horizontal velocity,

$$\mathbf{U}(\phi,\theta) = \int_{-H}^{\eta} \mathbf{u}(\phi,\theta,z)\,dz \qquad (3.2)$$

where (ϕ,θ) are longitude and latitude, $\mathbf{u}$ is horizontal velocity with eastward and northward components (u,v), z is height in the water column, $H(\phi,\theta)$ is the depth of the sea floor, and $\eta(\phi,\theta)$ is sea surface height. Here velocity refers to total velocity, not just the geostrophic component. Depth integrated velocity $\mathbf{U}$ is sometimes referred to as **transport** (unit: m^2/s) and has components (U,V). We define z so that the sea surface is near $z = 0$, so that η has typical magnitudes of about 1 m and H has typical magnitudes of thousands of meters (see Subsection 1.1.6).

Besides corresponding directly with ocean circulation theories, $\mathbf{U}$ has other desirable properties. Integrating vertically tends to hide details of the flow associated with a particular level in the water column, so it is a good way to give a general description of the horizontal flow. We can also simplify further. Consider any small region of the ocean, for instance any imaginary circle drawn on a map. The net flow into or out of this region must be balanced by compression of the water, by the rising or lowering of the sea surface in the region, or by flow through the surface by evaporation or precipitation. All these mechanisms are generally very small for the real ocean, so we can assume that the net exchange between the region and its surroundings are negligible: $\mathbf{U}$ is approximately non-divergent. In that case, as discussed in Subsection 2.1.4, (U,V) can be represented by a single scalar quantity, the streamfunction $\Psi(\phi,\theta)$.

Velocity vertically integrated over the entire water column is difficult to measure in the ocean. In principle, the surface height field (as in Figure 3.3) and the three-dimensional density field completely define the three-dimensional geostrophic velocity field (Subsection 2.1.2 and Subsection 2.2.2), and wind stress allows for calculation of the Ekman transport near the surface (Subsection 2.1.3). The last century of increasingly dense sampling of temperature and salinity (Figures 1.4 and 1.5) have been used to create global climatologies (see Appendix A) which can be used to compute density. Satellite altimetry allows for the more or less direct measurement of sea surface height (see Subsection 1.1.2). In practice, however, any small error in the surface height field produces velocity errors stretching all the way down through the water column (Equation 2.8). The wind-driven currents tend to be strongest in the top 10% of the ocean, so that an error in surface slope of only 10% of the average magnitude, when integrated over the entire water column, can create an error of the same order as the actual $\mathbf{U}$.

As mentioned in Subsection 1.2.2, numerical models can be used to constrain observations by statements of physical laws such as conservation of mass and conservation of momentum in the hope of producing a more accurate estimate of the ocean state (Wunsch, 2006). One such estimate is the Estimating Circulation and Climate of the Ocean (ECCO) project, which uses the Massachusetts Institute of Technology GCM to constrain ocean and atmosphere data from satellite altimetry, hydrography, Argo floats, scatterometer wind stress and other sources (Wunsch and Heimbach, 2007). A limit of the method is that errors in the model can produce inaccuracies in the estimate. For instance, the 1° horizontal resolution of the model used in the ECCO project cannot accurately capture the detailed structure of the western boundary currents, which are themselves barely 100 km wide. Put another way, a basic assumption of these models is that the transport properties are determined by the interior (near-)geostrophic flow plus constraints applied for property conservation.

The vertically integrated flow (Figure 3.6) from the ECCO estimate (Wunsch, 2011) shows similar patterns to the surface flow (Figure 3.4), but the gyres are somewhat more distinct. This is especially true in the tropical gyres, which were barely discernable in the surface flow. Typical volume transports in individual gyres are a few tens of Sverdrups. Subsection 3.2.4 compares volume transports in Figure 3.6a to the theoretical estimate shown in Figure 3.6b. The streamfunction gives some insight into why the tropical circulations are dominated by strong zonal jets: substantial zonal volume transports (especially in the Pacific) are confined to regions less than 10° wide in latitude. The volume transport $\Delta\Psi$ perpendicular to a length L is given by $\Delta\Psi = L\overline{U}$, where $\overline{U}$ is U averaged over the length. Therefore a relatively small L and relatively large $\Delta\Psi$ implies quite a large $\overline{U}$.

Observations with floats, hydrography, and current meters in the vicinity of the Gulf Stream have revealed a large increase in volume transport from about 30 Sv at 24°N to 100 Sv or more about 1000 km downstream of Cape Hatteras (good summaries appear in Richardson, 1985; Johns et al., 1995). The increase in velocity is almost entirely depth-independent, so that it is not apparent in the density structure of the flows. It appears to take the form of two depth-independent **recirculations** or miniature gyres in which westward flow just to the north and south of the Gulf Stream feed an increased flow in the Gulf

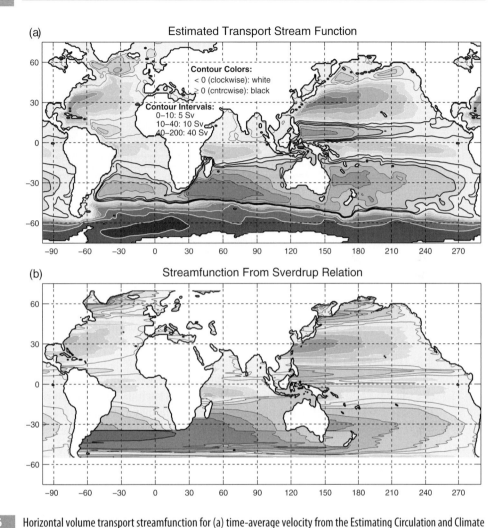

Figure 3.6 Horizontal volume transport streamfunction for (a) time-average velocity from the Estimating Circulation and Climate of the Ocean project (Appendix A; data courtesy of Carl Wunsch) and (b) estimate calculated from time-average SCOW wind using Sverdrup balance (Subsection 3.2.4). Contours are white for negative values (clockwise circulation) and black for 0 and positive values (counterclockwise). Streamfunction is measured with respect to reference value at eastern boundary of the Pacific at 29.5°N. Darker colors represent greater magnitude difference from reference value. Contour intervals are 5 Sv for 0–10 Sv, 10 Sv for 10–40 Sv, and 40 Sv for greater than 40 Sv. A black-and-white version of this figure (Figure 3.6a) appears in some formats. For the color version, please refer to the plate section.

Stream downstream of its separation point. The flow decreases further downstream as the Gulf Stream returns the water to the neighboring westward jets.

An estimate of the streamfunction based on altimetry and hydrography (Colin de Verdière and Ollitrault, 2016) shows a double recirculation in the Gulf Stream (Figure 3.7a) and a southern recirculation in the Kuroshio (Figure 3.7b), though the Wunsch (2011) state estimate (Figure 3.6a) does not show the recirculations, perhaps due to the coarse model resolution. Current meter measurements indicate that Kuroshio volume transport

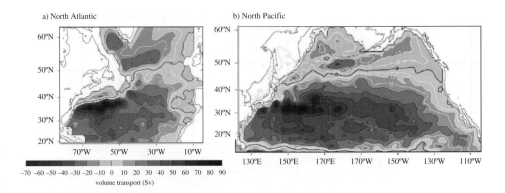

Observational estimate of horizontal volume transport streamfunction for (a) North Atlantic and (b) North Pacific, based on Colin de Verdière and Ollitrault (2016), Figures 3 and 7; ©American Meteorological Society. Used with permission. Contour interval is 10 Sv, with some 5 Sv intervals for white contours.

rises from 42 Sv where it separates from the coast of Japan to 114 Sv at about 146°E (Jayne et al., 2009). There is some evidence of recirculation to the south of the Kuroshio (Uehara et al., 2003; Kagimoto and Yamagata, 1997; Jayne et al., 2009) and to the north (Jayne et al., 2009).

While the discussion so far has emphasized flow within basins, flow between basins also occurs. The large continental landmasses such as the Americas and Eurasia–Africa can be thought of as islands which in principle may have a significant volume transport circling around them. In reality, the narrow, shallow Bering Strait between Alaska and Russia has only about 1 Sv of volume transport, which flows northward from the North Pacific to the Arctic (Coachman and Aagaard, 1966; Woodgate et al., 2012). This flow continues from the Arctic to the North Atlantic, though the precise pathway of its return to the Pacific is hard to measure among the strong currents in the Southern Ocean. The complex topography between Australia and southeast Asia allows communication between the Pacific and the Indian Ocean (Figure 1.9). Much of this region is also largely blocked by shallow continental shelf, but a circuitous pathway through Indonesia has narrow channels with depths of more than a kilometer (Figure 3.8). The flow in this region is difficult to measure, with the total **Indonesian Throughflow** recently estimated to be about 13–15 Sv from the Pacific to the Indian Ocean (Gordon et al., 2010). The volume transport from Pacific to Indian Oceans via Indonesia is balanced by a net southward transport out of the Indian Ocean which returns as a net northward transport into the South Pacific from the south.

The Wunsch (2011) horizontal streamfunction in Figure 3.6a also shows a roughly 15 Sv Throughflow and shows how this flow forms part of a global-scale circulation (see also Subsection 11.1.2). Streamlines emerging from Indonesia in the Indian Ocean (thick contours in figure) cross the Indian Ocean, and join its western boundary current along Africa. From the tip of South Africa, they continue westward across the Atlantic and join the Brazil Current flowing southward off the coast of South America. From the southern exit

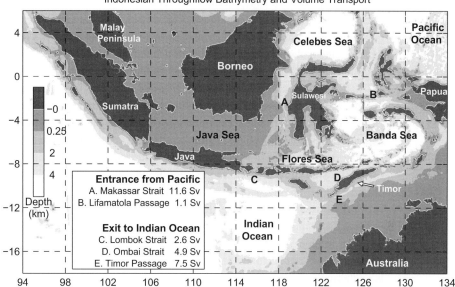

Figure 3.8 Indonesian passages bathymetry, nomenclature, and transport estimates from 2004–2006 measurements from the INSTANT program. Reprinted from Gordon et al. (2010), Figure 1, Copyright 2010, with permission from Elsevier.

of the Brazil Current, the water flows eastward all the way to the Pacific where it joins the South Pacific subtropical gyre, crosses the equator as a northward-flowing western boundary current along Papua New Guinea, and flows counterclockwise around the North Pacific tropical gyre, whose western boundary current carries it back to Indonesia to complete the path. (Note that, although Figure 3.6a depicts the time-average (Sverdrup) circulation, individual water parcels likely follow very different paths; see Subsection 7.1.1).

Consistent with Figure 3.6a, observations of the East Australian Current show it to be weaker than other western boundary currents. At approximately $30°$S, the current carries 27.4 Sv, but 18.4 Sv returns northward in a recirculation west of the longitude of New Zealand, for a net of only 9.0 Sv (Ridgway and Godfrey, 1994). Later current meter measurements showed the East Australian Current (not measuring the northward return flow) carrying 22.1 ± 4.6 Sv (Mata et al., 2000).

3.2 Concepts

3.2.1 Uniform-Density Models of the Gyres

The existence of ocean gyres, as illustrated by Figure 3.4, shows that there are interesting structures of horizontal velocity in the horizontal plane. The similarity in location and direction (cyclonic or anticyclonic) between the gyres and the wind curl (Figure 3.1b) also suggests that much of this horizontal flow is associated with the wind. Indeed, this circulation is often called *wind-driven*, meaning that the flow exists in a balance between wind forcing and dissipation.

Chapters 4 and 5 reveal important three-dimensional structures in the wind-driven ocean circulation, and Chapter 8 examines three-dimensional circulations associated with surface density variations. Here we discuss two-dimensional models of the horizontal circulation. Despite their simplicity, such two-dimensional models have been remarkably successful in explaining key qualitative and quantitative features of the horizontal circulation.

If the ocean had uniform density, a two-dimensional model would be a natural representation of the flow. The horizontal geostrophic flow $\mathbf{u}_G = (u_G, v_G)$ is proportional to horizontal pressure gradients, and the pressure at any point equals the weight of the water above (hydrostasy; Subsection 2.1.2). These pressure gradients are due to gradients in the water density and slopes in the sea surface height. For a uniform-density fluid, only the pressure gradient due to surface slope drives geostrophic flow, so (u_G, v_G) does not vary with depth and is therefore a function of latitude and longitude only. In other words, there is no thermal wind (Subsection 2.1.2).

The total flow includes top and bottom Ekman layers (Subsection 2.1.3) within which the velocity varies with depth. The Ekman speed in the bottom layer is approximately equal to, or less than, the geostrophic speed. For Ekman layer thickness H_{Ek} and total water thickness H, the contribution of the bottom Ekman layer to the vertical integral of velocity is negligible as long as $H_{Ek} \ll H$, which is a reasonable approximation for the ocean where H_{Ek} is typically tens of meters and H is thousands of meters (see (2.74)). As we discuss in Chapter 6, volume transport due to the surface Ekman layer is not negligible as we approach the equator, but outside the tropics it is also small compared to the geostrophic contribution.

In any case, we show in Subsection 3.3.3 (also Subsection 2.2.5) that the equations of motion can themselves be integrated vertically. The resulting two-dimensional equations do not allow us to solve exactly for $\mathbf{U}$ (Equation 3.2) for reasons that are described in detail in Subsection 3.3.3, but for much of the ocean they may give a good approximate solution for $\mathbf{U}$.

The correspondence between wind stress curl and the gyres is qualitatively similar to a shallow pan of water with a wind stress at the surface. The location of the gyres is about the same whether the pan is rotating or not (Stommel, 1948). We expect wind stress to drive the depth-average flow in the same direction as the wind, so that bands of parallel wind stress in alternating directions drive currents in the same general direction as the local wind (Figure 3.9a). The water is governed by the principle of continuity, so when the water reaches a wall of the pan, it cannot be squashed much (water is relatively incompressible; see Subsection 1.3.1), it cannot pile up much (gravity tends to smooth out hills of water), and it cannot stop (the water behind it is still moving), so it must turn to the side, flowing in a roughly oval trajectory as shown in Figure 3.9a. Now we consider wind stress that is all pointing in the same direction (Figure 3.9b), but with the same shear ($\partial \tau / \partial y$ in this example, assuming a conventional (x, y) plane with wind stress $\tau(y)$ in the x direction) as in (a). At first glance we might expect a very different circulation pattern, with all the flow

a) 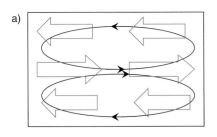   b)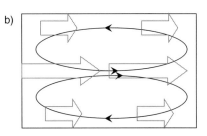

Figure 3.9 Schematic top view of pan of water showing location of walls (thick black lines), wind stress (gray arrows) and gyres (thin black curves) for case of (a) wind with alternating directions and (b) all wind in the same direction.

towards the right (the same direction as the wind stress). Such a flow would not satisfy continuity, and hence is impossible. All the water flowing to the right must turn around and flow to the left again. It is plausible that the strongest wind stress would succeed in pushing the water to the right, but that water under regions of weaker wind stress would be forced by continuity to flow back to the left. This results in Figure 3.9b having the same gyre circulation as Figure 3.9a. It is the gradient in the wind stress that determines the gyre flow, rather than the strength of the wind. There are likely to be differences in the surface height field between Figures 3.9a, b, but the flow is the same.

The thought experiment in Figure 3.9 can be crudely reproduced with a shallow pan of water. If one blows across the middle of the pan, two opposite gyres form, with water flowing in the same direction as the wind in the middle of the pan. To the left and right of the central current, currents flow in the opposite direction even though no wind is blowing in the opposite direction. The flow can be visualized by opening a tea bag and sprinkling the leaves on the surface. The "wind stress" curl associated with blowing in the center of the pan generates two gyres similar to those shown in Figure 3.9.

3.2.2 Vorticity Dynamics of the Gyres

A more precise statement of the behavior described in the experiment in the previous subsection is that the wind stress curl determines the water's depth-averaged relative vorticity (Subsection 2.1.4, Subsection 2.1.5). For $\mathbf{U}$, which is the horizontal velocity integrated over a uniform depth H, the depth-average relative vorticity is defined as

$$\zeta = \hat{\mathbf{z}} \cdot (\nabla \times \mathbf{U})/H. \tag{3.3}$$

Thus, the location and rotation sense (clockwise or counterclockwise) of the gyres discussed in Subsections 3.1.2 and 3.2.1 can be summarized as: relative vorticity of the gyres has the same sign as the wind stress curl at the same location. Note that vorticity refers to the "spin" of a water parcel but that the average vorticity of a region implies flow around the region (see Subsection 2.1.4). As described in Subsection 2.1.5, potential vorticity $q = (f + \zeta)/H$ has a budget which, for uniform depth H, can be written

$$\frac{D(qH)}{Dt} = W - r\zeta + v_h \nabla_h^2 \zeta \tag{3.4}$$

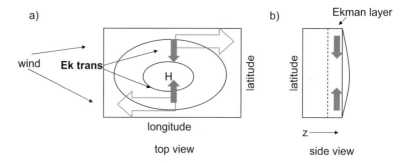

Figure 3.10 Schematic (a) top and (b) side views of counterclockwise rotating basin of water showing wind stress (unfilled arrows), Ekman transport (filled arrows) and surface height (contours).

where W is the wind forcing, r is a coefficient for bottom friction, and v_h is horizontal viscosity.

We can specify W in two different ways. One way is

$$W = \frac{1}{H}\hat{\mathbf{z}} \cdot (\nabla \times \mathbf{X}) \tag{3.5}$$

the curl of the wind stress, where we use $\mathbf{X} = \boldsymbol{\tau}/\rho$. In that case, (3.4) holds even if $f = 0$ as in the pan experiment in Subsection 3.2.1.

The careful reader may notice a problem with relating the wind stress curl to the vorticity of the entire water column in a rotating frame. The direct effect of wind stress is to create an Ekman transport at right angles to the stress, rather than to push water in the same direction as the stress (Figure 3.10a). Moreover, the Ekman transport is confined to a layer typically tens of meters thick, compared to wind-driven circulations reaching to hundreds or even thousands of meters. How is the deeper water driven by the wind? This is how the second way of specifying W enters. The wind curl is associated with horizontal divergence (or convergence) of Ekman transport; this implies Ekman suction (or pumping), that is, vertical velocity w_{Ek} at the bottom of the Ekman layer (see Subsection 2.1.3). If (3.4) refers to the depth range of geostrophic flow, which excludes top and bottom Ekman layers, then flow into or out of a layer modifies q by squashing/stretching (as discussed in Subsection 2.2.3 and Subsection 2.2.5). The forcing on such a geostrophic layer is given by

$$W = qHw_{Ek} = qH\nabla \cdot \left(\frac{\hat{\mathbf{z}} \times \mathbf{X}}{f}\right). \tag{3.6}$$

As the Ekman pumping is due to the wind curl, we expect that the vorticity input would be the same sign for Ekman pumping as for wind stress. Is it? Zonal wind stress increasing to the north (as in Figure 3.10a) has negative curl and hence, if W refers to the effect of wind stress curl on the entire water column, the wind inputs negative vorticity into the layer. In the northern hemisphere, this wind profile is associated with converging Ekman transport (Figure 3.10a), which generates Ekman downwelling (Figure 3.10b). For large-scale circulation, generally ζ is much less than f (as discussed in Subsection 3.3.1), so q has the same sign as f even if ζ has a different sign. If W designates Ekman pumping into the purely geostrophic layer, then for Earth's northern hemisphere ($f > 0$), downwelling ($w_{Ek} < 0$) also inputs negative vorticity. Thus the wind stress curl itself and the Ekman pumping due to the stress curl both input the same sign vorticity.

The remaining terms on the right-hand side of the potential vorticity equation (3.4) represent friction with the bottom and side walls, respectively (see Subsection 2.2.5). As we show in Subsection 3.3.5, (3.4) implies that in steady state, the integral over the entire domain of the forcing and friction terms on the right-hand side sums to zero. We can think of this as the wind injecting vorticity into the ocean and friction removing it (transferring it to the solid boundaries of the domain). Because both friction terms have the opposite sign of ζ, ζ must be the same sign as the wind stress curl for the domain-integrated vorticity balance to hold. This supports the claim made in Subsection 3.2.1 that the gyre should have the same vorticity as the wind. As the magnitude of the friction increases with the magnitude of ζ as well, a circulation starting from rest should increase over time until it is large enough to generate enough frictional vorticity loss to balance vorticity input from the wind.

Even if we see consistency between the vorticity input and the circulation direction, what actually "pushes" the water around the gyre if the wind stress causes water to flow as Ekman transport? It is a secondary effect of Ekman transport convergence, which creates a slight bulge in the sea surface height (Figure 3.10, not shown in Figure 3.9). The surface bulge causes a horizontal pressure-gradient force, which, according to the geostrophic balance, must be clockwise (Figure 3.10). Thus the geostrophic circulation is just as we expect from the vorticity argument.

3.2.3 Vorticity Dynamics of the Western Boundary Currents

Having related the circulation direction of the gyres to the vorticity balance, we now turn to the circulation strength and the existence of western boundary currents. Here the details about the rotation are crucial. We return to the potential vorticity equation, (3.4), which we can interpret as a statement of the rate of change in vorticity of a water column circulating in the gyre. In steady state, every time a column makes a complete circuit, q of the column must return to its value at the beginning of the circuit. An analysis of the changes in q during a circuit (Figure 3.11) leads to the conclusion that a few essential features of the gyre circulation on Earth demand the existence of a western boundary current.

We consider a clockwise gyre (such as a northern hemisphere subtropical gyre) on the Earth as an example. We assume that $\zeta \ll f$, and we remember that the gyres are large enough that f varies substantially from the northern to southern gyre boundaries. Thus, when a water column changes latitude, it changes f and q as well. For ζ small and H constant, potential vorticity is approximately a function of latitude.

When the water column is moving southward (in the eastern part of the gyre), f must decrease, hence q must decrease (Figure 3.11). This is a kinematic argument that a certain trajectory implies a certain change in q, but the terms on the right-hand side of the vorticity equation, (3.4) must be able to make the change. As the wind input of vorticity is negative for a clockwise gyre, and the friction input is positive, the decrease in q implies that the wind input of vorticity must have a greater magnitude than the friction input.

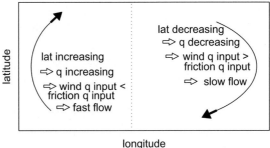

latitude

lat increasing
⇨ q increasing
⇨ wind q input <
   friction q input
⇨ fast flow

lat decreasing
⇨ q decreasing
⇨ wind q input >
   friction q input
⇨ slow flow

longitude

Figure 3.11 Illustration of potential vorticity, q, budget for clockwise ocean gyre, with arrows showing flow direction in east and west sides of basin, and text describing changes in q and their implications for current speeds.

When the water column is travelling northward (in the western part of the gyre), q is increasing (Figure 3.11), implying that friction must dominate wind in the vorticity budget there. This difference in the vorticity budget between east and west occurs even if the wind curl is independent of longitude (hence the same in the east and west), so it is the friction contribution that must be different in the two regions.

As the frictional vorticity input is proportional to flow speed, the way to achieve wind-dominance in the east and friction-dominance in the west is to have the current moving faster in the west than in the east. The relative width of each part of the flow follows from the relative speeds: because the northward volume transport must equal the southward volume transport (at least for a basin completely bounded by walls), the fast northward current must be narrower than the slow southward current.

Thus the existence of western boundary currents follows from the vorticity budget given relatively slow flow ($\zeta \ll f$) and meridional variations in f due to the sphericity of the Earth. The vorticity budget for a counterclockwise gyre yields a similar **western intensification** of the gyre. Note that the sign of f does not enter the analysis, which produces a western boundary current as long as f increases (grows more positive) as one goes to the north. Over most of the gyre, the wind inputs vorticity which allows water to flow to a different latitude. In a thin region in the western part of the gyre, friction removes the vorticity input by the wind, thus allowing water to return to its original latitude.

In contrast, uniform f (f-plane), such as in a rotating laboratory tank, does not demand any change in q as the water circumnavigates the gyre. There is no *a priori* reason to assume that one term dominates another in any part of the gyre. The input of vorticity by the wind has a *local* balance with its removal by friction. In that case the flow is symmetrical and has no western intensification.

Stommel (1948) first showed that a simple mathematical model would have a symmetric gyre for f constant or zero (Figure 3.12a; see also Winton, 1996) and that including the variation in latitude of f could produce a western boundary current (Figure 3.12b). Friction in the Stommel solution is only provided by bottom friction; Munk (1950) found similar behavior with side friction. Subsection 3.3.3 derives approximate solutions for both cases. The models of Henry Stommel and Walter Munk are triumphs of physical oceanography;

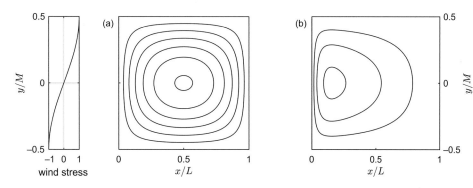

Figure 3.12 Stommel (1948) streamfunction for uniform-density fluid driven by zonally uniform zonal wind stress (shown at far left) for (a) constant f and (b) f increasing with latitude. The streamfunction contours start at zero at the edge and are spaced by 0.2π times (a) $X_0 L^2/(rM)$, and (b) $X_0 L/(\beta M)$, respectively (see Exercise 3.3 and Subsection 3.3.2; r is the vorticity damping rate from Subsection 2.2.5). See also Exercise 11.7.

they neglect topography, stratification, many aspects of the wind forcing, and the details of the current profile along the boundary; yet they capture the essential dynamics of western boundary currents.

3.2.4 Gyre Strength: The Sverdrup Balance

One might expect the total volume transport in the gyre to be controlled by both the wind stress curl and friction, but in this simple model, the strength of friction, as measured by the eddy viscosity (Subsection 1.3.1), is theoretically irrelevant to the flow except in the western boundary current. The main effect of increasing the eddy viscosity is to increase the width and decrease the speed of the western boundary current. Over most of the gyre, the meridional flow is just fast enough to make the planetary vorticity change at the same rate that vorticity is being input by the wind. This balance was derived and compared to tropical Pacific observations by Harald Sverdrup (Sverdrup, 1947) and therefore is known as the **Sverdrup balance**, one form of which is

$$\textbf{Sverdrup balance:} \qquad \beta V = \hat{\mathbf{z}} \cdot (\nabla \times \mathbf{X}), \qquad (3.7)$$

where β is the gradient of planetary vorticity ((2.63), Subsection 2.2.3). Various versions of the Sverdrup balance are derived in Subsection 3.3.1 and Subsection 3.3.2. The right-hand side of the equation is the wind input of vorticity, and the left-hand side is the rate at which the vorticity (well-approximated by the planetary vorticity, f) changes due to meridional motion.

The Sverdrup balance shows how to calculate the meridional transport V from the wind stress field. At each latitude θ, V can then be integrated westward along longitude ϕ to find a streamfunction $\Psi(\phi, \theta)$. The Sverdrup balance does not apply to the western boundary area, but we can estimate the value of Ψ along the western boundary if we know the net meridional transport between the eastern and western boundaries. Without knowledge of the details of the western boundary currents, we can still approximate the streamfunction by assuming it goes from its Sverdrup value to its value at the western boundary in a narrow region in the west.

For a closed basin such as the idealized case in Figure 3.12, the streamfunction has the same boundary value all around the basin. As described in Subsection 3.1.4, in the real world, only about 1 Sv flows from the Pacific to the Atlantic via the Arctic, so the coasts of America and the Eurasian-African landmass have about the same value of Ψ. We use the observational estimate of 15 Sv in the Indonesian Throughflow (Gordon et al., 2010) from the Pacific to the Indian Ocean to define the streamfunction value for the Australian coast. The value of the Indonesian Throughflow and the transports around other large islands can be predicted by an **Island Rule** (Godfrey, 1989) based on geostrophic and Ekman dynamics (Subsection 3.3.5). The Island Rule gives plausible estimates of observed transport (Godfrey, 1989; Godfrey and Masumoto, 1999), but friction may reduce the transport from the Island Rule prediction (Pratt and Pedlosky, 1998; Wajsowicz, 1993) and bathymetry may also be important (Pedlosky et al., 2009).

Using these boundary values of streamfunction, the Sverdrup balance, and the same annual wind stress data shown in Figure 3.1, we produce a nearly global estimate of the depth-integrated circulation (Figure3.6b). A comparison of estimated horizontal streamfunction (Figure 3.6a) with the Sverdrup prediction (Figure 3.6b) shows that the Sverdrup balance gives volume transports that are the right magnitude for the real ocean. It also reproduces some detailed features correctly, such as the subtropical gyre transport being greater in the North Pacific than in the North Atlantic. Our knowledge of the accuracy of the Sverdrup balance is limited by inaccuracies in the wind field and especially in wind stress on the ocean. While Leetmaa et al. (1977) concluded that the Sverdrup balance held at 24°N in the Atlantic, Wunsch and Roemmich (1985) argued that this conclusion strongly depended on weak assumptions, such as a 1000 m level of no motion and weak currents just east of the Gulf Stream. Hautala et al. (1990) argued that the Sverdrup balance approximately held at 24°N in the Pacific. Lu and Stammer (2004) compared a coarse resolution model (2° grid), which incorporated ocean measurements through data assimilation, with the Sverdrup balance prediction and found qualitative agreement but discrepancies that exceeded 10 Sv in many places. Saunders et al. (1999) found somewhat greater agreement between Sverdrup and an eddy-permitting model (1/4° resolution). More recently, Gray and Riser (2014) find agreement between the absolute geostrophic flow from Argo data and the predicted Sverdrup circulation in the tropics and subtropics, but not at higher latitudes or near boundaries.

The original paper on the Sverdrup balance (Sverdrup, 1947) applied a version of the formula to the eastern tropical Pacific geostrophic velocity, and found a rough correspondence between wind stress curl and V based on a small amount of data and neglect of currents below 1000 m. Several decades later, Landsteiner et al. (1990) found poor agreement between Sverdrup balance predictions and tropical Pacific V but argued that this could be due to uncertainty in the data. Using more extensive data available a decade later, Kessler et al. (2003) found that within about 8° of the equator, the Sverdrup balance predicted V to be weaker than observations, especially east of the date line.

3.2.5 Limitations of the Sverdrup Balance

Why would the Sverdrup balance *not* hold? A number of simplifications made in deriving the theory may be inaccurate. These simplifications are discussed more fully in Section 3.3, but we summarize them here.

Sverdrup theory neglects any interactions between bottom currents and bathymetric slopes. These interactions potentially alter the vorticity balance from which the Sverdrup balance is derived. In fact the uniform density model implies a very strong role for bottom topography, because even moderate variations in H (from continental slopes and mid-ocean ridges) contribute as much to variations in q as variations in f do. If the ocean had uniform density, these bathymetric effects would significantly alter Ψ from what the Sverdrup balance predicts. Because the ocean is stratified and the wind-driven circulation is largely confined to the upper ocean (Chapter 5), it is plausible that bathymetric variations are much weaker than the variation in f. In other words, if we approximate the stratified wind-driven circulation over topography with a uniform-density model, a flat bottom is probably better than realistic topography. Nevertheless, the real ocean has *some* deep flow from the wind-driven circulation and from the deep meridional overturning circulation (Chapter 8). The coastal sides of western boundary currents are also frequently found in water whose total depth is less than 1000 m. These may produce significant depth-integrated flow not predicted by the Sverdrup balance.

We expect such effects to be strongest in the subpolar regions, where the stratification is weaker and the gyre flow reaches more deeply than at lower latitudes. The Atlantic subpolar gyre has a prominent qualitative difference from the Sverdrup prediction. The East Greenland Current is a western boundary current flowing southward along Greenland's east coast. When this current reaches the southern tip of Greenland, according to the standard Sverdrup/western boundary current theory, it should flow westward to join the Labrador Current, which is the western boundary current flowing southward along the Canadian coast (such a circulation pattern is just barely visible in Figure 3.6b). In reality, the strong current continues northwestward along the coast as the West Greenland Current (Figure 3.13), in other words behaving as an *eastern* boundary current in contradiction to linear, flat-bottom theory. The current does eventually join the Labrador Current by flowing along shallower isobaths to the north in Davis Strait. It is possible that this is a topographic effect due to the bowl-like shape of the Labrador Sea's bottom. Another possible topographic effect is the intensification of the northward flow in the vicinity of the Reykjanes Ridge (Figure 3.13), which is the part of the Mid-Atlantic Ridge extending southward from Iceland. Finally, it is unclear if the high latitudes can attain Sverdrup balance because of the long time taken to reach a steady state (the gyre adjustment timescale decreases as stratification and β increase). Figure 3.13 shows surface currents rather than transports, but geostrophic calculations show the deep currents to be qualitatively similar. Coarse resolution numerical models (Bryan et al., 1995) and eddy-resolving models (Treguier et al., 2005) also indicate substantial discrepancies from Sverdrup theory in the subpolar gyre.

In Subsection 3.2.2, where we consider how a parcel of water changes its potential vorticity as it travels around the gyre, one influence we neglect is the transport of relative

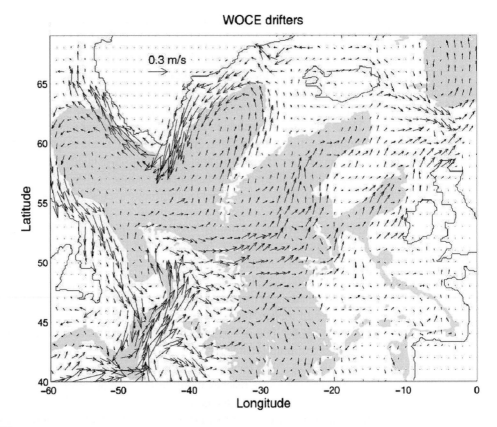

WOCE drifters

North Atlantic subpolar gyre surface currents based on a climatology of drifters, with shading representing water deeper than 2000 m. From Treguier et al. (2005), Figure 2a; ©American Meteorological Society. Used with permission.

vorticity ζ by the flow. Such transports are known as **inertial** effects and become important for currents which are strong enough and which vary over a small enough distance. Subsections 3.3.1 and 3.3.4 discuss just how fast and small such currents must be for inertial effects to be important. The inertial terms in the equation governing vorticity make the equation difficult to solve because they are **nonlinear** functions of Ψ.

The narrow zonal currents near the equator are one place where relative vorticity transport is significant. Kessler et al. (2003) analyzed the importance of these terms in a numerical model of the upper 400 m of the tropical Pacific. Their results imply that inertial terms strengthen the circulation in the tropics compared to what the Sverdrup balance predicts, but that friction partially counteracts the ζ contribution while broadening the eastward flow near the equator.

Outside the tropics, inertial terms are large in western boundary currents. The recirculation gyres associated with the separation of the Gulf Stream and Kuroshio from the coast (Subsection 3.1.4) may be related to these terms. Pioneering numerical solutions of the uniform-density model in an idealized configuration similar to Figure 3.12 (Bryan, 1963; Veronis, 1966) show that as inertia becomes stronger (either through increasing the current

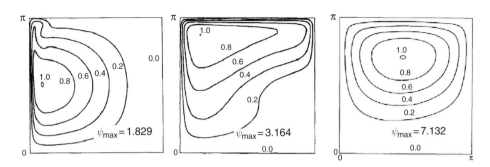

Figure 3.14 Streamfunction for numerical solution to nonlinear uniform-density model of wind-driven gyre with bottom friction, for increasing value of nonlinearity parameter L_I (see (3.13) in Subsection 3.3.1). Ψ_{max} printed in figure refers to maximum streamfunction scaled by a measure of the gyre transport calculated by the Sverdrup balance. Reprinted from Veronis (1966), Figures 7, 8, and 9, Copyright 1966, with permission from Elsevier.

speed by strengthening the wind or narrowing the western boundary current by reducing friction) a recirculation region develops in the part of the subtropical gyre where the current exits the western boundary. Maximum volume transport in the recirculation region, and the area covered by the recirculation region, grows as inertial effects become stronger (Figure 3.14). Subsection 3.3.4 discusses some further theoretical aspects of the recirculation gyres.

In numerical models that seek to reproduce realistic stratification and currents, the horizontal friction is usually set at a minimum value that will allow western boundary currents to be resolved by the model (as discussed in the previous section, western boundary current width increases with viscosity). Models with horizontal grid spacing of $1°$ resolve the large-scale features of the gyres but produce western boundary currents that are too broad (typically around 200 km) and consist of a jet that is only 1–2 gridpoints wide, that is not enough resolution to capture the detailed velocity profile across the jet. Decreasing the grid-spacing allows the model to form a western boundary current that is narrower, faster-flowing, and therefore has greater relative vorticity. A series of simulations of the North Atlantic with resolutions of $1°$ (Chassignet et al., 1996), about $1/3°$ (Bryan et al., 1995), and $1/10°$ (Smith et al., 2000; Bryan et al., 2007) show that recirculation gyres grow in strength as resolution increases, with the $1/10°$ model showing both northern and southern recirculation gyres of several tens of Sverdrups (Figure 3.15a). Similar recirculation gyres develop in high-resolution models of the North Pacific (Figure 3.15b, from Jayne et al., 2009).

In three-dimensional models such as that of Bryan et al. (2007), decreasing the grid spacing to less than about $1°$ allows mesoscale eddies to form (Chapter 7). Thus the recirculations which emerge in Figure 3.15 may be related to the dynamics of the eddies as much as to reproducing a high-ζ western boundary current. In general, the role of eddies on the general circulation is not well understood. As the large-scale subtropical gyre transports seem to be roughly captured by the Sverdrup balance, which ignores eddies, it is plausible that the large-scale dynamics can be approximated by neglecting eddies. Eddies

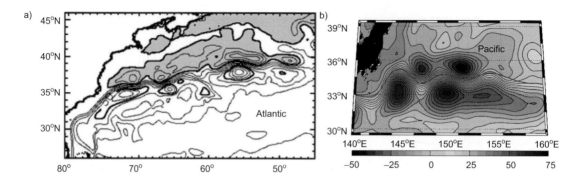

Figure 3.15 Streamfunction of vertically integrated velocity (multi-year time average) for 0.1° resolution three-dimensional simulation of (a) North Atlantic. Reprinted from Bryan et al. (2007), Figure 5a, Copyright 2007, with permission from Elsevier. (b) North Pacific. Reprinted from Jayne et al. (2009), Figure 5a, Copyright 2009, with permission from Elsevier. Contour interval is 10 Sv for (a) and 5 Sv for (b).

may be more important to the dynamics of western boundaries and recirculation regions and may have a smaller but non-negligible influence on the behavior of the rest of the gyre.

Our final example of a discrepancy from the Sverdrup/western boundary current view of the gyres is the Agulhas retroflection (Figure 3.5d). South of South Africa, there is no land boundary between the Atlantic and Indian Oceans, and south of New Zealand there is no land boundary between the Indian and Pacific Oceans (Figure 1.9). According to Sverdrup theory, the southern regions of the southern hemisphere subtropical gyres (Figure 3.1b) can be thought of as a single **Supergyre** (Ridgway and Dunn, 2007). The Supergyre stretches from the east coast of South America, through almost all longitudes, to the west coast of South America. If the South Atlantic western boundary current is fed by the entire Supergyre, the western boundary current volume transport should be equal and opposite to the Sverdrup transport integrated over the entire ocean at the same latitude (Figure 3.6b). However, observations and realistic numerical models show separate gyres in the Atlantic and Indian Oceans (Figure 3.6). If Sverdrup dynamics were the whole story, than all the Agulhas Current would continue westward when it reached the southern tip of the South African continental shelf. Such flow would cross the Atlantic, link to the western boundary current there, and return eastward somewhat to the south. Instead, the Agulhas flows southward from Africa and forms the retroflection which allows most of the Agulhas transport to return eastward near South Africa.

Several explanations have been proposed to explain this behavior. De Ruijter (1982) argued that nonlinear effects cause the Agulhas to "overshoot" the tip of South Africa far enough to reconnect to the eastward Sverdrup flow several degrees to the south. An idealized numerical model of the Agulhas region (Boudra and Chassignet, 1988) generates a retroflection when the inertial terms are included (Figure 3.16c). The solution consists of Sverdrup flow and western boundary current when the inertial terms are removed (Figure 3.16a). Interestingly, the same study shows that the current can also retroflect even in the linear case, if horizontal viscosity is raised to high enough values (Figure 3.16b). In

(a)   (b)

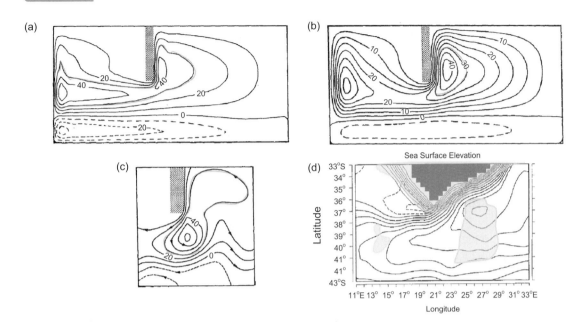

Sea Surface Elevation

(c)   (d)

Figure 3.16   Numerical models of an idealized Agulhas retroflection. (a), (b), and (c) show streamfunction from one-layer model (contour interval 10 Sv) for case of (a) linearized equations, horizontal viscosity $\nu_h = 330\,\text{m}^2/\text{s}$, (b) linearized equations, $\nu_h = 3300\,\text{m}^2/\text{s}$, and (c) nonlinear equations. From Boudra and Chassignet (1988) Figures 2, 3, 5a, ©American Meteorological Society. Used with permission. (d) Sea surface elevation (contours) and regions of relatively shallow model topography (depths < 4 km shaded gray) for multilevel model. From Matano (1996) Figure 7, ©American Meteorological Society. Used with permission.

each case we can think of the retroflection as being caused by the western boundary layer being wide enough to reach southward to the eastward flowing region of the gyre. The widening of the boundary layer can be caused by either making inertia large enough or by making friction large enough.

Another possible explanation involves topography (Matano, 1996). An idealized numerical model also develops a retroflection when realistic topography is added to the model (Figure 3.16d). Two features on the sea floor, the Agulhas Ridge near 10°E and the Agulhas Plateau near 26°E, rise to less than 2000 m depth. Though deeper than most of the flow, this may influence flow above the ridge crest enough to act as a barrier redirecting some of the Agulhas.

Despite a systematic parameter study (Dijkstra and de Ruijter, 2001), the relative importance of all these mechanisms in the real world is still unclear. The Agulhas Retroflection has impacts on larger issues than the local currents (for a review, see Beal et al., 2011). The amount of water flowing from the Indian to the Atlantic (in other words, the water that does *not* retroflect) depends on the meridional distribution of wind, which responds to changes in climate. Because the Agulhas water is warmer and saltier than the Atlantic, the degree of retroflection may in turn influence the water mass properties of the Atlantic and its heat flux with the atmosphere. Thus the Indian–Atlantic exchange in the vicinity of the Agulhas is influenced by, and in turn influences, climate variations.

3.3 Theory

3.3.1 Scale Analysis of the Vorticity Equation

In deriving the potential vorticity equation in Subsection 2.2.5, we replaced the real strati-fied ocean with a simpler uniform-density layer. The resulting equation is still very difficult to solve without a numerical model, even for uniform layer thickness H, because of the nonlinear terms such as $\mathbf{u} \cdot \nabla \zeta$. Therefore we will now simplify the equations by neglect-ing terms that are likely to be small for gyre-scale motion. To do this, we perform a scale analysis in which we estimate the relative magnitude of the terms in the equation.

We can write the material derivative of absolute vorticity as

$$\frac{D}{Dt}(f + \zeta) = \frac{D\zeta}{Dt} + \mathbf{u} \cdot \nabla f = \frac{D\zeta}{Dt} + \beta v. \tag{3.8}$$

so for uniform H, the single-layer vorticity equation from (2.94) becomes

$$\frac{\partial \zeta}{\partial t} + \mathbf{u} \cdot \nabla \zeta + \beta v = \frac{f + \zeta}{H} w_{Ek} - \frac{1}{2} \delta_{Ek}(f + \zeta)\zeta + \nu_h \nabla_h^2 \zeta, \tag{3.9}$$

where $\delta_{Ek} = H_{Ek}/H$ (to see where this term comes from, consult (2.40) and Exercise 3.6). This so-called **barotropic** (or depth-integrated) **vorticity equation** is the master equation for understanding the homogeneous two-dimensional circulations we discuss in this chap-ter. All of the conceptual and theoretical models are approximations to this equation. The focus throughout is on steady solutions, so the vorticity tendency on the left-hand side drops out. The βv term represents the rate at which the absolute vorticity is changing due to meridional flow changing the latitude of a water column. In Subsection 3.2.2 we identi-fied this as the main source of change in absolute vorticity. Therefore we will assume that this will be one of the largest terms in the equation. At least one other term must balance βv (unless the equation implies that $v = 0$). As we have identified wind stress as the driver for the motion, it is plausible that the w_{Ek} term is about the same size as βv. If we can neglect all the other terms, then the steady state form of (3.9) is a version of the Sverdrup balance:

$$\beta v H = f w_{Ek}. \tag{3.10}$$

We want to see under what circumstances we can neglect any of the other terms in (3.9).

The first step in the scale analysis estimates the magnitudes of the variables in the equa-tion. We assume that over most of the gyre, u and v are of equal magnitude given by U (not the depth-integrated zonal speed in this subsection), and that the horizontal length scale over which $\mathbf{u}$ changes by a significant amount in either the x or y direction is L. The mag-nitudes of the Coriolis parameter and β are approximated by f_0 and β_0, their values at the average latitude θ_0. First, compare the relative magnitude of the two terms in $f + \zeta$:

$$\zeta/f \sim \frac{U}{fL} = Ro, \tag{3.11}$$

where we have approximated the horizontal shear terms in ζ (see (3.3)) as U/L and the "$\sim$" means that the left-hand side has a typical magnitude given approximately by the right-hand side. Thus the magnitude of ζ/f is the Rossby number Ro. Chapter 2 shows that $Ro \ll 1$ for large scale flow in the ocean, at least away from the equator, so our scale analysis supports the assumption (Subsection 3.2.2) that relative vorticity is much smaller than planetary vorticity over most of the ocean.

Given that $\zeta \ll f$, the relative strength of the relative vorticity advection terms and the divergence of planetary vorticity advection is given by

$$\frac{\mathbf{u} \cdot \nabla \zeta}{\beta v} \sim \frac{U^2/L^2}{\beta_0 U} = \frac{U}{\beta_0 L^2} = \left(\frac{L_I}{L}\right)^2, \tag{3.12}$$

where we have defined the **inertial length scale** (also called the Rhines scale, see (7.46)),

$$L_I \equiv \sqrt{U/\beta_0} \tag{3.13}$$

to compare to the gyre length scale L. If $L_I \ll L$, then we can neglect the advection terms in the vorticity equation. Although the inertial term is based on the advection of relative vorticity, it may be significant even if relative vorticity is much smaller than f. We can rewrite (3.12) in terms of the Rossby number and use the definitions of f and β (Equations 2.1 and 2.63) to obtain

$$\frac{U}{\beta_0 L^2} = \frac{U}{f_0 L} \frac{f_0}{\beta_0 L} = \frac{Ro R_E \tan\theta_0}{L}, \tag{3.14}$$

where θ_0 is the average latitude. This number can exceed unity even for small Ro if L/R_E is even smaller (R_E is the Earth's radius).

The relative strength of the bottom friction term is given by

$$\frac{r\zeta}{\beta v} \sim \frac{rU/L}{\beta_0 U} = \frac{r}{\beta_0 L} = \frac{L_S}{L}, \tag{3.15}$$

where we have defined a frictional parameter

$$r \equiv \frac{f_0}{2} \delta_{Ek} \tag{3.16}$$

and a bottom friction length scale which is also known as the **Stommel length scale**

$$L_S \equiv \frac{r}{\beta_0} = \frac{1}{2} \delta_{Ek} R_E \tan\theta_0 \tag{3.17}$$

because Henry Stommel's paper (Stommel, 1948) was the first to use bottom friction for the gyre problem. Finally, the relative strength of the horizontal friction term is given by

$$\frac{\nu_h \nabla_h^2 \zeta}{\beta v} \sim \frac{\nu_h U/L^3}{\beta_0 U} = \frac{\nu_h}{\beta_0 L^3} = \left(\frac{L_M}{L}\right)^3, \tag{3.18}$$

where the horizontal friction length scale is

$$L_M \equiv (\nu_h/\beta_0)^{1/3}, \tag{3.19}$$

which is also called the **Munk length scale** since Walter Munk (Munk, 1950) first included horizontal friction in a gyre model.

The preceding scale analysis shows that the magnitude (relative to the largest terms) of the inertial, vertical friction, and horizontal friction terms are given by nondimensional parameters $(L_I/L)^2$, L_S/L, and $(L_M/L)^3$, respectively. Under what circumstances are these small?

To calculate L_I, we need an estimate of U and L. Observations (Subsection 3.1.3) show that each gyre consists of relatively slow flow over most of the gyre and a narrow western boundary region with much faster flow. Following the assumption that the βv and $f w_{Ek}/H$ terms have the same magnitude, we equate the scaling for both terms to get

$$U = \frac{f_0}{\beta_0} \frac{w_{Ek}}{H}. \tag{3.20}$$

This can also be written in terms of the wind stress by using (2.13), the definition of Ekman pumping:

$$U = \frac{\tau_*}{\beta_0 \rho H L}, \tag{3.21}$$

where τ_* is a typical wind stress magnitude. In the real world, $\rho \approx 1000$ kg/m^3, $\beta_0 = 2 \times 10^{-11}$ m^{-1}s^{-1} at about 30° latitude, and from Figure 3.2 we estimate that wind stress variations are about $\tau_* = 0.1$ N/m^2 over a length scale of $L = 2000 \times 10^3$ m. We take $H = 1000$ m, which is appropriate to the depth range of the strongest wind-driven flow (see Chapter 5), rather than the full ocean depth (see Exercise 3.8). Inserting (3.21) with these numbers into (3.13), we obtain $L_I \approx 10 \times 10^3$ m. Thus over basin scales of thousands of kilometers, we should be able to safely neglect the nonlinear terms because $(L_I/L)^2 \approx 10^{-2} \ll 1$. However, the narrow western boundary current has speeds that are roughly (M/L_W) faster than speeds over most of the gyre, where L_W is the width of the western boundary current and M is the gyre width. For L_W about 100 km and M for various gyres ranging from 5000 to over 10,000 km, L_I is roughly 100 km instead of 10 km. In the western boundary current, the current width (about 100 km) is a more appropriate choice for L than the gyre scale, and so it is plausible that the inertial terms are indeed important near the western boundary.

Scale estimates for the friction terms are independent of the velocity scale; larger velocities make both the friction term and the planetary vorticity term bigger by the same amount. The Stommel length scale can be estimated from (3.17) though it is problematic to apply to the real ocean because the bottom friction representation in (3.9) is based on a uniform-density ocean with depth-independent geostrophic velocity; in the real world the bottom velocity is much smaller than the depth-average. If we apply the formula with $H_{Ek} = 30$ m, $H = 4000$ m, and $R_E = 6400$ km, we get $L_S = 24$ km and hence $L_S/L \approx 10^{-2} \ll 1$. The Munk length scale is also difficult to calculate because v_h is not well known for the ocean. A mixing-length argument based on mesoscale eddies (Subsection 7.3.1) suggests that v_h is not much more than about 10^6 m^2/s, which according to (3.19) implies $L_M \approx 400$ km for $\beta_0 = 2 \times 10^{-11}$ m^{-1}s^{-1}, and hence $(L_M/L)^3 \approx 10^{-1} < 1$.

Thus our scale analysis implies inertial and friction terms are all negligible for gyre-scale motion, but because observations show that the velocity scale is larger and the length scale is smaller near the western boundary than they are for the gyre as a whole, these terms are potentially important for the western boundary current dynamics.

3.3.2 Sverdrup Balance and the Need for a Western Boundary Current

If we neglect all the terms that Subsection 3.3.1 found to be small for gyre-scale motion, the barotropic vorticity equation (3.9) reduces to a simplified form that looks similar to (3.7), the Sverdrup balance. Two different forms of the Sverdrup balance are derived from the simplified equations of motion.

First we derive the Sverdrup balance based on the geostrophic flow only. We start with the geostrophic relation, (2.3),

$$
\begin{aligned}
\rho_0 f v_G &= \frac{\partial p}{\partial x} \\
\rho_0 f u_G &= -\frac{\partial p}{\partial y}
\end{aligned}
\tag{3.22}
$$

and, as in Subsection 2.2.5, take the curl to get

$$
f \left(\frac{\partial u_G}{\partial x} + \frac{\partial v_G}{\partial y} \right) + \beta v_G = 0.
\tag{3.23}
$$

Combine this result with the continuity equation (2.83) to form the linear, inviscid form of the vorticity equation,

$$
\beta v_G = f \frac{\partial w}{\partial z}
\tag{3.24}
$$

(sometimes called the Sverdrup relation, Pedlosky, 1996; see Subsection 2.2.3). Integrating from the bottom to the top of the purely geostrophic part of the water column and assuming that $w = 0$ at the bottom and $w = w_{Ek}$ from (2.13) at the top, we get the geostrophic form of the Sverdrup balance,

Geostrophic Sverdrup balance: $\beta V_G = f w_{Ek}$

$$
= f \hat{\mathbf{z}} \cdot \nabla \times \left(\frac{\mathbf{X}}{f} \right),
\tag{3.25}
$$

where V_G is the geostrophic analogue of V. Neither (3.24) nor (3.25) assume much about density, so these equations apply to a stratified ocean as well as the uniform-density ocean discussed above. Equation 3.25 is less general than (3.24), because it includes the additional assumption of $w = 0$ at the bottom, which is not necessarily valid for flow along a sloping ocean bottom or for large bottom Ekman pumping.

Often it is desirable to include the surface Ekman velocity as well as the geostrophic flow. In that case, we should include vertical viscosity, represented here by a diffusion law:

$$
\begin{aligned}
-fv &= -\frac{1}{\rho} \frac{\partial p}{\partial x} + \frac{\partial}{\partial z} \left(\nu_v \frac{\partial u}{\partial z} \right), \\
fu &= -\frac{1}{\rho} \frac{\partial p}{\partial y} + \frac{\partial}{\partial z} \left(\nu_v \frac{\partial v}{\partial z} \right),
\end{aligned}
\tag{3.26}
$$

where v_v is the vertical eddy viscosity. Integrating from the bottom to the surface (including the Ekman layers), and using the fact that at the surface,

$$\mathbf{X} = v_v \frac{\partial \mathbf{u}}{\partial z}, \tag{3.27}$$

the momentum equations become

$$-fV = -\frac{1}{\rho}\frac{\partial P}{\partial x} + X,$$

$$fU = -\frac{1}{\rho}\frac{\partial P}{\partial y} + Y, \tag{3.28}$$

where $\mathbf{X} = (X, Y)$ and

$$P = \int_{-H}^{0} p \, dz \tag{3.29}$$

(bottom drag is neglected). The vertical integral of the continuity equation is

$$\frac{\partial U}{\partial x} + \frac{\partial V}{\partial y} = 0, \tag{3.30}$$

assuming that we can neglect w at the surface (due to evaporation and precipitation) and at the bottom (due to Ekman pumping and flow along bottom slopes). Taking the curl of (3.28) and combining with (3.30), we get

$$\beta V = \frac{\partial Y}{\partial x} - \frac{\partial X}{\partial y}, \tag{3.31}$$

which is a restatement of (3.7) in Subsection 3.2.4, the Sverdrup balance for the full water column. Like (3.25), the geostrophic version, (3.31) is valid for quite general density distributions but assumes that bottom velocities (and hence bottom drag) are negligible. It can be shown that the difference between the two forms of the Sverdrup balance is exactly the meridional component of Ekman transport. Notice that there is a term proportional to βX missing from (3.31). It arises because f varies with latitude and w_{Ek} depends on the curl of $\mathbf{X}/f$, not just $\mathbf{X}$ in (3.25) and (2.13). This missing term is sometimes important, for example in the tropics, and is discussed in Subsection 6.2.2.

Given V, U can be calculated from the continuity equation:

$$U(L, y) - U(x, y) = \int_{x}^{L} \frac{\partial V}{\partial y} \, dx', \tag{3.32}$$

where we integrate from an arbitrary value of x to the eastern boundary location $L(y)$. For the geostrophic Sverdrup balance, a similar procedure can obtain U_G from V_G and the Ekman pumping. The value of U at the eastern boundary can be obtained from the condition that there is no flow through the boundary. If we define the boundary by a vector $\hat{\mathbf{n}}$ normal to the boundary, the no-flow condition can be written

$$\mathbf{U} \cdot \hat{\mathbf{n}} = 0. \tag{3.33}$$

In the simplified case of a boundary which is a line of longitude, we have $U(L) = 0$.

One convenient feature of using $\mathbf{U}$ rather than $\mathbf{U}_G$ is that $\mathbf{U}$ is non-divergent and therefore has a streamfunction. Given V, the streamfunction can be calculated with (see (6.39)):

$$\Psi_{Sv}(L, y) - \Psi_{Sv}(x, y) = \int_x^L V \, dx', \tag{3.34}$$

which defines the **Sverdrup transport streamfunction**, $\Psi_{Sv}(x, y)$. The eastern wall at $x = L$ is impermeable, so the streamfunction there must be a constant, which we take to be zero. Hence, using (3.31),

$$\Psi_{Sv}(x, y) = \frac{1}{\beta} \int_x^L \left(\frac{\partial Y}{\partial x} - \frac{\partial X}{\partial y} \right) dx'. \tag{3.35}$$

If variations in f are small across the gyre, the geostrophic flow $\mathbf{U}_G$ is also non-divergent, but this is a dangerous approximation to make for the gyres. For one thing, it should be remembered that even if the divergence of $\mathbf{U}_G$ is small, this divergence is essential because it connects to the Ekman pumping w_{Ek} driving the flow. For another, on length scales appropriate for the gyre, the divergence is not necessarily small. Doing a scale analysis of (3.23), with velocity scaled by U and vorticity scaled by U/L, we get

$$\frac{\nabla \cdot \mathbf{u}_G}{\hat{\mathbf{z}} \cdot (\nabla \times \mathbf{u}_G)} = \frac{\frac{\partial u_G}{\partial x} + \frac{\partial v_G}{\partial y}}{\frac{\partial v_G}{\partial x} - \frac{\partial u_G}{\partial y}} = \frac{\beta v_G / f}{\frac{\partial v_G}{\partial x} - \frac{\partial u_G}{\partial y}} \sim \frac{\beta L}{f} = \frac{L}{R_E} \cot \theta. \tag{3.36}$$

For $L < R_E$, we assume that divergence is small. In that case, u_{Gx} and v_{Gy} must have opposite signs and largely cancel each other. For the gyres, L is not so much smaller than R_E, and the divergence may not be negligible.

Both forms of the Sverdrup balance are remarkably simple. The equations of motion include a partial differential equation (continuity), but we are able to combine them in such a way that the single resulting equation is an algebraic equation for V. This simplicity contains the seeds of the partial failure of the solution. For a closed basin, or an almost-closed basin with almost no net volume transport import or export to other basins (such as the Atlantic or North Pacific), we have

$$\int_0^L V \, dx' = 0. \tag{3.37}$$

Nothing in the Sverdrup balance guarantees that (3.37) is satisfied. For the idealized case in which $\mathbf{X}$ depends only on latitude, (3.37) is obviously *not* true because V has the same sign across the entire width of the basin.

The problem with the Sverdrup solution comes about because removing the friction terms from the barotropic vorticity equation (3.9) is a **singular perturbation** (Bender and Orszag, 1987, Chapter 7). Perturbation theory concerns finding an approximation to a problem that depends on a parameter ϵ based on knowing the solution for the case in which $\epsilon = 0$. A singular perturbation is one in which the solutions for $\epsilon \to 0$ do not converge to the $\epsilon = 0$ solution. For our circulation problem, the friction coefficient plays the role of ϵ. Mathematically, the qualitative change in our solution occurs because the differential equation is simplified by removing terms with the highest order derivatives in the equation (Bender and Orszag, 1987, Chapter 9). The highest order derivative determines how many boundary conditions a given differential equation satisfies. The bottom friction term in the vorticity equation is proportional to ζ, which contains first derivatives of the velocity components, and the side friction term is proportional to $\nabla^2 \zeta$, which contains higher order derivatives. For the full water-column version of the Sverdrup balance, the problem with satisfying (3.37) is equivalent to not being able to satisfy the boundary condition that Ψ is constant along the boundaries of the domain.

The solution to singular perturbation problems is often approximately correct over most of the domain but completely incorrect in part of the domain. A classic example in fluid dynamics is the problem of flow around an object in a non-rotating system (Kundu et al., 2012, Chapter 9). If friction is neglected by dropping the $\nabla^2 \mathbf{u}$ terms from the equations of motion, the resulting solution is approximately true everywhere except near the object. This solution cannot satisfy the no-slip condition, which states that there is no fluid flow relative to the object at the surface of the object. The complete solution includes a narrow **frictional boundary layer** in which the flow past the object goes to zero as one approaches the object. The gyre circulation also produces a boundary layer, although it is a fast current, not a slow one.

3.3.3 Depth-Integrated Gyres with Western Boundary Currents

To illustrate the relationship between the western boundary current and the Sverdrup solution, we now derive and solve a linear, full water-column vorticity equation including friction terms neglected in the previous section. In the real world, it is doubtful that the nonlinear terms can be neglected in the western boundary region, but the less realistic problem is still interesting because it shows how a western boundary current can develop even in the absence of inertial effects. The linear case is also relevant to numerical studies of the large-scale circulation and climate, for these often use artificially high values of ν_h which tend to weaken the flow and hence suppress the nonlinear terms.

When we include bottom and side friction, the vertically integrated form of the momentum equations (3.28) become

$$-fV = -\frac{1}{\rho}\frac{\partial P}{\partial x} + X + \nu_h \nabla^2 U - \frac{1}{2}fH_{Ek}(u_B - v_B),$$

$$fU = -\frac{1}{\rho}\frac{\partial p}{\partial y} + Y + \nu_h \nabla^2 V - \frac{1}{2}fH_{Ek}(u_B + v_B), \qquad (3.38)$$

where the last terms in each equation represent the bottom stress $\rho\nu_v\mathbf{u}_z$ in accord with Ekman theory (see Pedlosky, 1987a, eq. 4.3.30; see Exercise 3.6), with (u_B, v_B) representing geostrophic velocity at the bottom. We assume the seafloor is at a constant depth H. In Subsection 2.2.5, it was important to only consider the vertically uniform geostrophic velocity in an unstratified fluid, so that for the nonlinear terms, vertical integration was equivalent to multiplying by H:

$$\int_{-H+H_{Ek}}^{-H_{Ek}} \mathbf{u} \cdot \nabla\zeta \; dz \approx H\mathbf{u} \cdot \nabla\zeta. \qquad (3.39)$$

If **u** varies with depth, (3.39) is not true. Dropping the nonlinear terms opens up the possibility that we can consider depth-varying **u** as well, including the Ekman layers and geostrophic shear. The remaining complication is that our friction law is based on $\mathbf{u}_B = (u_B, v_B)$ rather than **U**. In the following we will replace $\mathbf{u}_B$ with $\mathbf{U}/H$, which will give a realistic bottom stress for a uniform-density fluid with negligible surface Ekman transport. For the more general case, this bottom friction term is somewhat unrealistic.

As before, we form a vorticity equation by taking the curl of the momentum equations and use continuity to eliminate $U_x + V_y$. While we allow ρ to vary in space, we assume that horizontal gradients in ρ are small compared to the other terms. The resulting vorticity equation is

$$\beta V = \left(\frac{\partial Y}{\partial x} - \frac{\partial X}{\partial y}\right) + v_h \nabla^2 \left(\frac{\partial V}{\partial x} - \frac{\partial U}{\partial y}\right) - \frac{1}{2} f \delta_{Ek} \left(\frac{\partial V}{\partial x} - \frac{\partial U}{\partial y}\right). \tag{3.40}$$

Now the fact that $\nabla \cdot \mathbf{U} = 0$ is especially convenient, because we can replace U and V with a streamfunction using (2.19), so that the vorticity equation becomes

$$\beta \frac{\partial \Psi}{\partial x} + \frac{1}{2} f \delta_{Ek} \nabla^2 \Psi - v_h \nabla^4 \Psi = \frac{\partial Y}{\partial x} - \frac{\partial X}{\partial y}. \tag{3.41}$$

Stommel Gyre Model

Stommel (1948) finds an exact analytic solution to this equation for a rectangular domain, constant β (β-plane, (2.64)) with idealized wind stress and $v_h = 0$. Solutions for more general cases can be found by making a boundary layer approximation. To calculate such a solution, we first find the "interior solution" $\Psi_S(x, y)$ away from the boundary layer. This is the solution when we neglect the friction terms; the solution to the Sverdrup balance (Subsection 3.2.4). We then find a solution for a region of width δL (the product of δ and L) near the western boundary, where L is the zonal length scale of the gyre and δ defines the boundary layer width. We assume $\delta \ll 1$ but we will not know its value until we compute the streamfunction. Assuming there is no flow through the outer boundaries of the gyre, Ψ must have the same value on the eastern and western boundaries. We can set the boundary value of Ψ to an arbitrary value, so for convenience we choose $\Psi = 0$. Outside the boundary layer region, Ψ must approach Ψ_S. Inside the boundary, we retain the friction terms but we neglect some terms based on a scale analysis in the boundary layer. We know that Ψ must vary from 0 to Ψ_S across the width of the boundary and between a central latitude and the northern and southern boundaries of the gyre, so it follows that the relative size of derivatives is given by

$$\frac{\partial \Psi / \partial y}{\partial \Psi / \partial x} \sim \delta, \tag{3.42}$$

which means that we can neglect y derivatives compared to x derivatives. In the region $x < \delta L$, we also have

$$\frac{\partial \Psi_S / \partial x}{\partial \Psi / \partial x} \sim \delta. \tag{3.43}$$

This implies that we can neglect the wind stress curl terms in the boundary layer. That is because the wind stress curl is of the same magnitude as $\beta \partial \Psi_S / \partial x$, so in the boundary layer $\beta \partial \Psi / \partial x$ must be much bigger. In the boundary layer, neglecting relatively small terms (and using subscripts to represent differentation) allows the vorticity equation (3.41) to be written

$$\beta \Psi_x + \frac{1}{2} f \delta_{Ek} \Psi_{xx} - \nu_h \Psi_{xxxx} = 0. \tag{3.44}$$

This is the **boundary layer equation**. At first glance it may not look much simpler than (3.41), but there is an important difference: (3.41) is a partial differential equation in (x, y), whereas (3.44) is an ordinary differential equation in x, which is much easier to solve. It looks like the resulting equation is independent of the forcing, but the wind stress enters through the condition that Ψ must match the Sverdrup solution outside the boundary layer.

It is illustrative to solve the $\nu_h = 0$ case. The boundary layer equation becomes

$$\beta \Psi_x + \frac{1}{2} f \delta_{Ek} \Psi_{xx} = 0, \tag{3.45}$$

which is solved by

$$\Psi = A(y)(1 - e^{-x/L_S}), \tag{3.46}$$

where the Stommel length scale (3.17) is

$$L_S = \frac{1}{2} \frac{f}{\beta} \delta_{Ek}. \tag{3.47}$$

This solution (3.46) automatically satisfies the boundary condition $\Psi = 0$ at $x = 0$. For $x \gg L_S$, $\Psi \to A(y)$, which satisfies the matching condition as long as $A(y) = \Psi_{Sv}(0, y) \equiv \Psi_S(y)$. It may seem strange to match a boundary layer solution at *large* x with the Sverdrup solution at $x = 0$. Nevertheless, the boundary layer occupies such a small interval in x that the streamfunction at large x/L_S is well-approximated by the interior solution at $x = 0$ (recall that Ψ_S varies over the much longer scale L).

The solution is illustrated by Figure 3.12b in Subsection 3.2.2. We calculate the meridional current in the boundary layer from (3.46) using (2.19):

$$V = (\Psi_S/L_S)e^{-x/L_S}, \tag{3.48}$$

which attains a maximum value at $x = 0$. V has the opposite sign to the Sverdrup value V_S and a magnitude that is a factor of $1/L_S$ larger.

Munk Gyre Model

It is also interesting to consider the pure horizontal-friction case, which is arguably a more realistic model of the real ocean. The boundary layer equation is then

$$\beta \Psi_x - \nu_h \Psi_{xxxx} = 0, \tag{3.49}$$

again for constant β. Because this is a higher order equation, when we solve it we need an extra constant $B(y)$ which implies an extra boundary condition. The general solution is

$$\Psi = \Psi_S(y)\left[1 - e^{-x/2L_M}\cos\left(\frac{\sqrt{3}x}{2L_M}\right) + B(y)e^{-x/2L_M}\sin\left(\frac{\sqrt{3}x}{2L_M}\right)\right], \qquad (3.50)$$

where the Munk scale L_M is defined in (3.19). There has been debate about which additional boundary condition is most appropriate for the ocean given small-scale turbulence, but for laminar conditions such as would occur in a laboratory experiment, the most natural boundary condition is no-slip (zero flow adjacent to the wall). This solution is given by

$$\Psi = \Psi_S(x, y)\Theta(x), \qquad (3.51)$$

and

$$v = v_S\Theta + \Psi_S\Theta_x, \qquad (3.52)$$

(where $v_S = \partial\Psi_S/\partial x$ is the meridional Sverdrup current) with

$$\Theta = 1 - e^{-x/2L_M}\left[\cos\left(\frac{\sqrt{3}x}{2L_M}\right) + \frac{1}{\sqrt{3}}\sin\left(\frac{\sqrt{3}x}{2L_M}\right)\right], \qquad (3.53)$$

and

$$\Theta_x = \frac{2}{\sqrt{3}L_M}e^{-x/2L_M}\sin\left(\frac{\sqrt{3}x}{2L_M}\right). \qquad (3.54)$$

(Note that in this solution (3.51) and (3.53) approximate the true solution, which does not have a simple closed form. The approximation deteriorates as L_M/L ceases to be small.) This solution is illustrated by Figure 3.17. The western boundary current is similar to the bottom-friction case, but the flow structure is slightly more complicated, with some recirculation seen most clearly in Figure 3.17b as southward flow, stronger than the Sverdrup circulation, centered at $x/L = 0.1$ in this example.

In both the bottom-friction and horizontal-friction solutions, the width of the boundary current is proportional to the frictional length scale (L_S for bottom friction and L_M for horizontal friction). Often in numerical experiments, v_h is not based on physical principles but is adjusted to give the smallest western boundary current width that can be resolved by the grid spacing of the model.

3.3.4 Nonlinear Recirculation Dynamics

We can get some insight into the effect of inertial terms by studying their effect on the potential vorticity $q = (f + \zeta)/H \propto f(y) + \nabla^2\Psi$ of the gyre (where depth H is constant). The distribution of q in a numerical model of a subtropical gyre with a recirculation confined to the northeastern corner serves our purpose (Figure 3.18a). In Subsection 3.2.2, we assume that the potential vorticity is completely defined by the latitude of the water, so that contours of constant potential vorticity are latitude circles. Even in the moderately nonlinear case shown in Figure 3.18, this is still true for much of the gyre where the Sverdrup balance holds. As a water parcel moves southward in this region, the motion is slow enough

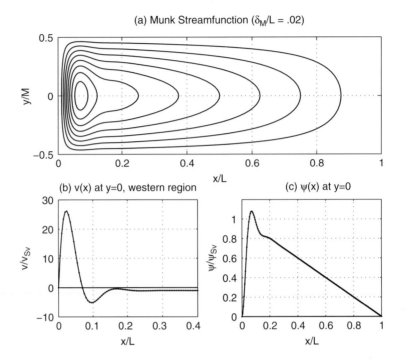

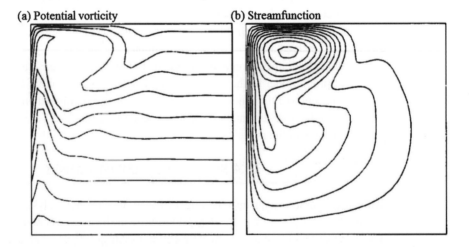

Figure 3.17 Numerical solution for the Munk gyre showing (a) streamfunction, (b) $v(x, y)$ at $y = 0$, and (c) $\Psi(x, y)$ at $y = 0$ for solution to (3.49) for $\hat{\mathbf{z}} \cdot \nabla \times \tau/(\beta\rho) = V_0 \cos(\pi y/M)$ and $L_M = 0.02L$.

Figure 3.18 Contours of (a) potential vorticity and (b) streamfunction in a nonlinear uniform-density model of a wind-driven gyre. Reprinted from Böning (1986), Figure 8, Copyright 1986, with permission from Elsevier.

that the input of vorticity by the wind keeps q at the value f/H appropriate for the parcel's latitude (Figure 3.18b). In the western boundary current, where friction is increasing q as it goes northward, the water is moving fast enough that it carries some anomalously low-q water northward. This is visible in the northward excursion of the q contours along the western boundary (Figure 3.18b).

In the recirculation region, q is nearly uniform over much of the area of closed streamlines. Here, advection dominates other processes; water travels around each streamline fast enough that the entire streamline keeps the same value of q. This does not explain why the entire recirculation region has the same q, rather than constant-q contours aligned with the recirculating streamlines. Nevertheless, it can be shown that the weak friction equalizes q between neighboring streamlines (see Subsection 5.2.3).

The effect of uniform q along a streamline can be seen as follows. Consider the analytic solution for a limiting case where, rather than ignoring inertial terms, we *only* keep the inertial terms and neglect forcing and friction. In this approximation, all that remains of the vorticity equation is

$$J(\Psi, \nabla^2\Psi + \beta y) = 0, \qquad (3.55)$$

where $q \propto \nabla^2\Psi + \beta y$ stands for the potential vorticity if we approximate $f = f_0 + \beta y$ with constant f_0. In this equation we use the Jacobian J to write the nonlinear inertial term, where

$$J(a, b) = \frac{\partial a}{\partial x}\frac{\partial b}{\partial y} - \frac{\partial a}{\partial y}\frac{\partial b}{\partial x} \qquad (3.56)$$

in Cartesian coordinates. If $J(a, b) = 0$, then contours of field a coincide with those of field b (see Exercises 3.9 and 5.9). Hence, in this case, q is uniform along a streamline, which implies

$$\nabla^2\Psi + \beta y = Q(\Psi), \qquad (3.57)$$

where Q is an arbitrary function of Ψ. There may be many different forms of $Q(\Psi)$ that give valid solutions. Fofonoff (1954) picked a particularly simple form and found a solution to the (linear) equation:

$$\nabla^2\Psi + \beta y = \beta(\Psi/U + y_0), \qquad (3.58)$$

where y_0 and U are free parameters. For a rectangular basin ($0 \le x \le L, 0 \le y \le M$), a boundary layer approximation gives this solution:

$$\Psi = U(y - y_0)\left(1 - e^{-x'} - e^{x'-L'}\right) - U(M - y_0)e^{y'-M'} + Uy_0e^{-y'}, \qquad (3.59)$$

where primes represent division by L_I see (3.13). This solution is called a **Fofonoff gyre** after Nicholas Fofonoff.

As Figure 3.19 shows, the resulting solution consists of westward flow everywhere except in boundary currents along all the edges of the basin. The westward flow returns eastward along the northern and southern boundary, with y_0 determining the bifurcation

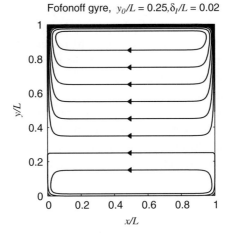

Fofonoff gyre, $y_0/L = 0.25, \delta_I/L = 0.02$

Figure 3.19 Fofonoff gyre solution from (3.59), with $y_0/L = 0.25, L_I/L = 0.02$. See also Exercise 3.9.

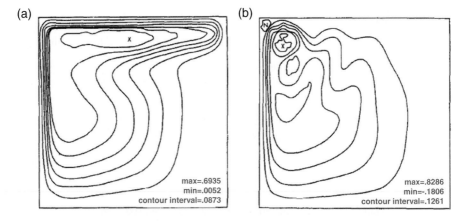

Figure 3.20 Streamfunction for numerical solution to nonlinear uniform-density model of wind-driven gyre with (a) no-stress boundary condition and (b) no-slip boundary condition. From Blandford (1971) Figure 6, Copyright 1971, with permission from Elsevier.

latitude for meridional flow along the boundaries. The Fofonoff gyre somewhat resembles the highly nonlinear solution (Figure 3.14). It also somewhat resembles confined recirculation gyres found in other numerical experiments.

Returning to wind-forced gyres, the size, shape, and strength of the recirculation depend on the boundary condition (Subsection 1.3.1) used for the solution. For a free-slip (no-stress) boundary condition, the current remains strong and narrow even downstream of the point of detachment from the western boundary, and a recirculation gyre forms adjacent to the eastward flowing jet (Figure 3.20a). For a no-slip boundary condition, the recirculation region does not reach as far from the western boundary, and the exiting current forms stationary meanders further downstream (Figure 3.20b).

3.3.5 Integral Vorticity Balances

Basin-Wide Input and Output

Integral constraints tell us useful and interesting things about the wind-driven circulation without even solving the equations.

One version of the vorticity equation was stated in Subsection 2.2.5 (Equation 2.90) and is equivalent to

$$\frac{\partial \zeta}{\partial t} + \nabla \cdot [\mathbf{u}_h(f + \zeta)] = \delta_T \hat{\mathbf{z}} \cdot \nabla \times (\mathbf{X}/h) - \delta_B r\zeta + \nu_h \nabla^2 \zeta. \tag{3.60}$$

If we integrate this over (stationary) area A of the domain, using the divergence theorem to write the advection term as a contour integral, we get

$$\frac{\partial}{\partial t} \int_A \zeta \, dA = - \oint_C [\mathbf{u}_h(f+\zeta)] \cdot \hat{\mathbf{n}} \, ds + \int_A \left[\delta_T \hat{\mathbf{z}} \cdot \nabla \times (\mathbf{X}/h) - \delta_B r\zeta + \nu_h \nabla^2 \zeta \right] dA, \tag{3.61}$$

where the bounding perimeter of the domain, shown in Figure 3.21, is C and $(\hat{\mathbf{n}}, \hat{\mathbf{t}})$ are unit vectors on C normal to (pointing out) and tangent to (pointing anticlockwise) the boundary, and s measures distance around it.

If the region of integration encompasses the entire ocean, then $\mathbf{u}_h \cdot \hat{\mathbf{n}} = 0$ and the advection term is zero. Therefore, the rate of change of the relative vorticity integrated over the entire domain is equal to the integrated difference between vorticity input by the wind stress and vorticity loss by friction. Once the system reaches steady state, we can also drop the left-hand side term and the wind and friction terms balance.

Using Stokes' theorem on wind stress and bottom friction terms and the divergence theorem on the horizontal friction term, we are left with

$$\oint_C \delta_T (\mathbf{X}/h) \cdot \hat{\mathbf{t}} \, ds = \oint_C \delta_B r\mathbf{u} \cdot \hat{\mathbf{t}} \, ds - \oint_C \nu_h \nabla \zeta \cdot \hat{\mathbf{n}} \, dA. \tag{3.62}$$

Written this way, we see that the wind input of vorticity occurs when the stress integrated around the boundary of the entire basin is nonzero. Consider (3.62) for the Stommel model (subtropical gyre, northern hemisphere). Then we have $\nu_h = 0$ and the entire domain consists of the subtropical gyre (so that walls bound all sides of the gyre). The wind stress curl integrated around C is negative because τ blows eastwards to the north and westwards to the south, opposite to $\hat{\mathbf{t}}$ in both regions. That means we must have $\oint_C \mathbf{u} \cdot \hat{\mathbf{t}} \, ds < 0$, meaning that the integrated current around C must be clockwise. In general, the bottom stress can

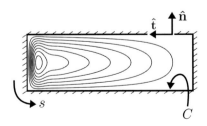

Figure 3.21 Schematic diagram of the domain A, enclosed by perimeter C, occupied by the homogeneous model of subtropical circulation.

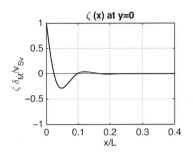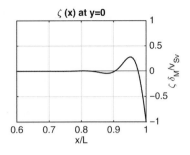

Vorticity in the northward-flowing jet for a western (left) and eastern (right) boundary current. The eastern boundary current cannot satisfy the domain-integrated vorticity balance for a steady state (3.62). The case of no-slip boundary conditions is shown (see Figure 3.17).

only strike a balance if the water circulates around the basin in the same direction as the wind.

The Munk model ($r = 0$, $v_h \neq 0$) has a similar balance. It is not immediately clear from (3.62) that balancing the wind curl with the horizontal friction term implies that the ocean currents have the same sense of circulation as the wind stress. This does in fact happen in numerical and analytic solutions for a no-slip boundary condition. This is what we expect if we think of vorticity entering the ocean from the wind stress and leaving due to friction with the domain boundary.

We can confirm the integral balance with horizontal friction by looking at the solution for the Munk model. The vorticity gradient is negligible everywhere except in the northward flowing boundary current. Figure 3.22 shows the vorticity in the western boundary current (left) and for a putative eastern boundary current (right). The eastern boundary current looks the same as the western boundary current in Figure 3.17, except with opposite sign and flipped in the x direction (it is instructive to check Figure 3.22 starting from the $v(x)$ plot in Figure 3.17). In both cases, the vorticity gradient at the wall in the x direction is negative. But the outward pointing normal direction is different. Only the western boundary current exhibits a positive vorticity gradient in the direction of the outward normal, so only the western boundary current can satisfy the steady-state domain-integrated vorticity balance (3.62).

The domain-integrated (potential) vorticity budget is a powerful constraint for understanding the circulation. Notice how it follows from the un-approximated shallow water equations. In particular, no assumption of weak flow (linearity) is necessary. It is also reassuring to grasp a theoretical statement that explains the sense of circulation. Despite the unfamiliar effects of rotation – geostrophy, Ekman, and Sverdrup dynamics – the gyre goes round the way we expect it to, driven by the swirl of the wind.

Circulation Integral from Momentum Equations

A similar integral to (3.62) can be deduced more directly from the momentum equations. We start with the horizontal momentum equations from the primitive equations, using equation (2.86) from Subsection 2.2.5 to write the advection terms:

$$\frac{\partial \mathbf{u}}{\partial t} + \frac{1}{2}\nabla(\mathbf{u} \cdot \mathbf{u}) - \mathbf{u} \times (\nabla \times \mathbf{u}) = -\frac{1}{\rho_0}\nabla p - f\hat{\mathbf{z}} \times \mathbf{u} + \nu_h \nabla^2 \mathbf{u} + \nu_v \frac{\partial^2 \mathbf{u}}{\partial z^2}. \quad (3.63)$$

Rather than taking the curl to get a vorticity equation and integrating, we integrate in z and then around a closed contour. If we reverse the order of the integrals in each of the two gradient terms, the inner integration is the closed contour integral of a gradient, which vanishes. Then the integral is

$$\oint \left[\int \frac{\partial \mathbf{u}}{\partial t}\, dz\right] \cdot \hat{\mathbf{t}}\, ds - \oint \left[\int \mathbf{u} \times (f\hat{\mathbf{z}} + \nabla \times \mathbf{u})\, dz\right] \cdot \hat{\mathbf{t}}\, ds =$$
$$\oint \left[\int \nu_h \nabla^2 \mathbf{u}\, dz\right] \cdot \hat{\mathbf{t}}\, ds + \oint \left[\int \nu_v \frac{\partial^2 \mathbf{u}}{\partial z^2}\, dz\right] \cdot \hat{\mathbf{t}}\, ds. \quad (3.64)$$

As $\mathbf{u}$ is parallel to the boundary, a cross product with $\mathbf{u}$ is perpendicular to the boundary, which has zero dot product with $\hat{\mathbf{t}}$. In steady state, the first term on the left is also zero. Integrating the last term in z we end up with the stress $\mathbf{X}_T$ at the top and $\mathbf{X}_B$ at the bottom of the water column. Thus the equation becomes

$$\oint \mathbf{X}_T \cdot \hat{\mathbf{t}}\, ds = \oint \mathbf{X}_B \cdot \hat{\mathbf{t}}\, ds - \oint \left[\int \nu_h \nabla^2 \mathbf{u}\, dz\right] \cdot \hat{\mathbf{t}}\, ds, \quad (3.65)$$

which is very similar to (3.62). The bottom stress term can be related to the vorticity as in (3.62), but will only get the same answer if the vorticity is the same throughout the water column, which is not necessarily true if the water does not have uniform density.

Island Rule

The contour integral of the momentum equation provides another constraint that is not accessible via the Stommel and Munk models. In previous sections, we assume no net flow into or out of the gyre, so that $\int V\, dx = 0$ across the gyre. In the real ocean, this is not always the case. For instance (Figure 3.23a), the South Pacific has net northward flow which leaves the Pacific as the Indonesian throughflow and returns to the South Pacific at about 44°S (the southernmost latitude of Tasmania, which connects to southern Australia by a shallow shelf). The momentum integral allows us to estimate the strength of this flow based on a simple geostrophic plus Ekman dynamics which implies an Island Rule for the transport (Godfrey, 1989).

A simple model of the circulation in this region considers an ocean basin divided into western and eastern basins by an island (Figure 3.23b). We want to know the flow around the island, which is the meridional volume transport passing through the basin at the latitudes of the island's northern (DC in figure) and southern (AB in figure) extremities. Neglecting evaporation, precipitation and runoff, the volume transports through both zonal sections are the same.

To calculate the flow through the basin, we integrate the momentum equation along the contour ABCD in Figure 3.23. The contour does not exclusively follow solid boundaries, so we cannot neglect the $\mathbf{u} \times (f\hat{\mathbf{z}} + \nabla \times \mathbf{u})$ terms as above. Instead, we linearize and drop the $\mathbf{u} \times \nabla \times \mathbf{u}$ term. Thus, the Island Rule is a linear theory like the Stommel

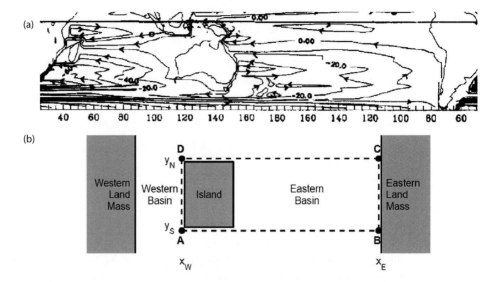

Figure 3.23 (a) Island Rule calculation. The computed transport around Australia is 16 ± 4 Sv, similar to the observed value in Figure 3.8. From Godfrey (1989), Figure 4; used with permission of Taylor & Francis Ltd. (b) Schematic for Island Rule calculation showing land masses (gray shading) including island at left and eastern boundary of the basin at right, and integration contours (dashed lines).

and Munk models. The contour avoids western boundaries, however, so from the previous analysis we assume that friction is negligible. Hence, for steady state flow, (3.64) becomes

$$- \oint \left[\int \mathbf{u} \times (f\hat{\mathbf{z}}) \, dz \right] \cdot \hat{\mathbf{t}} \, ds = \oint \mathbf{X}_T \cdot \hat{\mathbf{t}} \, ds. \tag{3.66}$$

Next, integrate in z and split the vertical integral of $\mathbf{u}$, $\mathbf{U}$, into components parallel to and perpendicular to the contour:

$$\mathbf{U} = U_n \hat{\mathbf{n}} + U_t \hat{\mathbf{t}}. \tag{3.67}$$

Equation (3.66) becomes

$$- \oint f[U_n(\hat{\mathbf{n}} \times \hat{\mathbf{z}}) + U_t(\hat{\mathbf{t}} \times \hat{\mathbf{z}})] \cdot \hat{\mathbf{t}} \, ds = \oint \mathbf{X}_T \cdot \hat{\mathbf{t}} \, ds. \tag{3.68}$$

Doing the cross product and then dotting with $\hat{\mathbf{t}}$, we now have

$$\oint f U_n ds = \oint \mathbf{X}_T \cdot \hat{\mathbf{t}} \, ds. \tag{3.69}$$

Applying this to the contour in Figure 3.23 shows that only segments AB and CD of the contour have nonzero U_n. It is convenient to define the volume transport passing through the basin:

$$T = - \int_A^B U_n \, ds = \int_{x_W}^{x_E} V(x, y_S) \, dx \tag{3.70}$$

where V is the northward vertically integrated velocity. Remember that $\hat{\mathbf{n}}$ points outward everywhere, so along AB, U_n gives the southward flow and along CD it gives northward flow. Using S to represent segment AB and N to represent segment CD, (3.69) becomes

$$f_N \int_{x_W}^{x_E} V(x, y_N) \, dx - f_S \int_{x_W}^{x_E} V(x, y_S) \, dx = \oint \mathbf{X}_\tau \cdot \hat{\mathbf{t}} \, ds, \qquad (3.71)$$

which gives

Island Rule: $\qquad T = \dfrac{1}{f_N - f_S} \oint \mathbf{X}_\tau \cdot \hat{\mathbf{t}} \, ds. \qquad (3.72)$

The Island Rule allows us to estimate the net flow around an island from the wind field. Compare it to the Sverdrup balance: instead of the wind stress curl, the Island Rule is based on the wind stress contour integral around a region encompassing the island and the ocean east of the island. Instead of this wind determining the meridional depth-integrated velocity times β, it determines meridional volume transport times the difference in f over the region. The additional circulation implied by the Island Rule must occur in the Western Boundary Current, as the Sverdrup balance still applies away from western boundaries.

Godfrey (1989) uses observations of wind stress to estimate 16 ± 4 Sv transport around Australia, which is consistent with the observed flow through the Indonesian Passages (Subsection 3.1.4). Although the derivation here treats a simplified geometry, the rule can also be derived for multiple islands, including influences of friction and topography (Wajsowicz, 1993).

3.4 Excursions: Pre-Scientific Knowledge of the Ocean Circulation

The scientific discipline of Oceanography began in the mid-nineteenth century. Yet, people have fished and travelled the seas for millennia. What did these prescientific observers know about the ocean circulation?

One interesting case concerns the medieval North Atlantic Vikings (Haine, 2008, 2012). They operated a colony for nearly five hundred years in southwest Greenland, starting around AD 1000. The enabling technology of these Greenland Norse colonists was their wooden longboat (knarr). Longboat seamanship requires navigation with primitive instruments, appreciation of tidal ranges and currents, plus knowledge of marine meteorology and sea ice. Evidently, Greenland Norse sailors knew about nontidal ocean circulation too. For instance, medieval depictions of Greenland emphasize driftwood floating along the coast. Greenland has no usable trees, so wood was a premium commodity for construction of ships and buildings. Greenland driftwood comes from Siberia, Canada, and Alaska, and is carried thousands of kilometers by the surface currents and sea ice to Greenlandic beaches. The strong surface fronts between the cold, fresh, icy polar water and the warm, salty Atlantic water are also readily apparent to careful observers. The fronts are associated with the western boundary currents, such as the East and West Greenland Currents. They presumably guided Norse navigators. Furthermore, it has been suggested that the Norse knew of the mesoscale eddies in the Iceland–Faroes front (Rossby and Miller, 2008).

These reasons support the idea that the existence and basic structure of the surface circulation of the western subpolar Atlantic Ocean, and the boundary current system in particular, were known 1000 years ago.

Exercises

3.1   This question follows Exercise 2.11.

(a) Load the SCOW wind stress data (see Appendix A). Make a plot of the annual average wind stress components over the basin you chose in Exercise 2.11.

(b) Compute (or download) the wind stress curl and plot it. Hence compute and plot the Ekman pumping speed. Remember to use the appropriate formula for curl in spherical polar coordinates from (B.22).

(c) Compute and plot the Sverdrup streamfunction along the average latitude of the zonal section you chose in Exercise 2.11. Do so by integrating the wind stress curl westwards from the eastern boundary of your ocean basin.

(d) Discuss whether the pressure-integrated geostrophic streamfunction you computed from the CTD section in Exercise 2.11 satisfies Sverdrup balance with the average wind.

3.2   Select one of the following regions:

- Subtropical North Atlantic,
- Subtropical South Atlantic,
- Subtropical North Pacific,
- Subtropical South Pacific,
- Subpolar North Atlantic,
- Subpolar North Pacific,
- Antarctic circumpolar current

(coordinate with your classmates to avoid replicates). Then, using altimetric sea-surface height data from a suitable source (see Appendix A),

(a) Make maps of time-average sea-surface height for your region corresponding to Figures 3.4 and 3.5.

(b) Explain how the surface circulation is inferred from your sea-surface height maps. Hence, estimate the time-average speed of the surface circulation and make a map of it corresponding to your map in part (a).

(c) Investigate seasonality in sea-surface height and speed for your region. Map, describe, and comment on it.

(d) Compute, map, and describe the variance of sea-surface height and speed over time for your region.

(e) Hence, identify the dominant source of surface speed variability for your region.

Hint: See also Section 7.1 and Exercises 7.1 and 7.2.

3.3   Consider the Stommel (1948) gyre shown in Figure 3.12 and:

   (a) Write down the equation for the streamfunction Ψ, making clear where it comes from.

   (b) Hence derive the two scaling formulae for Ψ shown in the caption to Figure 3.12.

   (c) Explain the argument that the streamfunction solution (a) for constant f applies for both rotating and non-rotating cases.

   (d) Explain why scaling (b) for variable f is independent of the damping rate.

   (e) **Optional:** Compute and graph the surface height field for these cases: (i) non-rotating (equatorial f-plane; $f = 0$), (ii) constant rotation (mid-latitude f-plane; f is a nonzero constant), (iii) linearly increasing rotation (mid-latitude β-plane; $f = f_0 + \beta y$).

3.4   Assuming that the Sverdrup solution, $\Psi_{Sv}(x, y)$ is already known, attempt to solve the Stommel problem for (i) a western boundary layer and (ii) an eastern boundary layer. What is the problem with the eastern boundary layer solution?

3.5   Purely geostrophic plus Ekman flow is governed by the Sverdrup balance, with no western boundary current. Does that mean the flow in the western boundary current is **not** geostrophic? To answer this question, look at the full depth-integrated x and y momentum equations for the Gulf Stream, pretending that we can ignore nonlinear terms and bottom friction. What is the relative strength of the Coriolis and side-friction terms in each momentum equation? What does that tell you about the accuracy of geostrophy? Assume that we are looking at a location about 30°N, that the Gulf Stream is about 100 km wide, and that the Sverdrup transport at that latitude is around 30 Sv.

3.6   Consider a homogeneous flat-bottomed ocean of depth H on a mid-latitude f-plane.

   (a) Write equations satisfied by the horizontal velocity vector, assuming that the sea floor has a no-slip Ekman layer. The velocity vector should be a function of position (x, y, z) and time t. Make simplifying assumptions, that you clearly state.

   (b) Find the solution to the linear, steady, stress-driven flow in the Ekman layer in terms of the interior geostrophic flow.

   (c) Write an expression for the bottom stress applied by the sea-floor on the water. Estimate a typical magnitude for the bottom stress and compare to the typical wind stress applied at the ocean surface.

   (d) Write an expression for the vertical velocity between the Ekman layer and the interior. Estimate a typical magnitude of the vertical velocity.

   (e) Write an equation for the evolution of the vertical component of the fluid vorticity in the interior of the flow, making necessary assumptions that are clearly stated.

   (f) Hence show that vorticity anomalies are damped by bottom friction over a timescale proportional to $1/(\delta_{Ek}f)$, where δ_{Ek} is the non-dimensional thickness of the Ekman layer. Compare and contrast this timescale with the timescale of viscous spin-down of vorticity anomalies for typical oceanic values. (See also Exercise 8.4.)

Hints: You may assume the flow is Boussinesq, hydrostatic, and incompressible. Review Subsection 2.2.4. Read Vallis (2006) Section 2.12, and especially 2.12.3, for help on how to solve the z-dependent Ekman theory.

3.7 Consider an idealized midlatitude basin with walls at $x = 0$, $x = L$ and $y = \pm M/2$. Wind stress consists of a zonal component only, which is given by

$$X = A \sin \pi y / M, \tag{3.73}$$

where L, M, and A are constants, and L and M are of the same order.

(a) Assuming the planetary vorticity gradient, β, is uniform over the basin, what is the Sverdrup depth-integrated meridional velocity V_S everywhere? Include the total velocity over the entire water column.

(b) Assume the west has a boundary layer of width δL (the product of δ and L), with nondimensional parameter $\delta \ll 1$, in which the depth-integrated meridional current, V, is nearly independent of y. Compute $V(x)$ in this boundary layer, making suitable assumptions that you clearly state.

(c) Calculate an expression that approximates the streamfunction $\Psi(x, y)$ of the depth-integrated flow over the entire basin, including the western boundary layer. Make a contour graph of (suitably nondimensionalized) $\Psi(x, y)$ for the case $\delta = 0.05$.

(d) Compute the depth-integrated zonal velocity, U, everywhere in the basin.

3.8 (a) Write down the equation for the geostrophic streamfunction, Ψ, for a flat-bottomed ocean on a midlatitude β-plane driven by surface wind stress (X, Y) (normalized by density, so the units are $m^2 s^{-2}$) and dissipated by bottom friction with (weak) viscosity ν. Explain the physical meaning of each term in your equation.

(b) An ocean lies between east–west boundaries at $y = 0$ and $y = b$ and is subject to a purely zonal surface stress, of form $-F \cos(\pi y / b)$, where F is a positive constant. Calculate the steady geostrophic velocity field that results, in the absence of friction and inertia, if the ocean has an *eastern* boundary at $x = b$. Sketch the streamlines.

(c) Using your solution from (b) write an expression for the correction to Ψ due to bottom friction. Estimate the fractional size of this correction for typical oceanic parameters.

Hint: You may find Exercise 3.6 useful here.

(d) Write an expression for the Ekman transport in the surface Ekman layer, and hence find an expression for the angle between the geostrophic transport and the total transport averaged over the surface Ekman layer. Estimate this angle for typical oceanic parameters.

(e) Write an expression connecting the maximum height of the sea surface to the wind stress. If the maximum height is 1 m, estimate F for the North Atlantic ocean. Is this value reasonable?

(f) **Optional:** Now consider *frictional flow near a western boundary* at $x = 0$, subject to the same wind stress as before. Seek forced solutions of the form $\Psi_B = $

$A(x)B(y)$, with suitable boundary conditions at $x = 0, x \to \infty, y = 0, y = b$. Show that $A(x)$ satisfies

$$A'' + \frac{\beta}{r}A' - \frac{\pi^2}{b^2}A = -\frac{\pi F}{rb}. \tag{3.74}$$

Hence find Ψ_B, and sketch the streamlines.

Show that as $r \to 0$ the northward flow becomes confined to a narrow layer close to the western boundary (assume the solution above still applies). Estimate the width of this layer when $r^{-1} = 10$ days.

Hint: This optional question uses the method of separation of variables (see, for example, Zwillinger, 1989 or Haberman, 1987).

3.9 (a) Referring to the Jacobian form of the inertial term (3.55), demonstrate that steady purely inertial flow has streamlines parallel to potential vorticity contours, as asserted.

(b) Consider Figures 3.18 and 3.20. Are these solutions purely inertial? Explain your answer.

(c) Compute and plot the potential vorticity for the Fofonoff gyre shown in Figure 3.19. Plot the $Q(\Psi)$ relation in this case.

(d) Explain why the circulation strength in the Fofonoff gyre is arbitrary (for this reason, it is sometimes called a mode of free flow). Suggest ways to resolve this indeterminacy using ideas from Chapter 3 (or otherwise; Marshall, 1986 is helpful too).

(e) **Optional**: Write a computer code to solve the steady inertial circulation problem (3.57) for arbitrary $Q(\Psi)$ functions. Hence, make plots of steady inertial gyres for several cases and describe the space of possible solutions. Consider positive definite potential vorticity only.

4

Surface and Mixed Layer Properties

The surface of the ocean has the most direct interaction with the atmosphere, the greatest variability of properties in time and space, the greatest concentration of biological activity, and arguably the greatest influence on human activity. For these reasons its properties interest scientists and engineers in many disciplines. As we discuss in Chapters 5 and 8, surface properties also play a role in determining the subsurface structure of the ocean and the large scale circulation.

A key organizing principle for studying the upper ocean is the concept of the mixed layer. Although the entire ocean interacts with the ocean's surface properties, the mixed layer has a particularly intimate relationship to the surface. Therefore it is useful to discuss surface and mixed layer properties together.

4.1 Observations

4.1.1 Distribution of Surface Properties in Space and Time

We start with temperature, perhaps the most significant physical property of water. The distribution of **sea-surface temperature** (SST) (Subsection 1.1.2) directly influences the atmosphere's temperature and indirectly its circulation, cloudiness, and precipitation. Temperature is a major component of seawater density and thus helps to determine ocean dynamics. It is of major chemical and biological importance as well.

Surface temperature means slightly different things in different contexts. Water temperature measured in the top meter of the ocean is sometimes referred to as **bulk** (or bucket) temperature. **Skin temperature** represents the top 0.01 mm or so of the ocean. This razor-thin layer (actually much thinner than a razor) is the source of upward infrared radiation from the ocean surface, and is best inferred from infrared measurements from satellites. The skin temperature is often 0.1–0.3°C cooler than the bulk surface temperature. Precise definitions of surface properties should be made with reference to the actual sea surface, which surface gravity waves undulate on scales of seconds and meters. As wind speed gets relatively high, wave breaking produces increasing amounts of foam so that the sea "surface" itself becomes ill defined. In what follows, we ignore these complications.

2006–2010 Average Sea Surface Temperature

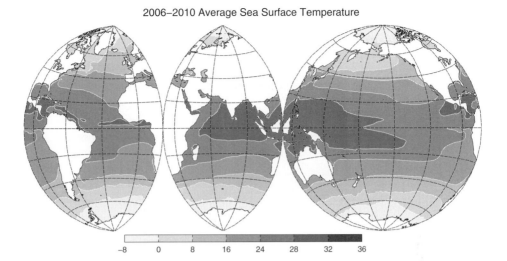

-8   0   8   16   24   28   32   36

Figure 4.1 Shading and thick contours represent global 2006–2010 average sea surface temperature (SST) climatology from the OISST satellite dataset (see Appendix A, Reynolds et al., 2007), 1/4° horizontal resolution. Contour intervals below and above 24°C are 8°C and 4°C, respectively.

Annual average global bulk SST ranges from over 28°C in the tropics to below 0°C (Figure 4.1). High latitude temperatures go no lower than $-1.9°C$ the freezing point of sea water at 1 atmosphere of pressure (see Figure 10.5). Generally speaking, SST is warmest in the tropics and coldest at high latitudes, and the SST contours are zonal, but there are other interesting features that are not as simple.

Parts of the equator, particularly in the eastern third of the Pacific and to a lesser extent in the eastern Atlantic, are colder than adjacent higher latitudes. In the Pacific, this is often referred to as the **cold tongue**, in contrast to the **Warm Pool** of high temperatures in the west. Much of the eastern edge of the ocean is colder than broader areas to the west. Notable examples of this include the regions off the west coast of South America, the California coast, and the southwest coast of Africa. At higher latitudes, particularly off the coast of Argentina, the Pacific coast of Siberia, and the east coast of Canada, it is the *western* part of the basin that is particularly cold.

Smaller-scale tongues of warm water extend poleward along the western boundaries of the North and South Atlantic and Pacific in the vicinity of 30° latitude (Figure 4.2). These are located where western boundary currents flow through strong meridional temperature gradients (see Figure 3.5 in Subsection 3.1.3). At the tip of South Africa (Figure 4.2d), the retroflection of the Agulhas back into the Indian Ocean (see Subsection 3.2.5) causes a substantial zonal surface temperature gradient between the warm water in the Indian Ocean and the colder South Atlantic water.

The **thermal equator** – the latitude which has the warmest zonal average SST – is slightly north of the geographic equator. The Southern Ocean is generally colder than similar latitudes in the northern hemisphere. The subpolar North Pacific is colder than similar latitudes in the North Atlantic. The tropical Indian Ocean is warmer than the tropical Pacific and Atlantic.

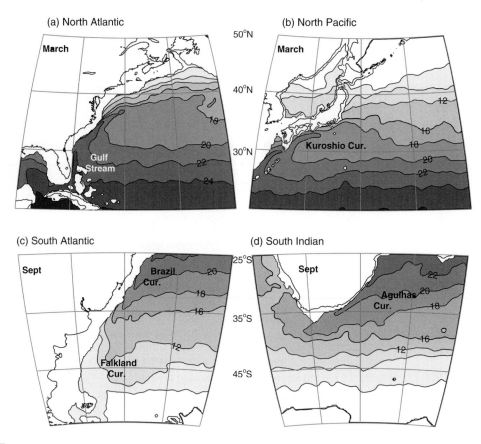

Figure 4.2 Global 2006–2010 average sea surface temperature (SST) climatology from the OISST satellite dataset (Appendix A) for March (top panels) and September (bottom panel). Contour intervals below and above 16°C are 4°C and 2°C, respectively. See also Figure 3.5.

Due to the geometry of the Earth's orbit and spin, the annual minimum in solar radiation occurs in December in the northern hemisphere and in June in the southern hemisphere. Typically the seasonal SST signal lags the insolation, with minima in March north of 10°N and in September south of 10°S. The maximum temperature in each hemisphere occurs about a month or less before the minimum in the opposing hemisphere. Equatorward of latitude 10°, the seasonal signal is small and can have a different phase than the rest of the ocean.

The SST differences between September and March are up to about 12°C, or greater than a third of the spatial variation in the annual average (Figure 4.3). The annual cycle reaches its largest values in midlatitudes. As with the annual mean, there are interesting geographical variations. In the tropics, the region off the west coast of South America has a relatively large annual cycle. The cold tongue feature is most prominent around September and nearly disappears around March. The southern hemisphere has a smaller annual cycle than the northern hemisphere. In the northern hemisphere the cycle is largest

SST, September Minus March, 2006–2010 Average

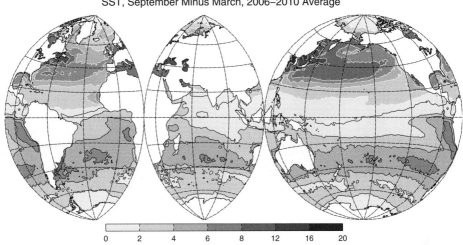

0   2   4   6   8   12   16   20

Figure 4.3 September SST minus March SST, 2006–2010 average, from same dataset as in Figure 4.1, smoothed with averaging using 5×5 array of neighboring grid points. Contour interval $= 2°C$. Shading represents absolute value of temperature difference. Contours are black for negative or zero values, and white for positive values.

near the western edges of the basin. At southern midlatitudes this western enhancement of the seasonal cycle occurs in the Atlantic but not the other basins.

The climatologies discussed here are processed in a way that simplifies the temporal variability. Surface temperature has a diurnal cycle, not shown here, with an amplitude that is typically of order $0.1°C$ but often reaches $1°C$ or higher in the tropics and during local summer in midlatitudes (Kawai and Wada, 2007). Irregular variations occur on a wide range of time and space scales. Some key timescales are associated with synoptic weather systems, like storms (timescales of a few days), mesoscale eddies (weeks, see Chapter 7), and El Niño (years). (For an introduction to atmospheric dynamics, consult Andrews, 2010; Holton and Hakim, 2013, for example.)

Sea surface salinity (SSS) (Subsection 1.1.2), along with temperature, determines the density of seawater (Subsection 1.2.3), so it is another important physical property to map. It also serves as a useful tracer of the subsurface flow: the distribution of salinity maxima and minima give clues to the circulation. The dominant pattern of surface salinity is one of maxima in the subtropics of both hemispheres and fresher water poleward of these and along the equator (Figure 4.4). These features broadly correspond to the regimes of atmospheric evaporation (E) and precipitation (P). In the subtropics, evaporation dominates E–P, while precipitation dominates elsewhere. The low-latitude fresh region is centered at the thermal equator rather than the geographical equator. In the atmosphere above, this region is called the **Intertropical Convergence Zone**, where the moist surface Trade Winds converge and upwell. Large-scale variations also show a trend from a relatively salty North Atlantic to a somewhat fresher South Atlantic, then to an even fresher South Indian and South Pacific, and finally to the freshest water in the North Pacific. Some of the semi-enclosed small seas (Mediterranean, Red Sea, Persian Gulf) are especially salty. There are

1955–2012 Average Sea Surface Salinity

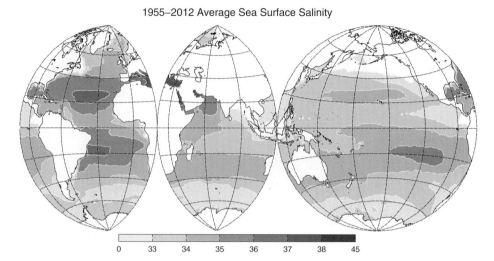

| 0 | 33 | 34 | 35 | 36 | 37 | 38 | 45 |

Figure 4.4 Like Figure 4.1, for global annual average sea surface salinity. Contour interval is 1 g/kg between salinities of 33 and 38. Data is from World Ocean Atlas 2013 (see Appendix A).

SSS, September Minus March, 1955–2012 Average

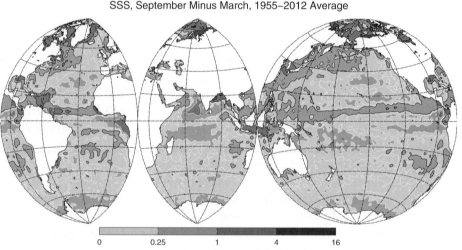

| 0 | 0.25 | 1 | 4 | 16 |

Figure 4.5 September minus March sea surface salinity, from same dataset as in Figure 4.4, with contours at 0, 0.25, 1, 4, 16°C. Contours are black for negative or zero values, and white for positive values.

also some especially fresh regions such as the seas around Indochina and eastern Canada, and the mouth of the Amazon River. Sea surface salinity has a less prominent annual cycle (Figure 4.5) than SST.

We calculate the surface density from temperature and salinity using the equation of state for seawater (Subsection 1.2.3). In fact, the subsurface density is most strongly influenced by the end-of-winter surface values: March in the northern hemisphere and September in

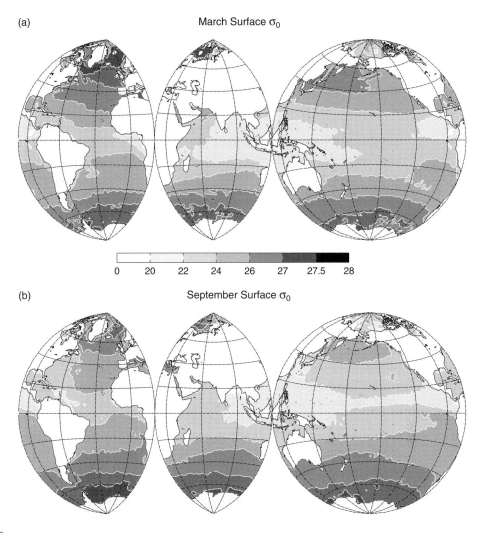

Figure 4.6 Like Figure 4.1, for global sea surface σ_0, (a) March average and (b) September average.

the southern hemisphere (see Subsection 5.2.4). Therefore we will examine those months rather than the annual mean (Figure 4.6).

Surface density is dominated by temperature variations, so to first order the surface is lightest in warm tropical regions and densest in cold polar regions. The major exception to this rule are the Mediterranean, Arabian Sea, and Persian Gulf, where high salinity makes winter density higher than most of the ocean's surface. Excluding these regions (where σ_0 can approach 29 kgm^{-3}), and some small regions where fresh river outflows can make surface water unusually light, the surface density of the ocean ranges from $\sigma_0 = 20-28$ kgm^{-3}.

Many detailed features of SST are also reflected in the density field. These include a density minimum at the thermal equator, lower densities at mid and high latitudes during local summer, and higher densities in the eastern cold regions and in September in the cold tongue (Figure 4.6b).

The influence of salt on the global distribution of surface density is generally more subtle than that of temperature. Perhaps the most important effect is that the northern North Atlantic is denser than the northern North Pacific. As we will see in Chapter 8, this relatively small inter-basin density difference has a large influence on the circulation. Excluding the warm, salty, semi-enclosed basins, the densest local-winter surface water in both hemispheres occur at around 0° longitude. In the north, the density maximum is in the vicinity of the Nordic Seas, north of Iceland. In the south, it is in the Weddell Sea to the east of the Antarctic Peninsula, though there is a smaller region of coastally trapped high-density water in the Ross Sea, which is in the Pacific sector of the Southern Ocean. In the Arctic Ocean proper (visible in the Pacific sector of Figure 4.6), very low salinities associated with annual ice melt (Figure 4.4) makes the surface water rather light even though it is very cold (Figure 4.1). Subsection 10.2.1 discusses this issue in detail.

At high latitudes, large regions of the ocean are covered by **sea ice** (frozen sea water, see Subsection 10.2.1) during at least part of the year. In the northern hemisphere, winter ice covers the Arctic, parts of the Nordic Seas between Greenland and Norway, and Davis Strait, Hudson Bay and parts of the Labrador Sea (Figure 4.7). Summer (hence perennial) ice is restricted largely to the Arctic Ocean. In the southern hemisphere, perennial ice encircles Antarctica, greatly expanding during austral winter (Figure 4.7). Northern hemisphere summertime ice has undergone a dramatic decrease in the last few decades. The summer minimum **sea ice extent** (the area in which at least 15% of the ocean's local surface is ice-covered) had a 1979–2012 average of about 7×10^6 km^2 and a 1979–2012 linear trend estimated at $-6.1\% \pm 0.8$ per decade (Vaughan et al., 2013). In September 2012 the sea ice extent dropped to 3.4×10^6 km^2 (Figure 4.7), a record minimum at the time of writing. This decrease is thought to be caused by anthropogenic global warming (Notz and Marotzke, 2012), which may result in the Arctic being largely ice free in the summers of the near future. Antarctic sea ice area does not show a statistically significant trend in recent decades.

4.1.2 Mixed Layer Structure

In general, ocean properties such as temperature, salinity, and density vary with depth. However, near the surface over much of the ocean, the vertical variation of these properties is much smaller than at other depths. The reason for this vertical uniformity is turbulent mixing, which is orders of magnitude stronger near the surface than it is over most of the water column. For this reason, the surface layer is often referred to as the **mixed layer**.

Vertical profiles of density clearly show the reduced stratification near the surface (Figure 4.8a). We have already seen in the previous subsection that surface density varies with season and latitude. Here we see that density variability decreases with depth. It is difficult to define the mixed layer or the location of the bottom boundary exactly, as will be discussed below, but in the profiles in Figure 4.8a for which the surface is at least 1 kg/m^3 lighter than at 200 m, there is a sharp contrast between the weak stratification in the top 30–40 m and the stronger stratification below. A simple criterion for the bottom of the mixed layer is the depth z at which $\sigma_\theta(z) - \sigma_\theta(0)$ reaches a certain value, $\Delta\sigma$. The appropriate value for $\Delta\sigma$ can be debated, but for a given choice of $\Delta\sigma$ we can see that mixed layer

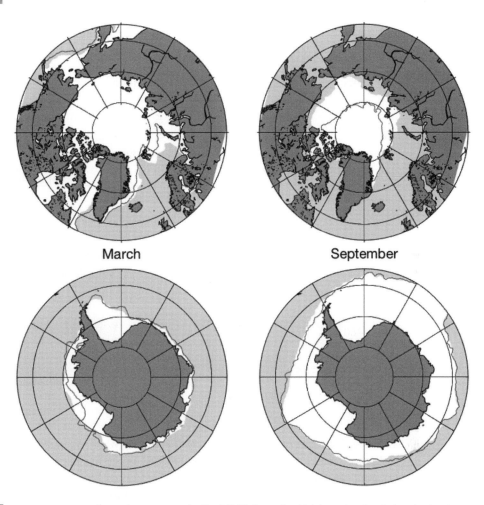

March   September

Figure 4.7   Average 1978–2015 observed sea ice extent for March (left), September (right), northern hemisphere (top) and southern hemisphere (bottom). The gray line shows the sea ice extent for 2012. Data are from the National Snow and Ice Data Center (see Appendix A).

thickness (commonly referred to as **mixed layer depth**) varies with latitude and month (Figure 4.8b). Generally speaking the mixed layer is deeper when the surface is denser, for instance at high latitudes or during the local winter.

During the course of a year, the mixed layer typically undergoes a cycle of deepening when the surface water grows colder and denser, and shallowing when the surface water grows warmer and less dense (Figure 4.8c,d). The emergence of shallow stratification during the spring and summer months is known as a **seasonal pycnocline**. The cycle is more pronounced at higher latitudes (Figure 4.8d) than at lower latitudes (Figure 4.8c). The time-depth plots show that in the mixed layer the annual cycle is much more pronounced than below. Strong vertical mixing within the mixed layer makes density and other properties respond quickly (compared to deeper water) to surface forcing. The difference is most dramatically seen at mid and high latitudes (Figure 4.8d). During spring, the mixed layer

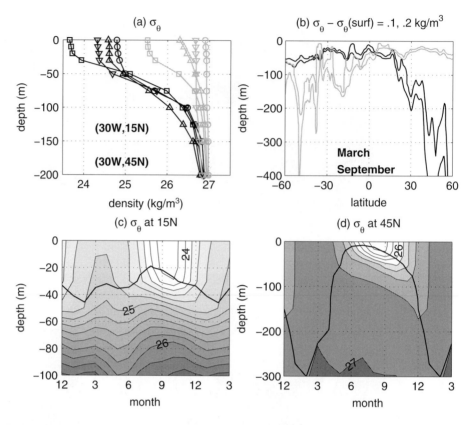

Figure 4.8   Mixed layer climatological distribution of σ_θ in the North Atlantic (30°W). (a) Function of depth at 15°N (black) and 45°N (gray) for March (circles), June (upward-pointing triangles), September (squares), and December (downward-pointing triangles). (b) Depth at which σ_θ is 0.1 and 0.2 kg/m³ denser than the surface value as a function latitude for March (black) and September (gray). σ_θ as a function of month and depth (contour interval = 0.2 kg/m³) and mixed layer depth defined by $\Delta\sigma_\theta = 0.1$ kg/m³ (thick line) at (c) 15°N and (d) 45°N. For (c) and (d), note different scales and gray shading scales.

shallows rapidly as the top of the mixed layer becomes warm and light. Throughout the summer (June through September), while the mixed layer depth undergoes relatively small deepening, the lightening of the water is stronger in the mixed layer than below (about 0.8 kg/m³ at the surface, 0.5 kg/m³ at 50 m depth, and 0.2 kg/m³ at 100 m). Starting in September, the mixed layer water cools and grows more dense, but the water below does not become significantly denser (below 50 m it even continues to get lighter) until the mixed layer depth dramatically grows in autumn and winter, making deeper water part of the mixed layer again.

In addition to the annual cycle, there can be a diurnal cycle corresponding to the diurnal SST cycle, with a **diurnal pycnocline** (and associated **diurnal thermocline**) developing in the top few meters of the ocean during daylight hours (Kawai and Wada, 2007). Day-time surface warming can be relatively strong and shallow for high insolation and low wind. As the wind increases, the temperature signal is mixed downward so that the diurnal

(a) March Mixed Layer Depth

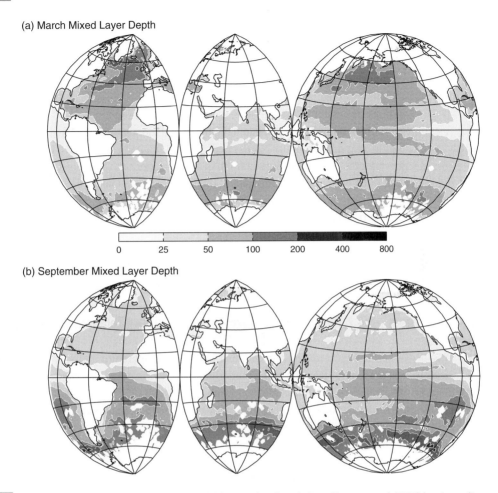

0   25   50   100   200   400   800

(b) September Mixed Layer Depth

Figure 4.9   Mixed layer depth climatology for (a) March and (b) September, from de Boyer Montegut et al. (2004) (see Appendix A), with contour values jumping by factors of 2 from 25 m to 800 m. Isolated grid points with no data have been replaced by the average of neighboring grid points.

pycnocline becomes broader but the density difference becomes smaller. As we discuss in Subsection 4.2.2, the mixed layer also tends to deepen when wind strength increases, and so variations in wind speed on all timescales can affect the mixed layer depth.

The mixed layer depth is an important parameter for determining how the ocean and atmosphere interact, so it is useful to know its geographical variation. There are a number of subtle issues which can make significant quantitative differences in the resulting clima-tology. To name a few, we can define the mixed layer based on a temperature or density criterion, or a combination. We can measure the mixed layer depth from a density (or tem-perature) difference from the surface or from the vertical gradient. Finally, climatological averages can produce subtle changes in the vertical profile of density or temperature, so that calculating the average of mixed layer depths from individual CTD profiles may be different than taking the mixed layer depth based on an average of all the profiles.

A climatology of mixed layer depth shows the local-winter deepening (and local-summer shallowing) in midlatitudes (Figure 4.9). Typical values are 20 to 50 m in summer and 60 to 200 m in midlatitude winter. As with other surface properties, there are more complicated geographical features as well. There is some sign of a mixed layer depth maximum in the tropics, as well as minima in the cold tongue and cold eastern boundary regions discussed in the previous section. The Southern Ocean has a band of high mixed layer depth during austral winter (Figure 4.9b). The wintertime mixed layer is deeper in the North Atlantic than in the North Pacific. More detailed studies have revealed winter mixed layer depths of around 1000–2000 m in the Labrador Sea (between Greenland and Canada) and to the bottom (approximately 2000 m) in small regions of the central Nordic Seas and off the Mediterranean coast of France. The Weddell Sea is also thought to sporadically have very deep winter mixed layers.

4.2 Concepts

4.2.1 How Surface Fluxes Influence Surface Temperature and Salinity

When trying to understand the distribution of temperature and salinity in the ocean, it is often useful to take the surface distribution of these quantities as given. It is legitimate to ask, however, what determines SST and SSS.

The evolution of temperature in any region in the ocean is related to how much thermal energy leaves and enters the region. The **heat transport** (Subsection 6.3.3) is the net transfer of thermal energy across a given imaginary surface. The **heat flux** is the heat transport per area. The **surface heat flux** Q is the heat flux into the ocean from the atmosphere. The surface heat flux Q includes electromagnetic radiation, latent heat release from evaporation, and sensible heat transfer between the atmosphere and ocean. As we will demonstrate in Subsection 4.3.1, the surface heat flux can be approximated as a **restoring boundary condition**

$$Q = \lambda(T_* - T) \tag{4.1}$$

(Haney, 1971), where T is SST and T_* is generally referred to as the **target temperature** or **restoring temperature**. The **restoring constant** λ has units of W m^{-2}K^{-1} (see Subsection 8.2.1). The target temperature and restoring constant are based on local atmospheric conditions such as temperature, humidity, and solar radiation near the sea surface and cloudiness. Note that T_* is *not* the atmospheric temperature (see Subsection 4.3.1).

The surface heat flux represents a tendency for the ocean temperature to change over time. There can also be heat fluxes within the ocean due to warmer or colder water flowing into a particular region. If there is no such ocean heat flux, then the water will be warmed if colder than T_* and cooled if warmer than T_*. In other words, the surface water will be restored toward the target value of T_*. The target temperature may vary in time due to weather or a seasonal cycle. In that case, T will lag behind T_*, because it takes a finite time for the mass of water to reach the target temperature. The smaller λ and the thicker the layer affected by Q, the longer it will take to approach a given T_*. Thus we can interpret T_* as the SST the ocean would reach for given atmospheric conditions and with negligible ocean heat flux and ocean depth. This is sometimes referred to as the SST for a **swamp ocean**.

The relevant depth to consider under conditions of heating and cooling is somewhat ambiguous. The mixed layer depth is often used, as temperature is vertically homogenized throughout the mixed layer. This is often a good assumption for seasonal and shorter timescales. For longer timescales such as decadal variability or global warming, weaker mixing or ocean currents may make a larger depth range relevant.

The restoring "constant" actually varies with location and time, but a typical value is $\lambda = 50$ W m^{-2} K^{-1} (Rahmstorf and Willebrand, 1995). In Subsection 4.3.1 we calculate the e-folding time t_Q (time for a reduction by a factor of $1/e$, where Euler's constant is $e = 2.718...$) for a departure from the target temperature. For a mixed layer depth of 30 m, the t_Q is about one month.

As a first approximation, we can estimate the ocean's temperature without knowing anything about the ocean. We can use atmospheric conditions based on observations or a theoretical or numerical model. This does not mean that the ocean is irrelevant to its own SST (see Subsection 4.2.2). For places and timescales where SST is close to the target temperature, Q will be small. In other regions, or looking at short timescales, mixed layer characteristics are crucial in determining seasonal SST evolution and Q will be large. Warm and cold currents can also make Q large and alter SST. In Subsection 4.1.1 we remarked on a number of interesting SST features (Figure 4.1) such as western tongues of warm water in the subtropics and eastern tongues of cold water at the equator, in the tropical Pacific, and in the subtropical Atlantic. The warm regions are due to strong western boundary currents carrying warm water to higher latitudes (Chapter 3). The cold regions are due to the coastal upwelling systems, which bring cold water up from below (Chapter 6). Ocean circulation is thought to have a role in other, more subtle features of the temperature field, such as the North Atlantic being warmer than the North Pacific (Chapter 8). In general SST has small but significant departures from T_* over large regions of the ocean.

Numerical ocean models can be run with a restoring surface heat flux to represent the effect of the sun and the atmosphere. Sometimes observed SST is used for T_*. This practically guarantees that the model SST will be close to observations, but it also guarantees that there will be some discrepancy (typically a few °C) in regions with significant ocean heat transport. The model can be forced to have a smaller discrepancy from observations by making λ artificially large. On the other hand, this can be counterproductive if we are interested in how SST responds to oceanic behavior, because we are nearly specifying SST.

The target temperature and λ depend on atmospheric conditions, but remember that these conditions themselves are influenced by the ocean. In fact, one motive for studying physical oceanography is to learn how ocean dynamics influence atmospheric temperature through SST. Regions in which large fluxes of heat are entering or leaving the ocean surface are by definition regions where large fluxes are leaving or entering the atmosphere. The direct effect of these fluxes is to change the atmospheric surface temperature as well as T_*. This effect of the ocean on the atmosphere may be parameterized by setting λ to 2 or 3 rather than 50 W m^{-2}K^{-1}, where T_* is now defined as the swamp ocean temperature not for actual atmospheric conditions but for the atmospheric conditions we would expect if there really were a swamp instead of a deep and circulating ocean (Rahmstorf and Willebrand, 1995). Such a weak value of λ reveals a much greater SST sensitivity to ocean dynamics than the classical value. Ocean dynamics may also affect Q in other ways not as easily parameterized. For instance, a spatial pattern of SST anomaly (relative to T_*) may produce a change in atmospheric circulation and hence in atmospheric properties such as temperature and cloudiness that affect T_*.

While temperature is influenced by the transfer of heat across the ocean surface, salinity is *not* influenced by the transfer of salt across the surface, as this transfer is negligible. Instead, it is altered by the surface **freshwater flux** F, which can be written as a mass flux (kg m^{-2} s^{-1}) or a volume flux (m^{3}m^{-2} s^{-1} = m/s). Evaporation increases salinity by removing freshwater and precipitation decreases salinity by adding freshwater. We can interpret the volume flux version of F to be the speed that the surface of water in a bucket would rise or fall based on evaporation and precipitation. Because these speeds are so slow, F is sometimes measured in m/yr. Typical ocean values are O(1 m/yr), as discussed in Subsection 11.1.4 (Figure 11.6).

Surface transport of freshwater affects both temperature and salinity, but through different mechanisms. When water evaporates, heat is removed from the surface because of the phase change of water from liquid to gas (see (4.7)). This should not be confused with the relatively small heat transport due to the fact that evaporating or precipitating water may be a different temperature than the mixed layer. F also affects volume transport within the ocean, because water that leaves the surface in one location generally returns to the ocean somewhere else. The circuit of water transfer through the atmosphere (as water vapor) and over land (as runoff) must be closed by flow in the ocean. However, volume transport associated with evaporation and precipitation is small (O(1 Sv) between net evaporation regions and net precipitation regions), and is sometimes ignored in numerical or theoretical models. In that case it is convenient in models to define a **virtual salt flux**, which is the flux of salt at the surface that would make the same changes in salinity as a given F. Subsection 4.3.1 shows the formula for this flux.

There is no physical justification for expressing the surface freshwater flux in a restoring law. There is no target salinity intrinsic to atmospheric conditions to which the SSS will tend. The amount of evaporation depends on local ocean and atmospheric conditions but the precipitation depends on the general circulation of the atmosphere. As will be discussed in Chapter 11, this can have important consequences for the number of possible ocean circulation states which are possible for given atmospheric conditions. The SSS does not directly affect the atmosphere, but it can indirectly affect the atmosphere by influencing ocean circulation which in turn can change SST or gas exchange.

Despite the lack of a physical justification, a restoring law for SSS is often used in numerical ocean models. In practice, this is simply a way to force SSS to be close to a desired distribution, such as observations or a theoretical profile. Such a numerical experiment cannot tell us about how SSS is set by ocean–atmosphere interaction, but it does allow us to investigate the consequence of a given SSS for the circulation and subsurface properties.

4.2.2 Mixed Layer Response to Atmospheric Forcing

Now think about what sets the mixed layer properties and how it responds to atmospheric forcing. The task is to explain the seasonal cycle and its regional variation shown in Figures 4.8 and 4.9. To do so, we consider how the upper ocean interacts with the atmosphere, how it stores and releases heat and freshwater, and how it is mixed by the wind and buoyant convection. The mixed layer can deepen by **entrainment** of the water below, meaning that the quiescent deeper water is moved into the turbulent mixed layer. Alternatively, the mixed layer can shallow by **detrainment**, which moves water in the opposite direction. The focus here is on a one-dimensional (vertical) **mixed layer model** with seasonal (but not diurnal) variations (see Figure 4.10). The horizontal structure in the mixed layer is also addressed (and in Subsection 5.2.4). The details of the mixed layer model are in Subsection 4.3.3; now we think about its conceptual basis and some key results.

Balance between Wind-Driven Deepening and Insolation-Driven Shoaling

The main idea in the one-dimensional mixed layer model is that a balance exists between competing processes to deepen and shoal the mixed layer. Most important, the wind adds mechanical energy to the mixed layer causing it to deepen and solar radiation (insolation) warms the mixed layer causing it to shoal. The balance between these two processes governs the mixed layer depth over vast areas of the ocean in summer. In particular, solar radiation stratifies the upper ocean by preferentially warming at the surface, which reduces potential energy. The wind opposes this tendency by mixing up the stratified water, raising potential energy (by conversion from turbulent kinetic energy).

To quantify this balance, let the rate of addition of thermal energy (mainly from the sun) be Q (in W m^{-2}). The rate of mechanical energy input from the wind is $\mathcal{G}$ (in W m^{-2}). The

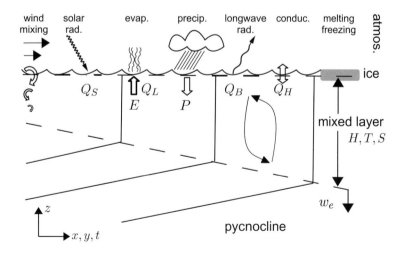

Figure 4.10   Schematic diagram of the surface mixed layer. The layer has thickness H, temperature T, and salinity S. It deepens by entraining pycnocline water at speed w_e. The layer exchanges heat (Q_S, Q_L, Q_B, Q_H) and freshwater (E, P) with the atmosphere via the processes shown, mediated by ice. Solid lines are isopycnals.

energy balance argument invokes thermal energy lowering gravitational potential energy. Potential energy depends on the mass distribution in the water column. Therefore we expect that gravitational acceleration g, specific heat capacity c_p, and thermal expansion coefficient α must feature. These parameters combine to yield a length-scale as $c_p/(\alpha g)$. Hence, we anticipate that the mixed layer depth should scale as:

$$H \sim \frac{c_p \mathcal{G}}{\alpha g Q}. \tag{4.2}$$

This formula shows how mixed layers that are strongly mixed relative to the stratifying tendency (large $\mathcal{G}/Q$) are deep, and vice versa. This formula for the mixed layer depth is (essentially) the **Monin–Obukhov depth**, which is defined precisely in (4.40). The complete argument in Subsection 4.3.3 uses an energy budget for the mixed layer along these lines (see also Subsection 1.3.3).

In this balance the mixed layer gains heat from the sun. Its temperature therefore increases, even though the depth is constant. The mixed layer model assumes that the mixed layer is homogeneous in its properties; namely, mixed layer depth H and temperature T (salinity S can be added, but it does not change the main idea). To account for the temperature change a mixed layer heat budget is constructed, and the rate of warming is

$$\frac{dT}{dt} = \frac{Q}{\rho c_p H}, \tag{4.3}$$

which shows that the mixed layer temperature simply changes as the heat flux divided by the mixed layer heat capacity, as expected (see (4.42)). For instance, if $Q = 100$ W m^{-2} and the mixed layer depth is 50 m, we have $dT/dt = 4.9 \times 10^{-7}$ °Cs^{-1}, or about 1.3°C per month.

Mixed Layer Seasonal Cycle and the Seasonal Thermocline

The rates of wind mixing and solar warming vary with a seasonal cycle, which affects the mixed layer (as seen in Figure 4.8). The main idea of a competition between wind mixing and solar warming, expressed as budgets of mechanical and thermal energy, is readily extended to cover this case. The increased solar warming in summer (and decreased wind mixing) decreases the Monin–Obukhov depth, so the mixed layer shallows and warms. As it does so, a stratified seasonal thermocline is left behind by detrainment of the mixed layer. The reverse process occurs in autumn and winter: the mixed layer deepens by entrainment of the stratified thermocline water below.

More precisely, Q changes sign (becomes negative) because the solar radiation warming the mixed layer is overwhelmed by the heat loss from the ocean, due to longwave radiation, evaporation, and conduction (the sensible heat flux). The ocean loses heat, which tends to deepen the mixed layer through **buoyant convection**. In winter, therefore, both wind mixing and net ocean heat loss act together to deepen the mixed layer. The rate of deepening depends on the strengths of these two effects relative to the temperature (buoyancy) difference between the mixed layer and the pycnocline water below. Again, the argument is quantified using energetics. With a large temperature difference relative to the supply of mechanical energy, the entrainment rate is low, and vice versa. Equation 4.39 defines the relation.

These unsteady effects are expressed in two rules for the mixed layer model. First, the mixed layer depth changes so as to approach the Monin–Obukhov depth, H_{MO} (see (4.40)). While Q is negative (net ocean cooling), the mixed layer continues to deepen (see Figure 4.13). Second, the mixed layer temperature evolves according to

$$\frac{dT}{dt} = \frac{Q}{\rho c_p H} \left(2 - \frac{H_{MO}}{H} \right). \tag{4.4}$$

This expression just generalizes (4.3) by adding a term that accounts for the mixed layer temperature tendency due to entrainment when the mixed layer depth is shallower than the Monin–Obukhov depth.

Figure 4.11 shows results from this simple one-dimensional mixed layer model for the tropical and subtropical eastern North Atlantic stations shown in Figure 4.8. The model is initialized with the observed hydrographic profiles in December and then uses estimates for the net air/sea heat flux Q and wind speed to give the mechanical energy injection rate, and hence H_{MO}. The basic pattern of mixed layer deepening and densification during fall and winter, then shallowing and density decrease in spring and summer is well described by the one-dimensional mixed layer model. The model also deposits a seasonal pycnocline as the mixed layer shallows in spring. The model predicts a summer mixed layer that is too shallow and light (warm) compared to the data, however, and the end-of-winter mixed layer depths are too large, in particular at the 15°N tropical station. Evidence for pycnocline processes also exists in the data Figure 4.8c,d, which are missing from the model in Figure 4.11b,c. For example, the vertical meanders of the isopycnals over time at 15°N are absent from the model in Figure 4.11b. (They are probably caused by variations in pycnocline circulation, or by Rossby waves, Subsection 1.2.1.) Also, the deepening isopycnals in the seasonal pycnocline at 45°N are absent from the model in Figure 4.11c. This deepening is probably caused by solar radiation penetrating below the mixed layer to warm the water in summer (this process is omitted from the simple model here, but see Exercise 4.4).

The mixed layer depth H tracks the Monin–Obukhov depth H_{MO} during the warming season (Figure 4.11b,c). As spring progresses into summer H_{MO} rapidly shallows, mainly because of increasing Q although the summer windspeed also weakens. The mixed layer depth shows the same behavior, with a delay of only a few weeks. In September, Q becomes negative (ocean cooling) and therefore the mixed layer depth increases. It lags behind H_{MO} because the mixed layer has to excavate the stratification of the seasonal pycnocline and then the permanent pycnocline. The mixed layer deepening continues until Q and H_{MO} change sign in March of the following year.

Exchange Between the Mixed Layer and Permanent Thermocline

With this understanding of the seasonal cycle in the one-dimensional mixed layer model, it is now straightforward to understand an important way that water exchanges between the mixed layer and the permanent thermocline. Consider Figure 4.11b,c again. They both show that the deepest mixed layer depth is greater in the second year of simulation than the first. Therefore, at the end-of-winter deepening phase in the second year, permanent

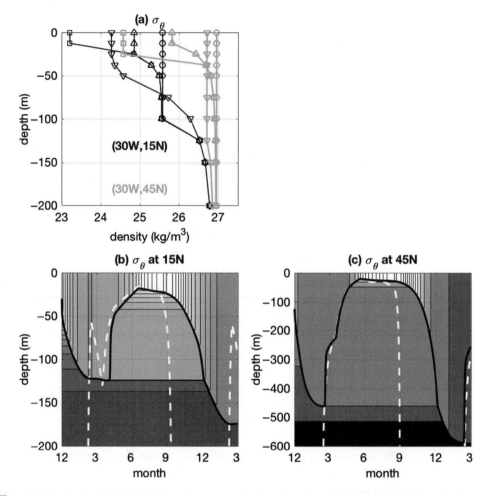

Figure 4.11 Results from the simple one-dimensional mixed layer model of Subsection 4.3.3 corresponding to the observations in Figure 4.8 (note the different scales). The dashed white lines in (b) and (c) are the Monin–Obukhov depths H_{MO}. The air/sea heat flux and evaporation rate are taken from the OAFlux product; the precipitation is taken from the GPCP product (see Appendix A).

pycnocline water is entrained into the mixed layer. This process is called **obduction**. Once the permanent pycnocline water is entrained into the mixed layer its properties (temperature, salinity, dissolved gas content) are rapidly reset by the intense mixing and air/sea exchange. The time taken for this property reset is much shorter than the time taken for properties to be changed by the weak interior mixing in the permanent pycnocline (see Subsection 7.1.2).

The obduction can occur another way too. Imagine that the horizontal axis in Figure 4.11b,c is time following a drifting parcel in the upper ocean. That means we view the time series as Lagrangian, rather than Eulerian (Subsection 1.3.1). We also ignore the fact that real currents vary in direction and strength with depth (Chapter 5). An upper ocean water column moving through the Gulf Stream and North Atlantic Current, for example, experiences deeper winter mixed layers in successive years. This fact is clearly seen in Figure 4.9a, which shows maximum (March) mixed layer depth increasing from the subtropical North Atlantic into the subpolar North Atlantic. Therefore, water in the permanent pycnocline below the deepest depth of mixing one year is entrained into the mixed layer in subsequent years. It thus obducts. This Lagrangian obduction process applies even if the deepest mixed layer depth is the same for all fixed (Eulerian) locations, which is mainly true (and assumed in Figure 4.9).

The reverse process, called **subduction**, is also easy to see. Subduction is the transfer of water from the mixed layer into the permanent pycnocline. It can occur whenever successive deepest mixing depths are shallower, either at a fixed station, or following a Lagrangian pathway (see, for example, Woods and Barkmann, 1986).

Hence, subduction and obduction occur whenever there is geographical (or interannual) variation in deepest H. This mode of exchange with the pycnocline is called **lateral induction**. Obduction (subduction) also occurs because of Ekman suction (pumping), as explained in Subsection 5.2.4 (see Figure 5.14). Technically, subduction and obduction are quantified as the rate (in meters per year) of equivalent vertical speed across the base of the winter mixed layer (see Figure 5.15). It is also common to talk of the ocean pycnocline being **ventilated** by obduction and subduction (see also Subsection 5.1.3). In this sense, the ocean interior is refreshed with oxygen-rich, and carbon-poor, mixed layer water that replaces oxygen-poor, and carbon-rich interior water. Körtzinger et al. (2004) show a particularly striking example of this ventilation in the Labrador Sea, in which oxygen measurements from Argo floats (Subsection 1.1.3) show the interior ocean taking a "deep breath." Note that ventilation in this sense is distinct from the ventilation of Subsection 5.2.2, albeit related.

Limits of the Mixed Layer Model

There are some limits of the one-dimensional mixed layer model, as follows. First, it assumes that the upper ocean consists of a vertically homogeneous mixed layer characterized by a single temperature T (and salinity S). The tacit assumption is that the timescale for turbulent mixing within the mixed layer is much shorter than the timescale for the mixed layer properties to change. This assumption is not strictly true and, in reality, the upper ocean is seen to be weakly stratified (and occasionally statically unstable), but it does not show zero stratification. Also, the presence of very weak surface stratification does not necessarily indicate active turbulent mixing. The region of active turbulence may not coincide with the layer of near homogeneous properties. In this way, the Ekman layer (Subsection 2.1.3 and Subsection 2.2.1) is distinct from the mixed layer. Moreover, the concept of the "mixed layer" obscures many interesting and important phenomena, such as surface waves, Langmuir circulation, and convection (as well as much of biological oceanography). Therefore, the "mixed layer" is an emergent concept (like the Ekman layer) that provides a convenient placeholder to refer to the complexity of the real upper ocean (see, for example, Thorpe, 2005 for more details).

Second, the model assumes that beneath the mixed layer lies a stably stratified pycnocline. The pycnocline interacts with the mixed layer via detrainment and entrainment. Within the pycnocline itself, no processes occur. Given that the main aim of this book is to identify and explain the global circulation, this is hardly a promising start! Nevertheless, neglecting all interior physics gives a useful reference point because it helps to distinguish between effects caused by mixed layer interaction and those that are not.

Third, the mixed layer model is limited because it has no realistic long-term steady state. If the annual-average air/sea heat flux is zero, then the mixed layer will deepen (and cool) year on year as the wind mixing erodes the pycnocline. If the annual-average mixed layer depth is constant, then the mixed layer must warm (see Figure 4.13). In this sense, the one-dimensional model is incomplete. There are two essential missing elements: The pycnocline below the mixed layer cannot be replenished in this model. In reality, water circulates below the mixed layer from remote locations. Chapters 5–6 explore this issue in detail. And there are mechanisms by which dense surface water can slide beneath lighter waters without mixing, in contradiction to the assumptions of the mixed layer model. Chapter 8 explores this issue in detail. Both these missing elements allow the pycnocline below the mixed layer to re-stratify, which therefore permits the upper ocean to achieve a long-term steady state.

4.3 Theory

4.3.1 Restoring Surface Temperature and Virtual Salt Flux

In order to analyze SST, it is convenient to consider a small region of water at the sea surface with thickness H, surface area A, and average temperature T. As will be shown in Chapter 11, the average temperature in this region can change due to the transport of thermal energy into and out of the region. If the heat transport into the region from the rest of the ocean is given by $\mathcal{H}$, then the rate of change of temperature in the region is given by

$$Hc_p\rho\frac{dT}{dt} = Q + \frac{\mathcal{H}}{A}, \tag{4.5}$$

where ρ is density and c_p is the specific heat capacity of sea water. c_p is roughly a constant, ranging from about 3996 J kg^{-1} K^{-1} at the surface to 3849 J kg^{-1} K^{-1} at 5000 m, with up to about ± 8 J kg^{-1} K^{-1} variations due to ocean temperature variations (Gill, 1982; IOC, SCOR, and IAPSO, 2010). Equation 4.5 follows from (1.8) applied to the heat $c_p\rho T$ (for water near the surface, we can neglect differences between in-situ temperature and potential temperature). There are no source terms and the integral of property flux over the bounding surface of the volume is given by AQ for the top surface and $\mathcal{H}$ for side and bottom surfaces. To be more precise, AQ and $\mathcal{H}$ are the negative of the surface integrals in (1.8), representing the flux convergence rather than divergence.

If the region we are examining is confined to the mixed layer (that is, H is the mixed layer depth or less), then T is SST. The surface heat flux into the ocean is given by

$$Q = Q_S - Q_L - Q_H - Q_B, \tag{4.6}$$

where Q_S is the net short wave radiation into the ocean (solar insolation that is not reflected by clouds, dust, or the ocean surface), Q_L is latent heat flux due to evaporation, Q_H is the sensible heat flux, and Q_B is the net long wave radiation. For Q_L, Q_H, and Q_B, (4.6) shows that positive values mean oceanic cooling Here we can take Q_S as a given function of time of day, time of year, latitude, and albedo. The other terms can be calculated based on local atmosphere ocean properties (Haney, 1971; see also Talley et al., 2011). Note that Q_L is directly related to evaporation at rate E, which also affects surface salinity (Subsection 4.3.2):

$$E = \frac{Q_L}{L\rho}, \tag{4.7}$$

for latent heat of vaporization L and density ρ. Subsection 11.1.3 shows the global distribution of heat flux components and the net heat flux, but here we explore their dependence on local variables in order to see how they act like a restoring boundary condition.

The long wave radiation is given by the **Stefan–Boltzmann law** of ideal black body radiation (available in many textbooks, for instance Peixoto and Oort, 1992; see also Subsection 11.1.1),

$$Q_B(T) = Q_* \sigma T^4, \tag{4.8}$$

where $\sigma = 5.67 \times 10^{-8}$ W m^{-2} K^{-4} is the **Stefan–Boltzmann constant** (derivable from other physical constants such as the speed of light, not to be confused with the density anomaly), and Q_* is a factor representing the reduction in upward radiation by water vapor and clouds. In (4.8), T is the absolute temperature in Kelvin. The remaining heat flux terms are based on the the turbulent exchange between the ocean and atmosphere of water vapor (for latent heat) and sensible heat, or

$$Q_L(T) = c_w UL \left[q_S(T) - q_A \right], \tag{4.9}$$

$$Q_H(T) = c_w U c_p \rho \left(T - T_A \right), \tag{4.10}$$

where T_A is the atmospheric temperature at a specified height z_A above the sea surface (usually 10 m), L is the latent heat of vaporization, q_A is specific humidity at z_A, q_S is saturation specific humidity, U is wind speed at z_A, and c_w is a measure of the turbulent mixing near the sea surface and depends on the wind speed and other parameters (see Fairall et al., 1996, 2003). Typical values of c_w are roughly 0.001 at $U < 10$ m/s. These equations, (4.9) and (4.10), are sometimes called the **bulk formulae** for air/sea exchange. They are a parameterization of the processes actually involved.

We can write $T = T_A - (T_A - T)$ and for the ocean $|T_A - T| \ll T_A$. Therefore, all the terms in (4.6) can be written as a function of $T_A - T$ and T_A, and we can linearize nonlinear terms such as the T^4 dependence in Q_B and the $q_S(T)$ term in Q_L. Then (4.6) can be approximated as

$$Q = Q_O + \lambda(T_A - T), \tag{4.11}$$

where

$$Q_O = Q_S - Q_B(T_A) - Q_L(T_A), \tag{4.12}$$

$$\lambda = 4Q_* \sigma T_A^3 + c_w U c_p \rho + \frac{\partial Q_L}{\partial T} \tag{4.13}$$

depend on top-of-atmosphere insolation, cloudiness, albedo, T_A, relative humidity near the surface, and near-surface atmospheric turbulence. We can further rewrite (4.11) as a **restoring equation** (mathematically, a Robin boundary condition),

$$Q = \lambda(T_* - T), \tag{4.14}$$

with **restoring temperature** or target temperature

$$T_* = T_A + Q_O/\lambda. \tag{4.15}$$

Notice that this framework allows a nonzero heat flux Q_O when the SST equals the air temperature, which is a desirable property.

Combining the restoring equation (4.14) and the heat balance equation (4.5), we can write the heat balance as

$$Hc_p\rho\frac{dT}{dt} = \lambda(T_* - T) + \frac{\mathcal{H}}{A}. \tag{4.16}$$

The solutions to this equation display the behavior discussed in Subsection 4.2.1. For constant T_* and $\mathcal{H} = 0$, (4.16) is solved by an exponential decay from the initial value of $T = T_I$ to a final value of T_*:

$$T = T_* + (T_I - T_*)e^{-t/\tau}, \tag{4.17}$$

where the decay time scale is given by

$$\tau = Hc_p\rho/\lambda. \tag{4.18}$$

Using typical values found at the ocean surface gives us $\lambda = 40$ W/m^2 (see Dickinson, 1981; Schopf, 1983), which for a layer that has $H = 30$ m gives $\tau = 4 \times 10^6$ s ≈ 45 days. Thus the mixed layer – in places where it is not very deep – responds to a change in T_* on a timescale of months. This idea is further developed in Subsection 4.3.3, where we also allow for changes in H.

In steady state with $\mathcal{H} \neq 0$, we have $T - T_* = Q/\lambda$. As Chapter 11 discusses, the regions of maximum annual-average heat flux have $Q \sim 100$ W/m^2, so for $\lambda = 40$ W m^{-2}K^{-1}, T differs from T_* by 2.5°C. Much of the ocean has Q, and hence $T - T_*$, several times smaller.

4.3.2 Virtual Salt Flux from Freshwater Flux

As discussed in Subsection 4.2.1, there is no restoring condition for sea surface salinity. Here we derive an expression for virtual salt flux. For a given surface volume of seawater containing mass per surface area m of salt and mass per surface area M of freshwater, the salinity is

$$S = \frac{m}{M + m}, \tag{4.19}$$

(see Subsection 1.2.3 and (1.7)). The surface freshwater flux changes M:

$$\frac{dM}{dt} = F. \tag{4.20}$$

The units of F here are kg m^{-2} s^{-1}, but sometimes we report F/ρ_w with units of ms^{-1} instead, for example in Figure 11.6, where $\rho_w = 1000$ kgm^{-3} is the density of fresh water. We want to know the effect of F on S, so differentiate (4.19):

$$\frac{dS}{dt} = -\frac{m}{(M + m)^2}\frac{dM}{dt} = -\frac{S}{M + m}\frac{dM}{dt}. \tag{4.21}$$

If the volume has thickness H and density ρ, $M + m = \rho H$, and we can rewrite (4.21) as

$$\rho H \frac{dS}{dt} = -SF \equiv \mathcal{S} \tag{4.22}$$

where $\mathcal{S}$ is the virtual salt flux. Over most of the ocean, variations in S are relatively small (10% or less; Figures 4.4, 4.5), while variations in F are around 100% (since F changes sign from one region to another), so the virtual salt flux can be roughly approximated as a constant times the freshwater flux.

4.3.3 One-Dimensional Mixed-Layer Model

Here we construct a simple mathematical model for the one-dimensional mixed layer, including the processes in Figure 4.10. The argument is based on assuming the mixed layer is homogeneous in its properties and accounts for the budgets of heat and mechanical energy (see Subsection 4.2.2). It is based on the pioneering model of Kraus and Turner (1967). For simplicity, we neglect the effects of precipitation, evaporation and salinity; see Subsection 10.3.2 for a variation on the model described here that focuses on freshwater and ice.

Begin by writing the equation for temperature in the mixed layer from (1.9),

$$\rho c_p \frac{\partial T}{\partial t} = -\frac{\partial \Phi}{\partial z} + \mathcal{F}, \tag{4.23}$$

where $\Phi(z, t)$ is the vertical turbulent flux of heat within the ocean and $\mathcal{F}(z, t)$ is the diabatic heat source/sink. In (4.23) T is a function of depth z and time t, but variations with depth within the mixed layer are small and averaged out below. We also assume that the convergence of horizontal heat fluxes is negligible. The diabatic term is due to the heat fluxes across the sea surface:

$$\mathcal{F}(z, t) = Q(t)\delta(z), \tag{4.24}$$

where the total heat flux Q is the sum of the incoming shortwave and longwave radiation, latent heat, and the sensible heat flux (due to conduction), as in (4.6). In (4.24), $\delta(z)$ is the Dirac delta function which focuses all the diabatic terms at the sea surface. Thus, we assume that the shortwave radiation is absorbed entirely at the surface; relaxing this assumption to account for sunlight penetration through the **euphotic zone** is a straightforward extension (see Exercise 4.4 and Subsection 6.1.2).

At the base of the mixed layer the vertical turbulent heat flux is due to entrainment of cooler water from below when the mixed layer deepens:

$$\Phi(-H) = w_e \rho c_p \left(T_+ - T \right) \Lambda(w_e). \tag{4.25}$$

Here, T_+ is the pycnocline temperature at depth H, immediately below the mixed layer,

$$w_e = \frac{dH}{dt} \tag{4.26}$$

is the entrainment speed, and

$$\Lambda(x) = \begin{cases} 1 & \text{if } x > 0, \\ 0 & \text{otherwise} \end{cases} \tag{4.27}$$

is the Heaviside function. This flux formula (4.25) models the effect on the mixed layer temperature when the mixed layer deepens and therefore entrains water from the pycnocline. When the mixed layer is not deepening, $w_e \leq 0$ and this flux switches off, which is the purpose of the Heaviside function. Notice that w_e is similar to the vertical cross-interface speed w_{*B} from Subsection 2.2.5. Be aware that the representation of the vertical stratification is different, however (variable density homogeneous layer on top of continuous stratified pycnocline here; stack of homogeneous density layers there).

Now integrate (4.23) through the mixed layer to find:

$$\rho c_p H \frac{dT}{dt} = - [\Phi(z)]^0_{-H} + Q, \tag{4.28}$$

$$= w_e \rho c_p \left(T_+ - T\right) \Lambda(w_e) + Q \tag{4.29}$$

from (4.25) and noting that there is no turbulent heat flux through the sea surface, $\Phi(0) = 0$.

To progress further we must add another equation that allows us to find w_e and close the system of equations. Here, we construct a mechanical energy budget that is generic: in Subsection 10.3.2, we specify a parameterization for the entrainment speed more precisely. The mechanical energy budget for the mixed layer is

$$\frac{\alpha g}{c_p} \int_{-H}^{0} \Phi \, dz + \mathcal{G} = \mathcal{D}. \tag{4.30}$$

Here $\mathcal{G}$ (units of W m^{-2}) is the turbulent kinetic energy input by the wind (for example, due to surface wave breaking) and $\mathcal{D}$ is the turbulent kinetic energy dissipation (for example, due to turbulent mixing and ultimately viscosity). Both $\mathcal{G}$ and $\mathcal{D}$ are positive quantities. The first term in (4.30) is the source of turbulent kinetic energy released by **convective overturning** in the vertical (α is the thermal expansion coefficient; see (1.4)). To see where this term comes from, think of the vertical exchange of two parcels of water in the mixed layer, as shown in Figure 4.12. (This is an example of a **parcel theory** argument; see Haine

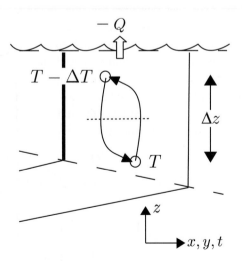

Figure 4.12   Schematic diagram of two parcels of water being exchanged in the mixed layer by convective overturning. Heat loss to the atmosphere, as shown, means that Q is negative.

and Marshall, 1998, for instance). Imagine the upper parcel is slightly cooler, therefore denser, than the lower one, with a (positive) temperature difference of ΔT. The parcels are **gravitationally unstable** so that exchanging them as shown releases potential energy. Compute the potential energy release this way: the heat ΔH carried up across the mid-depth dashed line by the parcel exchange is

$$\Delta H = \rho V c_p \Delta T, \tag{4.31}$$

for parcel volume V (ΔH is not related to the depth H here). The mass carried down is

$$\Delta m = \alpha \rho V \Delta T. \tag{4.32}$$

And the potential energy released is

$$\Delta PE = \Delta m g \Delta z = \alpha \rho V \Delta T g \Delta z = \frac{\alpha g \Delta z}{c_p} \Delta H. \tag{4.33}$$

The release of potential energy by convection per unit time is therefore $\alpha g \Phi \Delta z / c_p$ and so the total rate of release of potential energy for the whole mixed layer is

$$\frac{\alpha g}{c_p} \int_{-H}^{0} \Phi \, dz, \tag{4.34}$$

as in (4.30).

Now we substitute for Φ from (4.23):

$$\begin{aligned}
\frac{(\mathcal{D} - \mathcal{G})c_p}{\alpha g} &= \int_{-H}^{0} \int_{z'}^{0} \left(\mathcal{F} - \rho c_p \frac{dT}{dt} \right) dz \, dz', \\
&= -\int_{-H}^{0} \left(Q + \rho c_p \frac{dT}{dt} z' \right) dz', \\
&= -HQ + \frac{\rho c_p H^2}{2} \frac{dT}{dt}.
\end{aligned} \tag{4.35}$$

Hence, we find dT/dt and w_e separately:

$$\frac{dT}{dt} = \frac{2}{\rho c_p H^2} \left[\frac{(\mathcal{D} - \mathcal{G})c_p}{\alpha g} + HQ \right], \tag{4.36}$$

$$\Lambda(w_e)w_e = \frac{1}{\rho c_p H (T_+ - T)} \left[\frac{2(\mathcal{D} - \mathcal{G})c_p}{\alpha g} + HQ \right], \tag{4.37}$$

using (4.29). Finally, rewrite these two equations as

$$\frac{dT}{dt} = \frac{Q}{\rho c_p H} \left(2 - \frac{H_{MO}}{H} \right), \tag{4.38}$$

$$\Lambda(w_e)w_e = \frac{Q}{\rho c_p (T - T_+)} \left(\frac{H_{MO}}{H} - 1 \right), \tag{4.39}$$

where

$$H_{MO} = \frac{2(\mathcal{G} - \mathcal{D})c_p}{\alpha g Q} \tag{4.40}$$

defines the Monin–Obukhov depth (Thorpe, 2005; Turner, 1973). For reference, the net turbulent kinetic energy injection rate can be parametrized as

$$\mathcal{G} - \mathcal{D} = \rho m_0 \left(\frac{c_D \rho_a}{\rho} \right)^{3/2} U_{10}^3. \tag{4.41}$$

In this expression, $\rho_a \approx 1.2$ kgm^{-3} is the surface air density, $c_D \approx 1.1 \times 10^{-3}$ is a nondimensional drag coefficient, $m_0 \approx 1.25$ is a nondimensional dissipation factor, and U_{10} is the windspeed at 10m above the sea surface. See also Subsection 10.3.2. The cube in (4.41) shows that H_{MO} is sensitive to the surface windspeed. In practice, variations in Q are typically more important than in U_{10}, at least for the seasonal cycle. The reason is that Q changes sign over the year, whereas variations in U_{10} are more muted (see Subsection 4.2.2).

Equations 4.38 and 4.39 are revealing about the behavior of the mixed layer. Think of these cases (and see Figure 4.13, which organizes them):

Case 1: The mixed layer depth is constant in time. Hence, $w_e = 0$ and therefore $H = H_{MO}$. This balance is possible when the mixed layer is gaining heat $Q > 0$ and being mixed by the wind $\mathcal{G} > \mathcal{D} \geq 0$. If the mixed layer depth is not initially equal to H_{MO}, then it will tend towards it (either deepening or shallowing). Therefore, the mixed layer depth approaches the Monin–Obukhov depth whenever the mixed layer is gaining heat from the atmosphere. This circumstance applies over large areas of the low-latitude ocean and in summer (see Figure 4.11). In this case, the mixed layer temperature increases at a rate (see (4.3))

$$\frac{dT}{dt} = \frac{Q}{\rho c_p H_{MO}}, \tag{4.42}$$

which is always positive. This rate is larger for low net wind mixing and thus shallower mixed layers with smaller heat capacity (which has a natural scaling $\rho c_p H_{MO}$).

Case 2: The mixed layer temperature is constant in time, $dT/dt = 0$. From (4.38) and (4.39) we have:

$$H = \frac{H_{MO}}{2} \tag{4.43}$$

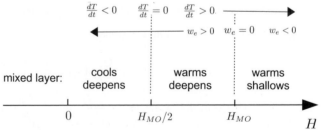

Figure 4.13 Schematic diagram of mixed layer behavior according to the mixed layer depth H and the Monin–Obukhov depth H_{MO} (4.40). If $H_{MO} < 0$ (Q is negative, cooling), then the mixed layer cools and deepens.

and the entrainment speed is

$$w_e = w_{e*} \equiv \frac{Q}{\rho c_p \, (T - T_+)}, \tag{4.44}$$

which is always positive (deepening mixed layer). In this case, the air/sea heat flux Q (warming) exactly balances the entrainment heat flux (cooling). The entrainment speed in this case (4.44) provides a natural scaling for w_e, which we call w_{e*} (see the end of this section).

Case 3: The mixed layer cools and deepens, that is, $dT/dt < 0$ and $w_e > 0$. From (4.38) and (4.39) we have:

$$H < \frac{H_{MO}}{2} \text{ or } Q < 0. \tag{4.45}$$

These inequalities say that the layer deepens and cools when the net wind mixing overcomes the tendency of the mixed layer to warm because of net heating or when the ocean loses heat to the atmosphere ($Q < 0$, which gives a negative H_{MO}). This circumstance applies, for example, in mid and high latitudes and in winter (see Figure 4.11).

Case 4: Sometimes it is possible for the mixed layer to deepen and warm. That requires $dT/dt > 0$ and $w_e > 0$, therefore:

$$H_{MO} > H > \frac{H_{MO}}{2}. \tag{4.46}$$

In words, that means that the (positive) heating rate Q is strong enough to overcome the cooling from entrained water beneath the mixed layer.

Case 5: The mixed layer shallows and warms: $w_e < 0$ and $dT/dt > 0$, for which we need

$$H \geq H_{MO}. \tag{4.47}$$

In this case the wind mixing is too weak to overcome the warming and shallowing influence of the heating Q. (Strictly, (4.39) guarantees that $H = H_{MO}$ whenever the mixed layer is shallowing. In practice, a short, but nonzero, timescale for adjustment is more physical and better-behaved computationally. This variant is used to produce Figure 4.11.)

Link to SST Restoring

To close these ideas about the time-varying one-dimensional mixed layer model, we connect to the theme of restoring SST conditions from Subsection 4.3.1. Specifically, write the heat flux from (4.14)

$$Q = \lambda(T_* - T), \tag{4.48}$$

with restoring rate λ given by (4.13) and target temperature T_* given by (4.15). In deriving this restoring equation in Subsection 4.3.1 we assumed that the mixed layer depth was fixed. That led to the characteristic timescale $\tau = Hc_p\rho/\lambda$, (4.18), for decay of SST anomalies. How does this idea change when we also allow H to vary?

Consider that the mixed layer depth changes according to

$$\frac{dH}{dt} = w_e \tag{4.49}$$

from (4.26). Therefore, τ_H the characteristic timescale for H to change, is

$$\tau_H = \frac{H}{w_e}. \tag{4.50}$$

We expect that the SST restoring arguments of Subsection 4.3.1 apply if $\tau \ll \tau_H$. This condition amounts to

$$c_p \rho \ll \frac{\lambda}{w_e}, \tag{4.51}$$

and thus, using $w_e = w_{e*}$ as the natural scaling for entrainment speed from (4.44),

$$T_* - T \ll T - T_+. \tag{4.52}$$

This formula says that the SST restoring ideas apply for SST perturbations that are small compared to the size of the temperature jump across the mixed layer base. Then the entrainment speed is small and the SST anomalies are damped by air/sea interaction faster than the mixed layer deepens. This makes sense physically and is a useful rule of thumb.

Exercises

4.1 Search for a long time series of hydrographic profile data at a fixed position (CTD data is best). The Bermuda Atlantic Time-series Study and the Hawaii Ocean Time-series are notable examples. Download the data in a suitable form and then:

(a) Plot the time series of upper ocean temperature, salinity, and density (similar to Figure 4.8c,d, but for the entire record).

(b) Compute and plot the climatological seasonal time series of upper ocean temperature, salinity, and density (as in Figure 4.8c,d).

(c) Suitably define a criterion for the mixed layer depth, and explain your choice. Compute the mixed layer depth over time for your data and add this information to the plots from parts (a) and (b).

(d) Estimate the climatological seasonal air/sea fluxes of heat and freshwater using the changes in temperature, salinity in the upper ocean over time. Explain your reasoning.

(e) Compare and contrast your results from part (d) with the estimates for air/sea heat and freshwater fluxes in the vicinity of your profile data from Chapter 11. Is it reasonable to assume that all changes are due to air/sea exchange?

4.2 Select one of the following regions:

- Subtropical North Atlantic,
- Subtropical South Atlantic,
- Subtropical North Pacific,
- Subtropical South Pacific,
- Subpolar North Atlantic,
- Subpolar North Pacific,
- Antarctic circumpolar current.

Coordinate with your classmates to ensure you make a unique choice. Download select Argo profile data (see Appendix A) for 1–5 floats from this region that travel a large distance. For example, pick long-lived floats that travel through the western boundary current and then circulate in the gyre interior. Pick floats that allow you to answer all the questions below.

(a) Make a map showing the locations of the Argo float profiles and identify the float and the period it samples.

(b) Make a (Θ, S_A) diagram marking all the profiles and label them by season (e.g., by using four different colors). Label the mixed layer, seasonal pycnocline, permanent pycnocline, and deep water on your diagram.

(c) Make (distance, pressure) sections of potential density anomaly (σ_θ), conservative temperature Θ and absolute salinity S_A from the Argo profiles. Label the x-axes with time as well as distance.

(d) For each of your floats, compute the mixed-layer depth using a suitable criterion, and make a plot of it versus distance (and time, as in part (c)).

(e) On your (distance, pressure) sections from part (c) label the periods of *obduction* and *subduction*, when water transfers from the permanent thermocline to the mixed layer and vice versa, respectively. Hence, estimate the equivalent vertical speed of water leaving and entering the permanent thermocline. Compare and contrast with the Ekman suction/pumping velocity in the vicinity of your data.

Hint: To see both obduction and subduction you may need to look at multiple floats.

(f) Estimate the air/sea fluxes of heat and freshwater using the changes in Θ and S_A between profiles for your floats. Compare and contrast with the estimates for air/sea heat and freshwater fluxes in the vicinity of your floats from Chapter 11. Assume that all changes are due to air/sea exchange. What other processes might cause the changes you see?

4.3 Here is an idealized model of time-average SST along the equator in the Pacific ocean: Assume the restoring temperature T_* is uniform along the equator. Let $x = 0$ be the western boundary of the equator and let sea surface temperature T be T_2 ($T_2 < T_*$) at $x = L$. East of $x = L$ there is upwelling which brings up cold water and west of $x = L$ there is no upwelling. Near the equator $v = 0$, and $u = -u_0(x/L)$ where u_0 is a constant. Treat the near-surface water as a mixed layer with uniform depth H.

(a) Derive a heat budget equation for the mixed layer for $0 < x \leq L$ as follows. Use the conservative tracer equation (1.9) with $\chi = c_p \rho T$, $\Phi = \mathbf{u}\chi + \mathbf{D}$, and $\mathcal{F} = 0$. Integrate vertically, and assume that $\mathbf{D}$ is negligible except at the surface where it is given by a restoring heat flux law (4.1) in order to get an ordinary differential equation for the steady state $T(x)$.

(b) Solve your equation from (a) and give a physical interpretation of any nondimensional parameters included in the solution.

(c) Plot the solution $T(x)$ for $L = 10{,}000$ km, $H = 50$ m, $u_0 = 1$ m/s, $\lambda = 50$ W m^{-2}K^{-1}, $T_* = 29.5°$C, and $T_2 = 27°$C. Also show the solution for

$\lambda = 5$ W m^{-2}K^{-1} and for $u_0 = 0.5$ m/s. Try to give a physical explanation of why the solution changes the way it does when you change each parameter.

4.4　Modify the one-dimensional mixed layer model of Subsection 4.3.3 to consider solar radiation that penetrates the sea surface. That is, let

$$\mathcal{F}(z, t) = Q_S \exp(kz) - (Q_L + Q_H + Q_B)\, \delta(z) \tag{4.53}$$

replace (4.24) for shortwave absorbtion coefficient k. How are the results concerning warming/cooling and deepening/shallowing of the mixed layer changed (Figure 4.13)?

4.5　Is it ever possible for the one-dimensional mixed layer model of Subsection 4.3.3 to simultaneously shallow and cool? Explain why.

4.6　Consider a mixed layer with positive net air/sea heat flux ($Q > 0$) and an initial mixed layer depth that is (a) much greater than H_{MO}, (b) much less than H_{MO}. By considering suitable approximations to (4.39), or otherwise, describe the evolution of the mixed layer depth over time until a steady state is reached. Quantify for the (approximate) net change in mixed layer temperature in each case.

4.7　Consider the rate of deepening of an ocean mixed layer under constant buoyancy loss B that commences at time $t = 0$. Explain how B relates to heat flux and freshwater flux (see, for example, (10.18)). Assume that the ocean is very deep and is initially uniformly stratified with Brunt–Väisälä frequency $N > 0$. Show by budgeting the buoyancy lost that the depth of the well-mixed layer H is approximated by:

$$H = \frac{\sqrt{2Bt}}{N}. \tag{4.54}$$

Neglect any horizontal variations in any property and clearly state any other physical assumptions you have made.

5

Depth-Dependent Gyre Circulation

In Chapter 3, we examine the surface geostrophic circulation of the ocean and relate it to the torque induced by horizontal variations in the wind stress (the wind stress curl). Chapter 3 discusses how the gyres are explained by the two-dimensional theory of a uniform-density ocean driven by the wind.

In the real ocean, density varies with depth. This allows geostrophic flows to vary with depth, and so poses the questions of describing and explaining the three-dimensional distribution of geostrophic currents. Because of the thermal wind relation, these currents in turn help to determine the density distribution within the ocean. Here we extend our exploration of the wind-driven gyres to include these depth-dependent currents.

5.1 Observations

5.1.1 Topography of the Pycnocline

At the crudest approximation, the shape of the thermocline is determined by the fact that deep water is cold and surface water is also cold at high latitudes but warm in the tropics. Density variations are qualitatively similar because over most of the world, alpha oceans dominate over beta oceans (Stewart and Haine, 2016). The pycnocline is essentially a lens of warm, light, near-surface water centered on the equator and cradled by denser water around it (Figure 5.1). Individual isopycnals **outcrop** (intersect the surface) with denser (and mostly deeper) isopycnals outcropping at higher latitude. Thus contours in a surface density map (Figure 4.6) mark the outcrop locations.

Closer inspection reveals other ubiquitous features of the pycnocline. In both hemispheres of the Atlantic and Pacific and in the South Indian, many of the isopycnals make a bowl shape, with the sides of each isopycnal rising up to both the north and south of the bottom of the "bowl." This bottom occurs at higher latitudes for deeper isopycnals; for instance in the North Pacific, the maximum depth of the $\sigma_\theta = 26.5$ isopycnal occurs at 30°N, while the maximum depth of $\sigma_\theta = 25.0$ occurs at about 20°N. There is also a sign of an upside-down bowl – a hill – in the northern North Pacific and northern North Atlantic. The bowl is a true three-dimensional structure, as shown by longitude–depth sections at 30°N and 30°S (Figure 5.2) and a longitude–latitude map of the depth of the $\sigma_\theta = 26.8$ isopycnal in the North Atlantic (Figure 5.3). The bottom of the bowl is near the western

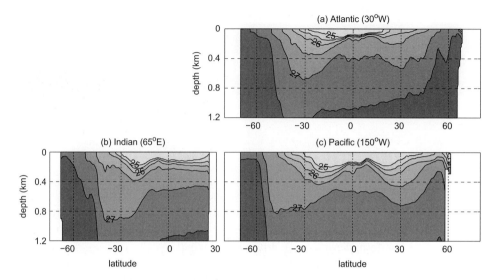

Figure 5.1 Latitude–depth contours of annual average σ_θ for (a) Atlantic (30°W), (b) Indian (65°E), and (c) Pacific (150°W), Oceans. Contour interval 0.5 kg/m^3 starting at 25 kg/m^3. Data from CARS (see Appendix A).

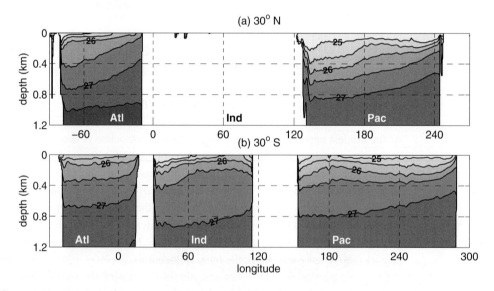

Figure 5.2 Longitude–depth contours of annual average σ_θ for (a) 30°N and (b) 30°S. Contour intervals as in Figure 5.1. Data from CARS (Appendix A).

boundary of the ocean, with a sharp upward slope to the west and a gentle upward slope to the east.

Though the subpolar regions have weaker stratification than the subtropics, they still have significant density features. The westward deepening of isopycnals in the subtropics (Figure 5.2) is reversed in subpolar regions (Figure 5.4), with most of the lightest water

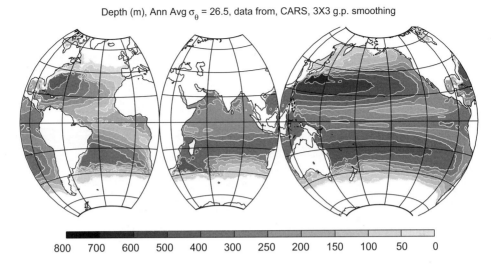

Depth (m), Ann Avg σ$_\theta$ = 26.5, data from, CARS, 3X3 g.p. smoothing

800   700   600   500   400   300   250   200   150   100   50   0

Figure 5.3   Depth (m) of $\sigma_\theta = 26.5$ isopycnal from climatology of annual-average temperature and salinity. Depths are smoothed with 7×7 gridpoint running mean. The thick contour is 300 m which divides 50 m contour interval for lower values and 100 m interval for higher values. Data from CARS (Appendix A).

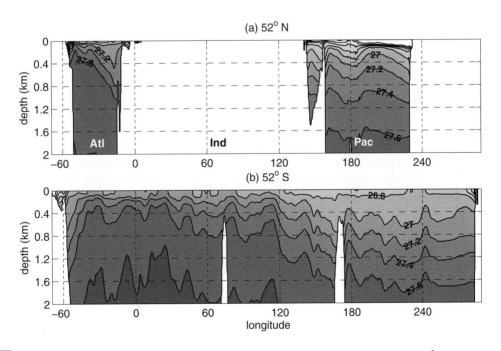

Figure 5.4   Longitude–depth contours of annual average σ_θ for (a) 52°N and (b) 52°S. Contour interval is 0.2 kg/m^3 ($\sigma_\theta \leq 26$) with an additional contour at $\sigma_\theta = 25$, and thin contours at 0.1 kg/m^3 intervals for $\sigma_\theta > 27$. Note that depth range, as well as longitudes of eastern and western boundaries of figure, are different from Figure 5.2. Data from CARS (Appendix A).

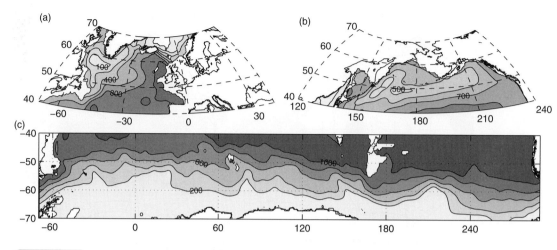

Figure 5.5 Depth (m) of (a) $\sigma_\theta = 27.6$ isopycnal in North Atlantic, (b) $\sigma_\theta = 27.2$ in North Pacific, and (c) $\sigma_\theta = 27.6$ in Southern Ocean. Depths are smoothed with 7×7 gridpoint running mean. Contour interval in (a) is 200 m with an extra 100 m contour, in (b) is 100 m, and in (c) is 400 m with an extra 200 m contour. Data is from CARS (Appendix A).

occuring in the east rather than in the west. This is true not only for the North Atlantic and North Pacific (Figure 5.4a), but also in the Southern Ocean (Figure 5.4b), where the "basin" consists of the domain whose western boundary is the east coast of South America and whose eastern boundary is the west coast of South America. Isopycnals plunge westward in a narrow region associated with western boundary currents, especially in the southern hemisphere. This is consistent with the view that the dense water forms a hill with a peak near the western boundary.

The topography of this hill can be seen in the depth of isopycnals in the North Atlantic (Figure 5.5a) and the North Pacific (Figure 5.5b). The shapes of these hills are somewhat irregular, probably due to Greenland in the Atlantic and the Aleutian Islands in the Pacific. In the Southern Ocean, isopycnals slope upwards as one goes poleward, but they do not come down again (Figure 5.5c). Much of the O(1000) km zonal structure seen in the Southern Ocean isopycnal depth may be eddy activity aliased by the relatively poor data sampling in the region, but the overall southeastward tilt of depth contours corresponds to the northeastward tilt of depth contours in the eastern half of basin in the northern hemisphere.

5.1.2 Depth Distribution of Currents

Tilting isopycnals are associated with vertical shear in the geostrophic velocity due to the thermal wind relation (Subsection 2.1.2, Subsection 2.2.2). An isopycnal that is low in the middle implies that the flow below the isopycnal is cyclonic relative to the flow above it. As we saw in Chapter 3, the surface flow in the subtropical gyre is anticyclonic. Thus the flow above the bowl is anticyclonic, and the bowl implies that this anticyclonic flow

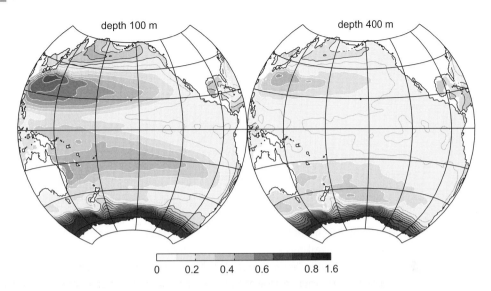

Figure 5.6 Dynamic height normalized by gravitational acceleration (m) for the Pacific Ocean. The dynamic height is computed with respect to 500 db for (left) 100 m and (right) 400 m depth, and smoothed with 7×7 gridpoint averaging. Shading represents magnitude and line color is black (positive), white (negative), and gray (zero). Data is from World Ocean Atlas 2013 (see Appendix A).

weakens with depth. Similarly, the upside-down bowls seen at higher latitudes imply that the cyclonic subpolar gyres weaken with depth. The strongly tilted density front which extends to great depths at roughly subpolar latitudes in the southern hemisphere of all three oceans (Figures 5.1 and 5.3) indicates a weakening of the Antarctic Circumpolar Current (ACC) with depth.

Dynamic pressure reveals the geostrophic flow at different depths in the Pacific (Figure 5.6). Pressure is displayed as $p'/\rho_0 g$, the dynamic height, which can be interpreted as the height of a given isobar or as the sea surface height anomaly needed to generate the same pressure (for a uniform-density ocean; see (2.48)). The dynamic height at a given level z_* is calculated by adding the sea surface height from satellite altimetry (see Subsection 1.1.2 and Figure 3.3) to the vertical integral between $z = 0$ and $z = z_*$ of the density anomaly divided by a representative density ρ_0.

Dynamic height contours in Figure 5.6, interpreted as geostrophic streamlines, display the northern and southern subtropical gyres, subpolar gyre, and ACC. The subtropical currents are weaker (as measured by pressure differences) at 400 m (right panel) than 100 m (left panel), and by 1000 m (not shown) are barely distinguishable except for the ACC. The circulation patterns in the tropics are less similar to the surface patterns, however. The dynamic height difference between the center and edge of the subtropical and subpolar gyres are about two times smaller at 400 m than they are at 100 m; in the ACC the deep speeds are closer to the surface values.

In general the gyre flow weakens as one descends through the pycnocline and is a small fraction of its surface value throughout the abyss. Direct measurements of velocity

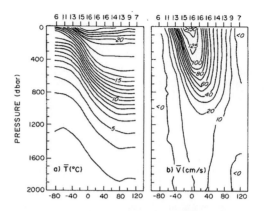

Figure 5.7 Section through the Gulf Stream showing temperature (left) and velocity (right) as a function of pressure (db) and cross-stream distance (km). The section is synthesized from repeated temperature and drifter-velocity measurements across the Gulf Stream in the vicinity of 36°N and 74°W. Float measurements eliminate the need to assume a level of no motion. See also Figure 3.5. From Halkin and Rossby (1985), Figure 10, ©American Meteorological Society. Used with permission.

with current meters and floats support this view. For example, geostrophic calculations combined with float data yield a cross section through the Gulf Stream, which shows a similar decrease of velocity with depth (Figure 5.7). The weakness of the flow in the abyss compared to the pycnocline is a common pattern in the ocean, but we see in Chapter 8 that there are important exceptions. Remember that weak flow does not necessarily imply weak volume transport, because (for instance) a 10 cm/s flow in the top 0.5 km will be associated with the same volume transport as a 1 cm/s flow in the bottom 5 km. Observational, numerical, and theoretical results indicate that most of the volume transport of the wind-driven gyres occurs in the pycnocline (less so in the ACC, see Chapter 9).

5.1.3 Density Distribution of Currents

Ventilated and Unventilated Flow

The density field provides an important tool, in addition to its contribution to geostrophic flow, for analyzing the three-dimensional gyre circulation. If a water parcel does not change its density, it remains, by definition, on the same isopycnal while it drifts. As the equations of motion show, changes of density occur due to interaction with the atmosphere at the sea surface, solar radiation (principally in the top few tens of meters of the ocean), and mixing. Below the mixed layer, if mixing is small enough, then the density of a water parcel changes little as it circles the gyre. Observations indicate that vertical mixing in the pycnocline due to small-scale turbulence is about 10^{-5} m^2/s (see Chapter 7). As we demonstrate in Subsection 5.3.1, this seems to be small enough to treat pycnocline flow as approximately isopycnic. This behavior, to the extent that it holds in the gyres, gives another constraint

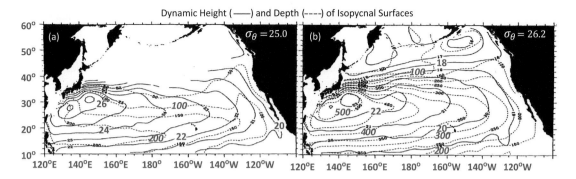

Figure 5.8 Dynamic height (solid lines, referenced to 2000 m, also called the acceleration potential) in $m^2\,s^{-2}$, and depth contours (dashed lines) in meters, for (a) $\sigma_\theta = 25.0$ and (b) $\sigma_\theta = 26.2\,kgm^{-3}$ for annually averaged climatology. From Huang and Qiu (1994), Figure 3, ©American Meteorological Society. Used with permission.

on the flow. The density and horizontal and vertical components of velocity must arrange themselves so that the parcel stays on an isopycnal.

For flow that follows isopycnals up and down through the gyres, the velocity is a better indicator of trajectories if it is calculated on isopycnals rather than at a given depth. For instance, climatological views of the dynamic height at $\sigma_\theta = 25.0$ and $\sigma_\theta = 26.2$ show the Pacific subtropical gyre flow and western intensification at both densities (Figure 5.8).

As pointed out in Chapter 3, for geostrophic flow, the dynamic pressure and dynamic height are streamfunctions of $f(\mathbf{u} - \mathbf{u}_0)$ (where $\mathbf{u}_0$ is the velocity at the deep level, sometimes presumed to be small and referred to as the level of no motion; see Subsection 2.2.2).

On the shallower surface (Figure 5.8a), these streamlines appear from nowhere in the northeastern corner of the subtropical gyre. The streamlines cannot be traced back further because the 25.0 isopycnal does not exist upstream, where all the water is denser than $\sigma_\theta = 25.0$. The streamline start-points do not actually represent flow appearing from nowhere, but water that changes its density as it flows and enters the given density range from a different density range. There are also streamline start-points and end-points near the western boundary which are an artifact of the data failing to resolve the narrow western boundary current.

If flow is generally along isopycnals, then there is an important qualitative difference between flow on deep isopycnals and shallow isopycnals. Shallow isopycnals (such as $\sigma_\theta = 24.5$ in Figure 5.1a) outcrop (touch the mixed layer) within the subtropical gyre, while deeper isopycnals (such as $\sigma_\theta = 27$, same figure) outcrop at higher latitudes. Water parcels on shallow isopycnals directly interact with atmospheric forcing as they flow around the subtropical gyre. This is referred to as **ventilated** flow. Water at the outcrops in the subpolar gyre does not have a direct isopycnic path to the subtropical gyre, so the lower pycnocline water in the subtropical gyre can be considered unventilated. As we see in Subsection 5.2.2, ventilation is one of the key organizing principles to understand the three-dimensional pycnocline circulation. In addition, ventilation is a potentially interesting characteristic because we expect ventilated water to have its properties (such as salinity, chemical tracers, and heat transport) influenced by surface conditions in the subtropical gyre.

The location of isopycnal outcrops changes with season, with outcrop latitudes typically moving equatorward in the winter and poleward in the summer (Chapter 4). For reasons explained in Subsection 5.2.4, the winter outcrop position is most relevant for determining ventilation.

Potential Vorticity as a Tracer

When we analyze the structure of the pycnocline in Sections 5.2 and 5.3, a revealing quantity is the planetary potential vorticity (Subsection 2.1.5), $q = f/h$. Recall that f is the Coriolis parameter and h is the vertical distance between given isopycnals (namely, the layer thickness if we approximate the pycnocline as a stack of layers of different densities). The potential vorticity includes a relative vorticity term which is ignored here because it is small away from western and equatorial boundary currents. It is also small on large scales (planetary potential vorticity is sometimes called the large-scale potential vorticity for this reason). As explained in Subsection 2.2.5, q is conserved following the flow for each layer in the absence of friction and water property transformation (diabatic processes). These requirements are met to reasonable accuracy in the pycnocline away from the surface (Subsection 5.3.1). Therefore, we expect q contours in each layer to align with streamlines. This is a powerful constraint that is exploited in Subsection 5.2.2.

The potential vorticity shows an interesting horizontal and vertical structure when mapped on a given isopycnal. (Talley, 1985) In the Pacific at $\sigma_\theta = 26$ there are regions of high q values near where the isopycnals reach the surface (Figure 5.9a). These appear to be sources of "tongues" of high q fluid which flow equatorward and westward in both hemispheres, as if being wrapped around the subtropical gyres by the circulation. There are also broad regions with relatively little variation in q, such as on the $\sigma_\theta = 27$ isopycnal in the Pacific poleward of 20°N (Figure 5.9a). As discussed in Subsection 5.2.3 and Subsection 5.3.4, the idea that dynamics may force large regions to have uniform q has greatly influenced theories of the subtropical circulation.

Density Partition of Volume Transport

Because of the importance of ventilation, Huang and Qiu (1994) summarize the vertical variation in the flow by separating the Ekman, ventilated, and unventilated components in the North Pacific subtropical gyre. They use isopycnal depths and geostrophic velocity fields such as those in Figure 5.8, along with wind data, to estimate the amount of flow in each layer. Huang and Qiu (1994) choose $\sigma_\theta = 26.2$ as the boundary between ventilated and unventilated flow because it is the densest isopycnal that outcrops in the subtropical gyre during the winter. They also separate the ventilated layer into a seasonal thermocline, which includes all geostrophic flow in the seasonal thermocline and the mixed layer, and a permanent thermocline. Their calculation only includes the Sverdrup interior region (see Subsection 3.2.4 and Subsection 3.3.2), and so also shows exchange with the western

(a) Normalized f/h, $\sigma_\theta = 26.0$ (b) Normalized f/h, $\sigma_\theta = 27.0$

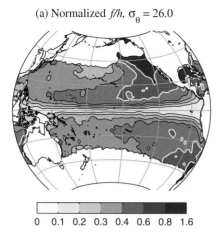
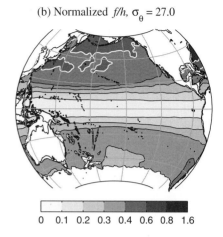

0 0.1 0.2 0.3 0.4 0.6 0.8 1.6   0 0.1 0.2 0.3 0.4 0.6 0.8 1.6

Figure 5.9 Distribution of normalized planetary potential vorticity $q = f/h$, where h is vertical distance between annual averaged isopycnals (σ_θ) of (left) 25.75 and 26.25, and (right) 26.75 and 27.25. The normalization factor is $2\Omega/H_0$, where Ω is the Earth's rotation rate and H_0 is a typical value of h for each layer, (a) 100 m and (b) 400 m. Density is taken from local winter values (March and September), with larger contour intervals marked by white contours. Thick black contours show winter surface location (March for northern hemisphere, September for southern hemisphere) of isopycnals. At each point, q is averaged over a $4.75^\circ \times 4.75^\circ$ square centered at the point. Data is from World Ocean Atlas 2013 (Appendix A).

boundary current. It also shows exchange with the tropics, which is discussed in Chapter 6. The resulting budgets are not in perfect balance, due to sampling errors, but the differences are mostly tolerably small.

The transport budget has several notable features (Figure 5.10). The total gyre strength is about 41 Sv (as measured by inflow from the Kuroshio in the northwest and outflow to the Kuroshio in the southwest), which is about what the Sverdrup transport implies. Of this 41 Sv, nearly two-thirds occurs in the ventilated thermocline. Within the ventilated thermocline, inflow from the Kuroshio is evenly divided between seasonal and permanent thermocline, but outflow to the Kuroshio is dominated by water in the permanent thermocline. The difference is due to the fact that the seasonal thermocline generally thickens as one goes poleward (Chapter 4); this difference in transport implies that as water flows northward in the Kuroshio, it is transferred from the permanent thermocline to the seasonal thermocline. This process is called obduction (Qiu and Huang, 1995; Subsection 4.2.2). The transfer occurs because as the water flows northward, it enters regions where the winter mixed layer deepens to include deeper and denser water, so that some water that is below the mixed layer at 20°N (for instance) may be within the mixed layer at 40°N.

Ekman transport is often assumed to be negligible, with geostrophic flow dominating (which is sometimes dubious, and is discussed further in Subsection 6.2.2). Figure 5.10 shows that the Ekman pumping for the North Pacific subtropical gyre is about 29 Sv, which is not that much smaller than the Sverdrup transport. When looking at depth-average flow, the gyres do not appear to exchange fluid between each other. However, there is a large

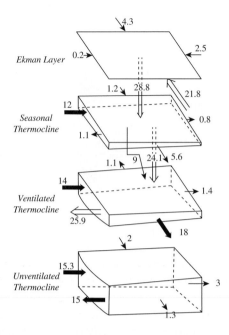

Figure 5.10 Volume transport (Sv) flow through boundaries of the North Pacific subtropical gyre (excluding western boundary currents), based on climatological wind and density data. North is into the page, east is to the right. From Huang and Qiu (1994), Figure 8, ©American Meteorological Society. Used with permission.

Ekman inflow from the tropics, and smaller Ekman inflows from the subpolar region and the eastern boundary (coastal North America). Similarly, about 24 Sv of ventilated water flows out of the tropical boundary of the gyre, with smaller amounts leaving other boundaries. This exchange of Ekman transport and geostrophic flow is part of the North Pacific subtropical cell, which is described in more detail in Subsection 6.1.1 and Subsection 6.2.3.

A similar budget for the South Pacific does not balance very well, perhaps indicating a more important role for deep (below the thermocline) flows (Huang and Russell, 1998). It is also harder to apply this analysis to the Atlantic, where flows associated with the deep meridional overturning circulation are large (Chapter 8, but see also Huang, 1990).

5.2 Concepts

5.2.1 The Paradox of Wind-Driven Flow in the Pycnocline

Observations show that a nearly ubiquitous feature of the ocean is the presence of a thermocline and a pycnocline. The reasons *why* there is a thermocline and pycnocline are discussed in Chapter 8. For the remainder of this section we refer to the pycnocline, though much of the same discussion could address the thermocline instead. Some further questions about the structure of the pycnocline arise. Why does the subtropical pycnocline have a bowl shape? Why does the bowl have the magnitude (vertical excursion of isopycnals from the center of the bowl to the edges) that it has?

The presence of the pycnocline allows the geostrophic flow of the subtropical gyres to vary with depth. Why is this flow strong in the pycnocline and weak below it? Can we understand how the flow varies within the pycnocline? The vertical variation within the pycnocline is interesting precisely because properties such as density, temperature, salinity, and chemical tracers vary dramatically from the top to the bottom of the pycnocline. Taking temperature as an example, the meridional heat transport depends on the difference in temperature between northward flowing water and southward flowing water in the gyre (Chapter 11). Because of temperature variations within the pycnocline, even if we know that both northward and southward flow occur somewhere in the pycnocline, we cannot understand the heat transport unless we know the depth (or more accurately, the temperature) distribution within the pycnocline of the flows. Similar reasoning applies to other properties, whose transports are less well understood.

The question of pycnocline structure and the question of subtropical gyre structure are linked via the thermal wind law (Subsection 2.1.2), which relates vertical variations in velocity to horizontal variations in density. This potentially complicates the task of explaining each structure, but also provides clues to finding the explanation.

The most basic features of the subtropical gyres – bowl-shaped pycnocline and wind-driven flow faster in the pycnocline than in the abyss – can be explained relatively easily. We can idealize the ocean as having two layers, a relatively deep abyssal layer with a uniform density and a relatively thin pycnocline with a lower density (Figure 5.11a). Given the density cross sections shown in Figure 5.1, characterizing the top layer with a single density is a rather crude approximation, but it is the simplest way to bring the pycnocline into the circulation problem.

Chapter 3 analyzes the gyre circulation in terms of potential vorticity. Ekman pumping at the surface gives water columns anticyclonic potential vorticity, which implies equatorward flow. To maintain equilibrium, this potential vorticity must be removed by friction, which occurs in the western boundary current and implies poleward flow. In the two-layer case, all that vorticity injection by Ekman pumping goes into the top layer. Because there is no pumping into the bottom layer, each water column in that layer apparently keeps a constant vorticity and hence does not circulate. This simple model shows us why it is reasonable that the gyre-flow is largely confined to the pycnocline.

The bowl shape follows directly from geostrophy. If the abyss is motionless, there can be no horizontal pressure gradients in that layer. The weight of the water above a given depth in the abyss must be the same everywhere in the subtropical gyre. This only can occur if the interface between the two layers has a shape which compensates for the surface slope of the water. In other words, comparing the center of the subtropical gyre to the edges, the raised sea surface contributes a high pressure to the abyss, while the thicker light, upper-layer water contributes a low pressure to the abyss.

This model runs into trouble when we try to extend it to more layers to get a more detailed view of the pycnocline, however. Consider a three-layer model, with the pycnocline represented by two layers of equal mean thickness (Figure 5.11b). Now, the same reasoning implies that all the flow must be in the top half of the pycnocline, with none in

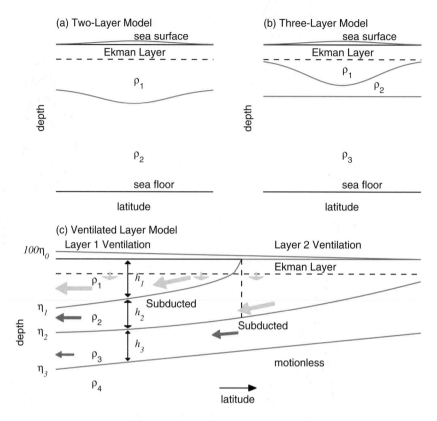

Figure 5.11 Schematic latitude–depth cross sections of layer models showing interface(s), sea floor, sea surface position (exaggerated in the vertical), Ekman layer, and densities for (a) two layers, (b) three layers, and (c) ventilated layers. In (c), layer thicknesses, Ekman pumping, and interior (away from western boundary current) flow directions are indicated.

the bottom half. Adding more layers to the representation of the pycnocline confines the circulation to an increasingly small depth range near the surface. As the density stratification of the real ocean is continuous, this reasoning seems to imply that all the flow occurs in an infinitesimal layer just below the Ekman layer. Such a vertical distribution seems rather strange, and it contradicts the observational evidence of Subsection 5.1.3 which shows significant flow throughout the pycnocline.

What has gone wrong with this model? In the 1980s, researchers at Woods Hole Oceanographic Institution proposed two complementary mechanisms which together can solve the apparent contradiction. The simpler idea to understand, which was actually proposed second, hinges on the concept of ventilation (Luyten et al., 1983, henceforth LPS after the three authors). It is humbling to note that the same Henry Stommel who first explained the existence of western boundary currents in 1948 helped formulate this fundamental extension of the theory of gyres. The more subtle idea, proposed by Rhines and Young (1982) and Young and Rhines (1982), shows how deeper, unventilated layers can be put into motion. See Pedlosky (2006) for an interesting historical account of the ideas.

5.2.2 Ventilated Pycnocline

Assumptions

Figure 5.11b portrays the pycnocline as a stack of uniform-density layers, one on top of another, so that only the lightest one is exposed to the surface. Density in the real ocean (Figure 5.1) looks more like Figure 5.11c. Each layer is in contact with the sea surface (and Ekman layer) in some latitude range (Subsection 5.1.1). As explained below, this outcropping allows many layers to be driven by Ekman pumping and thus ventilated.

LPS use this picture of what they call the **ventilated thermocline** to extend the Sverdrup balance to multiple layers. Just as the Sverdrup balance predicts the vertically integrated flow, so LPS theory predicts how the flow varies with depth. Neither theory applies to the western boundary current, which demands more complex dynamics. In addition, LPS theory applies to the subtropical gyres but not the subpolar gyres. Because geostrophic flow is intimately related to density gradients, it also predicts the shape of the pycnocline.

Subsection 5.2.1 makes a tacit assumption: water does not move from one layer to another. For a model with a small number of layers, this assumption seems obvious, because it would appear rather artificial for the water density to jump as it flows into a different layer. For many layers, which becomes a good approximation for the continuous stratification of the ocean, the assumption is not only not obvious, it is obviously wrong. That is because we know that in the real ocean, and in realistic numerical models, water changes density. In a multiple-layer model in which density is by definition set to a constant and uniform value within each layer, such density changes are represented by water crossing the interface from one layer to another. We refer to flows in which water parcel density does or does not change as **diabatic** or **adiabatic**, though these terms more properly refer to potential temperature, not density (see Subsection 1.2.3).

The importance (or unimportance) of diapycnal mixing is a fundamental and often unstated assumption in general circulation studies. In the ventilated thermocline theory (at least in its earliest and best-known forms), mixing is neglected below the mixed layer. In theories of the deep meridional overturning (Chapter 8), mixing is considered one of the primary driving mechanisms. One may wonder how mixing can be unimportant and important at the same time. A partial answer is that mixing is thought to be much smaller in the pycnocline than in the abyss (see Chapter 7). Stratification and flow in the pycnocline is an important component of the deep meridional overturning circulation, however, so this is not a very satisfying explanation. A better explanation considers the timescale of the phenomenon under consideration. As mentioned in Subsection 5.1.2, in the time it takes a water parcel to traverse the subtropical gyre in the thermocline, mixing probably cannot change its density enough to make the trajectory depart much from an isopycnal (Subsection 5.3.1). On longer timescales associated with the deep meridional overturning, mixing can have a first order influence on the dynamics. The interaction of this idealized, adiabatic gyre flow with diffusively driven dynamics is a subtle issue to which we return later.

LPS idealize the stratification with ventilated layers over a motionless, uniform-density abyss (Figure 5.11c). The density of each layer is specified, as is the latitude at which it outcrops. This is equivalent to specifying the ocean's surface and abyssal density. For simplicity, outcrops are assumed to be latitude circles, which is a somewhat crude approximation for the real world.

The theory predicts the distribution of density with depth by predicting the thickness h_n of each layer. As explained below, h_n of the lowest non-motionless layer at the eastern boundary must also be assumed. The theory assumes that the Ekman layer is so thin that it carries an insignificant geostrophic volume transport. Below the Ekman layer, water parcels cannot change their density, so water does not cross the interfaces between layers. The theory can be applied to $N\frac{1}{2}$ layers (N moving layers, with "$\frac{1}{2}$" referring to the deep stagnant layer), though the algebra becomes very complicated as N increases.

Each layer consists of a ventilated region where it is exposed to Ekman pumping, and a **subducted** region, equatorward of the ventilated region, where the layer is covered by shallower layers. The term "subduction" is borrowed from the study of plate tectonics, where it refers to the phenomenon of one plate of the Earth's crust sliding under another. Flow in the ventilated region is directly forced by Ekman pumping. All the geostrophic flow is equatorward (remember, this theory excludes the western boundary current), as in the barotropic subtropical gyre. When this flow reaches the subducted region, it must continue flowing in order to satisfy continuity (Figure 5.11c). This is a solution to the conundrum of what forces the deeper thermocline layers. They are not driven by the Ekman pumping directly above, but by the Ekman-induced geostrophic flow from the poleward border of the subducted region.

Results of the Theory

As the flow below the Ekman layer is assumed to be geostrophic, we can calculate the velocities once we know all the layer thicknesses h_n as a function of latitude and longitude. For a single layer, a single equation – the Sverdrup balance – suffices to calculate the flow (Chapter 3). For additional layers, we need additional expressions. Recalling that the Sverdrup balance is a special case of the budget equation for potential vorticity, it is not too surprising that analogous expressions can be found for each individual subducted layer. As the subducted regions have no local Ekman pumping, and as we are assuming (as in Sverdrup theory) that friction is negligible outside the Ekman layer, these additional equations are statements that potential vorticity is constant for each water parcel. LPS use these potential vorticity constraints, geostrophy, and the assumption of inviscid, adiabatic flow to construct an elegant theory that predicts the density and velocity structure of the pycnocline in the subtropical gyre. We reproduce the theory for the case of two active layers (Figure 5.11c without layer 3) in Subsection 5.3.3.

Here we summarize some key results of the two-layer theory. The streamfunction for each of the two layers looks qualitatively similar to the barotropic streamfunction (Figure 5.12a). Flow is not defined in the upper layer poleward of its outcrop latitude. Two subregions of the subducted flow display noteworthy behavior (Figure 5.12a):

First, in the eastern and tropical edges of the domain, there is no layer 2 flow. This region is inaccessible to water in the ventilated part of layer 2 (poleward of the outcrop line). It is known as the **shadow zone**, referring to this disconnect from the ventilation region. The

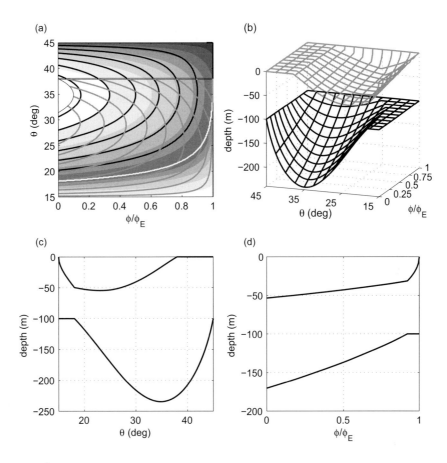

Figure 5.12   Solution of $2\frac{1}{2}$-layer LPS model. (a) Streamlines for layer 1 (gray) and layer 2 (black) showing pool (white region) and boundary of shadow zone (white contour), along with layer 2 potential vorticity (shading). (b) Perspective view of interfaces. (c) Meridional cross section of interfaces near western boundary. (d) Zonal cross section of interfaces at 25° N. For details of solution, see Subsection 5.3.3.

reason for the disconnect can be guessed by regarding the easternmost streamline that flows from the outcrop (shown by white curve in Figure 5.12a). This streamline bends westward going equatorward. As none of the water equatorward of this streamline can be driven by subducting water, it is motionless (or possibly driven by some other process).

Second, the westernmost subducting streamline that intersects the outcrop line may bend eastward (though this depends on the latitude of the outcrop line; Figure 5.12a). This shields a western region, known as the **pool**, from ventilation. This region is connected to flow from the western boundary current, so motion is possible. The ventilation concept does not specify this flow, however.

Finally, the layer 2 potential vorticity map (Figure 5.12a) reflects the dynamical differences between the three different regions (subduction region, shadow zone, and pool). In the subduction region, potential vorticity contours follow streamlines, whereas in the shadow zone, they have stronger gradients which reflect gradients in f and h_1. These features are reminiscent of the observed potential vorticity field (Figure 5.9), particularly the North Pacific. A perspective rendering of h_1 and $h = h_1 + h_2$, the depth of the interfaces (Figure 5.12b), shows the bowl shape that we saw in observations (the western slope of the bowl is not visible because the theory does not include the western boundary current). Note that in the shadow zone, where both layer 2 and the abyss are motionless, the thermal wind relation constrains the lower interface to be flat.

Cross sections of the layers give us a different perspective on the predicted density structure. The meridional section (Figure 5.12c) reproduces the observed feature (Figure 5.1) that the bottom of the bowl is closer to the equator for shallower layers. The zonal section (Figure 5.12d) is also reminiscent of observed density (Figure 5.2), and illustrates an important point about the eastern boundary condition. At the eastern boundary, the condition of no flow normal to the boundary implies that all the interfaces are flat along the boundary. For layer 1, we have $h_1 = 0$ at the outcrop latitude, so it must have zero thickness everywhere along the eastern boundary. For layer 2, the eastern boundary is in the shadow zone, so the boundary value of H_2 is a free parameter that must be specified. Something about the internal dynamics of the ocean must control this parameter, but that something is outside the scope of LPS theory. Similarly, although the pool is circumscribed by a q_2 contour and a streamline, its dynamics are not yet explained.

5.2.3 Unventilated Pycnocline

There is more to say about the ventilated pycnocline, but let us first describe the other major paradigm for explaining wind-driven flow in the pycnocline. We start with the observation that there are isopycnals that have a bowl shape, indicating their participation in the gyre flow, but which are not ventilated in the subtropical gyre (see the $\sigma_\theta = 27$ isopycnal in Figure 5.1). Rhines and Young (1982) and Young and Rhines (1982) analyzed the potential vorticity within such unventilated layers to propose a forcing mechanism. Recall that, for a (linear) uniform-density ocean with uniform thickness H, equatorward-flowing water must change its value of $q = f/H$ as it flows. Ekman pumping provides the mechanism for changing q appropriately. For an ocean which consists of a stack of uniform-density layers, the potential vorticity of each layer, $q_n = f/h_n$, depends on both f and layer thickness h_n. As these deep layers do not contact the surface, Ekman pumping cannot inject vorticity into them as it does in either the uniform-density or ventilated case.

Peter Rhines and William Young noticed that the upper layer circulation can qualitatively change the q_n of deeper layers. Consider again the two-active-layer model in Figure 5.11b. If the change in thickness of layer 2 is small across the bowl (for instance, 10%), contours of f/h_2 are only changed a small amount (Figure 5.13a). If there is no source of vorticity in layer 2, then q_2 for any water parcel must remain constant, which can only happen if the parcel travels along a q_2 contour. Such trajectories are impossible, however, because each q_2 contour starts and ends at a solid wall which prevents flow.

For a larger change, such as 50% (Figure 5.13b), the isolines of constant q_2 change by a much larger amount. More importantly, a local extremum forms in q_2, so the q_2 isolines form closed paths around the extremum. Flow along such closed paths *is* possible because those q_2 contours never encounter a solid boundary.

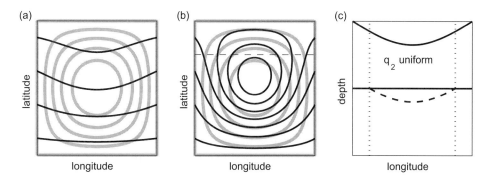

Figure 5.13 Layer thickness and potential vorticity for idealized three-layer model, with $0 \leq x \leq L$ and $-M/2 \leq y \leq M/2$, upper layer thickness given by $h_1 = D_1 \sin(\pi x/L) \cos(\pi y/M)$, and lower layer thickness given by $h_1 + h_2 = D_2$, where D_1 and D_2 are constants. For (a) $D_1/D_2 = 0.1$ and (b) $D_1/D_2 = 0.5$, gray curves show contours of constant h_1 (different contour intervals used for the two cases), and black curves show contours of constant q_2 (same contour interval for both cases). In (c), cross section of interface positions (solid lines) along dashed line in (b), with dashed curve showing distortion of interface needed to make q_2 uniform in the closed-q_2-contour region.

So far, we have shown that the presence of a bowl can eliminate a major barrier to deep flow, namely the need for a source of vorticity to allow water parcels to cross q_n contours. We still need to show what actually determines the strength of the flow along the closed q_n contours. Rhines and Young (1982) identified a constraint that at first glance seems a rather unlikely force for driving the circulation. They argued that the deep flow could be driven by the horizontal **homogenization of potential vorticity**. In the example in Figure 5.13b, we show q_2 given the layer 1 flow and given the assumption that layer 2 is motionless. Potential vorticity homogenization instead removes the closed black q_2 contours in Figure 5.13b. Thus the unventilated layer 2 circulates. The flow is determined by assuming that this region has a uniform q_2 equal to the value in the motionless fluid bordering the closed-contour region; Subsection 5.3.4 gives an idealized example. A cross section of the interfaces through the homogenized-q_2 region gives some idea of the flow. The dashed curve shows the interface below layer 2 forming a bowl shape (and hence supporting flow) so that f/h_2 is uniform.

What is the justification for assuming that q_n is homogenized? Potential vorticity homogenization has not been rigorously proven, but several factors support the assumption. Conceptually, we can think of the q_n homogenization as being carried out by mesoscale eddies (depth-varying winter mixed layers can also contribute; Williams, 1991). Chapter 7 looks more closely at eddies; here we just point out that we expect eddies to mix properties laterally along isopycnals. One such property is q_n. Generally, the net effect of mixing is to reduce gradients in properties. There is no source of potential vorticity in the middle of the closed-contour region, so it is plausible that the gradients vanish: homogenized q_n.

There is some evidence that this is true. Eddy-resolving quasi-geostrophic models of the circulation driven by idealized subpolar-subtropical winds show homogenized potential vorticity in both gyres (Holland et al., 1984). Quasi-geostrophic ocean models assume that the flow is approximately geostrophic and that the layer thicknesses do not vary by large amounts across the domain, so that ventilation as in Figure 5.11c is not allowed. Observations give somewhat mixed evidence of potential vorticity homogenization. The North Pacific, South Indian, and South Atlantic show relatively small gradients, but the North Atlantic does not.

Chapter 7 shows how potential vorticity homogenization by eddies can be thought of as a kind of vertical friction in a stratified, geostrophic flow (see discussion after (7.53)). Vertical friction does gives a more intuitive mechanism than potential vorticity homogenization for driving deep flow: directly forced shallow motion pushes on deeper flows via friction. While we have claimed to exclude both friction and mixing in these pycnocline theories, friction has crept back into the solution.

The ventilated and the unventilated pycnocline are two paradigms which can coexist in the real ocean as well as in theoretical and numerical models. The ventilated theory refers to the shallower density ranges, while the unventilated theory refers to deeper densities, with closed q_n contours and the concomitant homogenized q_n found in the pool region and below.

5.2.4 Mixed Layer Variations and Subduction

The theory of subduction described in Subsection 5.2.2 and Subsection 5.3.3 portrays the water injected into the pycnocline as coming from a thin mixed layer, as shown schematically in Figure 5.14a. As discussed in Chapter 4, we assume the Ekman layer is roughly coincident with the mixed layer. Therefore in the geometry shown in Figure 5.14a, the only transport between the mixed layer and pycnocline occurs through the vertical velocity at the base of the mixed layer. The vertical velocity is given by the divergence of the horizontal velocity in the Ekman layer.

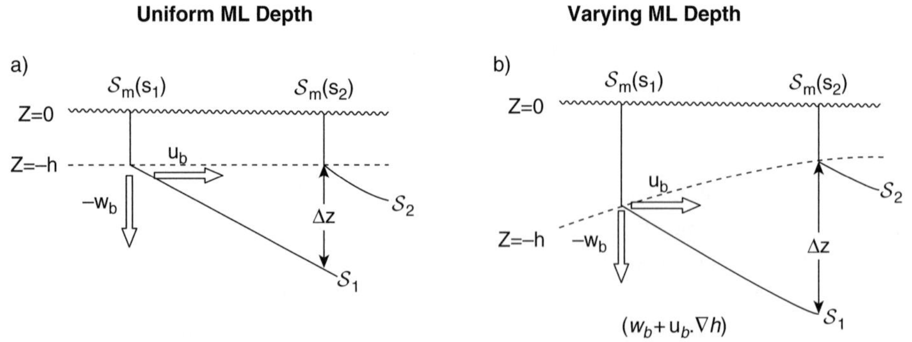

Figure 5.14 Schematic showing mixed-layer (ML) subduction from (a) uniform-depth mixed layer (by vertical downwelling due to Ekman pumping) and (b) variable-depth mixed layer (by lateral induction). Dashed curve shows bottom of mixed layer, solid curves (except for top one showing sea surface) represent isopycnals, arrows show horizontal (u_b) and vertical (w_b) speeds at the base of mixed layer, which has depth h. From Williams (1989), Figure 1, ©American Meteorological Society. Used with permission.

Horizontal velocity in the Ekman layer is mostly composed of Ekman velocity and geostrophic velocity. While the horizontal divergence of the Ekman transport is the Ekman pumping, the geostrophic transport can also have horizontal divergence. This is usually small compared to the divergence of the Ekman transport, but in some locations it is not negligible. Thus the vertical transport is only approximately given by Ekman pumping. Generally speaking, the thicker the mixed layer, the bigger the geostrophic transport.

The flow across the bottom of the mixed layer should not be confused with flow across isopycnals. Because (by definition) the density is roughly uniform through the depth of the mixed layer at any latitude–longitude location, isopycnals that are tilted in the pycnocline are vertical in the mixed layer (see Figure 4.10). As Figure 5.14a shows, water can flow into or out of the mixed layer without changing density as long as its horizontal velocity allows it to slide along an isopycnal. It is in the mixed layer where water typically flows across isopycnals through its horizontal motion. The change in density implied by this flow is accomplished by surface fluxes which heat, cool, freshen or salinify the water column in the mixed layer (see Subsection 4.2.2).

Chapter 4 discusses the large variations in mixed-layer depth that occur in time and space. Horizontal gradients in mixed-layer depth allow another mode of exchange between the mixed layer and the pycnocline, as explained in Subsection 4.2.2. This lateral induction mode has horizontal flow crossing the tilted mixed-layer base (Figure 5.14b). In that case an exchange can occur even if $w = 0$ at the mixed-layer base, with flow from thick to thin mixed-layer depths producing subduction. Climatological distributions of mixed-layer depth (Figures 4.8 and 4.9) show the mixed layer thinning towards the equator in the polar half of the subtropical gyre at the end of the given hemisphere's winter. The Sverdrup flow is also equatorward in these regions, so the large changes in mixed-layer depth there enhance subduction.

Huang and Qiu (1994) calculate the strength of subduction in the North Pacific Ocean from these mechanisms. Ekman pumping produces vertical velocities of 25–50 m/yr over most of the gyre (Figure 5.15a). The lateral induction, though generated by horizontal motion, can be quantified as an equivalent vertical velocity and reaches about 25 m/yr over the northern half of the subtropical gyre (Figure 5.15b). The total subduction rate is significantly enhanced by non-Ekman mechanisms (Figure 5.15c), reaching a high of over 75 m/yr just south of the Kuroshio Extension, where subduction due to Ekman pumping should be small. The subtropical North Atlantic has a similarly enhanced subduction rate of up to about 100 m/yr (Marshall et al., 1993).

Regions of deep winter mixed layer have a large annual cycle in mixed-layer depth. Water that exits a relatively shallow mixed layer in the summer may be re-entrained when the mixed layer deepens in the winter (Stommel, 1979). Thus water being pumped downward permanently separates from the mixed layer only if it is pumped out at a time when the bottom of the mixed layer is deep enough that it will not overtake the water parcel later. Because the properties (density, temperature, etc.) of water are reset by interaction with the atmosphere when the water is entrained into the mixed layer, it is the permanently subducted water that determines the characteristics of the pycnocline.

As discussed above, the equatorward Sverdrup flow in the subtropical gyres tends to push water parcels into regions of shallowing mixed layer, which facilitates permanent

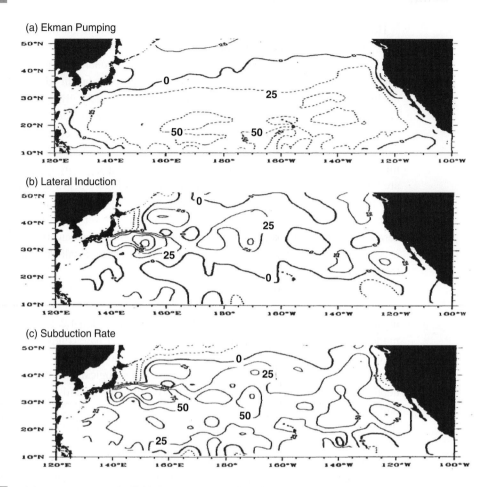

Figure 5.15 Subduction rate associated with (a) Ekman pumping, (b) lateral induction, and (c) all sources. Subduction rate (c) is the sum of (a) and (b) plus a small contribution from the divergence of the geostrophic flow within the mixed layer. Contour interval 25 m/yr. From Huang and Qiu (1994), Figure 5. ©American Meteorological Society. Used with permission.

escape from the mixed layer. A plot of mixed-layer depth and water parcel depths as a function of time (Figure 5.16a) shows the main processes determining which water avoids re-entrainment. Re-entrainment depends on the magnitude of the depth-cycle of the mixed layer, the downward velocity of the water, and the difference in local mixed-layer depth between winter of one season and winter a year later in the parcel's new location. A numerical model of the Atlantic with realistic topography and forcing (Williams et al., 1995) shows that "drifters" placed within the model subtropical gyre to follow water parcels do in fact get re-entrained into the mixed layer multiple times before Ekman pumping carries them deeply enough to leave the mixed layer for the last time (Figure 5.16b).

The Williams et al. (1995) model shows that for much of the subtropical North Atlantic, all the water in the permanent thermocline left the mixed layer for the last time during

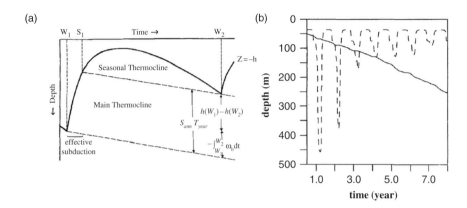

Figure 5.16   Illustration of Stommel's Demon. Depth of subducting water parcel and of mixed layer over time following parcel's horizontal trajectory in subtropical gyre for (a) schematic and (b) numerical model (depths of mixed layer and water parcel given by dashed and solid curves, respectively). From Williams et al. (1995), Figures 6 and 5, ©American Meteorological Society. Used with permission.

a brief (1–2-month) period near the end of winter. They refer to the process of selecting winter time water as **Stommel's Demon**, because Henry Stommel (Stommel, 1979) likened the selection of winter-only water for permanent subduction to the process of Maxwell's Demon sorting gas molecules between two connected containers. Stommel's Demon gives us some justification for neglecting the seasonal cycle in the permanent pycnocline: the surface properties from late winter are more important than the annual average.

5.2.5 Some Open Questions

Subpolar Gyres

The subpolar gyres pose several challenges to theorists trying to explain them. Therefore, theories of the subpolar gyre are less advanced than for the subtropics. The reasons are as follows.

In LPS theory, surface density is imposed via zonal outcrop lines for the layers. Observationally, surface isopycnals are more zonal in the subtropics than in the subpolar gyre, where there is a strong zonal gradient, causing isoypcnals to tilt northwards (Figure 4.6a). This complicates the LPS theory. In fact, the difficulty is more fundamental. The density outcrop distribution is intrinsic to the nature of subpolar and subtropical gyres. As we see in Subsection 5.1.1, subtropical gyre isopycnals form a bowl, and subpolar ones form a hill. The bowl can get as deep as the wind pushes it, short of reaching the sea floor, a limit outside the parameter range of the real ocean. For strong enough wind, the hill can pierce the surface, causing the "hilltop" region in the west to be denser than the hill's flanks in the east. This can be seen even in idealized shallow water models (numerical and analytical), in which the top-layer flow skirts a "hole" in the western half of the basin where top-layer thickness goes to zero (for instance, see Huang and Flierl, 1987; Chassignet et al., 1995). Therefore, the surface density distribution appears to be part of the solution to the circulation problem, rather than an externally imposed boundary condition.

The vertical Ekman speed is upwards (suction) in the subpolar gyre (Figure 3.1, Figure 3.2). Pedlosky (1996) argues that the upper ocean would retain the imposed surface density pattern and develop a diffusive boundary layer near the surface. In such a putative boundary layer, surface-intensified mixing would transform the density of upwelling water in order to match the imposed boundary condition. Even so, the Ekman suction means that isopycnic trajectories emerge from deep water towards the surface, rather than the other way round, as in the subtropical gyres. What sets the properties, such as potential vorticity, on the isopycnals is much less clear in the subpolar gyres than in the subtropical gyres.

The North Atlantic subpolar gyre is especially complex. Winter mixed layers are greater than 400 m thick (Figure 4.9a), with small regions reaching thousands of meters. Consequently, one premise of the subtropical theories – that most of the flow occurs below the mixed layer – is dubious there. The low stratification of the Atlantic (and the Pacific, to some extent) also means that interaction with bottom topography is more important than in the subtropics. This calls into question the validity of the Sverdrup balance itself.

Wind-Driven Pycnocline Flow and Thermocline Theory

It is tempting to view the theories described in this chapter as solving what is called the **thermocline problem** (or pycnocline problem). The thermocline problem is to explain the observed three-dimensional hydrographic structure given the surface forcing. We emphasize that the theory is far from complete, however, even conceptually. In Subsection 5.2.4 we described how the ventilated flow is influenced by horizontal and temporal gradients in mixed-layer thickness. These variations should themselves be explained by the theory rather than given from observations. Also, the mixed layer characteristics are plausibly influenced by the intensity of the western boundary current. But if true, that would violate an assumption of all variants of Sverdrup theory, namely that the interior circulation can be calculated without knowing any details of the western boundary. The tacit assumption is that the western boundary current just recirculates the water within each isopycnic layer (see Subsection 3.3.4).

Much of the vertical thickness of the pycnocline can be linked to the wind-driven bowl: isopycnals that are close to the surface at the eastern edge of the basin descend to their greatest depths just east of the western boundary current. In other words, the pycnocline thickness can be estimated by seeing how much geostrophy tilts the isopycnals (see Exercise 5.4). Figure 5.2 shows that this is not the whole story. For some isopycnals, such as $\sigma_\theta = 26.5$ in the Atlantic (Figure 5.2a), this is true, but much of the vertical variation in density occurs independent of this tilt. For instance, there is significant vertical stratification at the eastern boundary, so that even if we "shut off" the wind and allow the isopycnals to lay flat in longitude, it seems that there would be a pycnocline that was a few hundred meters thick. This is partly reflected by the free parameter of eastern boundary lower-layer thickness in LPS theory. Similarly, the unventilated theory of Rhines and Young (1982) assumes a given background stratification that cannot be derived from the theory itself. As described in Chapter 8, the missing ingredient for completing the explanation of pycnocline structure probably lies in diapycnal mixing and other global-scale forcing associated with the deep meridional overturning circulation. A full conceptual explanation for the thickness of the pycnocline is still missing.

5.3 Theory

5.3.1 Condition for Neglect of Mixing

In the following sections, we assume that flow below the mixed layer follows isopycnals, which is equivalent to saying that the density does not change for a water parcel. Below the euphotic zone (Subsection 6.1.2), water density only changes where there is mixing. Therefore, we now ask under what circumstances can mixing be neglected? The criterion for neglect of mixing is that water parcels do not drift far from their initial isopycnal in the time it takes them to circulate around the gyre.

In steady state, conservation equations for temperature T and salinity S, away from surface forcing, are:

$$\mathbf{u} \cdot \nabla T = \kappa_v T_{zz},$$
$$\mathbf{u} \cdot \nabla S = \kappa_v S_{zz} \tag{5.1}$$

(from (1.16) with vertical diffusivity κ_v). Here we have ignored lateral diffusion, which has a more subtle influence on density (see Chapter 7). For turbulent mixing, it is plausible that κ_v is the same for T and S (although for molecular mixing their values differ; see Subsection 1.3.1). If we also use the linear equation of state (1.6) and add β times the salinity equation above to $-\alpha$ times the temperature equation, we produce an advection-diffusion equation for density:

$$\mathbf{u} \cdot \nabla \rho = \kappa_v \rho_{zz}. \tag{5.2}$$

Now write the velocity in terms of its components in the horizontal and vertical directions, $\mathbf{u} = u\hat{\mathbf{x}} + v\hat{\mathbf{y}} + w\hat{\mathbf{z}}$, and divide the vertical velocity w into two components, $w = w_I + w_*$. The component w_I is the vertical speed needed for the water parcel to stay on an isopycnal (Figure 5.17). Similarly, w_* is the cross-isopycnal (or diapycnal) vertical velocity, which measures how fast the water is rising above or sinking below its original isopycnal (Figure 5.17). Mathematically, w_I is defined by

$$w_I \rho_z + u \rho_x + v \rho_y = 0, \tag{5.3}$$

so that density advection vanishes moving at velocity (u, v, w_I). Substituting the definition of w_I into the density equation, we are left with

$$w_* \rho_z = \kappa_v \rho_{zz}. \tag{5.4}$$

In a time Δt, a water parcel drifts a vertical distance $\Delta z = w_* \Delta t$ from its original isopycnal, so that

$$\Delta z = \Delta t \kappa_v \rho_{zz} / \rho_z. \tag{5.5}$$

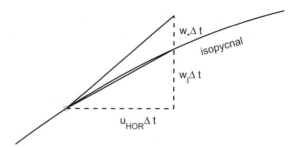

Figure 5.17   Particle trajectory illustrating definitions of w_I and w_*.

The value of Δz is estimated by scale analysis. The vertical length scale associated with vertical derivatives is the pycnocline thickness D, hence

$$\Delta z = \Delta t \kappa_v / D. \tag{5.6}$$

In the subtropical gyre pycnocline, it is reasonable to set $D = 500$ m, and $\kappa_v = 10^{-5}$ m^2/s. To estimate the timescale for flow in the gyre interior, we take $\Delta t = L/v_S$, where $L = 2000$ km is the meridional length scale for the gyre and v_S is a typical Sverdrup velocity. Taking the North Atlantic as an example, the volume transport is $\Psi \approx 30$ Sv (Chapter 3), and $v_S \approx \Psi/(MD) = 0.01$ m/s, where $M = 6000$ km is the width. This yields a timescale of about 6 yr, and inserting into (5.6), we arrive at an estimate of $\Delta z = 4$ m. This demonstrates that our assertion is plausible: the scale analysis shows that neglect of mixing across isopycnals makes a relatively small error in the trajectories.

5.3.2 Elements of Layer Models

To construct a multiple-layer model of the Sverdrup circulation, we need several basic elements. We consider an $N\frac{1}{2}$-layer model, consisting of N layers with nonzero velocities overlaying a bottom layer with zero velocity. Such an arrangement, also known as a reduced gravity model (because of the resting layer), is shown in Figure 5.11c.

Our first element consists of the horizontal momentum equations. We neglect the advection terms and momentum mixing, and we get the geostrophic relations (2.2), which in spherical coordinates (Appendix B) are:

$$f v_n = \frac{1}{\rho_0 R_E \cos \theta} \frac{\partial p_n}{\partial \phi},$$

$$f u_n = -\frac{1}{\rho_0 R_E} \frac{\partial p_n}{\partial \theta}. \tag{5.7}$$

Here θ is latitude (not potential temperature) and ϕ is longitude, R_E is the radius of the Earth (approximately 6400 km, Appendix C), f is the Coriolis parameter, ρ_0 is the reference density, and (u_n, v_n) are zonal and meridional geostrophic velocity components for layer n. In the Ekman layer, vertical mixing of momentum *is* important and cannot be neglected. Therefore, these geostrophic equations only apply to the ocean below the Ekman layer.

The next element is the hydrostatic relation, which connects the interface depths and pressure. As in Subsection 2.1.2, the hydrostatic pressure is given by (2.5). We are only interested in the dynamically active part of the pressure, which varies in (x, y), because the geostrophic velocity is proportional to horizontal gradients in p. Hence, following the arguments in Subsection 2.2.4 leading to (2.70), we find that

$$\nabla p_n = \nabla p_{n-1} + \rho_0 \gamma_{n-1} \nabla \eta_{n-1}, \tag{5.8}$$

where the gradient is in the horizontal plane and

$$\gamma_n = \frac{g}{\rho_0}(\rho_{n+1} - \rho_n) \tag{5.9}$$

is the reduced gravity ((2.68); see also Exercise 2.7).

So far, we have not used the fact that the deepest layer is at rest. By geostrophy, a motionless fluid has uniform dynamic pressure, and we can set $\nabla p_{N+1} = 0$. From (5.8),

$$\nabla p_{N+1} = \nabla p_N + \rho_0 \gamma_N \nabla \eta_N, \tag{5.10}$$

which can be rewritten

$$\nabla p_N = -\rho_0 \gamma_N \nabla \eta_N. \tag{5.11}$$

This means that the tilt in the deepest interface tells us the pressure gradient (hence the geostrophic velocity) in the deepest active layer. Thus the relationship between the isopycnal position and velocity is particularly simple in this layer. For the next deepest layer, we can use (5.8) again to find the pressure gradient based on the deepest two interfaces, and repeat all the way to the top of the water column. In general, the pressure gradient in the reduced gravity model is given by

$$\nabla p_n = -\rho_0 \sum_{m=n}^{N} \gamma_m \nabla \eta_m \tag{5.12}$$

(compare to (2.71)). Above the deepest active layer, the pressure gradient is based on a series of interface positions, and so the relationship is not as simple as in the deepest layer. However, (5.12) shows that given all the interface depths, we can calculate all the pressure gradients and thus the geostrophic velocities.

Another element we use is the Sverdrup balance for a layered model. As for a uniform-density fluid, the divergence of the geostrophic momentum equations and the continuity equation (2.62) yields

$$\beta v_n = f w_z \tag{5.13}$$

(β is the gradient of planetary vorticity, not the haline contraction coefficient). We integrate (5.13) vertically to obtain the layer-model version of the Sverdrup balance,

$$\beta \sum_{n=1}^{N} v_n h_n = f w_{Ek}, \tag{5.14}$$

where

$$h_n = \eta_{n-1} - \eta_n \tag{5.15}$$

is the layer thickness, and we assume that $w = w_{Ek}$ at the bottom of the Ekman layer, $w = 0$ in layer $N + 1$, and the geostrophic volume transport in the Ekman layer is negligible. A more subtle assumption is that w_z is defined at all the layer interfaces. This is only true if w is continuous at the interfaces. This is not completely obvious, as we know that $\mathbf{u}$ is *not* continuous. In steady state, just below the interface η_n, the condition that there is no flow through the interface implies

$$w_n = \mathbf{u}_n \cdot \nabla \eta_{n-1} \tag{5.16}$$

while just above the interface, the same condition gives

$$w_{n-1} = \mathbf{u}_{n-1} \cdot \nabla \eta_{n-1}. \tag{5.17}$$

It turns out that $w_n = w_{n-1}$, and so the integral (5.14) is valid (see Exercise 5.5).

The Sverdrup balance as defined in (5.14) can be rewritten purely in terms of the layer depths. We consider a $2\frac{1}{2}$-layer model. For that case, the pressure gradients for the two layers are given by

$$\nabla p_1 = -\rho_0 \gamma_2 \nabla \eta_2 - \rho_0 \gamma_1 \nabla \eta_1,$$
$$\nabla p_2 = -\rho_0 \gamma_2 \nabla \eta_2. \tag{5.18}$$

Inserting these expressions into the geostrophic relations, we obtain

$$v_1 = -\frac{1}{f R_E \cos \theta} \frac{\partial}{\partial \phi} (\gamma_2 \eta_2 + \gamma_1 \eta_1),$$
$$v_2 = -\frac{1}{f R_E \cos \theta} \frac{\partial}{\partial \phi} (\gamma_2 \eta_2). \tag{5.19}$$

The two-layer case of the Sverdrup balance is

$$\beta(v_1 h_1 + v_2 h_2) = f w_{Ek}. \tag{5.20}$$

Note that η_0 is much smaller than all other η_n because η_0, which represents sea surface height variations, is generally O(1) m in the ocean, while the other interfaces vary by O(100) m. Thus $h_1 = -\eta_1$ and $h_2 = \eta_1 - \eta_2$ to good accuracy. Inserting (5.19) for velocities and (5.15) for the thicknesses, finding that two of the terms cancel, and using the identity $\eta \eta_\phi = \frac{1}{2}(\eta^2)_\phi$, the Sverdrup equation becomes

$$\frac{\beta}{f R_E \cos \theta} \frac{\partial}{\partial \phi} \left(\frac{1}{2} \gamma_1 \eta_1^2 + \frac{1}{2} \gamma_2 \eta_2^2 \right) = f w_{Ek}. \tag{5.21}$$

This is a differential equation for η_1 and η_2. It is easy to turn it into an algebraic equation for η_1 and η_2 instead. Integrating from an arbitrary ϕ to the eastern boundary ϕ_E, we can write

$$\frac{1}{2} \gamma_1 (E_1^2 - \eta_1^2) + \frac{1}{2} \gamma_2 (E_2^2 - \eta_2^2) = (f^2/\beta) R_E \cos \theta \int_\phi^{\phi_E} w_{Ek} \, d\phi, \tag{5.22}$$

where E_n is η_n at the eastern boundary. It is convenient to define the parameter $\Gamma = \gamma_1/\gamma_2$, and to note that $x = \phi R_E \cos \theta$, so that the Sverdrup balance can be written as

$$\eta_2^2 + \Gamma \eta_1^2 = E_2^2 + \Gamma E_1^2 - \frac{2 f^2}{\beta \gamma_2} \int_x^M w_{Ek} \, dx, \tag{5.23}$$

where the eastern boundary is at $x = M$. This formula can also be written in terms of layer thickness, as in (6.57). Applying geostrophy at the eastern boundary tells us something about E_n. Equation 5.7 says that u_n, the flow into the boundary, is proportional to $\partial p_n / \partial \theta$. For the deepest active layer, (5.11) tells us p_N is proportional to η_N. Because there is no flow into or out of the wall at the eastern boundary, η_N must not vary with latitude, so E_N is a constant. And because E_N is flat along the wall, the layer above it must also be flat. Therefore, in general for any number of layers, each E_n is a constant. Chapter 6 shows that this rule must be modified when there is significant Ekman transport towards or away from the wall, but here we exclude such cases.

If we consider the situation where layer 1 outcrops (as in Figure 5.11c), then poleward of the outcrop latitude, $\eta_1 = 0$, and the Sverdrup balance simplifies to

$$\eta_2^2 = E_2^2 - \frac{2f^2}{\beta \gamma_2} \int_x^M w_{Ek} \, dx. \tag{5.24}$$

Alternatively, this is the Sverdrup balance for a $1\frac{1}{2}$-layer model (if we redefine the subscripts so that layer 2 $\rightarrow$ 1 in (5.24)). For the subtropical gyre, $w_{Ek} < 0$, so the Sverdrup balance becomes a description of how the top-layer thickens as we get further from the eastern boundary.

For a $1\frac{1}{2}$-layer model, the Sverdrup balance completely defines the solution. Two of the parameters, γ_2 and E_1, have an ambiguous status, however, because it is unclear how to specify them. γ_2 represents the density jump between the active and inactive layer. If we take the $1\frac{1}{2}$-layer model as a crude representation of the pycnocline (top layer) and abyss (bottom layer), then γ_2 can be estimated from the surface density difference between some central latitude of the subtropical gyre and some latitude poleward of the gyre. This density difference is determined by surface temperature and salinity balances, which are set by processes beyond the scope of this theory. It is less obvious how E_2 should be set. As we are postulating that the bottom layer is motionless, then our single active layer should not outcrop within the subtropical gyre. Such an outcrop would expose the "motionless" layer to Ekman pumping and then force motion within it. The theory gives us no further guidance on the value of E_2, and so we are free to choose a value for it.

For a $2\frac{1}{2}$-layer model, we can allow the top layer to outcrop within the subtropical gyre. The outcrop latitude θ_1 and the buoyancy jump γ_1 again must be imposed because they represent surface processes outside the scope of the theory. In reality surface isopycnals are not strictly latitude circles, and a more general theory would allow θ_1 to vary with longitude. Here we make the approximation that θ_1 is a constant. In any case, the fact of outcropping gives us $\eta_1 = 0$ at the outcrop latitude, and because E_1 must be a constant, we can set $E_1 = 0$. For models with more layers, $E_n = 0$ for all layers that outcrop within the subtropical gyre.

Once we have more than one active layer, the Sverdrup balance and the accompanying parameters are no longer enough to completely define the solution. For each successive layer, we must have one more equation constraining the system. In the following subsections, we look at these constraints and the resulting solutions.

5.3.3 Model of Ventilated Pycnocline

Continue to consider the $2\frac{1}{2}$-layer example, for which we solved the flow poleward of the outcrop latitude in Subsection 5.3.2. In the poleward (1-layer) part of the gyre, equatorward-flowing fluid in layer 2 eventually reaches the outcrop latitude, where layer 1 sits above layer 2. We need an additional equation to solve the system in the two-layer domain. It is the conservation of potential vorticity (Subsection 2.2.5). For layer n, the planetary potential vorticity (2.22) is

$$q_n = \frac{f}{h_n}. \tag{5.25}$$

Conservation implies that q_2 is uniform along a streamline so that $\mathbf{u}_2 \cdot \nabla q_2 = 0$. Geostrophy, (5.7), implies that

$$\frac{\partial p_2}{\partial \theta} \frac{\partial q_2}{\partial \phi} = \frac{\partial p_2}{\partial \phi} \frac{\partial q_2}{\partial \theta}, \tag{5.26}$$

which means that contours of p_2 and q_2 coincide (see Exercise 5.9 and (3.56)). Equation 5.11 shows that p_2 is proportional to η_2 and (5.15) relates h_n to the interface depths. Therefore, conservation of potential vorticity for layer 2 can be written

$$q_2 = \frac{f}{\eta_1 - \eta_2} = Q_2(\eta_2) \tag{5.27}$$

for function $Q_2(\cdot)$, which is to be determined.

Now we use a trick that perhaps appears too good to be true at first. At the outcrop latitude θ_1, where we define $f_1 = f(\theta_1)$, the upper layer thickness is zero (by definition of the outcrop layer), so that $\eta_1 = 0$. Then at that latitude, (5.27) is

$$\frac{f}{\eta_1 - \eta_2} = \frac{f_1}{-\eta_2}. \tag{5.28}$$

Equation 5.27 shows that the left-hand side equals some function of η_2, but does not say what that function is. The right-hand side of (5.28) happens to be a function of η_2 (f_1 is a constant), and so at the outcrop latitude, the function Q_2 that we are looking for is given by this expression. However, if $Q_2(\eta_2) = -f_1/\eta_2$ is true at the outcrop, it is true *everywhere* in the subduction region. We have found our additional constraint on η_1 and η_2. (Mathematically, this argument solves the hyperbolic equation $\mathbf{u}_2 \cdot \nabla q_2 = 0$ using the method of characteristics. The characteristics emanate from the outcrop latitude into the ventilated interior.)

In the subduction region (where layer 2 is beneath layer 1), (5.28) is

$$\eta_1 = \left(1 - \frac{f}{f_1}\right)\eta_2. \tag{5.29}$$

We can write the individual layer thicknesses in terms of the total pycnocline thickness, $h \equiv h_1 + h_2$. Using (5.29) and (5.15) relating the h_ns to the η_ns, we obtain

$$h_1 = \left(1 - \frac{f}{f_1}\right)h,$$

$$h_2 = \frac{f}{f_1}h. \tag{5.30}$$

Note that

$$\frac{h_1}{h_2} = \frac{f_1}{f} - 1, \tag{5.31}$$

is independent of longitude, density step, and forcing amplitude.

We still do not know the solution, because we have not solved for h. We use (5.23), the Sverdrup balance in terms of η_1 and η_2. If we write the η_ns in terms of the h_ns, and use (5.30), we can rewrite the Sverdrup balance as a solution for the total layer thickness:

$$h^2 = \frac{E_2^2 - (2f^2/\beta\gamma_2)\int_x^M w_{Ek}\,dx}{1 + \Gamma(1 - f/f_1)^2}. \tag{5.32}$$

We have used the fact that $E_1 = 0$, which was demonstrated in Subsection 5.3.2. Another way to write (5.32) is

$$h^2 = \frac{E_2^2 + d_0^2}{1 + \Gamma(1 - f/f_1)^2}, \tag{5.33}$$

where d_0 is the integral term in the numerator of (5.32):

$$d_0^2 = -\frac{2f^2}{\beta\gamma_2}\int_x^M w_{Ek}\,dx. \tag{5.34}$$

Note that d_0 represents the thickness of a $1\frac{1}{2}$-layer wind-driven gyre for the case of layer thickness going to zero at the eastern boundary (from (5.24)).

The solution can also be written in a way that emphasizes the dependence of h on ϕ, θ and the parameters. Here we use the special case that $w_{Ek}(\phi, \theta) = w_{Ek}(\theta)$, though the more general case can be analyzed in a similar way. Using the definitions of f and β, (2.1) and (2.63), and setting $w_{Ek} = W w'_{Ek}$, where W represents the amplitude of the Ekman pumping, we get

$$d_0^2 = -D_0^2(1 - \phi/\phi_E)w'_{Ek}\sin^2\theta, \tag{5.35}$$

where the constant

$$D_0 = -\frac{d_0(\phi = 0, \theta)}{w'_{Ek}\sin^2\theta} = \sqrt{4\Omega R_E^2 \phi_E W/\gamma_2}, \tag{5.36}$$

is a measure of how deep the interface would be at the western boundary (and at $\theta = \theta_1$) for $E_2 = 0$.

The solution portrayed in Figure 5.12 is calculated with

$$w'_{Ek} = -\cos\pi\frac{\theta - 30^o}{30^o}, \tag{5.37}$$

where θ is in degrees latitude. This example sets Ekman pumping to zero at the northern and southern boundaries of the domain, which are at 15°N and 45°N, similar to the real northern-hemisphere subtropical gyres. (As shown in Chapter 3, however, w_{Ek} does not go to zero at the equatorward border of the subtropical gyres). Parameters used for Figure 5.12 are $\theta_1 = 38^o$, $\Gamma = 2$, $E_2 = 100$ m, and $D_0 = 400$ m. These are all plausible parameters for the real ocean.

The solution given by (5.33), (5.35), and (5.36) can be used to define the location of the shadow zone discussed in Subsection 5.2.3. Each constant-h contour represents a layer 2 streamline in Figure 5.12. Therefore, if we take h to be given, we can define a constant-h curve by solving (5.33), (5.35), and (5.36) for $\phi(\theta)$:

$$\frac{\phi}{\phi_E} = 1 - \frac{(E_2/D_0)^2 - (h/D_0)^2[1 + \Gamma(1 - f/f_1)^2]}{w'_{Ek}\sin^2\theta}. \tag{5.38}$$

The border of the shadow zone is the layer-2 streamline that intersects the outcrop line $(\theta = \theta_1)$ at $\phi = \phi_E$. From (5.33), (5.35), and (5.36), $h = E_2$ at (ϕ_E, θ_1), and so the boundary of the shadow zone is given by

$$\frac{\phi}{\phi_E} = 1 + (E_2/D_0)^2\Gamma\frac{(1 - f/f_1)^2}{w'_{Ek}\sin^2\theta}. \tag{5.39}$$

Similar arguments define the location of the pool (see Exercise 5.10).

5.3.4 Model of Unventilated Pycnocline

In Subsection 5.3.3, we investigate the subtropical gyre flow for a $2\frac{1}{2}$-layer case in which layer 2 is ventilated by flow from higher latitudes. Here we solve the equations for a $2\frac{1}{2}$-layer flow in which the lower active layer is completely unventilated, as in Figure 5.11b. As discussed in Subsection 5.2.3, the unventilated layer is driven by potential vorticity homogenization. We want to find the interface depths in both layers and the boundary between the motionless and moving regions in layer 2.

In Subsection 5.2.3, we saw that those parts of layer 2 that are linked to a solid boundary by a curve of constant q_2 are motionless. Figure 5.13 showed that there was also a region in which the q_2-contours formed closed loops, allowing flow to move there.

To find the boundary between the two regions, we start by assuming that near the eastern boundary, layer 2 is motionless. In that region, the bottom interface is flat, so that $h = H = -E_2$, the eastern boundary value. We want to find contours of f/h_2, so we just need to find h_1 and use the fact that $h = h_1 + h_2$. We have already solved the one-layer problem (see (5.24)). Here we assume that $w_{Ek} = w_{Ek}(\theta)$ and use the notation in Subsection 5.3.3, and we get

$$h_1^2 = H_1^2 - (D_0^2/\Gamma)(1 - \phi/\phi_E)w'_{Ek}\sin^2\theta, \tag{5.40}$$

where $H_1 = -E_1$ is the eastern boundary value of h_1. It follows that the potential vorticity is given by

$$q_2 = \frac{2\Omega\sin\theta}{H - \sqrt{H_1^2 - (D_0^2/\Gamma)(1 - \phi/\phi_E)w'_{Ek}\sin^2\theta}} \tag{5.41}$$

or, setting $\delta_1 \equiv H_1/H$,

$$q_2 = \left(\frac{2\Omega}{H}\right)\frac{\sin\theta}{1 - \sqrt{\delta_1^2 - (D_0/H)^2(1 - \phi/\phi_E)w'_{Ek}\sin^2\theta/\Gamma}}. \tag{5.42}$$

While we could plot $q_2(\phi, \theta)$ contours, it is more revealing to calculate the curve in the ϕ-θ plane corresponding to each contour. It is straightforward to solve (5.42) for $\phi(\theta)$. Each

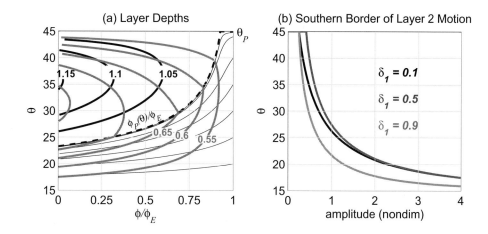

Figure 5.18 (a) $2\frac{1}{2}$-layer unventilated solution for $\delta_1 = 0.5, D_0/H = 1.5$, and $\Gamma = 1$. Contours of q_2 (thin), border of motionless-layer-2 region (dashed), h_1/H (thick gray), and h/H (thick black). (b) Latitude of $\phi_p = 0$, the intersection of the border with the western boundary, as function of D_0/H, for $\delta_1 = 0.1$ (black), 0.5 (dark gray), and 0.9 (light gray).

contour is defined by its value of q_2, but it is more convenient to define it by the latitude θ_E at which it intersects the eastern boundary. At the eastern boundary, (5.42) becomes

$$q_2 = \left(\frac{2\Omega}{H}\right)\frac{\sin\theta_E}{1-\delta_1}, \qquad (5.43)$$

which upon substitution back into (5.42) gives

$$\frac{\phi}{\phi_E} = 1 - \Gamma\left(\frac{H}{D_0}\right)^2\frac{\delta_1^2 - [1-(1-\delta_1)\sin\theta/\sin\theta_E]^2}{w'_{Ek}\sin^2\theta}. \qquad (5.44)$$

We illustrate these $\phi(\theta)$ contours with the thin curves in Figure 5.18a. The Ekman pumping w_{Ek} is the same as in Subsection 5.3.3. The q_2 contour $\phi_p(\theta)$ (dashed curve) that intersects the poleward eastern corner of the subtropical gyre ($\phi = \phi_E$, $\theta_E = \theta_p$) is the boundary of the region in which layer 2 is motionless. (Actually as $\theta_E \to \theta_p$, the q_2 curves asymptote to a curve that reaches θ_E offshore of the eastern boundary and continues eastward at θ_E) Contours of q_2 to the west of $\phi_p(\theta)$ intersect the western boundary of the Sverdrup interior and hence form closed loops when continued into the western boundary current (outside the scope of our solution). This is similar to Figure 5.13b, except with the western half of the domain squashed into a very narrow layer next to the western boundary. Therefore, west of $\phi_p(\theta)$, motion is allowed in layer 2.

West of $\phi_p(\theta)$, we assume that q_2 is homogenized by eddies and is therefore equal to its value on the boundary. In this moving-layer-2 region, we no longer have the constraint $h = H$, but we have a new constraint because $q_2 = f/(h - h_1)$ is known. Using the q_2 on $\phi_p(\theta)$, we get

$$h - h_1 = H(1-\delta_1)\frac{\sin\theta}{\sin\theta_P}. \qquad (5.45)$$

We can now combine this with the Sverdrup balance to solve for both h and h_1 in the moving-layer-2 region.

As in Subsection 5.3.2, we zonally integrate the two-layer Sverdrup balance (5.21) to create an equation in h and h_1. In this case, we do not integrate to the eastern boundary of the basin but to $\phi_p(\theta)$, the eastern boundary of the moving-layer-2 region. The Sverdrup balance then becomes

$$h^2 + \Gamma h_1^2 = H^2 + \Gamma h_p^2 - D_0^2(\phi_p/\phi_E - \phi/\phi_E)w'_{Ek}\sin^2\theta, \qquad (5.46)$$

where $h_p(\theta) = h_1(\phi_p, \theta)$ calculated in the motionless-layer-2 region. From (5.40) and (5.44), and setting $\theta_E = \theta_p$, $h_p(\theta)$ has a relatively simple form:

$$h_p = H\left[1 - (1 - \delta_1)\frac{\sin\theta}{\sin\theta_p}\right]. \qquad (5.47)$$

Inserting (5.47), (5.45) and (5.44) into (5.46), we get expressions for h/H and h_1/H as functions of the coordinates, w'_{Ek}, and the nondimensional parameters δ_1, Γ, D_0/H. Contours of h/H and h_1/H are plotted in Figure 5.18a.

For a given domain and w'_{Ek}, the size of the moving-layer-2 region depends on the parameters δ_1 and D_0/H. A measure of the size of that region is the latitude θ_0 at which the border of the moving-layer-2 region intersects the western boundary ($\phi_p = 0$). This can be found by setting $\theta_E = \theta_p$ and $\phi = 0$ in (5.44). It is difficult to solve for $\theta_0(\delta_1, D_0/H)$, but easy to plot D_0/H as a function of θ_0 as we do in Figure 5.18b for several values of δ_1. The stronger the forcing parameter D_0/H, the further south θ_0 moves, thus enlarging the region where layer 2 is in motion.

Using three active layers, it is possible to combine the model of the unventilated pycnocline with the ventilated model of Subsection 5.3.3 (Pedlosky and Young, 1983; Liu et al., 1993). As with two active layers, the Sverdrup balance provides one constraint on the interface depths, while potential vorticity rules appropriate to either ventilation, homogenization, or no motion are applied to individual regions. Motion in the "pool" of the ventilated theory can also be solved by potential vorticity homogenization. Using three active layers produces a more complicated solution with more subregions than two active layers. Similarly, we can subdivide the water column into more active layers in order to get a more accurate picture of the vertical structure. In practice, the growing complexity of the solution has deterred attempts at solution with more than three layers.

5.4 Excursions: Henry Stommel Worried That He Came Too Late to Oceanography

Henry Stommel (1920–1992) had an unparalleled influence on physical oceanography. Like many in the field, he drifted into it through a series of accidents. He joined the tongue-in-cheek "Society of Subprofessional Oceanographers" after a Naval Research Laboratory newsletter referred to scientific staff without PhDs as "subprofessionals." The leader of ocean theory for half a century was rejected by the Scripps Institute of Oceanography

(SIO) graduate program, a rejection that he attributed to the SIO Director's irritation at Stommel having written a book about oceanography while still a novice (Veronis, 1981).

A PhD is supposed to indicate that an individual has demonstrated an ability to identify a scientific question and answer it in a way that can stand up to the questioning of experts. One can have these skills without having a PhD.

The Scripps director was Harald Sverdrup (1888–1957). Sverdrup is best remembered now for the Sverdrup Relation (Subsections 3.2.2 and 3.3.2; Sverdrup, 1947), but it was Stommel who first found a more complete solution that included (and explained) the western boundary currents, a key result in Chapter 3 of this book. Stommel deduced the existence of deep western boundary currents (Subsection 8.2.4), by analogy with the wind-driven gyres, and did important work in establishing the idea of an advective-diffusive balance (Subsection 8.2.2). Thus large parts of Chapter 8 also owe a debt to Stommel. In his 60s, he collaborated with colleagues at Woods Hole Oceanographic Institution (WHOI) to use the idea of subduction to extend the Sverdrup relation to a stratified flow (Luyten et al., 1983), a central idea of Chapter 5. Even later, a fanciful essay (Stommel, 1989) helped launch the idea that became the Argo network (Chapter 1). Several sections of Chapter 11 discuss the idea of multiple ocean states caused by differing surface boundary conditions for freshwater and heat, an idea originated in Stommel (1961).

This text book might have been much shorter if Stommel had stuck to his original field, astronomy. He remarked that when he began in oceanography, Sverdrup, Johnson, and Fleming's monumental textbook *The Oceans* (Sverdrup et al., 1942) depressed him: it gave the impression that there was nothing left to discover in physical oceanography. Occasionally even Stommel could be wrong.

Exercises

5.1 Use suitable hydrographic climatologies (see Appendix A) for this question and include the Arctic Ocean.

   (a) Make global maps of the differences in potential temperature, potential density, and salinity between the surface water in winter and the bottom water.
   (b) Using the maps from part (a), identify and label the boundaries between the alpha oceans and the beta oceans. Explain your reasoning.
   (c) Identify the places in the global ocean with the strongest and weakest vertical stratification, based on the maps of potential density differences from part (a).
   (d) Identify and comment on the relationship between the alpha and beta oceans from part (b) and the stratification strength from part (c).

5.2 Consider Figures 5.6, 5.8, and 5.9. Using a suitable hydrographic climatology for the North Atlantic Ocean (see Appendix A):

   (a) Make maps of dynamic height on depth surfaces, corresponding to Figure 5.6.
   (b) Make maps of dynamic height on isopycnic surfaces, corresponding to Figure 5.8. Include maps of the depth of the isopycnic surfaces.

(c) Make maps of normalized planetary potential vorticity, corresponding to Figure 5.9.

(d) Hence, compare and contrast the upper pycnocline circulation in the North Atlantic and North Pacific subtropical gyres.

5.3 Use a suitable source of hydrographic profile data to answer the following questions. Include profiles at a wide range of latitudes.

(a) Compute the density stratification N^2 for your data. Compute the temperature stratification $\alpha\,\partial T/\partial z$ and salinity stratification $\beta\,\partial S/\partial z$. Explain (with a formula) how N^2 relates to these quantities defining all variables.

(b) Make a plot of your data in a space spanned by $\beta\,\partial S/\partial z$ on the x-axis and $\alpha\,\partial T/\partial z$ on the y-axis.

(c) Add contours of N^2 on your plot and identify the parts of the space that are statically (gravitationally) stable and unstable.

(d) An alpha (beta) ocean may be defined as one for which temperature (salinity) stratification, $\alpha\,\partial T/\partial z$ ($\beta\,\partial S/\partial z$), dominates the density stratification N^2. Mark on your plot the alpha and beta ocean regimes.

(e) Specifically, the degree of alpha- or beta-ness, K^2, may be quantified by contours orthogonal to the N^2 contours on your plot. Add these contours and write a formula that defines K^2 (sometimes called the **spice** stratification).

(f) Label the data points on your diagram that are alpha and beta oceans and state where they occur in the world ocean.

(g) Salt-fingers occur in alpha oceans where the salinity stratification is destabilizing. Label the salt-finger regime on your plot and state where in the world ocean your salt-fingers occur.

(h) Double diffusion occurs in beta oceans where the temperature stratification is destabilizing. Label the double diffusion regime on your plot and state where in the world ocean your double diffusion occurs.

Hint: Stewart and Haine (2016); IOC, SCOR, and IAPSO (2010), and the TEOS-10 software are helpful.

5.4 Given eastern edge thicknesses D_n ($n = 1, 2$ for top and bottom layers, respectively), horizontal length scale L, maximum Sverdrup transport Ψ, density difference $\Delta\rho$ between layers, and Coriolis parameter f, what is the difference in depth Δh between the deepest part of the bowl and the edges of the bowl? Given reasonable values based on observations of the real ocean, what is a reasonable prediction of Δh? How good an estimate is this for Δh in the gyres of the real ocean?

5.5 Use geostrophy to show that (5.16) and (5.17) in Subsection 5.3.2 are equal. Does the equality of vertical speeds either side of the interface still hold if the flow is *not* geostrophic?

5.6 Consider an idealized depth–longitude section through a subtropical gyre (excluding the western boundary current), as shown in Figure 5.19. The density structure is approximated by two layers, each of uniform density. Assuming there is no flow in the bottom layer, derive an algebraic formula for volume transport as a function of the parameters shown in the figure and any other parameters that are needed. For

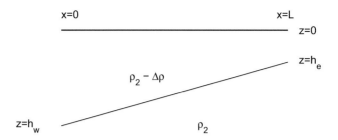

Figure 5.19 Idealized depth–longitude section through a subtropical gyre for Exercise 5.6.

parameter values appropriate for an approximation to the North Atlantic subtropical gyre density section (Figure 5.2a) what is the volume transport?

5.7 Consider a zonal re-entrant channel of zonal width $\Delta\phi$ and meridional width $\Delta\theta$ at average latitude θ_0. A meridional Gaussian ridge obstructs the channel. The ridge has zonal characteristic width $\delta\theta$ and rises a height ΔH above the otherwise flat sea floor at depth H.

(a) Write formulae for the water depth as a function of latitude and longitude and hence for the planetary potential vorticity of a homogeneous ocean filling the channel.

(b) Now consider that a two-layer ocean occupies the channel, with layer densities ρ_1, ρ_2 ($\rho_2 > \rho_1$) and average layer depths H_1 and H_2. Write expressions for the planetary potential vorticity in each layer. Under what circumstances can the channel carry a nonzero net zonal volume flux that is unforced and undamped? What restrictions on the zonal volume flux apply? Make sketches of the layer interfaces to explain your answer and assume the flow is geostrophic, adiabatic, inviscid, and that potential vorticity is homogenized within closed contours.

(c) Apply these ideas to the Antarctic Circumpolar Current and the Kerguelen Plateau (see Figure 9.1) and comment on what you find.

5.8 Given hydrostatic and geostrophic relations for a $2\frac{1}{2}$-layer ocean, prove that the Sverdrup balance can be written as (5.21), defining all the symbols carefully.

5.9 Consider (5.26) and using geometric arguments (or otherwise) demonstrate that p_2 and q_2 contours are aligned. Hence prove (5.27).

5.10 Using similar arguments to define the edge of the shadow zone, (5.39), find a parametric formula defining the edge of the pool.

5.11 Consider a $2\frac{1}{2}$-layer ocean in a rectangular basin with layer thicknesses $h_1(\phi,\theta)$, $h_2(\phi,\theta)$ and $h = h_1 + h_2$, with (longitude, latitude) (ϕ,θ).

(a) Use hydrostatic and geostrophic relations to write an expression for the meridional volume transport in each active layer. Your expression should be a function of h, h_1, and any relevant parameters.

Hint: These expressions will contain a zonal integral.

(b) For the case of an unventilated lower layer and Ekman pumping $w_{Ek}(\theta)$ (independent of ϕ), show that the integral(s) in the answer to (a) can be evaluated (that is, converted to expression(s) that do not contain integrals). Use any results from Chapter 5 that are relevant.

5.12 For a $2\frac{1}{2}$-layer ocean with lower layer ventilated in the north, write a computer program to make a contour graph of lower interface depth $h(\phi, \theta)$ with

$$w_{Ek} = -W \cos \pi \frac{\theta - \Delta\theta}{\Delta\theta}, \qquad (5.48)$$

where the domain is between $15°N$ and $45°N$ and $\Delta\theta = 30°$, and the subduction latitude at which $h_1 = 0$ is $\theta_1 = 30°N$. For your graph, normalize h by the eastern boundary value H and ϕ by basin width ϕ_E. Contours of h should include the regions both north and south of θ_1. What nondimensional parameters do you need to know to make the graph? Assign values that are reasonable for the real ocean when you do this.

5.13 Consider a $3\frac{1}{2}$-layer model of the ventilated pycnocline of Luyten et al. (1983) (see also Vallis, 2006, Chapter 16). Using the arguments explained in Subsection 5.3.3:

(a) Write formulae for the thicknesses of layers 1, 2, and 3, in the region where layer 1 is outcropping and layers 2 and 3 are submerged.
(b) Compute and hence sketch the solution for the depths of the interfaces between layers 1, 2, and 3 for: (i) a meridional section at mid-basin, and, (ii) a zonal section at mid-basin. Make the following assumptions:
   • The basin geometry is the same as in Exercise 3.8 with $b = 4000$ km.
   • The wind stress is of the same form as in Exercise 3.8 with $F\rho_0 = 0.1\,\mathrm{N\,m^{-2}}$.
   • Layer 1 has density $1026\,\mathrm{kgm^{-3}}$, and the layer 1/layer 2 outcrop is at $y/b = 0.33$.
   • Layer 2 has density $1026.2\,\mathrm{kgm^{-3}}$, and the layer 2/layer 3 outcrop is at $y/b = 0.66$.
   • Layer 3 has density $1026.5\,\mathrm{kgm^{-3}}$, thickness 400 m at $x = b$ where it outcrops, and the layer 3/layer 4 outcrop is at $y/b > 1$.
   • Layer 4 has density $1027\,\mathrm{kgm^{-3}}$ and is resting.
   Make other reasonable parameter choices, appropriate for the North Atlantic ocean.
(c) Sketch the (x, y) trajectories of a parcel of layer 3 water starting at $(x = b, y = 0.66b)$.
(d) Repeat (c) for the trajectory starting at $(x = 0, y = 0.66b)$. What is the significance of the trajectories in (c) and (d)?

5.14 This question concerns the so-called **beta spiral**. Consider steady flow in a subtropical gyre under thermal wind balance with material conservation of density. Assume also that the geostrophic divergence formula (2.62) applies.

(a) Write the corresponding equations for the velocity field $\mathbf{u} = (u, v, w)$ in terms of the height η of an isopycnal and other parameters that you define.
(b) Hence write a single equation for (u, v) in terms of horizontal derivatives of η.

(c) Deduce the change of direction of the horizontal flow with height z. Compare your result to Figure 5.12 and comment.

(d) Discuss how this idea might be used to infer the absolute geostrophic velocity from observations of η alone, and hence beat the level-of-no-motion problem (Subsection 2.2.2).

6 Equatorial Circulation, Shallow Overturning, and Upwelling

Equatorial ocean currents differ from currents at higher latitudes. The magnitude and direction of the flow, which is strong and mainly zonal, vary sharply with depth and latitude. In places the surface and subsurface water flows in opposite directions. The strong constraint of geostrophic balance breaks down at the equator. Surface Ekman transport is not necessarily negligible, and Ekman suction and pumping play a larger role than in the extra-tropics. Thus the circulations are fully three-dimensional, in the sense that the flow varies with both horizontal position and depth and the vertical current cannot be disregarded. In this chapter we explore low-latitude circulation and the associated vertical flow. The three-dimensional equatorial circulation is synthesized with the subtropical gyres. Strong upwelling regions at eastern boundaries, which are related, are also discussed.

6.1 Observations

6.1.1 Equatorial Currents and Subtropical Cells

Near the equator, the circulation is qualitatively different than the gyre flow discussed in Chapter 5 (and Chapter 3). Specifically, it is stronger, more zonal, and varies more with depth. These features are shown most clearly in the Pacific, although they apply to the equatorial Atlantic and Indian Oceans too. The surface flow (Figure 6.1a) consists of the eastward **North Equatorial Countercurrent (NECC)** sandwiched between westward flow known as the **North Equatorial Current (NEC)** and **South Equatorial Current (SEC)**. The South Equatorial Current spans the equator despite its name; the "South" refers to its position relative to the North Equatorial Countercurrent rather than to the equator. The subsurface flow within the pycnocline (Figure 6.1b) also has a South Equatorial Current, North Equatorial Current, and North Equatorial Countercurrent, but right on the equator there is a strong eastward current known as the **Equatorial Undercurrent (EUC)** (Cromwell et al., 1954; see also Figure 1.1).

Cross sections of the EUC (Figure 6.2) show that the climatological current is a strong and narrow jet, with peak speed of about 1 m/s and typical width of 3–4°. Stretching from New Guinea to South America, the Pacific EUC is one of the longest and strongest

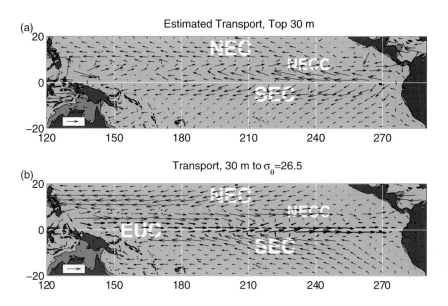

Horizontal velocity integrated between (a) surface and 30 m and (b) 30 m depth and the depth of the $\sigma_\theta = 26.5$ surface. Reference arrow over Australia shows 10 m^2/s. Velocity in (a) is from drifter-derived climatology of velocity at 15 m depth (Lumpkin and Garraffo, 2005) and in (b) is a blend of smoothed geostrophic velocity estimate from WOA density and 2000 m level of no motion (see Appendix A), and Johnson et al. (2002) near-equator zonal velocity measurements. Thicker arrows less than 5° from the equator are drawn half as long as the rest of the field.

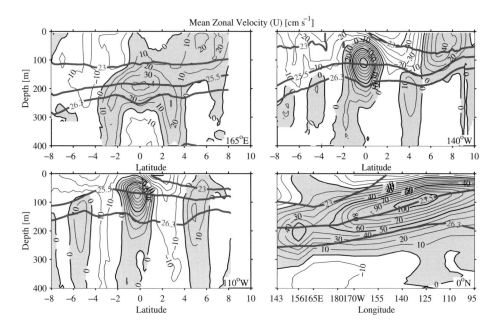

Equatorial Pacific zonal velocity (thin contours, cm/s; eastward flow shaded) and σ_θ (thick contours) from repeat ADCP and CTD sections. Panels show meridional sections except for lower-right panel showing equatorial section. From Schott et al. (2004), Figure 5, ©2004 American Geophysical Union. All rights reserved. Used with permission.

currents of the ocean, though it is mostly hidden from surface observations. The Atlantic has a similar EUC (Halpern and Weisberg, 1989; Johns et al., 2014). In the Pacific, the core of the current rises from about 200 m depth at 165°E to less than 80 m at 110°W. As the lower right panel of Figure 6.2 shows, the upward tilt in the zonal velocity field is similar to the upward tilt in density (and temperature; not shown), with the peak velocity at any longitude occuring at about $\sigma_\theta = 25.5$. The flow is mostly isopycnal, though there may be some cross-isopycnal flow associated with water warming and lightening due to mixing (Bryden and Brady, 1985; Brown and Fedorov, 2010).

Additional eastward jets a few degrees north and south of the equator extend downward from the flanks of the EUC (Figure 6.2). These are known as **Subsurface Counter Currents** (Tsuchiya, 1975) or **Tsuchiya Jets** (Rowe et al., 2000); the southern one was called a **South Equatorial Countercurrent** by Reid (1959), who first described it. The jets diverge from west to east: they border the EUC in the west (Figure 6.2, upper left), are detached from the EUC at middle longitudes (Figure 6.2, upper right), and (for the northern jet) merge with the NECC in the east (Figure 6.2, lower left). Like the EUC, the jets shallow and become warmer and lighter as they flow. Because the tropical gyre is stronger in the northern hemisphere than in the southern (Figure 3.6), the northern jet is harder to distinguish as a purely subsurface current than the southern jet.

Although one can imagine a closed cell along the equator in which the EUC and the westward surface flow replenish each other, the relatively high density of the EUC core suggests that this water comes from elsewhere. Solar heating warms and lightens water surfacing in the east. This warm, light, surface water must somehow become much denser in order to repeat the cycle, and there is no plausible mechanism to cool the water enough at the equator. Instead, the water comes from higher latitudes, where the surface properties are similar to those at the western end of the EUC. For instance, local winter $\sigma_\theta = 25.5$ at the surface around 30–40°N and about 30°S (Figure 4.6). Subsection 6.2.1 discusses the dynamics of the EUC.

Consistent with an extra-equatorial source for the water of the EUC, the large-scale currents in the pycnocline have an equatorward component (Figure 6.1b), in contrast to the generally poleward flow near the surface (Figure 6.1a). In the southern hemisphere, there appears to be a direct path to the EUC in the mid-Pacific (especially near 210°E) as well as flow to the western boundary. In the northern hemisphere, most of the flow appears to travel to the west. Pycnocline water also flows equatorward in **low latitude western boundary currents**, including the **Mindanao Current** in the North Pacific off the Philippines (Figure 6.3a) and the **New Guinea Coastal Undercurrent** in the South Pacific along the northern coast of New Guinea (Figure 6.3b).

The meridional flows close circulations left open in Chapter 5. In the subtropics, there is equatorward flow in the Sverdrup interior of the pycnocline. The equatorward and westward flow (as in Figure 5.8) feeds the EUC via the low latitude western boundary currents and some interior flow; this water flows eastward and upward until it reaches the surface and heads westward again. In Chapter 5 we mostly neglect the Ekman flow as small, but the magnitude of Ekman transport, $|\mathbf{U}_{Ek}| = |\mathbf{X}/f|$ from (2.11), becomes large in the tropics as $f \to 0$ (Figure 6.4). At the surface, water flows poleward towards the subduction

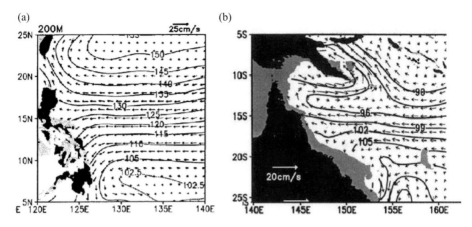

Figure 6.3 Geostrophic velocity (arrows) based on 0.5-degree resolution gridded density and 1200 m level of no motion. (a) Western North Pacific, 300 m depth. From Qu and Lukas (2003), Figure 8, ©American Meteorological Society. Used with permission. (b) Western South Pacific, 200 m depth. From Qu and Lindstrom (2002), Figure 6, ©American Meteorological Society. Used with permission. In both panels, contours represent dynamic pressure used to calculate geostrophic velocity (see Subsection 2.2.2).

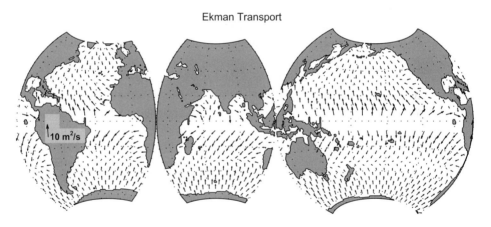

Figure 6.4 Surface Ekman transport. Arrow lengths are proportional to the square root of vector magnitudes. Data are from the SCOW wind climatology (averaged to $1°$ squares; see Appendix A).

regions as Ekman transport and (in the subtropical gyres) as shallow water in the western boundary currents. These circuits (one in each hemisphere) – subtropical subduction, pycnocline flow to the equator, equatorial upwelling, and return in the surface flow – are known as the **Subtropical Cells** (STCs).

6.1.2 Upwelling and Downwelling

Subsection 6.1.1 shows that basic features of the horizontal flow imply vertical flow as well. The vertical component, w, of the ocean velocity field is generally much smaller than the horizontal components. A relatively strong vertical speed may be only 3×10^{-6} m/s, or 100 m/yr. Such small speeds can be of great importance because ocean properties change quickly with depth. A water parcel excursion of tens of meters in the vertical may be as significant as hundreds of kilometers in the horizontal.

Wind and Current Measurements

Vertical speeds are too slow to directly measure with either mechanical or electronic current meters. To infer w, we measure the horizontal velocity $\mathbf{u}_h$ and relate it to w with the continuity equation (1.16),

$$\nabla \cdot \mathbf{u}_h + w_z = 0. \tag{6.1}$$

If we integrate (6.1) between depth h (assumed horizontally uniform here) and sea surface η (which we set to 0 under the approximation that $\eta \ll h$), and if $\bar{\mathbf{u}}_h$ is the depth average velocity over this depth range, we get

$$w(-h) = w(\eta) + h\nabla \cdot \bar{\mathbf{u}}_h, \tag{6.2}$$

where we have explicitly written the dependence of w on height. The $w(\eta)$ term is caused by vertical motion of the sea surface, flow up or down horizontal gradients in η, and flow through the surface due to evaporation and precipitation. All these terms are much smaller than subsurface w of interest, so we neglect $w(\eta)$ and estimate $w(-h)$ from $\bar{\mathbf{u}}_h$.

In the top 100 m, we expect the vertical velocity to be approximated by the divergence of Ekman transport. Over most of the ocean, the horizontal Ekman transport is dominated by meridional flow, with a strong tropical and subtropical flow away from the equator, and somewhat weaker equatorward flow at high latitudes (Figure 6.4). Near the eastern boundaries of basins, Ekman transport also has a strong zonal, offshore component. Assuming $h\bar{\mathbf{u}} = \mathbf{U}_{\mathbf{Ek}}$ and no normal flow through coasts, (6.2) gives the Ekman pumping/suction (Figure 6.5a; see also (2.13)). Note that equatorial w is based on Ekman transport at $\pm4°$ latitude, because the Ekman formula becomes inacurate near the equator. Ekman pumping predicts downwelling in the subtropical gyres, upwelling in the subpolar gyres, and intense upwelling near the equator in the Atlantic and Pacific as well as along some coasts, especially on the eastern boundaries of the basins (Figure 6.5a).

Ekman velocity, derived from the wind field, is not actual ocean near-surface velocity but a prediction of the ageostrophic flow (Subsection 2.2.1). Data from surface drifters attached to a drogue at 15 m depth (Subsection 1.1.3) comprise the basis of a climatology of near-surface velocity (Lumpkin and Garraffo, 2005). The resulting flow (Figure 6.6) is similar to the geostrophic near-surface circulation (Figure 3.3), with subtropical and subpolar gyres and the Antarctic Circumpolar Current. The drifter observations include both geostrophic, Ekman, and any other ageostrophic components. Flow clearly diverges from the equator in the Atlantic and Pacific, and (less clearly) the water flows away from the eastern boundary off Peru, California, the west coast of southern Africa, and elsewhere. The divergent flow is opposite the geostrophic flow (Figure 3.3), which is *toward* the equator and these eastern boundary regions.

(b) Curl of Wind Stress

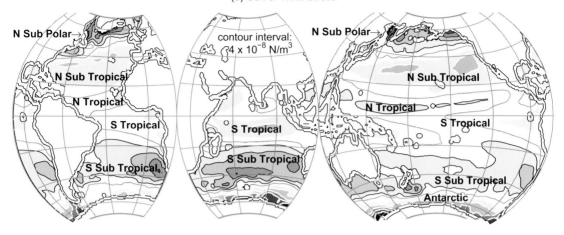

Figure 3.1b Annual average wind stress data from the SCOW product averaged on 1° cells. See Appendix A and Risien and Chelton (2008). (a) Wind stress, (b) wind stress curl. For (b), contour interval $= 4 \times 10^{-8}$ N/m^3, black and white indicating positive (counterclockwise) and negative (clockwise) values, respectively.

(a)   Estimated Transport Stream Function

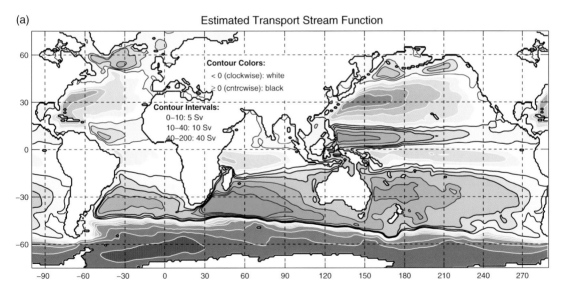

Figure 3.6a Horizontal volume transport streamfunction for (a) time-average velocity from the Estimating Circulation and Climate of the Ocean project (Appendix A; data courtesy of Carl Wunsch). Contours are white for negative values (clockwise circulation) and black for 0 and positive values (counterclockwise). Streamfunction is measured with respect to reference value at eastern boundary of the Pacific at 29.5°N. Darker colors represent greater magnitude difference from reference value. Contour intervals are 5 Sv for 0–10 Sv, 10 Sv for 10–40 Sv, and 40 Sv for greater than 40 Sv.

(a) Ekman Upwelling Velocity (m yr⁻¹)

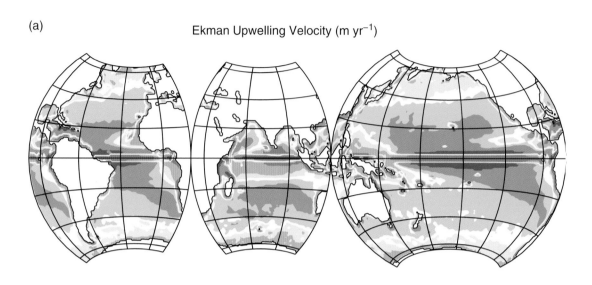

−160   −40   −10   0   10   40   160

(b) Surface Vertical Velocity from Drifters (m yr⁻¹)

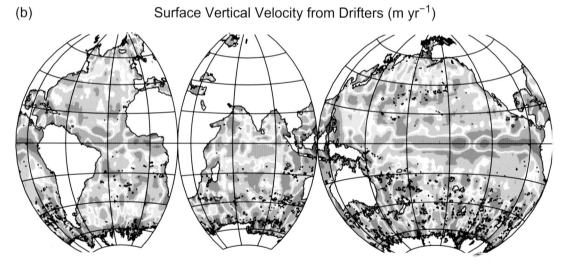

Figure 6.5 Vertical velocity for (a) Ekman suction/pumping smoothed with 3° × 3° running mean and (b) estimate at 30 m with 7° × 7° smoothing. Contour values ± 10, 40, 160, 640 m/yr. Within 3° of equator, (a) is calculated by interpolating values at 4° latitude. Data are from (a) SCOW wind climatology (averaged to 1° squares; see Appendix A) and (b) the Global Drifter Program (Appendix A).

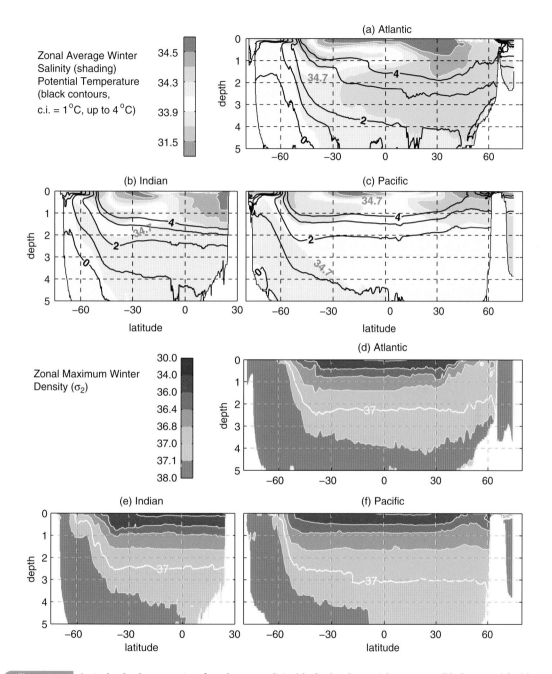

Figure 8.1 Latitude–depth cross section of zonal average salinity (shading) and potential temperature (black contours) for (a) Atlantic, (b) Indian, and (c) Pacific Oceans, and zonal maximum density (σ_2) for (d) Atlantic, (e) Indian, and (f) Pacific. Data from WOA 2013 (see Appendix A). Contour interval is $1\,^{\circ}$C for temperature and is given by color bars for salinity and density.

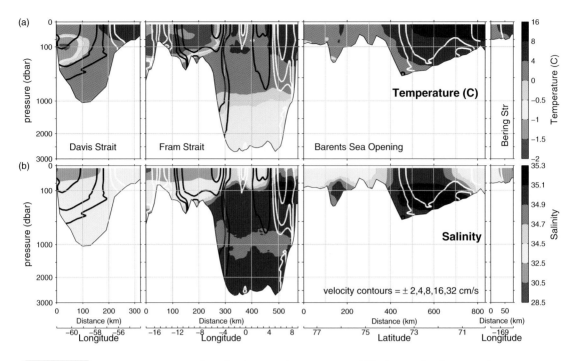

Figure 10.2 Arctic gateway sections with shading showing (a) temperature, and (b) salinity, and with contours showing geostrophic velocity into (white) and out of (black) the Arctic. Depths are proportional to square root of vertical distances. Adapted from Figures 5 and 9c of Tsubouchi et al. (2012).

Annual Average Net Heat Flux Into Ocean

contours at 0, ±20, 40, 80, 160, 320 W/m²

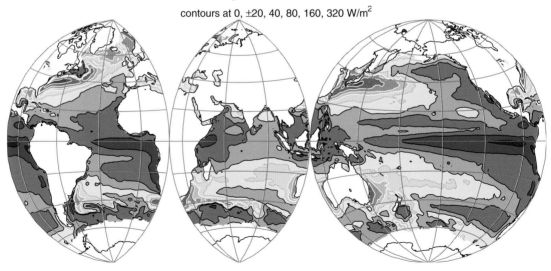

Figure 11.4 Climatological annual average net heat flux into the ocean. Contours at ± 20, 40, 80, 160, 320 W/m²; contours black for heat gain, white for heat loss, gray for zero line. Data from OAFlux (see Appendix A).

(a)   Ekman Upwelling Velocity (m yr^{-1})

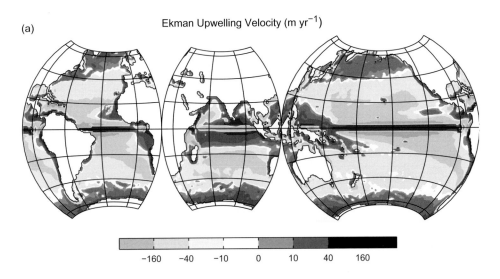

$$-160 \quad -40 \quad -10 \quad 0 \quad 10 \quad 40 \quad 160$$

(b)   Surface Vertical Velocity from Drifters (m yr^{-1})

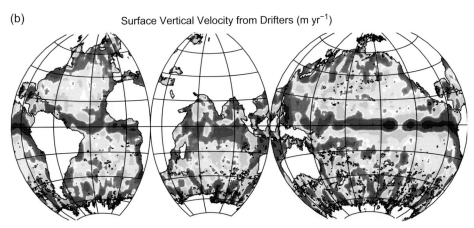

Figure 6.5 Vertical velocity for (a) Ekman suction/pumping smoothed with $3° \times 3°$ running mean and (b) estimate at 30 m with $7° \times 7°$ smoothing. Contour values $\pm$ 10, 40, 160, 640 m/yr. Within $3°$ of equator, (a) is calculated by interpolating values at $4°$ latitude. Data are from (a) SCOW wind climatology (averaged to $1°$ squares; see Appendix A) and (b) the Global Drifter Program (Appendix A). A black-and-white version of this figure appears in some formats. For the color version, please refer to the plate section.

We calculate vertical velocity (Figure 6.5) from (6.2) (in which we assume a uniform layer thickness $h = 30$ m). The w field shows strong equatorial upwelling. Elsewhere the field is noisy, even with heavy smoothing, but the tropics and subtropics tend to downwell, and subpolar latitude and eastern boundaries tend to upwell. The zonal average of Ekman w (Figure 6.7) clearly shows the large equatorial upwelling, tropical and subtropical downwelling, and polar upwelling. The drifter climatology shows a similar pattern but with large quantitative differences. The discrepancy near $60°$S may be at least partly related to data gaps in the Southern Ocean.

Surface Horizontal Velocity from Drifters

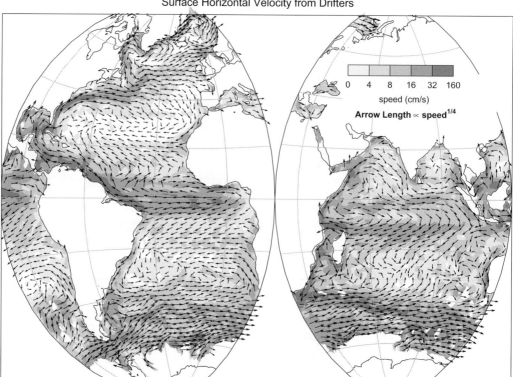

Impact on Tracers

The effect of w on property distributions is interesting because it links the circulation and other subjects such as marine biology, fisheries, and climate. Property distributions, in turn, give us additional evidence for near-surface patterns in w.

Vertical velocity has a strong impact on the abundance of phytoplankton, the base of the marine food chain. Upwelling is a key mechanism for bringing mineral nutrients, like nitrate, phosphate, and silicate (Sarmiento and Gruber, 2006, Chapter 4), to the surface. Phytoplankton incorporate nutrients using photosynthesis in the **euphotic zone**, which is the top few tens of meters illuminated by sunlight (see Subsection 4.3.3). The nutrients then slowly fall as a "rain" of fecal pellets, animal corpses, and other matter. In the annual mean over most of the ocean, surface phytoplankton abundance is limited by the supply of nutrients. Therefore upwelling regions have high **primary productivity** (rate of energy conversion by photosynthesis) and hence rapid phytoplankton growth, which attracts fish and other organisms and is partly responsible for several economically important fisheries.

Satellite photos reveal phytoplankton concentration which is measured in terms of **chlorophyll**. Chlorophyll concentration is relatively low in regions with little upwelling, such as subtropical gyres, and high in upwelling regions such as subpolar gyres, the equatorial Atlantic and Pacific, and the coasts (Figure 6.8). Note that coastal regions have higher concentrations associated with river outflows and strong tidal mixing (Mann and Lazier,

Surface Horizontal Velocity from Drifters

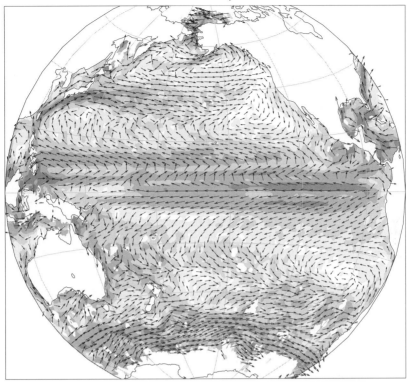

Figure 6.6 Drifter-derived climatology of horizontal velocity, $2.5° \times 2.5°$ smoothing. Arrow format as in Figure 6.4. Data is from the Global Drifter Program (Appendix A and Lumpkin and Garraffo, 2005).

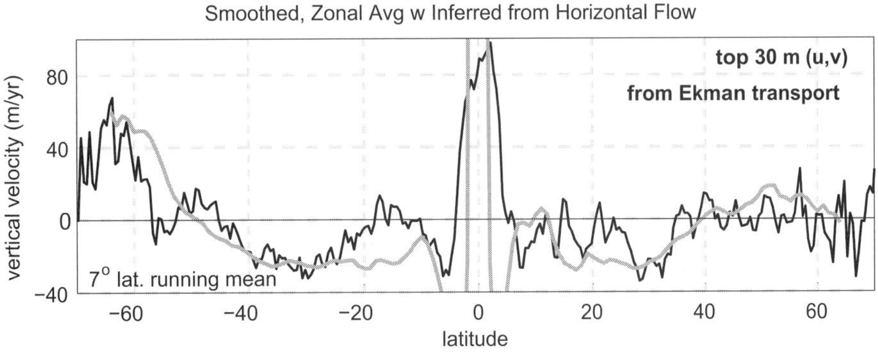

Smoothed, Zonal Avg w Inferred from Horizontal Flow

Figure 6.7 Zonal average vertical speed w based on drifter data (black) and SCOW climatology Ekman transport (gray) from Figures 6.4 and 6.6; $7° \times 7°$ smoothing.

2005, Chapter 4), but in regions where such processes are absent, high coastal chlorophyll concentrations indicate coastal upwelling regions. Prominent high-chlorophyll regions of coastal upwelling include the west coasts of South and North America, parts of the African west coast, and a band from the Horn of Africa to India.

Chlorophyll A Concentration (mg m^{-3})

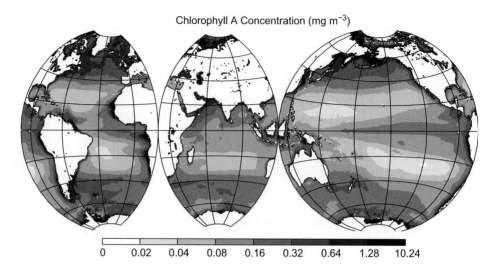

0	0.02	0.04	0.08	0.16	0.32	0.64	1.28	10.24

Figure 6.8 Surface ocean chlorophyll-a concentration (mg m^{-3}) from the SeaWiFS satellite instrument, September 1997 to July 2010 average (see Appendix A).

Sea surface temperature also indirectly measures upwelling, and the importance of vertical motion to climate. Strong upwelling brings relatively cold water to the surface. SST (Figure 4.1) shows prominent cold regions in the eastern equatorial Pacific, along the Pacific coast of South America, along the Atlantic coasts of southern Africa and northern Africa. More subtle cold features can be seen in the eastern equatorial Atlantic and along the eastern North Pacific off California. All these regions have intense localized upwelling.

6.1.3 Eastern Boundary Circulation

The eastern boundary upwelling systems mentioned in Subsection 6.1.2 can be seen better in regional maps (Figure 6.9). The surface flow away from the coast predicted by Ekman transport is clearly visible in the tropical South Pacific (Figure 6.9c), and subtropical South Atlantic (Figure 6.9d) and to a lesser extent in the midlatitude North Pacific (Figure 6.9a) and tropical North Atlantic (Figure 6.9b). For further evidence that the offshore flow is Ekman transport, we note that it is absent from the surface geostrophic flow field shown in Figure 3.4. The **cross-shore** (perpendicular to the shore) flow is also visible in a cross section in the South Pacific (Figure 6.10). Along the coast in the upwelling regions, a strip of water 100–300 km wide is colder than offshore water (Figure 6.9), most clear in the southern hemisphere (Figure 6.9c,d).

Seasonality in upwelling systems is strong (see Exercise 6.3). During the season of strongest upwelling in each region, seasonal average temperature differences between minimum SST and locations 500 km offshore can exceed 5°C (Carr and Kearns, 2003). Seasonality in wind forcing is strongest in the North Indian Ocean, where the wind reverses direction between summer and winter as part of the **monsoon** (Schott and McCreary Jr., 2001). The southerly winds blowing along the Horn of Africa drives the Somali upwelling

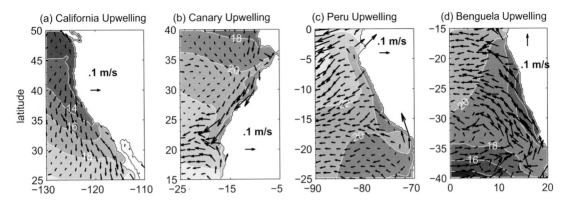

SST (shading) and near-surface velocity (arrows) for eastern boundary regions. SST is from satellite AVHRR gridded data averaged for mid-month March, June, September, and December for years 2004–2008. Velocities are surface drifter climatology (see Appendix A).

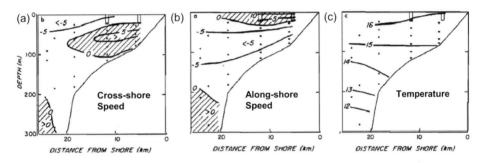

Cross sections at approximately 15°S in the South Pacific off the coast of Peru of quantities averaged over March–May 1977. (a) Cross-shore component of velocity, (b) alongshore, and (c) temperature. Units are cm/s for velocity and °C for temperature. From Brink et al. (1980), Figure 4, ©1980 American Geophysical Union. All rights reserved. Used with permission.

(Schott, 1983), which is a rare example of a strong *western* boundary upwelling zone. For these reasons, the time-average behavior shown in Figure 6.9 is less intense than the episodes of strong upwelling observed during wind events. The spatial scale of the most obvious upwelling structures can also be tens of kilometers in the cross-shore direction.

Small-scale surveys in all the upwelling areas (see, for instance, Lentz, 1992) have shown ubiquitous features (Hill et al., 1998). The strongest upwelling and surface cooling occurs near the shelf break the cooling is the surface manifestation of the isotherms being pulled upwards by the upwelling during a wind event (Figure 6.10). The shoreward flow has been observed at a depth of about 100–150 m at 15°S in the South Pacific upwelling zone (Strub et al., 1998) and inferred (based on temperature and salinity analysis) to originate at 200–300 m depth in the South Atlantic (Nelson and Hutchings, 1983). Off the west coast of South America, water appears to come from the southern Tsuchiya Jet (Subsection 6.1.1) at lower latitudes and directly from the subtropical gyre at higher latitudes.

The eastern boundary upwelling regions are associated with equatorward flow within about 20 m of the surface. The strongest (5–20 cm/s) and best-known parts of these currents occur over the continental shelf and slope. The surface current's name is associated with the upwelling system as a whole: **California Current** in the North Pacific, **Peru Coastal Current** in the South Pacific, **Benguela Current** in the South Atlantic, and **Canary Current** in the North Atlantic. These jets are tens of kilometers wide and have been measured in fine-scale surveys (Figure 6.10c). They are not visible in the drifter climatology of surface currents (Figure 6.9), though some of the regions have broader and weaker flows. The offshore boundary of the equatorward jet can be ambiguous, because in the subtropical gyres it is in the same direction as the equatorward Sverdrup flow that is offshore. Thus the name "California Current" has been applied to different widths, including all the southward flow within 1000 km of the American coast (Hickey, 1979). The meridional upwelling-favorable wind has a wind stress curl that strengthens and deepens the flow over the broader region. In the southern hemisphere, equatorward flow offshore of the Peru Coastal Current is referred to as either the **Peru Current** or the **Humboldt Current** (Strub et al., 1998).

A subsurface poleward flow accompanies the equatorward surface flow in the eastern boundary regions (Hill et al., 1998). The current is usually 20–100 km wide, extends to a few hundred meters below the surface equatorward flow, and sits over the continental slope (Figure 6.10). Typical speeds are 10–20 cm/s. In the South Pacific, there is also poleward flow offshore of the undercurrent and separated from it by a minimum in southward flow (Strub et al., 1998). In the North Pacific, poleward flow along the continental slope has been traced along the entire length of the United States (Pierce et al., 2000). North of about 35°N, this poleward flow reaches the surface during the winter as the **Davidson Current** (Hickey, 1979).

The South Indian Ocean provides an exception to the current pattern described above. Despite upwelling-favorable winds, surface coastal water along the Australian west coast is warm and flows *southward* in the **Leeuwin Current** (Cresswell and Golding, 1980), with an opposing subsurface flow (Waite et al., 2007). The contrary behavior appears to be driven by buoyancy gradients rather than wind (McCreary. et al., 1986).

6.2 Concepts

6.2.1 Equatorial Undercurrent

Dynamics without Rotation

The Equatorial Undercurrent (EUC) sits on the equator, so it is tempting to only consider the longitude–depth plane when explaining the existence and behavior of the current. At the equator, $f = 0$, so we neglect rotation. The trick is to explain how the westward wind at the equator can drive the EUC, which flows eastward. The explanation lies in two effects of the wind stress (Figure 6.11). One effect is to pile water up at the western end of the basin and hence to induce a slope in the sea surface. The other effect is

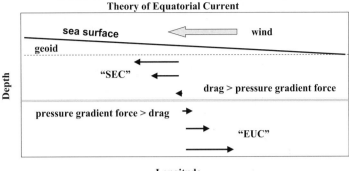

**Crude, Nonrotating, Unstratified, and Unsatisfactory
Theory of Equatorial Current**

Figure 6.11   Schematic showing elements of a longitude–depth theory for the equatorial undercurrent. Black arrows represent horizontal currents. The gray line represents boundary between two regions with different force balances. The sea surface slope height is exaggerated.

to induce a vertical shear in the horizontal velocity, as shown in Figure 6.11. The water moves under the influence of the drag associated with the wind stress and the horizontal pressure gradient due to the sea surface slope. The horizontal pressure gradient is independent of depth, assuming the fluid is hydrostatic. The stress decreases with depth. Near the surface, the pressure force is weaker than the drag, and the flow is westward, as is the surface flow in the equatorial Pacific and Atlantic. Further down, the pressure force is greater than the drag, so the flow is eastward. This eastward flow is the EUC we are trying to explain.[1]

The reasoning embodied in Figure 6.11 can be represented by a very simple theory (Subsection 6.3.1; see Arthur, 1960 and Stommel, 1960 for a slightly more complicated version). The resulting EUC is proportional to X/v, where X is equatorial wind stress and v is vertical eddy viscosity. The solution produces flows in the right direction (westward near the surface and eastward below), but there are several problems. One is that in the real world, the flow is not confined to the equator. As discussed in Subsection 6.1.1, pycnocline water must flow geostrophically from higher latitudes, upwell, and return to higher latitudes as Ekman transport of surface water. Such a circulation constitutes the subtropical cell discussed in Subsection 6.1.1.

Another problem with an EUC controlled by the balance between the zonal pressure gradient and vertical friction (see (6.12)), is that this balance does not seem to apply to the real ocean. Measurements in the equatorial Pacific near 150°W show that friction is important above 90 m depth, but below 90 m depth (where most of the EUC flows), vertical friction appears to make a relatively small contribution to the zonal momentum balance (Johnson and Luther, 1994).

---

[1] The first author's son, Benjamin Klinger, demonstrated this in a swimming pool in 2014. A strong wind blew along the length of the pool, which had a rectangular shape. A plastic cup drifted in the direction of the wind when floating at the surface (about 95% submerged) and in the opposite direction when floating near the bottom.

Dynamics with Rotation

Flaws in the two-dimensional frictional theory point to improvements. Following Fofonoff and Montgomery (1955), we think of water in the pycnocline converging on the equator. If friction can be neglected, then the water's potential vorticity is conserved (Subsection 2.1.5 and Subsection 2.2.5). Idealizing the pycnocline (for example the water between σ_θ of 23 and 26.3 in Figure 6.2) as a layer of uniform-density water of thickness h, the potential vorticity is $q = (f + \zeta)/h$ (see (2.22)). Because f vanishes as we approach the equator, we can no longer assume $\zeta \ll f$ as we did in Chapter 5. Figure 6.2a–c shows that it is plausible that changes in h are small compared to changes in f as water approaches the equator. In equatorward-flowing water parcels, ζ must greatly increase to compensate the great decrease in f. The generation of relative vorticity manifests itself as a narrow jet near the equator with

$$u = \frac{\beta}{2}(y - y_0)^2, \tag{6.3}$$

where y_0 is the initial distance of the water parcel from the equator. This jet is the Equatorial Undercurrent. The model allows u to attain realistically large values. For instance, for $y_0 = 250$ km, (6.3) gives $u = 1$ m/s at the equator.

This theory seems to replace forcing by the wind with conversion of planetary vorticity into relative vorticity. In reality, the wind forcing is "hiding" within the theory, because it is the wind that is responsible for dragging pycnocline water to the equator to compensate for the poleward Ekman transport at the surface.

Subsection 6.3.2 derives the strength of the EUC from the potential vorticity ideas of Fofonoff and Montgomery (1955). The theory is incomplete because it does not include variations in h and because the value of q in the EUC depends on its value at some arbitrary starting latitude where it begins to approach the equator. These gaps can be filled by connecting the equatorial flow to the wind-driven circulation further from the equator. Chapter 5 uses the idea of subduction to calculate the density and velocity structure in the subtropical gyre, but avoids discussion of the equatorial edge of the domain. In the current chapter, we find that the source water of the EUC appears to be the water that subducts in the subtropics.

Pedlosky (1987b), following the insights of the subduction theory of Luyten et al. (1983) (Chapter 5), shows how the problem of determining the EUC behavior can be solved by using the dynamics outside the equatorial region. He treats the equator as a **boundary layer** which must be attached to the tropical circulation. In the boundary layers needed to complete the Sverdrup circulation at the western edges of basins, the essential feature is that friction cannot be neglected (Chapter 3). In the equatorial boundary layer, the essential feature is that relative vorticity (associated with the nonlinear advection of momentum) cannot be neglected. (Relative vorticity and advection of momentum are also important in the western boundary currents, but as Chapter 3 discusses, key features of the western boundary currents are captured without these factors). More precisely, the equatorial boundary layer theory assumes that the zonal pressure gradient causes water to accelerate as it flows along the equator. As Subsection 6.3.2 shows, a scaling based on this assumption as well as the hydrostatic relation and continuity with dynamics outside the equatorial boundary layer predicts that the velocity scale U is given by

$$U = (\gamma_2 X_0 L)^{1/4}, \tag{6.4}$$

where $\gamma_2 = g\Delta\rho/\rho_0$ (mean density ρ_0, density difference between EUC layer and abyssal layer $\Delta\rho$, gravitational acceleration g), wind stress scale $X_0 = \tau_0/\rho_0$, and zonal distance scale L. For plausible Pacific values of $\Delta\rho = 4$ kg/m^3, $\tau_0 = 0.02$ N/m^2, $L = 15{,}000$ km, (6.4) gives a velocity scale of 1.9 m/s which is a little too big but not that far off for a scaling relation. Unlike the linear theory embodied by (6.15), this theory predicts that the EUC velocity scale is independent of viscosity ν. Speed U does depend on wind stress in the vicinity of the equator, but the $X_0^{1/4}$ dependence means that U is much less sensitive to wind stress than the linear theory predicts.

Numerical Tests of EUC Dynamics

Experiments with a primitive equation model test the sensitivity of an equatorial circulation to wind (Liu and Philander, 1995). The idealized wind stress is applied to a 60° wide basin. As shown in Figure 6.12 (top panels), the experiments vary the strength of the mid-latitude westerlies (Figure 6.12a,b), and a (uniform in latitude) equatorial wind stress (Figure 6.12a,c). The strength of the equatorial undercurrent is insensitive to the strength of the westerlies (Figure 6.12d,e) but weakens when equatorial wind is weakened (Figure 6.12d,f). The 50% reduction in equatorial wind leads to a roughly 25% reduction in EUC speed. The speed is less sensitive to X_0 than the linear, frictional prediction (6.15) but more sensitive than the nonlinear, inviscid prediction (6.4), perhaps because both friction and advection are important. The numerical model of Liu and Philander (1995) may have too much friction, so it is possible that the real world behavior is more like the inviscid model.

The upward and eastward flow in the EUC supplies relatively cool and dense water to the surface in the east. This is reflected in the model's equatorial temperature structure, which is also sensitive to the wind stress. The westerlies only affect the lower part of the pycnocline, in the density range corresponding to mid-latitude surface densities (Figure 6.12g,h). The easterlies affect the surface and upper pycnocline, with stronger wind leading to cooler water (Figure 6.12g,i) through a faster EUC and stronger upwelling.

The dynamics of the equatorial region, where $f \to 0$, are special, so it is unclear if the Sverdrup balance applies there. However, the tropical Pacific gyres deduced from the wind stress curl (Figures 3.1, 3.6b) show some sign of eastward flow at the equator. Kessler et al. (2003) devise an interesting diagnosis of the departures of the equatorial ocean from Sverdrup balance. They start by writing the momentum equations as

$$f\hat{\mathbf{z}} \times \mathbf{U} = -\nabla P + \mathbf{X}_* \tag{6.5}$$

where $\mathbf{U}$ is the depth-integrated horizontal velocity, P is the (scaled) depth-integrated dynamic pressure, and $\mathbf{X}_*$ is a **generalized wind stress**. In addition to the wind stress, $\mathbf{X}_*$ includes terms representing vertical integrals of momentum advection and horizontal friction. In this form, the momentum equations look linear even though they are not; the advection terms in the momentum equation, which are nonlinear functions of $\mathbf{u}$, are hidden in $\mathbf{X}_*$. When we drop the nonlinear and horizontal friction terms from $\mathbf{X}_*$, the curl of

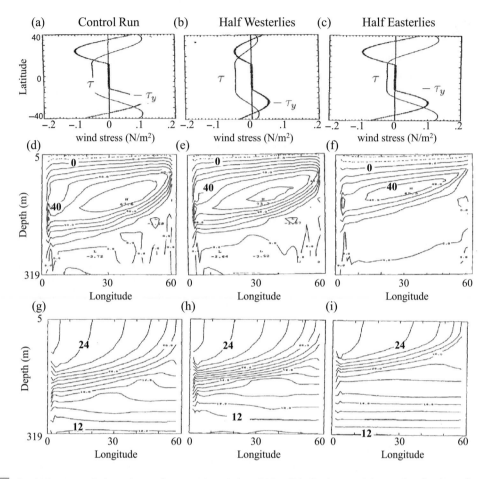

Figure 6.12 Results from numerical experiments demonstrating wind sensitivity of idealized equatorial ocean, based on Liu and Philander (1995), Figures 2, 5, and 7; ©American Meteorological Society. Used with permission. Top: wind stress τ (thin) and scaled wind shear $-\tau_y$ (thick). Middle: zonal speed at the equator (contour interval 10 cm/s). Bottom: temperature at the equator (contour interval 1°C). The experiments are characterized by different wind forcing: "Control" (left), "Half westerlies" (center), and "Half Easterlies" (right).

(6.5) allows us to calculate the Sverdrup transport. Similarly, if we retain the $\mathbf{X}_*$ instead of $\mathbf{X}$, the curl of (6.5) produces a **generalized Sverdrup transport** that depends on curl of $\mathbf{X}_*$ rather than curl of $\mathbf{X}$. Unlike the traditional Sverdrup balance, the generalized Sverdrup balance cannot be used to solve for the motion, since $\mathbf{X}_*$ includes functions of $\mathbf{u}$ such as the nonlinear advection terms. Instead, Kessler et al. (2003) use a primitive-equation numerical model to find the solution for realistic geography, forcing, and dynamics. They then use the solution to calculate the contribution of nonlinear terms and horizontal friction to the generalized Sverdrup transport.

Using continuity to calculate U as a function of longitude ϕ and latitude θ from the Sverdrup transport, we see (Figure 6.13a) that the wind stress does indeed drive an eastward flow along the equator (similar to the EUC). The eastward flow is sandwiched between

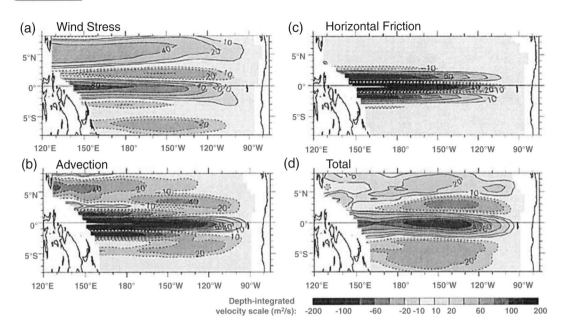

Figure 6.13 Depth-integrated zonal component of generalized Sverdrup velocity, U, based on (a) wind stress, (b) nonlinear advection, (c) horizontal friction, and (d) sum of (a)–(c), from Kessler et al. (2003), Figure 7, ©American Meteorological Society. Used with permission. In all panels, (solid, dashed) contours represent (eastward, westward) flow. Note the exaggerated aspect ratio.

regions of westward flow (South Equatorial Current), and there is an additional eastward flow near 5°N (North Equatorial Countercurrent). The advection terms strongly enhance the eastward equatorial flow (Figure 6.13b), consistent with the idea that advection is an important influence on the EUC. The narrow jets generate strong frictional transfer of momentum which tend to slow the jets (Figure 6.13c). The combined influence of all these terms (Figure 6.13d) looks qualitatively like the Sverdrup flow from wind stress, but with much stronger eastward U along the equator in the mid-Pacific.

6.2.2 Three-Dimensional Motion in Uniform-Density Water

Ekman Overturning Cells

Three-dimensional flow in uniform-density water may seem like an oxymoron, because the geostrophic flow is independent of depth. However, the surface Ekman flow introduces an important variation with depth. Our discussion will treat Ekman velocity as if it were vertically uniform and completely confined to the Ekman layer. This simplification is somewhat justified here because we are primarily interested in the effect of the Ekman flow on the entire water column rather than vertical variations within the Ekman layer.

As discussed in Chapter 3, the divergence or convergence of the surface Ekman transport is associated with upwelling or downwelling, respectively. It is useful to divide the sources

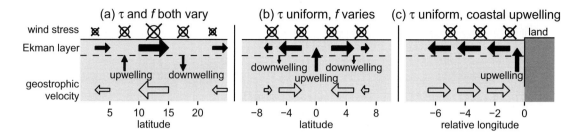

Figure 6.14 Schematic of upwelling and downwelling driven by Ekman transport divergence for the (a) subtropical, (b) equatorial, and (c) coastal ocean. Circles with ×'s represent wind stress *into* the page; note (c) assumes southern hemisphere.

of this Ekman pumping into three broad classes based on (2.11) for Ekman transport. Ekman pumping can be dominated by variations in wind stress **X** (Figure 6.14a) as in Chapter 3. It can be driven by the gradient in the Coriolis parameter f, which occurs most dramatically at the equator, where f changes sign (Figure 6.14b). Or, it can be caused by the presence of a land boundary (Figure 6.14c).

Large-scale zonal wind stress drives the gyres in the *horizontal* plane (Chapter 3) as well as circulation cells in the *latitude–depth* plane. Just as the horizontal streamfunction shows gyres that represent the vertically averaged flow, a meridional streamfunction shows **overturning cells**, which represent the zonally averaged flow (see Subsection 2.1.4 and Subsection 6.3.3). At a given latitude, the Ekman transport at the surface is balanced by a geostrophic flow in the opposite direction (Figure 6.14, unfilled arrows), which in a uniform-density ocean is distributed uniformly through the entire water column. Water travels vertically between the shallow and deep current via Ekman pumping.

A similar overturning exists in the case of Ekman pumping near a zonal boundary (Figure 6.14c), though defining a zonal streamfunction is problematic because there are no meaningful closed meridional boundaries. If the meridional coastal wind were independent of longitude, the zonal cell would extend to the Ekman pumping at the opposite coast. For example, eastern boundary upwelling would flow into westward Ekman transport, western boundary downwelling, and eastward geostrophic return flow. For the more realistic case of meridional wind varying with longitude, the zonal overturning is at least partially closed by mid-ocean Ekman pumping. Volume transports of coastal Ekman pumping are generally smaller than those associated with zonal wind stress, and so less work has been done on the basin-scale circulation associated with coastal w.

Ekman and Sverdrup Overlap

Outside the tropics, the Ekman pumping typically has the same sign as wind stress curl. Thus the regions of wind stress curl associated with the subtropical gyres drive downwelling throughout the gyres, while the oppositely-circulating subpolar gyres have upwelling. This correspondence breaks down at low latitudes because the variation of f with latitude becomes important. Thus the **tropical gyres** have the same sign of wind curl as the subpolar gyres, but have Ekman downwelling like the subtropical gyres. To see why, we consider a zonal wind stress X that depends on latitude θ. The gyre circulation is defined by the meridional velocity outside the western boundary current. The depth-integrated Sverdrup velocity (3.7) for zonal wind stress is, in spherical coordinates (B.22),

$$V_S = \frac{1}{\beta}\hat{\mathbf{z}} \cdot (\nabla \times \mathbf{X}) = \frac{1}{\beta R_E \cos\theta} \frac{\partial}{\partial\theta} (-X\cos\theta), \tag{6.6}$$

where R_E is the Earth's radius. Given the Ekman transport (2.12)

$$V_{Ek} = -\frac{X}{f}, \tag{6.7}$$

the Ekman pumping/suction is given by the divergence of the Ekman transport (2.13),

$$w_{Ek} = \frac{1}{R_E \cos\theta} \frac{\partial}{\partial\theta} \left(-\frac{X\cos\theta}{f} \right), \tag{6.8}$$

which can be rewritten using the fact that $f = 2\Omega\sin\theta$ as

$$w_{Ek} = \frac{1}{2\Omega R_E \sin\theta} \left(-\frac{\partial X}{\partial\theta} + \frac{X}{\sin\theta\cos\theta} \right). \tag{6.9}$$

The first term in this expression has the same sign as the wind stress curl ($-\partial X/\partial\theta > 0$ in the northern hemisphere) but the second term is proportional to $X/\sin\theta$ (< 0 in the northern hemisphere) and overcomes the curl term at low latitudes where $\sin\theta$ gets small.

The relationship between these different aspects of the circulation is illustrated by an idealized version of the observed zonal wind stress (Figure 6.15a) blowing over a northern hemisphere basin that is 60° wide in longitude. The wind stress curl (Figure 6.15b) displays the usual pattern, from the equator to the northern boundary, of positive values (tropical gyre), negative values (subtropical gyre), and positive values (subpolar gyre). The Ekman volume transport is similar to the wind stress except that the factor of $1/f$ makes the Ekman transport grow near the equator (Figure 6.15c). The Sverdrup transport is similar to the wind stress curl except that a factor of $1/\beta$ makes the Sverdrup transport weaken near the equator (Figure 6.15c). Note that Ekman volume transport refers to the zonal integral of the Ekman transport. The geostrophic component of the Sverdrup transport is simply the difference between the Sverdrup transport and the Ekman volume transport. Because of the opposing trends with latitude of Sverdrup and Ekman components, the geostrophic Sverdrup transport is nearly identical to the total Sverdrup transport in the subtropical and subpolar gyres, but quite different at low latitudes. In the tropical gyre, the Sverdrup transport is northward but the geostrophic Sverdrup transport is southward (Figure 6.15c).

The similar magnitudes of the Sverdrup and Ekman transports can be seen from a scale analysis. From (6.6) and (6.7), we find that the relative size of the transports is given by

$$\frac{V_{Ek}}{V_S} = -\frac{\beta X}{f\hat{\mathbf{z}} \cdot (\nabla \times \mathbf{X})}. \tag{6.10}$$

Using the definitions of f and β and assuming that the magnitude of curl $\mathbf{X}$ is given by X/M, where M is a typical horizontal distance between minimum and maximum magnitude of X, the scale of the ratio is

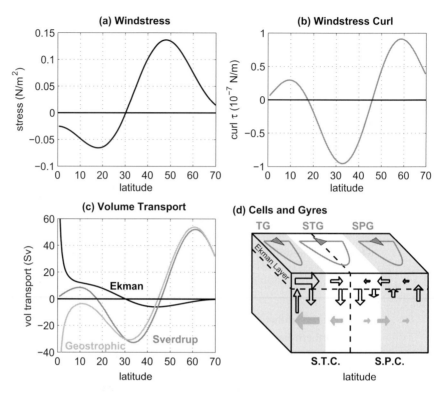

Figure 6.15 For idealized zonally uniform zonal wind forcing, (a) wind stress, (b) wind stress curl, (c) volume transport associated with Ekman transport, Sverdrup transport, and geostrophic component of Sverdrup transport (all marked in figure), and (d) schematic of overturning cells showing depth-average horizontal streamfunction (shown on top surface of basin), zonal-average meridional and vertical transport (shown on eastern edge of basin).

$$\frac{V_{Ek}}{V_S} \sim \frac{M}{R_E} \cot\theta. \tag{6.11}$$

Typical meridional scales of $M \approx 1500$ km give $V_{Ek}/V_S = O(0.1)$ at mid-latitudes and $O(1)$ values in the tropics, as we see in Figure 6.15c.

It is helpful to see the gyres and overturning cells illustrated together (Figure 6.15d). In this figure the surface of the basin shows the gyres, while the front cross section shows the cells. In the tropical gyre and southern half of the subtropical gyre the Ekman transport flows northward and decreases with latitude, thus feeding Ekman downwelling which in turn feeds a southward geostrophic flow. The geostrophic flow is depicted by arrows below the Ekman layer; in reality it extends into the Ekman layer, but most of its transport is below the Ekman layer if the Ekman layer thickness is much smaller than the total ocean depth. This circuit is known as the **subtropical cell**. The circuit is completed by equatorial upwelling.

The exact width in latitude of the equatorial upwelling is not immediately obvious. Its poleward edge is not defined by the wind stress. At the equator, where $f = 0$, zonal wind does not drive meridional transport. Close to the equator the expression for Ekman transport must lose its validity, so that the large transport values some distance from the equator can decrease to zero approaching the equator. In the real ocean, the upwelling region where this happens extends on the order of $1°$ of latitude from the equator.

From the subpolar gyre to the northern subtropical gyre, equatorward Ekman transport constitutes the surface limb of the **subpolar cell**. This cell links to poleward geostrophic circulation via upwelling in the subpolar gyre and downwelling in the subtropical gyre.

Building a More Realistic Circulation Picture

Neither the gyre nor the cell necessarily describes the flow at a given location, which will differ from the depth-average and the zonal average, and so both streamfunctions can be misleading. For instance, northward flow implied by the tropical gyre outside its western boundary current hides the fact that the flow beneath the Ekman layer is actually southward. Or consider the northward geostrophic flow of the subpolar cell. The northern part of the cell is in the subpolar gyre, where the western boundary current flows southward, and the southern part is in the subtropical gyre, where northward geostrophic flow is *only* in the western boundary current (WBC).

Still, gyres and cells are useful constructs. They are flow structures whose magnitude we can actually calculate given the wind stress and basin geometry, so they can be useful stepping stones to a complete solution. It is difficult to picture the full three-dimensional circulation, so each streamfunction isolates some key features of the flow. Finally, some ocean processes may be more sensitive to either the gyres or the cells, so it can be a useful approximation to consider only one or the other.

The subtropical cells are somewhat more prominent than the subpolar cells. One reason is that Ekman volume transport is larger compared to Sverdrup transport at low latitudes than at high latitudes. Another reason is that in the real ocean, the subtropical cells appear to play an important role in determining equatorial temperature, which in turn is important for El Niño and climate. As we see in Chapter 8, the subpolar region has important thermodynamic and climate implications, but the role of the subpolar cell (as opposed to other processes) may be more complicated than the role of the subtropical cells. Ekman transport in the subpolar southern hemisphere is usually considered part of a single circulation feature called the **Deacon Cell** (Subsection 9.2.1).

For a stratified fluid such as the real ocean, the qualitative features of the subtropical cells described here should be relevant for the flow in the top of the ocean (thermocline and mixed layer). Indeed, observations in the tropics and subtropics show flow generally away from the equator at the surface (Figure 6.6a), flow towards the equator in the thermocline (Figures 6.1 and 6.3), and upwelling at the equator (Figures 6.6b and 6.7). As discussed in Subsection 6.2.3, subtropical cells are also robust features of realistic numerical models.

6.2.3 Subtropical Cells in Numerical Models

Subsection 6.2.2 discusses subtropical cells in an idealized uniform-density ocean. Just as a stratified ocean carries most of its Sverdrup flow in the pycnocline, so the deep limb

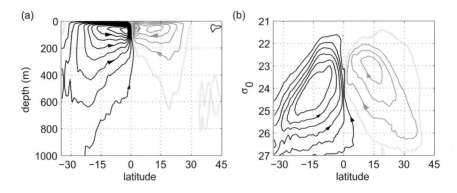

Figure 6.16 Indo-Pacific meridional overturning streamfunction for near-global simulation with the Modular Ocean Model forced by seasonally varying climatological atmospheric fields. (a) $\Psi_z(\theta, z)$, based on zonally averaged $v(\phi, \theta, z)$. (b) $\Psi_\sigma(\theta, \sigma_0)$, based on zonally averaged $v(\theta, \sigma_0)$. Contour interval is 4 Sv in both panels. See also Richards et al. (2009), Figure 3.

of the subtropical cells also occurs in the pycnocline. The subtropical cells can be seen most clearly in numerical models. The Indo-Pacific meridional overturning streamfunction $\Psi_z(\theta, z)$ (z referring to the depth coordinate) shows the poleward Ekman transport, subtropical sinking, equatorward subsurface flow, and equatorial upwelling (Figure 6.16a). The Indo-Pacific rather than Pacific basin is used here because the streamfunction is not defined where the Indonesian passages connect the Indian and Pacific basins. Similarly, Figure 6.16 only shows regions north of 35°S in order to exclude latitudes where the Indo-Pacific can exchange water with the Atlantic. Attempting to show the streamfunction for only the Pacific Ocean results in a similar velocity field except for an additional 10 Sv that enters the Pacific in the south and exits to the Indian Ocean at the equator.

The volume transport can be calculated as a function of density rather than depth. The resulting density-coordinate streamfunction, $\Psi_\sigma(\theta, \sigma_0)$ shows important differences from Ψ_z; Subsection 6.3.3 explores the mathematical properties of Ψ_σ. $\Psi_\sigma(\theta, \sigma_0)$ shows that despite being confined to shallow depths, the cells encompass a wide density range (Figure 6.16b). Equatorward flow is as dense as $\sigma_0 = 27$ and replaces poleward Ekman flow as light as $\sigma_0 = 22$, which spans most of the range of the entire ocean (Figures 4.6 and 1.6). The large density range and large volume transport (36 Sv for both hemispheres, Figure 6.16b), implies a substantial buoyancy transport associated with the subtropical cells. Because the density variations are primarily associated with temperature variations, the subtropical cells are also an important mode of meridional heat transport (see Subsections 11.1.1, 11.1.2, and 11.2.1). In both hemispheres, warm poleward flow and cold equatorward flow means the cells transport heat poleward. The most dramatic changes in density along the streamlines occur at the surface and at the equator. At the surface, the interaction with the atmosphere cools and "densifies" (makes denser) the water as it travels poleward, and at the equator, the water is warmed and made lighter as it upwells into the equatorial mixed layer. The deep limbs of the cells have smaller variations along streamlines, especially in the southern hemisphere.

There are several interesting differences between $\Psi_z(\theta, z)$ and $\Psi_\sigma(\theta, \sigma_0)$. Remember that these streamlines are not merely mapped differently in the vertical, but can have different values because v is integrated zonally along different paths (Subsection 6.3.3). Compared to the equatorward transport, the poleward transport is confined to a small *depth* range (Figure 6.16a) but a substantial *density* range (Figure 6.16b); at each latitude, the Ekman flow is distributed over a wide span of longitudes which has significant zonal density variation (as in Figure 4.6). The strength of the cells are somewhat weaker for Ψ_σ than for Ψ_z. Near the equator, shallow intensifications known as the **Tropical Cells** are visible in Ψ_z but not in Ψ_σ. However, Ψ_σ includes them in some simulations such as Hazeleger et al. (2001), and eddies have an important influence on the tropical cells (Chapter 7).

The horizontal extent of the subtropical cells is different for Ψ_z and Ψ_σ. In depth coordinates, the poleward boundary of each cell is approximately 30° latitude, where the meridional Ekman transport vanishes. In density coordinates, the cells extend almost to 40°N and to at least 35°S. Western boundary currents carry warm water beyond 30°. The density-coordinate cells exchange a substantial amount of water across ± 30° latitude, implying a transport of heat there as well as an additional transport of heat at lower latitudes due to the additional cold water from higher latitudes. Finally, the latitude of the upwards flux is closer to the equator than it is for the diapycnic flux (to lighter density), which has been linked to Tropical Instability Waves (see Subsection 7.1.1; Hazeleger et al., 2001; Richards et al., 2009).

6.2.4 Gyre and Cell Circulation Together

Figure 6.15d displays the locations of both the cells and gyres of the wind-driven circulation, yet it does not show how they combine to form the full three-dimensional wind-driven circulation. For instance, we want to know how the gyre circulation is altered by transport between the gyres, and what pathway water takes towards and away from the equator. In the following, we explore the horizontal structure of the flow, and how it varies between the thermocline and mixed-layer, with the $2\frac{1}{2}$-layer model of McCreary and Lu (1994).

In Chapter 5, the layers represent different density ranges in the pycnocline, below the Ekman layer. As Subsection 6.2.2 discusses, the flow in the Ekman layer is essential to the subtropical cells. So here we represent the surface layer (including both geostrophic and Ekman transport) with layer 1, and the pycnocline with layer 2 (Figure 6.17). These are also uniform-density layers, but their correspondence to various densities in real water is less straightforward. Layer 2 is similar to a layer in the LPS model (Subsection 5.2.1), but layer 1 does not correspond in a simple way to density ranges in the ocean. The basin is divided by the subduction latitude θ_d, which represents the equatorward boundary of the subduction region for layer 2 water. In the LPS model, layer 1 would have zero thickness poleward of θ_d (Figure 5.11). Instead, the McCreary and Lu (1994) model assigns layer 1 a constant nonzero thickness poleward of θ_d to represent the Ekman layer. Equatorward of θ_d, layer 1 represents water that is lighter than that in layer 2, but poleward of θ_d it represents water that would have density increasing poleward in the real ocean or in a more realistic model. In the real world, heat loss to the atmosphere makes water poleward of θ_d get denser as it drifts northward in the mixed layer, after which the water parcel

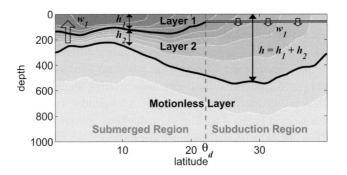

Figure 6.17 Schematic of interface depths h_1 and h as a function of latitude for the McCreary and Lu (1994) model, superimposed on cross section of March North Pacific σ_0 on the Date Line (c.i. = 0.5) from WOA13 (see Appendix A).

maintains its density as it subducts into the pycnocline. In the model, the water keeps the same density as long as it stays in layer 1, but becomes denser when Ekman pumping pushes it down into layer 2.

McCreary and Lu (1994) solve an analytic version of this model as well as a numerical version that includes more complete representations of nonlinear effects, friction, and fluid exchange between layers. We discuss results from the numerical version here and from the theory in Subsection 6.3.5.

Idealized Wind and Geometry

Two experiments illustrate the separate influence of wind *stress* and wind *curl* in the tropics. In both, the $2\frac{1}{2}$-layer model is driven by the familiar wind pattern of mid-latitude Westerlies and tropical Easterlies. One case has uniform tropical easterlies (Figure 6.18a), which drive a strong EUC but not a tropical gyre (Figure 6.18b,c). In the other case, the easterlies go to zero at the equator (Figure 6.18d), removing the EUC but gaining a strong cyclonic tropical gyre. In both cases, layer 2 has a clear subtropical gyre flow pattern and a somewhat less clear subpolar gyre (Figure 6.18c,f). These gyres are less clear in layer 1 (Figure 6.18b,e) because of the Ekman transport superimposed on the gyre flow.

The subtropical cell is also visible in the experiments. The cell's lower limb, in which pycnocline water flows to the equator, occurs in a southwestward layer 2 flow from the eastern subtropical gyre (Figure 6.18c,f). The current flows between the curves intersecting $\theta = \theta_B$ in the west and $\theta = \theta_d$ in the east. In both cases, the equatorward flow finishes its journey to the equator as a low-latitude western boundary current. The subtropical cell's poleward upper limb takes a somewhat different route in the two cases. With uniform tropical easterlies, layer 1 flow (Figure 6.18b) is nearly the opposite of layer 2, as one expects from the northward Ekman transport and zero Sverdrup transport. Similarly, the western boundary current is also northward. For the zero-equatorial-wind case, the northward flow in layer 1 (Figure 6.18e) circles the tropical gyre and is opposed by a southward western boundary current.

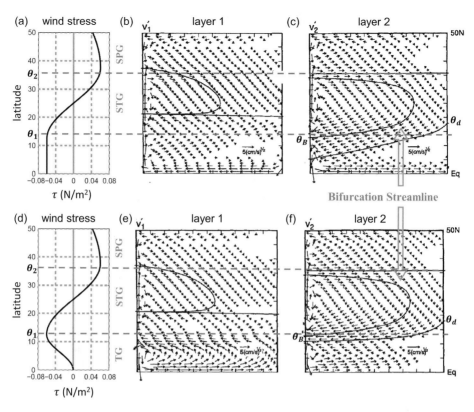

Wind stress (left panels), layer 1 velocity (center panels), and layer 2 velocity (right panels) for two numerical experiments (top and bottom panels) with $2\frac{1}{2}$-layer model. From McCreary and Lu (1994) Figures 5 and 7 (velocity) and Eq. 9 (wind stress). Dark gray dashed lines mark gyre boundaries. Arrow length is proportional to the square root of speed. Figures from McCreary and Lu (1994) are ©American Meteorological Society. Used with permission.

The EUC plays an important role in the upward limb of the subtropical cell. Water arriving at the equator in the west is carried eastward along the equator (Figure 6.18c) where it upwells into layer 1 (not shown). Flow in the opposite direction, reminiscent of the South Equatorial Current (SEC), returns the upwelled water to the west where it joins the northward flow. In the uniform-tropical-wind case, EUC transport carries all of the layer 2 water from the subtropical gyre. This presents a paradox. The volume transport of pycnocline equatorward flow out of the subtropical gyre is largely determined by the Ekman transport near the equatorward boundary of the subtropical gyre, as we demonstrate in Subsection 6.3.5. The strength of the EUC depends on equatorial wind stress (Subsection 6.2.1), which is independent of the subtropical wind. What is to prevent a mismatch between the amount of water flowing equatorward and the amount of equatorial upwelling? The resolution to the paradox lies in the the effect of tropical wind shear on both the EUC and the tropical gyre. For a given subtropical Ekman transport, a weaker equatorial wind implies a bigger tropical wind curl. To the extent that larger wind shear weakens upwelling in the EUC, it is able to upwell excess subtropical water in the tropical gyre.

What happens to pycnocline water in the subtropical gyre that does not immediately flow towards the equator? The western boundary of the equatorward flow in the layer 2 subtropical gyre is called the **bifurcation streamline** because it is the streamline that intersects the basin western boundary at latitude θ_B where the flow bifurcates between northward and southward western boundary currents. The curve is determined by calculating the path a water parcel would take (with velocity reversed) from the bifurcation point until it approaches the western boundary again in the north. Water that is pumped from layer 1 to layer 2 west of the bifurcation streamline recirculates within the subtropical gyre. The Ekman pumping implies a horizontal divergence, which means that water parcels spiral outward from the center of the gyre. Each time the water enters the western boundary current, it emerges a little further north, putting it on to a wider trajectory around the gyre, until it exits the WBC *outside* the bifurcation streamline. Some layer 2 water also leaks into the subpolar gyre as part of a subpolar cell as well. Such water upwells in the subpolar gyre and crosses back into the subtropical gyre via Ekman transport.

Pacific and Atlantic Cases

The real world wind (Figure 3.1) has both strong equatorial wind stress (like Figure 6.18a) and strong tropical wind curl (like Figure 6.18d), so the resulting currents share features of both solutions.

For Pacific Ocean wind stress and basin shape, the pycnocline in a layer model shows a similar flow from the subtropics to the EUC (Figure 6.19) as in Figure 6.18c. The model allows flow to reach the EUC both in a western boundary current (region between the figure's *Th* and *In* contours in northern hemisphere and between *Bi* and *In* contours in southern hemisphere) and within the Sverdrup region (between the *In* and *Sz* contours in both hemispheres). The model imposes flow into the basin from the southern border of the model domain and out of the basin as the Indonesian Throughflow. Some of the northern hemisphere flow to the equator (between contours *Bi* and *Th*) leaves the basin as part of the Throughflow rather than reaching the EUC. Realistic winds drive coastal upwelling along South America, and some of the EUC water feeds this upwelling. The tropical pycnocline circulation is similar to that of the real ocean (Figure 6.1) but lacks the strong westward-flowing North Equatorial Countercurrent which should be at about 8°N. The Countercurrent is strongest in the top 200 m, which at that latitude is mostly in layer 1. Layer 1 of the model does in fact have an eastward-flowing jet at about 8°N as well as other features of the surface flow shown in Figure 6.6a (see Lu et al., 1998, Figure 4a).

The Atlantic Ocean meridional overturning does not show obvious subtropical cells because of the large western boundary current from South Atlantic to North Atlantic in the top kilometer (see Chapter 8). However, a realistic model (Lazar et al., 2002) with many density layers in the pycnocline shows clear trajectories of flow to the EUC within the pycnocline (Figure 6.20).

Eastern Boundary Upwelling Systems

Some aspects of equatorial circulation are echoed by coastal flows, especially at upwelling zones along eastern boundaries. Wind blowing anti-cyclonically around a boundary

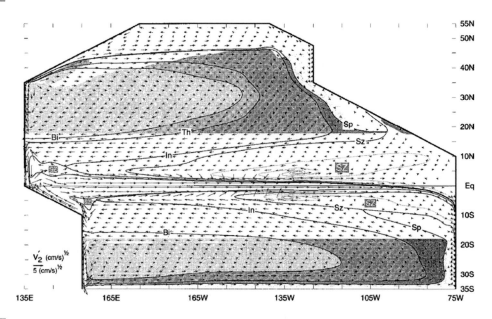

Figure 6.19 Layer 2 velocity for $3\frac{1}{2}$-layer model in Pacific-shaped basin driven by climatological annual-average wind stress. Arrows representing velocity have lengths proportional to the square root of speed. Contours represent streamlines dividing different flow regions. From Lu et al. (1998), Figure 7, ©American Meteorological Society. Used with permission.

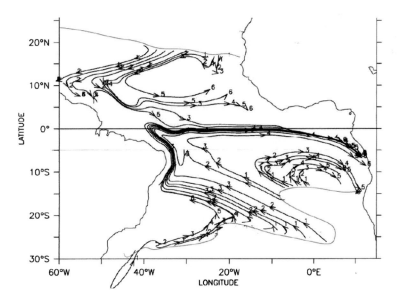

Figure 6.20 Trajectories of water parcels initially at $\sigma_\theta = 25$ and circulated by annual mean flow in multilayer models driven by climatological wind and surface buoyancy fluxes. Each number on each trajectory represents year since trajectory initialization at the location of the number. From Lazar et al. (2002), Figure 15, ©2002 American Geophysical Union. All rights reserved. Used with permission.

(clockwise in the northern hemisphere, counter-clockwise in the southern) drives Ekman transport away from the boundary whether that boundary is at the equator or a coast (Figure 6.4). In either case, water upwells in a narrow zone along the boundary (Figure 6.14). In fact, observations show such upwelling zones along the equator and coasts in the Atlantic and Pacific (Subsection 6.1.2 and Subsection 6.1.3).

Ventilated pycnocline theory (Subsection 5.2.2) predicts that the pycnocline in the eastern low-latitude corner of the subtropical gyre is a quiescent "shadow zone." The presence of a meridional wind implies coastal upwelling that draws water from the pycnocline in the shadow zone. Recall that the shadow zone exists because the easternmost streamline of subducted water curves away from the coast, leaving a region that is unreachable by subducted water. As Cessi (1992) shows, for upwelling-favorable wind, the easternmost streamlines curve eastward, carrying water into the coast. In this modification of the ventilated pycnocline theory, individual layers are not able to satisfy the no-flow condition at the eastern boundary, though the vertically integrated flow does. Higher order dynamics is needed to satisfy the boundary condition there.

The flow *along* the equatorial boundary also has coastal analogues. On the equator, the surface current is westward, the same direction as the wind, and the subsurface EUC is eastward. Similarly, the typical coastal currents in upwelling zones consist of surface water flowing with the wind (equatorward) and subsurface poleward-flowing water flowing against the wind.

6.3 Theory

6.3.1 Some Incomplete Theories of the Equatorial Undercurrent

Early theories of the EUC discussed in Subsection 6.2.1 were only able to account for some of the the current's observed features, but they describe some processes that are relevant to a more complete theory.

First we look at a rudimentary theory in which the EUC is determined by vertical friction, and then at an incomplete theory in which the EUC is independent of vertical friction. As Subsection 6.2.1 explains, the latter case is probably more realistic, but the former case illustrates the way the zonal pressure gradient caused by wind could drive an eastward current.

If we neglect nonlinearity and horizontal friction, then the zonal momentum equation becomes simply

$$0 = -\frac{1}{\rho_0}p_x + \nu u_{zz}. \tag{6.12}$$

If we further neglect zonal variations in p_x and u, (6.12) is simply an ordinary differential equation in $u(z)$ (we already noted that hydrostatic p_x is independent of z). The stress in the fluid is given by $\rho_0 \nu u_z$, so assuming we know the surface wind stress $\tau = \rho_0 X$ and the depth H of the wind-driven layer, we can apply two boundary conditions,

$$vu_z(0) = X \quad \text{and} \quad vu_z(-H) = 0. \tag{6.13}$$

Finally, if the flow is truly two-dimensional and we assume there is no flow through the eastern or western boundaries, we have

$$\int_{-H}^{0} u \, dz = 0. \tag{6.14}$$

Integrating (6.12) twice and applying the three conditions, we find a solution given by

$$u = \frac{XH}{\nu} \left[\frac{1}{2} \left(\frac{z}{H} \right)^2 + \left(\frac{z}{H} \right) + \frac{1}{3} \right]. \tag{6.15}$$

Alternatively, the inviscid limit allows us to derive the expression (6.3) for u based on conservation of potential vorticity (Fofonoff and Montgomery, 1955). We consider a layer in the pycnocline that is initially at meridional location $y = y_0$, corresponding to Coriolis parameter $f = f_0$, and layer thickness $h = h_0$. We assume that at this initial location, the relative vorticity ζ is negligible. Then conservation of potential vorticity says

$$\frac{f + \zeta}{h} = \frac{f_0}{h_0}. \tag{6.16}$$

If we assume that the zonal scale is much longer than the meridional scale (in other words, that the EUC looks like a boundary layer), then

$$\zeta = v_x - u_y \approx -u_y. \tag{6.17}$$

We make a somewhat arbitrary assumption that $h = h_0$ everywhere and approximate the region with an equatorial β-plane on which $f = \beta y$, then (6.16) can be rewritten

$$u_y = (y - y_0)\beta. \tag{6.18}$$

Assuming that u at $y = y_0$ is negligible compared to its values inside the EUC, we integrate (6.18) to obtain

$$u = \frac{\beta}{2}(y - y_0)^2, \tag{6.19}$$

which is a restatement of (6.3). A problem with this theory is that it does not say what y_0 should be. In Subsection 6.3.2, we examine a more complete scaling analysis which produces a less arbitrary estimate of u.

6.3.2 Equatorial Undercurrent Scaling

As discussed in Subsection 6.2.1, Pedlosky (1987b) formulated an inviscid theory for the EUC that combines the extra-equatorial flow structure of LPS with the Fofonoff and Montgomery (1955) mechanism for converting planetary vorticity into relative vorticity approaching the equator. Following Pedlosky (1987b), we perform a scale analysis of the equations of motion in the vicinity of the equator to estimate the speed (see (6.4)) and other parameters associated with the EUC.

We define scales for variables $u \sim U$, $h \sim H$, and $(x, y) \sim (L, M)$, where $M \ll L$. We want to estimate zonal velocity scale U, current width scale M, and layer thickness H.

The zonal length scale, L is given by the width of the basin at the equator. The continuity equation can be written

$$(hu)_x + (hv)_y = 0, \tag{6.20}$$

which yields

$$V \sim (M/L)U. \tag{6.21}$$

Here we have neglected flow through the top or bottom of the layer, but as long as such flow is not much larger than the other terms, our scaling for v is at worst an upper bound for the size of v.

At the edge of the EUC ($y \approx M$), the meridional momentum equation can be approximated by geostrophy, which for the equatorial β-plane is

$$\beta y u = -\frac{1}{\rho_0} p_y, \tag{6.22}$$

which gives a scale for the meridional pressure gradient based on the scales described above

$$p_y \sim \rho_0 \beta M U, \tag{6.23}$$

where we use the equatorial value of β throughout this subsection. For the zonal momentum equation, we include nonlinear terms but continue to neglect friction,

$$u u_x + v u_y - \beta y v = -\frac{1}{\rho_0} p_x. \tag{6.24}$$

At the equator the Coriolis term disappears because $y = 0$, and the nonlinear terms have magnitude

$$u u_x \sim v u_y \sim U^2/L, \tag{6.25}$$

which implies the magnitude for the zonal pressure gradient,

$$p_x \sim \rho_0 U^2/L. \tag{6.26}$$

If the total difference in pressure between eastern and western extremities of the EUC is the same order of magnitude as the difference in pressure between the equator and the poleward edges of the EUC, then we can relate the two pressure gradients,

$$\frac{p_x}{p_y} \sim \frac{M}{L}, \tag{6.27}$$

which can then be combined with the separate scalings (6.23) for p_y and (6.26) for p_x, we obtain

$$U = \beta M^2. \tag{6.28}$$

This is just the scaling implied by (6.19) from Fofonoff and Montgomery (1955) (see Subsection 6.3.1), but here the length scale M represents the width of the current rather than an arbitrary starting point.

The scaling is not complete yet because we need to solve for M as well. First we relate the vertical length scale to the pressure gradient via the hydrostatic relation,

$$\frac{1}{\rho_0}\nabla p = \gamma_2 \nabla h, \tag{6.29}$$

where $\gamma_2 = g(\rho_3 - \rho_2)/\rho_0$ as in Chapter 5 (see (5.18)). Combining this with (6.23) and (6.28), we get

$$H = \frac{1}{\gamma_2}\beta^2 M^4. \tag{6.30}$$

So far none of these scale relationships involve any forcing. We use the version of the Sverdrup balance for ventilated thermocline theory (Subsection 5.3.3) which relates h to Ekman pumping and hence to the wind forcing. The Sverdrup balance only holds outside the EUC, so we assume here that the depth scale is similar inside and at the edge of the EUC. The two-layer Sverdrup balance for the ventilated pycnocline is (from (5.33))

$$h^2 = \frac{E_2^2 + d_0^2}{1 + \Gamma(1 - f/f_1)^2}, \tag{6.31}$$

where, for longitude-independent Ekman pumping,

$$d_0^2 = -(4\Omega R_E^2 \phi_E/\gamma_2)(1 - \phi/\phi_E)w_{Ek}\sin^2\theta, \tag{6.32}$$

and w_{Ek} can be written in terms of X according to (6.9). Near the equator, the second term of (6.9) dominates, and the scaling for h based on (6.31) and (6.9) is

$$H = \sqrt{\frac{X_0 L}{\gamma_2}}, \tag{6.33}$$

where X_0 is the scale for X.

To summarize, we have assumed that the flow is hydrostatic, has similar pressure ranges in x and y, is geostrophic and Sverdrupian at the edge of the EUC, and has a balance between the zonal pressure gradient and the advection terms on the equator. This gives us the following scale relations for unknowns U, H, and M:

$$U = \beta M^2, \quad H = \frac{1}{\gamma_2}\beta^2 M^4, \quad H^2 = \frac{1}{\gamma_2}X_0 L. \tag{6.34}$$

Solving gives :

$$M = \left(\frac{\gamma_2 X_0 L}{\beta^4}\right)^{1/8}, \quad H = \left(\frac{X_0 L}{\gamma_2}\right)^{1/2}, \quad U = (\gamma_2 X_0 L)^{1/4}. \tag{6.35}$$

For $X_0 = (0.1 \text{ N/m}^2)/\rho_0$, $L = 3000$ km, $\gamma_2 = 0.01 \text{ m/s}^2$, we get plausible values: $H = 170$ m, $U = 1.3$ m/s, $M = 250$ km. The parameters have a weak dependence on X_0. However EUC volume transport is proportional to MHU:

$$MHU = \frac{(LX_0)^{7/8}}{\gamma_2^{1/8}\beta^{1/2}}, \tag{6.36}$$

which is nearly proportional to X_0.

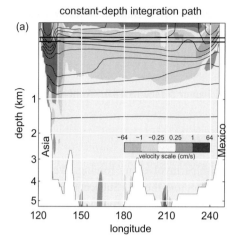
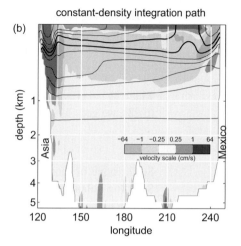

Longitude–depth section of potential density σ_0 (contours, c.i. $= 0.5$ kg m^{-3}) and meridional component of velocity (shading, scale shown in plots), for Pacific latitude 26°N in the numerical experiment displayed in Figure 6.16. The plot highlights integration path for (a) constant-depth overturning Ψ_z and (b) constant-density overturning Ψ_σ.

6.3.3 Varieties of Meridional Streamfunction

Alternative Meridional Streamfunctions

The subtropical cells (Subsection 6.2.2) are revealed most clearly by the meridional over-turning streamfunction Ψ (Subsection 6.2.3). As mentioned in Subsection 2.1.4, it is straightfoward to create $\Psi(y, z)$ based on the zonal integral of v or w. In analogy to the horizontal streamfunction, one expects that the integration in x is along a line of con-stant z, indicated by the two thick lines in Figure 6.21a. As the density increases with depth, there is a rough correspondence between z and σ and the streamfunction *approximately* tells us the magnitude and direction of volume transport in different density classes. However, since v and σ are complicated functions of both x and z (Figure 6.21), it is illuminating to see *exactly* how much transport is in each density class, that is, overturning streamfunction as a function of density. The depth-coordinate and density-coordinate streamfunctions can show us contrasting views of the overturning, as we see in Figure 6.16 and Chapter 9.

To find the density-coordinate streamfunction we integrate in x, but this time integrating along $z = z(x, y, \sigma)$ (thick contours in Figure 6.21b), where σ is a particular fixed density and $z(x, y, \sigma)$ defines the height of the σ contour at each horizontal location (x, y). For convenience we work in Cartesian coordinates here, but the same reasoning applies in spherical coordinates. The volume transport between zonal boundaries W and E and between densities σ_B and σ_T is

$$\Psi_\sigma(y, \sigma_T) - \Psi_\sigma(y, \sigma_B) = \int_W^E \int_{z(x,y,\sigma_B)}^{z(x,y,\sigma_T)} v(x, y, z(x, y, \sigma)) \, dz \, dx. \tag{6.37}$$

We use this to derive a relationship between zonal integral velocity and each of the stream-functions, Ψ_z and Ψ_σ, where the z and σ subscripts stand for "height" and "density," respectively (they are not derivatives). If we change the integration variable from z to σ, the substitution $dz = (\partial z/\partial\sigma)d\sigma$ for the inner integral gives us:

$$\Psi_\sigma(y,\sigma_T) - \Psi_\sigma(y,\sigma_B) = \int_W^E \int_{\sigma_B}^{\sigma_T} v(x,y,z(x,y,\sigma)) \frac{\partial z}{\partial\sigma} \, d\sigma \, dx. \tag{6.38}$$

Notice that for this change of variable to make sense we require $\partial\sigma/\partial z < 0$ everywhere so that $\partial z/\partial\sigma$ is bounded. In other words, we need a nonzero density stratification $N > 0$ everywhere so a one-to-one correspondence exists between σ and z in the vertical.

Now define zonally integrated velocities for the two coordinate choices as

$$V_z(y,z) = \int_W^E v(x,y,z) \, dx, \tag{6.39}$$

$$V_\sigma(y,\sigma) = \int_W^E v(x,y,z(x,y,\sigma)) \frac{\partial z}{\partial\sigma} \, dx. \tag{6.40}$$

Switching the order of integration,

$$\Psi_z(y,z) - \Psi_z(y,z_B) = \int_{z_B}^z V_z(y,z') \, dz', \tag{6.41}$$

$$\Psi_\sigma(y,\sigma) - \Psi_\sigma(y,\sigma_B) = -\int_{\sigma_B}^\sigma V_\sigma(y,\sigma') \, d\sigma'. \tag{6.42}$$

In other words, the zonally integrated velocities relate to the streamfunctions by,

$$\frac{\partial\Psi_z}{\partial z} = V_z, \tag{6.43}$$

$$\frac{\partial\Psi_\sigma}{\partial\sigma} = -V_\sigma. \tag{6.44}$$

The units are m^3s^{-1} for Ψ_z and Ψ_σ, m^2s^{-1} for V_z, and $\text{m}^6\text{s}^{-1}\text{kg}^{-1}$ for V_σ. Sometimes a different sign convention is used to define the streamfunctions in (6.41) and (6.42).

It is important to understand that Ψ_σ is not simply Ψ_z stretched or squashed by a mapping from z to σ. For example, in Figure 6.16, the extent *in latitude* for STCs is different in Ψ_z and Ψ_σ (Subsection 6.2.3).

It is better to think of the streamfunction and volume transports as two kinds of census. For Ψ_z calculated at a given meridional location y, we sort the cross section of the ocean into bins with different ranges of z values (for example, 0 to 50 m, 50 to 100 m, etc.), and we sum the volume transport in each bin. For Ψ_σ, we sort the section into bins with different σ ranges and sum the volume transports in *those* bins. The meridional overturning streamfunction defined in this way may have different magnitudes and even different signs.

Insight from Alternative Streamfunctions

Any three-dimensional ocean property field that is monotonic in depth can be used as an alternative "vertical" coordinate. A temperature-coordinate overturning streamfunction Ψ_T (usually using potential temperature, Subsection 1.1.2) is useful if we are more interested in the thermodynamics than in the dynamics, though Ψ_T and Ψ_σ are closely related when salinity variations are small.

Besides showing how much of each density of water is flowing, the density-coordinate streamfunction shows how density is transformed by the flow. In regions where contours of $\Psi_\sigma(y, \sigma_2)$ are horizontal (such as, approximately, 20°S, $\sigma_2 = 26$ in Figure 6.16b), density remains constant for water parcels in the flow. Tilted or vertical contours indicate changing density, for instance parcels becoming lighter near the equator and denser for the lightest water in the subtropics. In such regions some process such as mixing or air–sea interaction must be changing the density of the water. A similar analysis applies to Ψ_T which shows heating and cooling of water parcels.

Another useful property of Ψ_σ is its connection with the **meridional density transport** Q_σ

$$Q_\sigma(y) = \int_W^E \int_{z_B}^{z_T} v(x, y, z) \sigma(x, y, z) \, dz \, dx. \tag{6.45}$$

The density transport is related to surface density fluxes at different latitudes, and measures how strongly density is altered in the course of a water parcel's transit through the basin (see Chapter 11). A similar quantity is the **meridional heat transport** Q_T, which has the same definition as Q_σ except that σ is replaced by $c_p \rho T$ (c_p is the specific heat capacity at constant pressure of seawater; Appendix C). The heat transport is an important measure of thermodynamic ocean–atmosphere interaction and hence of the ocean's influence on the climate (Chapter 11).

A transparent form of (6.45) is derived as follows. Change the variable representing the vertical direction from z to σ,

$$Q_\sigma(y) = \int_{\sigma_B}^{\sigma_T} \int_W^E v(x, y, z(x, y, \sigma)) \sigma \frac{\partial z}{\partial \sigma} \, dx \, d\sigma, \tag{6.46}$$

and substitute (6.40) for the inner integral. Now we have an extra factor of σ, but as we are integrating x along contours of constant σ, each integral in x is independent of σ. Therefore, bring the σ term outside the x integral and find

$$Q_\sigma = \int_{\sigma_B}^{\sigma_T} V_\sigma \sigma \, d\sigma. \tag{6.47}$$

This can be further simplified by writing V in terms of Ψ_σ. From (6.44) and applying the product rule to rewrite the factor of $V_\sigma \sigma$, we find

$$V_\sigma \sigma = -\frac{\partial \Psi_\sigma}{\partial \sigma} \sigma = -\frac{\partial}{\partial \sigma} (\Psi_\sigma \sigma) + \Psi_\sigma. \tag{6.48}$$

Inserting this into (6.47) gives

$$Q_\sigma = \int_{\sigma_B}^{\sigma_T} \Psi_\sigma \, d\sigma - [\Psi_\sigma \sigma]_{\sigma_B}^{\sigma_T}. \tag{6.49}$$

Now think about the final term. If there is no net meridional volume transport through a basin, which is often a good approximation, then $\Psi_\sigma(\sigma_B) = \Psi_\sigma(\sigma_T)$. In that case, if

$\Psi_\sigma = 0$ at the density limits σ_B and σ_T, then the final term is zero. If $\Psi_\sigma = \Psi_0 \neq 0$ at those limits, then define a new streamfunction $\Psi' = \Psi_\sigma - \Psi_0$, and

$$Q_\sigma = \int_{\sigma_B}^{\sigma_T} \Psi' \, d\sigma + (\sigma_T - \sigma_B)\Psi_0 - \left[\Psi'\sigma\right]_{\sigma_B}^{\sigma_T} - (\sigma_T - \sigma_B)\Psi_0, \qquad (6.50)$$

which resembles (6.49) with Ψ' replacing Ψ_σ. Thus, in either case, the meridional density transport is

$$Q_\sigma = \int_{\sigma_B}^{\sigma_T} \Psi' \, d\sigma. \qquad (6.51)$$

Once we have calculated $\Psi_\sigma(\sigma)$ at a given latitude, we can get the density transport by integrating. The heat transport Q_T is similarly related to Ψ_T, see (11.4). In many cases, $\Psi_\sigma(\sigma)$ at a given latitude looks like the idealized case shown in Figure 6.22. This represents a 6 Sv northward flow of water lighter than $\sigma_2 = 34$ and an equal volume transport of denser water flowing southward. The volume transport $\Delta\Psi$ and density range $\Delta\sigma_2$ between mean northward and mean southward flow can be read off the graph and the density transport estimated as $\Delta\Psi \Delta\sigma_2 \approx 6$ Sv $\times 1.5$ kg m$^{-3} = 9 \times 10^6$ kg s^{-1}. Similar estimates can be made from more complex $\Psi_\sigma(y, \sigma)$ patterns such as in Figure 6.16b. Once one gets used to reading them, Ψ_σ and Ψ_T plots give an immediate impression of the meridional property fluxes associated with a given flow.

6.3.4 Layer Models of Extra-Equatorial Tropics

In order to derive the full three-dimensional wind-driven flow in the subtropics and tropics, we need to include both Ekman transport and western boundary currents. Essentially, we are trying to give a more complete description of the flow than the cartoon in Figure 6.15 for the case in which the wind-driven flow occurs in and above a pycnocline. In general this is a difficult problem but some simple cases can be solved analytically. Here we formulate

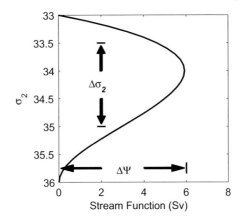

Figure 6.22 Idealized example of $\Psi_\sigma(\sigma)$ illustrating the estimate of property transport from meridional streamfunction in the appropriate coordinate system.

a model which is then solved in Subsection 6.3.5 to derive solutions such as the ones discussed in Subsection 6.2.4 and pictured in Figure 6.18.

We choose two different ways to represent the vertical structure of the wind-driven flow in the tropics. The simplest has a uniform density model as in Figure 6.15, with vertically uniform geostrophic flow and Ekman transport confined to a thin layer at the top. Here we use a slightly more complex model, a $1\frac{1}{2}$-layer arrangement in which all geostrophic flow is confined to the top layer and the Ekman transport is confined to a thin layer within the top layer. This allows us to represent the confinement of most of the wind-driven flow to the pycnocline. For a more realistic model that allows velocity in the lower pycnocline to be different from flow near the surface, we follow McCreary and Lu (1994) in using a $2\frac{1}{2}$-layer hydrostatic model which extends LPS theory (Chapter 5) to the western boundary layer and equator.

The McCreary and Lu (1994) model, shown in Figure 6.17, includes another important extension to the equations used in Chapter 5: there is an exchange of fluid between the two active layers, which is represented by a vertical cross-interface velocity w_1. (We assume that $w_2 = 0$ so there is no exchange between layer 2 and the motionless abyss.)

We choose w_1 based on the desired layer thicknesses h_1 and h_2. As shown in Figure 6.17, poleward of what McCreary and Lu (1994) call the **detrainment latitude** θ_d, layer 1 represents the Ekman layer and has uniform thickness $h_1 = H_1$. In this region, most of the divergence of Ekman transport pumps into layer 2. To the extent that the Ekman layer corresponds to the mixed layer, this transfer corresponds to detrainment of mixed-layer water (layer 1) into the pycnocline (layer 2). There is some geostrophic transport in layer 1 which in principle can cancel the Ekman divergence, but in practice the cancellation is small if H_1 is small compared to $h = h_1 + h_2$.

Equatorward of θ_d, layer 1 represents both the Ekman layer and the upper pycnocline. As Ekman downwelling at low latitudes enters the upper pycnocline and then follows isopycnals, such downwelling should remain in layer 1 rather than contribute to w_1. Ekman upwelling at the equator increases layer 1 thickness. We expect that when the upper layer becomes sufficiently thin, layer 2 water upwells into the mixed layer. This is implied in Figure 6.2d, which shows the EUC approaching the sea surface in the eastern Pacific where the upward-sloping velocities deposit water into the surface layer. Also, we avoid vanishing layer 2 thickness (and hence infinite stratification) by invoking vertical mixing of water from layer 1 to layer 2, when layer 2 is sufficiently thin.

For each layer n of the two-layer model of McCreary and Lu (1994), the variables are pressure p_n, horizontal velocity components (u_n, v_n), and those above: layer thickness h_n, and the cross-interface velocity component w_n. The solution is defined by the wind stress forcing $\mathbf{X} = (X, Y)$, layer densities ρ_n and detrainment latitude θ_d. Using the Kronecker delta function ($\delta_{nm} = 1$ if $n = m$, else $\delta_{nm} = 0$), in spherical coordinates (Appendix B) the equations read (see Subsection 5.3.2 and (2.76))

$$f h_n v_n = \frac{h_n}{\rho_0 R_E \cos\theta} \frac{\partial p_n}{\partial \phi} - \delta_{n1} X, \tag{6.52}$$

$$f h_n u_n = -\frac{h_n}{\rho_0 R_E} \frac{\partial p_n}{\partial \theta} + \delta_{n1} Y, \tag{6.53}$$

$$w_n = \frac{1}{R_E \cos\theta} \left[\frac{\partial}{\partial\phi}(h_n u_n) + \frac{\partial}{\partial\theta}(h_n v_n \cos\theta) \right], \tag{6.54}$$

where (ϕ, θ) are (longitude, latitude). These equations can be written entirely in terms of (u_n, v_n, h_n) by expressing pressure in terms of h_1 and h using the hydrostatic equation and the assumption of no flow in layer 3 (from (2.96)):

$$\frac{1}{\rho_0}\nabla p_1 = \gamma_1 \nabla h_1 + \gamma_2 \nabla h, \tag{6.55}$$

$$\frac{1}{\rho_0}\nabla p_2 = \gamma_2 \nabla h, \tag{6.56}$$

where reduced gravity $\gamma_n = g(\rho_{n+1} - \rho_n)/\rho_0$. The momentum equations, (6.52) and (6.53), are the same as in Chapter 5 except that the zonal wind stress is included and hence the velocity includes both the geostrophic velocity and Ekman transport.

In Chapter 5, we derive equations for geostrophic layers below an Ekman layer (see Subsection 5.3.2). Consideration of the Ekman transport in the top layer does not change the behavior of the geostrophic flow, so we can use the same model for geostrophic flow in the tropics (except within about 100 km of the equator). Chapter 5 finds that the Sverdrup balance can be written in terms of the layer thicknesses and forcing given by Ekman pumping w_{Ek}. Here we are interested in both the Ekman transport and the Ekman pumping, so we retain the relationship between w_{Ek} and wind stress **X**.

To solve for the velocity distribution in either case ($1\frac{1}{2}$ or $2\frac{1}{2}$ layers), start with the Sverdrup balance. It says (see (5.23) and use (2.65) to replace ηs with $h s$)

$$h^2 + \Gamma h_1^2 = H^2 + \Gamma H_1^2 - \frac{2f^2}{\beta\gamma_2} R_E \cos\theta \int_\phi^{\phi_E} w_{Ek}\, d\phi, \tag{6.57}$$

where $\Gamma = \gamma_1/\gamma_2$, and $(h_1, h) = (H_1, H)$ at the eastern boundary $\phi = \phi_E$. Taking advantage of the dependence of f and β on θ and substituting for w_{Ek} with (6.9), (6.57) becomes

$$h^2 + \Gamma h_1^2 = H^2 + \Gamma H_1^2 + \frac{2R_E}{\gamma_2}\int_\phi^{\phi_E}\left(\frac{\partial X}{\partial\theta}\sin\theta - \frac{X}{\cos\theta}\right)d\phi. \tag{6.58}$$

The stress X is independent of ϕ, so the integral is straightforward. Nondimensionalizing with $(h', h_1', H_1') = (h, h_1, H_1)/H$, $X' = X/X_0$, and defining the nondimensional parameter

$$D_0'^2 = \frac{2R_E X_0}{H^2\gamma_2}, \tag{6.59}$$

gives

$$h'^2 + \Gamma h_1'^2 = 1 + \Gamma H_1'^2 + D_0'^2(\phi_E - \phi)\left(\frac{\partial X'}{\partial\theta}\sin\theta - \frac{X'}{\cos\theta}\right). \tag{6.60}$$

For the $2\frac{1}{2}$-layer case, we need another equation to solve for both h and h_1. Poleward of θ_d, we have $h_1 = H_1$. Equatorward of θ_d (but away from the equator), we can use conservation of potential vorticity (f/h_2) for layer 2 flow, as in LPS theory (Subsection 5.3.3). The condition is somewhat different in this case, because instead of $h_1 = 0$ at the *subduction* latitude, we have $h_1 = H_1$ at the *detrainment* latitude. Because potential vorticity is conserved along streamlines and h are streamlines in layer 2, there must be some function $Q_2(\cdot)$ for which

$$f/h_2 = Q_2(h). \tag{6.61}$$

This is true at $\theta = \theta_d$, where a matching condition to the solution poleward of θ_d gives

$$Q_2(h) = \frac{f}{h_2} = \frac{f_d}{h - H_1} \tag{6.62}$$

and writing h_2 in terms of h and h_1, (6.62) is equivalent to

$$h_1 = (1 - f/f_d)\,h + (f/f_d)H_1. \tag{6.63}$$

Inserting this into the two-layer Sverdrup balance (6.60) gives

$$h'^2 + \Gamma\left[\left(1 - \frac{f}{f_d}\right)h' + \left(\frac{f}{f_d}\right)H_1'\right]^2 = 1 + \Gamma H_1'^2 + D_0'^2(\phi_E - \phi)\left(\frac{\partial X'}{\partial \theta}\sin\theta - \frac{X'}{\cos\theta}\right). \tag{6.64}$$

If $h_1 = 0$ ($1\frac{1}{2}$-layer model, in which case $H_1 = 0$ too) or $h_1 = H_1$ ($2\frac{1}{2}$-layer model poleward of θ_d) the Sverdrup balance (6.60) does not depend on h_1:

$$h'^2 = 1 + D_0'^2(\phi_E - \phi)\left(X_\theta'\sin\theta - X'/\cos\theta\right), \tag{6.65}$$

with $X_\theta' = \partial X'/\partial\theta$.

6.3.5 Gyre With Subtropical Cell

Here we apply the model developed in Subsection 6.3.4 to the idealized configurations shown in Figure 6.18. We illustrate the differences and similarities of the solutions to $1\frac{1}{2}$- and $2\frac{1}{2}$-layer models with a set of wind profiles which we can alter in the tropics. The wind stress is given by

$$X' = \begin{cases} \frac{1}{2}(1 - X_Q')\cos(\pi\frac{\theta}{\theta_1}) - \frac{1}{2}(1 + X_Q') & \theta < \theta_1 \\ -\cos(\frac{2\pi}{\theta_n}[\theta - \theta_2]) & \theta_1 \le \theta < \theta_2 \\ \cos(\frac{\pi}{\theta_n}[\theta - \theta_2]) & \theta \ge \theta_2 \end{cases} \tag{6.66}$$

and is displayed in the left panels of Figure 6.23. The parameter X_Q' controls the wind strength at the equator, so that we can look at examples with uniform wind in the deep tropics (top), a 25% decrease in wind to the equator (middle), and zero wind at the equator (bottom). This creates a positive (counterclockwise) wind curl that is zero (top), small (middle) and big (bottom). The wind in the top and bottom panels correspond to those in the top and bottom panels of Figure 6.18.

Pathways to the Equator

The geostrophic flow in the deepest active layer is along contours of h, so we can see the geostrophic flow in the $1\frac{1}{2}$-layer model (Figure 6.23, middle column of panels) and in layer 2 of the $2\frac{1}{2}$-layer model (Figure 6.23, panels in right column). The h contours in Figure 6.23 do not include the western boundary layer.

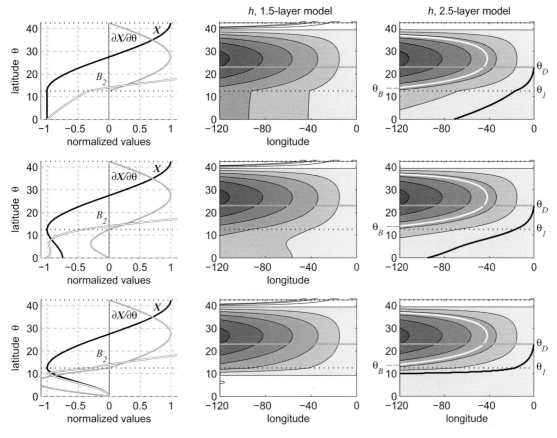

Figure 6.23   Wind stress X (black), $\partial X/\partial \theta$ normalized by maximum values (gray), and WBC transport B_2 (double line; left panels); h for $1\frac{1}{2}$-layer model (center), and h for $2\frac{1}{2}$-layer model (right) for $X'_Q = 1$ (top), $X'_Q = 0.75$ (middle), and $X'_Q = 0$ (bottom). Lighter shades show thinner h, thick gray line is at $\theta = \theta_d$, dotted line is at $\theta = \theta_1$, thick black contour represents boundary of shadow zone, and thick white contour is bifurcation streamline. The parameters for calculating the fields in figure are $\theta_1 = 12.5°$, $\theta_d = 23°$, $H'_1 = 0.25$, and $D_0'^2 = 0.8$. See also Exercise 6.9.

As the top panels show, adding the extra layer to the model introduces a significant change in the circulation. In the $1\frac{1}{2}$-layer model, the geostrophic flow is nearly due south because of the absence of a tropical gyre. This flow balances the northward Ekman transport. However the $2\frac{1}{2}$-layer model shows that the pycnocline flow has a strong westward component as well. This same component can be seen in Figure 6.18c. The depth-averaged velocity in the zero-curl region is zero, which is consistent with the layer 2 southwestward flow, because layer 1 has flow in the opposite direction (Figure 6.18b). If we add a weak cyclonic gyre in the tropics (Figure 6.23, middle panels), some of the $1\frac{1}{2}$-layer flow in the northwestern corner of the tropical gyre also flows southwestward. In the $2\frac{1}{2}$-layer model, the westward component is more pronounced than in the $X'_Q = 1$ case. As we reduce X'_Q to zero (and hence strengthen the tropical gyre), equatorward-flowing pycnocline water from the subtropical gyre is channeled into a small region in the north of the tropical gyre in both models (Figure 6.23, bottom panels).

In the $2\frac{1}{2}$-layer model, the southern boundary of the pycnocline flow from the subtropical gyre is the boundary of the shadow zone discussed in Chapter 5. This boundary is the $h = H$ streamline that extends southward from $(\phi, \theta) = (\phi_E, \theta_d)$ (Figure 6.23, right panels). As in Subsection 5.3.3, we can determine the path of this streamline by setting $h' = 1$ in (6.64),

$$\Gamma \left[1 - f/f_d + (f/f_d)H_1' \right]^2 = \Gamma H_1'^2 + D_0'^2 (\phi_E - \phi) \left(X_\theta' \sin\theta - X' \cos\theta \right), \tag{6.67}$$

and rearranging to find $\phi_E - \phi$ as a function of θ:

$$(\phi_E - \phi) = \frac{\Gamma \left\{ \left[1 - f/f_d + (f/f_d)H_1' \right]^2 - H_1'^2 \right\}}{D_0'^2 \left(X_\theta' \sin\theta - X' \cos\theta \right)}. \tag{6.68}$$

At $\theta = \theta_d$, $\phi_E - \phi = 0$ by design. If we have $\phi = \phi_W$ at any latitude, then the shadow zone boundary intersects the western boundary and the subtropical water cannot flow to the equator via the "interior" (Sverdrup) flow. Only the denominator of (6.68) depends on the wind stress profile. Near the equator, $\sin\theta \to 0$ and $\cos\theta \to 1$, so the denominator is approximately $-D_0'^2 X'$. The smaller $-X'$, the further westward the shadow zone reaches, which is consistent with Figure 6.23 when X_Q' is lowered from 1 to 0.75. Furthermore, the denominator can also get small away from the equator if X_θ' is a different sign than $-X'$, as it is in the tropical gyre (Figure 6.23, left panels). In that case, if there is any θ in the tropical gyre for which $X_\theta' \tan\theta = X'$, as there is for the $X_Q' = 0$ example (Figure 6.23, bottom panels), then the shadow zone streamline veers sharply westward near the subtropical gyre.

In the real world, the tropical gyre has stronger wind curl in the northern hemisphere than in the southern hemisphere, because the Intertropical Convergence Zone is north of the equator in the Pacific (Figure 3.1). Lu and McCreary (1995) argue that this explains why subtropical water flows to the western boundary in the North Pacific but some of it flows directly to the equator in the central South Pacific (Figures 6.1, 6.19).

Water that flows to the western boundary can also reach the equator if the western boundary current flows equatorward. The depth-average western boundary current direction changes sign at the zero-wind-curl line dividing the subtropical and tropical gyres, but it is not obvious that the geostrophic pycnocline flow changes direction at the same latitude. In order to find this **bifurcation latitude** dividing poleward subtropical gyre flow from equatorward tropical gyre flow in the western boundary current, we must extend our solution to the western boundary. First we define transport metrics, then we compute in turn the western boundary current transports poleward and equatorward of the detrainment latitude.

Circulation Transport Metrics

Start by defining the volume transport between longitudes ϕ_a and ϕ_b for layer n,

$$T_n(\phi_a, \phi_b, \theta) = R_E \cos\theta \int_{\phi_a}^{\phi_b} h_n v_n \, d\phi. \tag{6.69}$$

This can be written in terms of h and h_1 by way of the zonal momentum equation (6.52) and the hydrostatic relation (6.55):

$$T_1 = R_E \cos\theta \int_{\phi_a}^{\phi_b} \left[\frac{h_1(\gamma_1 h_{1\phi} + \gamma_2 h_\phi)}{f R_E \cos\theta} - \frac{X}{f} \right] d\phi, \tag{6.70}$$

$$T_2 = R_E \cos\theta \int_{\phi_a}^{\phi_b} \frac{\gamma_2(h - h_1)h_\phi}{f R_E \cos\theta} \, d\phi. \tag{6.71}$$

Identifying the Ekman volume transport,

$$T_{Ek} = -R_E \cos\theta \int_{\phi_a}^{\phi_b} \frac{X}{f} \, d\phi, \tag{6.72}$$

and using $hh_\phi = (h^2/2)_\phi$, rewrite the layer volume transport expressions as

$$T_1 = \frac{\gamma_1}{2f}[h_1(\phi_b)^2 - h_1(\phi_a)^2] + \frac{\gamma_2}{f} \int_{\phi_a}^{\phi_b} h_1 h_\phi \, d\phi + T_{Ek}, \tag{6.73}$$

$$T_2 = \frac{\gamma_2}{2f}[h(\phi_b)^2 - h(\phi_a)^2] - \frac{\gamma_2}{f} \int_{\phi_a}^{\phi_b} h_1 h_\phi \, d\phi. \tag{6.74}$$

The bifurcation point may occur either north or south of the detrainment latitude, depending on the details of the wind field and the value of θ_d. If $\theta \geq \theta_d$, $h_1 = H_1$, and the expressions simplify:

$$T_1 = \frac{\gamma_2 H_1}{f}[h(\phi_b) - h(\phi_a)] + T_{Ek}, \tag{6.75}$$

$$T_2 = \frac{\gamma_2}{2f}\left[h(\phi_b)^2 - h(\phi_a)^2\right] - \frac{\gamma_2 H_1}{f}[h(\phi_b) - h(\phi_a)]. \tag{6.76}$$

Two measures of volume transport are of special interest, the western boundary transport

$$B_n(\theta) \equiv T_n(\phi_W, \phi_W + \delta\phi, \theta), \tag{6.77}$$

where $\delta\phi$ is the western boundary current width, and the total layer transport

$$M_n(\theta) \equiv T_n(\phi_W, \phi_E, \theta). \tag{6.78}$$

Note that $h_n(\phi_W + \delta\phi)$ represents values just outside the western boundary current, so that h_n equals the Sverdrup value for $\phi \approx \phi_W$ which is given by expressions such as (6.60). The bifurcation latitude occurs where $B_2(\theta) = 0$. Similarly, we also define B_{Ek} and M_{Ek} for Ekman volume transport in the boundary layer and entire basin width, respectively.

Western Boundary Transport Poleward of Detrainment Latitude

First, poleward of θ_d, (6.76) and (6.77) imply

$$B_2 = \frac{\gamma_2}{2f}\left(h_S^2 - h_0^2\right) - \frac{\gamma_2 H_1}{f}(h_S - h_0), \tag{6.79}$$

where $h_S \equiv h(\phi_W + \delta\phi)$ is known. The height at the western coast, $h_0 = h(\phi_W)$, is not known, however, because it depends on the very western boundary current properties that we are trying to find. We can calculate h_0 by using the fact that we know that the net transport $M_1 + M_2$ is zero, and by adding (6.73) and (6.74) to get an expression for $M_1 + M_2$ in terms of h_0. (As discussed in Chapter 3, for the general case $M_1 + M_2 \neq 0$ if there is net flow around a continent, but we keep the simpler case here.) The sum of (6.75) and (6.76) gives

$$T_1 + T_2 = T_{Ek} + \frac{\gamma_2}{2f}\left[h(\phi_b)^2 - h(\phi_a)^2\right].\tag{6.80}$$

Substituting $\phi_b = \phi_E$, $h(\phi_E) = H$, $\phi_a = \phi_W$, and $h(\phi_W) = h_0$, and then noting that $T_1 + T_2 = M_1 + M_2 = 0$, gives

$$h_0^2 = H^2 + \frac{2f}{\gamma_2}M_{Ek}.\tag{6.81}$$

We can substitute this expression for h_0 back into (6.79) to close the problem.

In fact, there is a somewhat more illuminating expression for B_2. First define $\bar{h} = (h_S + h_0)/2$ and $\Delta h = h_S - h_0$, so that we can rewrite (6.79) as

$$B_2 = \frac{\gamma_2}{f}\left(\bar{h} - H_1\right)\Delta h.\tag{6.82}$$

We can find an expression for Δh in terms of known parameters by using (6.80) again, but now substituting $\phi_b = \phi_W + \delta\phi$:

$$\frac{\gamma_2}{f}\bar{h}\Delta h = B_1 + B_2 - B_{Ek}.\tag{6.83}$$

Based on the analysis in Chapter 3, we can assume that, to good accuracy,

$$B_1 + B_2 \approx -\frac{R_E\cos\theta}{\beta}\int_{\phi_W}^{\phi_E}\hat{\mathbf{z}}\cdot(\nabla\times\mathbf{X})\,d\phi = -\Psi_{Sv},\tag{6.84}$$

where Ψ_{Sv} is the Sverdrup transport. Then if we use (6.83) and (6.84) to replace Δh, (6.82) becomes

$$B_2 = -\left(1 - \frac{H_1}{\bar{h}}\right)(\Psi_{Sv} + B_{Ek}).\tag{6.85}$$

As B_{Ek} is only integrated over the narrow western boundary current, it is small, and in general $\bar{h} > H_1$. Therefore B_2 has the same sign as $-\Psi_{Sv}$ and so the bifurcation point occurs near the border between the subtropical and tropical gyres, where the Sverdrup transport vanishes.

Western Boundary Transport Equatorward of Detrainment Latitude

Equatorward of θ_d, in order to calculate B_2 we need to calculate M_1. This transport is interesting in its own right as a measure of the subtropical cell strength. Consider first latitudes poleward of θ_d; the argument is extended equatorward of θ_d momentarily. Using $h(\phi_E) = H$ and $h(\phi_W) = h_0$, (6.75) and (6.81) yield

$$M_1 = \frac{\gamma_2 H_1}{f}\left(H - \sqrt{H^2 + \frac{2f}{\gamma_2}M_{Ek}}\right) + M_{Ek}.\tag{6.86}$$

This expression simplifies when $2fM_{Ek}/(\gamma_2 H^2)$ is not very big (see Exercise 6.7). In that case

$$M_1 \approx \frac{\gamma_2 H_1 H}{f}\left[1 - \left(1 + \frac{fM_{Ek}}{\gamma_2 H^2}\right)\right] + M_{Ek}, \tag{6.87}$$

$$\approx \left(1 - \frac{H_1}{H}\right)M_{Ek}. \tag{6.88}$$

Not surprisingly, the subtropical cell volume transport is the Ekman volume transport minus a fraction associated with layer 1 geostrophic compensation. Because we are assuming that the combined volume transports of layer 1 and layer 2 is zero at any latitude, $M_2 = -M_1$.

Between the detrainment latitude and the equatorial boundary layer, there is no exchange between layers 1 and 2 (Subsection 6.3.4). Therefore M_1 and M_2 are the same at $\theta < \theta_d$ as they are at $\theta = \theta_d$. We can find B_2 by conserving volume, namely,

$$B_2(\theta) + T_2(\phi_W + \delta\phi, \phi_E; \theta) = M_2(\theta_d) \tag{6.89}$$

and calculating T_2 from the interior solution. Using the relationship between h_1 and h (6.63) we rewrite the T_2 equation (6.74) as

$$T_2(\phi_W + \delta\phi, \phi_E; \theta) = \frac{\gamma_2}{2f}\left(H^2 - h_S^2\right) - \frac{\gamma_2}{f}\int_{\phi_W+\delta\phi}^{\phi_E}\left[\left(1 - \frac{f}{f_d}\right)h + \frac{f}{f_d}H_1\right]h_\phi\, d\phi,$$

$$= \frac{\gamma_2}{2f_d}\left[H^2 - h_S^2 - 2H_1(H - h_S)\right]. \tag{6.90}$$

Insert this into (6.89)

$$B_2(\theta) = -M_{Ek}(\theta_d) - \frac{\gamma_2 H_1}{f_d}[H - h_{0d}] - \frac{\gamma_2}{2f_d}\left[H^2 - h_S^2 - 2H_1(H - h_S)\right], \tag{6.91}$$

where $M_2(\theta_d) = -M_1(\theta_d)$, $M_1(\theta_d)$ is given by (6.86), and

$$h_{0d} = h_0(\theta_d) = H\sqrt{1 + \frac{2f_d}{\gamma_2 H^2}M_{Ek}(\theta_d)}. \tag{6.92}$$

Hence,

$$B_2(\theta) = -M_{Ek}(\theta_d) + \frac{\gamma_2}{f_d}\left[\frac{1}{2}(h_S^2 - H^2) + H_1(h_{0d} - h_S)\right]. \tag{6.93}$$

From (6.81),

$$-M_{Ek}(\theta_d) = \frac{\gamma_2}{2f_d}(H^2 - h_{0d}^2) \tag{6.94}$$

and therefore, finally,

$$B_2(\theta) = \frac{\gamma_2}{f_d}(h_S - h_{0d})\left[\frac{1}{2}(h_S + h_{0d}) - H_1\right]. \tag{6.95}$$

With this expression, we now possess the solution for the western boundary current transport equatorward of the detrainment latitude.

For insight into the bifurcation latitude θ_B in this case, think about the following argument. Assume that the factor of $\frac{1}{2}(h_S + h_{0d}) - H_1$ is positive (see Exercise 6.8), then $B_2 = 0$ from (6.95) at latitude θ_B, where $h_S = h_{0d}$. Now,

$$(h_{0d}/H)^2 = 1 - \frac{2X(\theta_d)R_E \cos\theta\,\Delta\phi}{H^2\gamma_2} = 1 - D_0'^2 X'(\theta_d)\cos\theta\,\Delta\phi \tag{6.96}$$

using (6.72) to compute $M_{Ek}(\theta_d)$, $\Delta\phi = \phi_E - \phi_W$, and (6.59). From the formula for the pycnocline depth (6.60), and in the limit $(h_1, H_1) \to 0$,

$$(h_S/H)^2 \approx 1 + D_0'^2\Delta\phi\left(\frac{\partial X'}{\partial\theta}\sin\theta - \frac{X'}{\cos\theta}\right). \tag{6.97}$$

Because $h_S^2 = h_{0d}^2$ at the same latitude at which $h_S = h_{0d}$, equating (6.96) and (6.97) shows that $h_S = h_{0d}$ occurs close to the zero wind-curl latitude, where $\partial X'/\partial\theta = 0$, and X' is maximum towards the west. Relaxing the limit of $(h_1, H_1) \to 0$ makes a small correction (Figure 6.23, right panels). Thus the bifurcation latitude occurs near the border between the subtropical and tropical gyres (as we found above if θ_B is poleward of the detrainment latitude).

The **bifurcation streamline** reaches the western boundary at the bifurcation latitude, $\theta = \theta_B$. Layer 2 water to the west of the bifurcation streamline recirculates within the subtropical gyre, while water between the bifurcation streamline and the shadow zone boundary all flows to the equator, either through the equatorward-flowing western boundary current or through flow in the Sverdrup interior. Because the volume transport of equatorward flow in layer 2 at θ_d is largely determined by the Ekman transport there, the volume transport is about the same in the three cases shown in Figure 6.23. Only the division between western boundary current and interior flow in the tropics is different.

Exercises

6.1 Using data from the Ocean Sustained Interdisciplinary Timeseries Environment observation System (OceanSITES, Appendix A):

(a) Make plots showing the time-average zonal upper ocean temperature for the equatorial Pacific, Atlantic, and Indian Oceans. Compare and contrast the three ocean basins.

(b) Make plots at various longitudes of the time-average meridional upper ocean temperature for the equatorial Pacific, Atlantic, and Indian Oceans. Compare and contrast the three ocean basins.

(c) Make plots at various longitudes of the time-average vertical profiles of the zonal current at the equator for the Pacific, Atlantic, and Indian Oceans. Identify and label the main currents.

(d) El Niño events are characterized by anomalous warm surface water in the eastern equatorial Pacific Ocean. Using OceanSITES time series plots (or otherwise), identify El Niño events. Explore and describe the associated response in the thermocline structure and the upper ocean circulation in the equatorial Pacific.

6.2 Download the near-surface velocity climatology from surface drifters (see Appendix A and Laurindo et al., 2017). Also, download annual average chlorophyll data from ocean color satellite instruments (such as the SeaWIFS instrument, see Appendix A). In each case, make and state sensible choices.

(a) Make figures of the global average surface horizontal speed and vertical speed, similar to Figure 6.6. Explain your methods.

(b) Make a figure of the annual average surface ocean chlorophyll concentration, similar to Figure 6.8.

(c) Make a scatter plot of chlorophyll against vertical speeds. Describe and explain the relationship between the variables.

(d) A few upwelling zones show anomalously low chlorophyll. Identify them in your figures. What do you expect the surface nutrient concentration field to look like in these areas?

(e) Find, download, and map data of surface nutrient concentration (for example, nitrate). Test your expectation from part (d) and comment.

6.3 Choose one of the four eastern upwelling systems in Figure 6.9 (coordinate with your classmates to avoid replicates). Using observations of your choice (see, for example, Section 6.1 and Appendix A):

(a) Investigate seasonality in the strength of upwelling and anomalies in sea-surface temperature, chlorophyll, and nutrient concentration. Explain the relationships between these variables.

(b) Coordinate with your classmates to compare and contrast different upwelling systems. Identify common relationships between upwelling systems and speculate about the origins of their idiosyncrasies.

6.4 Consider the sea-surface temperature anomalies typical of eastern boundary upwelling systems (such as in Figure 6.9). Think about what sets the characteristic length scale of zonal SST anomalies in such regions, then:

(a) Write a formula for the zonal surface Ekman transport, defining all the terms.

(b) Write an equation for the SST as a function of zonal distance from the coast, neglecting meridional variations and nonlinear effects. Include air–sea heat exchange with an atmosphere of temperature T_A (use ideas from Subsection 4.3.1).

(c) Under some circumstances, the SST relaxes to T_A over the characteristic distance

$$\frac{Y \rho c_p}{f \lambda} \tag{6.98}$$

(in our standard notation). Derive this result, explaining suitable assumptions and defining all variables and parameters.

(d) Estimate the characteristic relaxation distance (6.98) for the upwelling systems in Figure 6.9. Comment on the agreement with the observed scale for SST anomaly decay.

Hint: This problem is based on Spall and Schneider (2016).

6.5 Consider the theories for the equatorial undercurrent in Subsections 6.3.1 and 6.3.2 (see also Subsection 6.2.1).

For the frictional non-rotating theory:

(a) Compute and plot the zonal speed, pressure field, and sea-surface height field as functions of longitude and depth for the equatorial Pacific. Use realistic parameter values.

(b) What value of vertical viscosity matches the data best? Comment on the realism of this value.

For the inviscid rotating theory:

(c) Using the continuity equation, write an expression for the meridional speed v in the equatorial undercurrent layer.

(d) Hence, sketch streamlines of the equatorial undercurrent layer. Assume the equatorial flow smoothly merges with the ventilated thermocline flow away from the equator. Comment on the realism of this flow.

6.6 Examine the provided files with annual-average potential temperature θ, salinity, potential density (σ_θ) and meridional velocity v on longitude–depth sections in the Atlantic and Pacific basins from a GCM simulation. For each basin:

(a) Use contour plots, or otherwise, to show the θ and v sections.

(b) Calculate and plot the zonal integral of v, at constant depth, z, as a function of z.

(c) Calculate and plot the zonal integral of $v dz/d\theta$, calculated along constant potential temperature θ, as a function of θ. There are a number of ways to estimate this from the data. Here is a suggestion: assume that each grid point is a box which is Δx m wide and Δz m tall (Δz is different for each level in the model), and that θ and v are uniform within each box. Divide the temperature axis into small intervals (you decide how small), and sort the grid boxes into these intervals. In other words, the temperature of the box determines to which θ that particular box is contributing its velocity.

(d) Calculate and plot the z-coordinate meridional overturning streamfunction as a function of z at this particular latitude.

(e) Repeat (d) for the θ-coordinate overturning streamfunction as a function of θ.

(f) How would you interpret the zonal average flow based on your z-coordinate integrals? How about based on your θ-coordinate intervals? Describe each in a few sentences for each ocean.

6.7 Consider the formula for the subtropical gyre strength (6.86). Make arguments about the realism of assuming $2fM_{Ek}/(\gamma_2 H^2) \ll 1$, and explain physically what the criterion means.

6.8   Is it reasonable to assume that $\frac{1}{2}(h_S + h_{0d}) > H_1$ in (6.95) in order to estimate the bifurcation latitude? Explain your answer.

6.9   Consider the theory for the $2\frac{1}{2}$-layer McCreary and Lu (1994) model in Subsections 6.3.4 and 6.3.5, then:

(a) Write down the final formulae for the full solution. Identify the variables and parameters.

(b) Write pseudo-code defining an algorithm to compute the solution.

(c) Hence compute the solution to reproduce the three cases in Figure 6.23. Test the reasonableness of the assumptions in Exercises 6.7 and 6.8.

(d) Using the observed SCOW wind stress from Figure 3.2 and reasonable choices for the other parameters, compute the solution for the $2\frac{1}{2}$-layer model.

(e) Repeat part (d), but customizing your solutions to the Pacific, Atlantic, and Indian Oceans. Comment on the realism of your results.

Eddies and Small-Scale Mixing

Ocean general circulation is primarily concerned with lengthscales of hundreds of kilometers and larger, and timescales of seasons and longer (Subsection 1.1.1). Here we discuss smaller and less persistent features because they influence the larger scales (Subsection 1.2.1).

Representing the effects of eddies and small-scale mixing through **parameterization** is an important theme. In fact, it remains an active research area, with many basic open questions. Ocean eddy parameterization is related to the great unsolved problem of fluid turbulence. One should therefore not forget that our ideas about the effects of eddies are provisional, as are the results from models and theories that depend on them.

We divide small-scale motion into two groups. One consists of structures that evolve significantly over days to weeks and have widths that are tens to hundreds of kilometers. These structures are generally known as **mesoscale eddies**, a term that is derived from the Greek word for "middle" due to the intermediate size of eddies. To obtain a gut feeling for these features see, for instance, the animations at `https://earth.nullschool.net/`. There is less ocean activity on scales that are a little faster and smaller than the mesoscale, but the activity increases again on scales of minutes or shorter and hundreds of meters or smaller. This is the realm of **internal gravity waves**, a disturbance in isopycnals that is similar to the small, fast surface gravity waves on the surface of the ocean (see Subsection 1.2.1). Just as surface waves can break on the beach in an explosion of turbulence, internal waves can also break in mid-ocean, generating turbulence on scales from meters to centimeters.

7.1 Observations

7.1.1 Mesoscale Eddies

Rings

Chapters 3, 5, and 6 portray ocean velocity as a smooth, unvarying field of vectors, as in the surface current climatology shown in Figure 7.1a. The trajectories of individual surface drifters tell a different story (Figure 7.1b). While the mean flow looks consistent

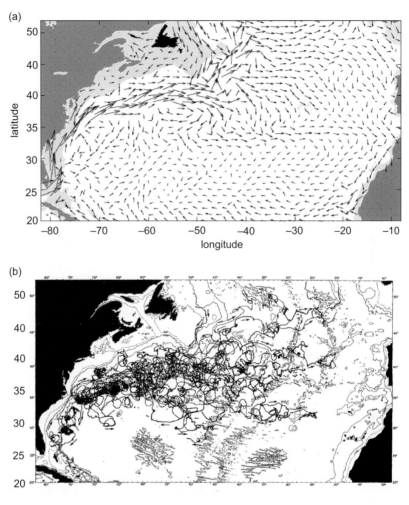

Figure 7.1 North Atlantic surface drifter (drogue at 15 m depth). (a) Velocity climatology for 1992–2011 drifters from the Global Drifter Program (see Appendix A and Lumpkin and Garraffo, 2005). (b) Individual trajectories of 35 instruments from Richardson (1981), Figure 1a, ©American Meteorological Society. Used with permission.

with the general direction in which individual instruments drifted, the trajectories are more complex. In particular, many of the drifters made one or more loops of (very roughly) 100 km radius.

The loops in the trajectories are due to mesoscale eddies, roughly circular geostrophic current systems that wander through the ocean. An instantaneous snapshot of the circulation would reveal numerous mesoscale eddies in each ocean basin. We can approximate such a snapshot with a weekly view of sea surface height (SSH). In the Atlantic, the SSH snapshot captures climatological features such as the Gulf Stream, high SSH in the subtropical gyre and low SSH in the subpolar gyre (Figure 7.2a). The snapshot reveals the mesoscale variability missing in Figure 7.1a (see also Figure 3.4 and Exercise 3.2),

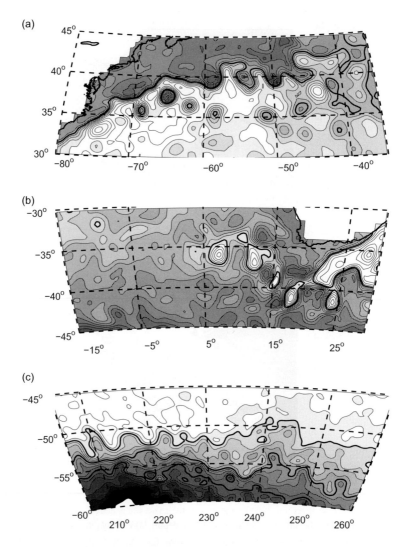

Figure 7.2 Sea surface height for week 4 of 2008 showing (a) Gulf Stream Extension region of the North Atlantic, (b) Agulhas Retroflection and South Atlantic, and (c) Indian Ocean sector of the Southern Ocean. Contour interval is 10 cm, lighter shading represents higher surface, and contours for selected heights are thickened to highlight key features. Data are from the AVISO sea surface height product (see Appendix A). See also Figure 9.4.

including at least six anticyclonic eddies to the south of the Gulf Stream system with central SSH at least 40 cm lower than surroundings, a handful of SSH highs east of 50°E, and numerous weaker highs and lows. The Gulf Stream extension has dramatic meanders east of 60°E and is narrower than the climatological Gulf Stream. Similarly, the Agulhas region (Figure 7.2b) includes several eddies to the east of the Agulhas Retroflection. A snapshot of the Indian Ocean sector of the Antarctic Circumpolar Current (ACC; Figure 7.2c) shows that the relatively broad ACC can be divided into three separate jets which have meanders and eddies associated with them.

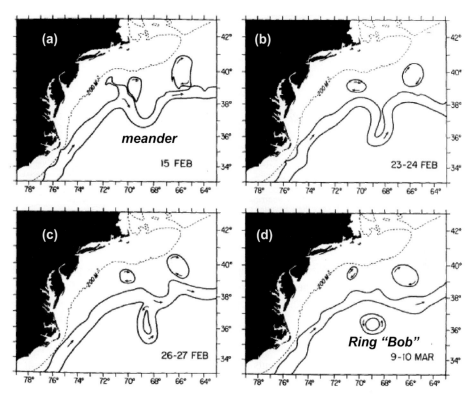

Figure 7.3 Boundary between warm and cold water based on SST from infrared satellite pictures from four different days in 1977 showing formation of Ring Bob from a Gulf Stream meander. From Richardson (1980b) Figure 1a, ©American Meteorological Society. Used with permission.

The strongest eddies are known as **rings** because the fastest flow speeds in the eddy occur in a ring-like band surrounding the eddy's center. Rings are generally formed from growing **meanders** (Figure 7.3a) in strong, jet-like currents such as the Gulf Stream, Kuroshio, Agulhas Current, and ACC.

The meander elongates (Figure 7.3b), forms a closed loop (Figure 7.3c), and drifts away from the formation site (Figure 7.3d). A western boundary current will create 3–10 of each type of ring in a year, and the ACC generates more due to its great length.

A ring encloses a core of water and carries it as it drifts. This can be seen, for instance, in SST fields, which show **cold core rings** carrying cold water from the northwest of the Gulf Stream into the subtropical gyre, and **warm core rings** carrying warm water northwestward from the subtropical gyre (Figure 7.3d). Drifters have been seen traveling around a ring for many circuits as the ring drifts hundreds of kilometers from an initial location (Richardson, 1980b).

Strong rings have swirl speeds comparable to the current which formed them (up to about 1 m/s), drift speeds of a few centimeters per second, and life times of up to several years. Agulhas rings traverse the entire Atlantic from south Africa to south America (Figure 7.4). The rings decay over time, with decreasing swirl speed and temperature

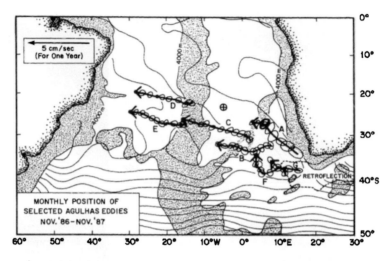

Figure 7.4 Trajectories of eddies, labeled A through G, climatological sea surface dynamic pressure relative to a level of no motion of 1 km, and 4000 m isobath. Small circles show approximate location and size of eddies at successive times. From Gordon and Haxby (1990), Figure 3, ©1990 American Geophysical Union. All rights reserved. Used with permission.

anomalies, but they usually "die" by interacting with another current. Many rings from western boundary currents and the ACC are reabsorbed by the current that formed them, while Agulhas rings may be absorbed by western boundary currents along South America.

Cold core rings are cyclonic, with a low in sea-surface height at the center of the ring. Warm core rings are anticyclonic. As swirl currents in rings are generally greatest near the surface and decrease to small values below the thermocline, isotherms and isopycnals slope in the opposite direction as the sea surface. Isotherms and isopycnals take on the shape of a "hill" for a cold core ring (Figure 7.5a) or a "bowl" for a warm core ring (Figure 7.5b). While the greatest currents in a ring are found at the same depths as the current that spawned it, the ring may reach deeper, as in Figure 7.5b where the warm core ring extends to the bottom.

Other Eddies

Rings are the most prominent eddies, but other O(100) km wide, swirling structures abound in the ocean. The MODE experiment in the North Atlantic (The MODE Group, 1978) showed that the ocean is filled with eddies with pressure excursions equivalent to O(10) cm of sea surface height anomaly (compared to several times that for the isolated rings). When the equatorial cold tongue (Subsection 4.1.1) is most prominent (generally September–November), the surface temperature front at its edge shows a wave-like pattern (Legeckis, 1977). These **Tropical Instability Waves** (also called Legeckis Waves) are westward-propagating eddies largely confined to the top 100 m of the ocean. Lenses of high salinity water, apparently from the Mediterranean, have been detected at about 1 km, the depth at which the Mediterranean Outflow travels around Iberia (Subsection 8.1.4) and across the Atlantic (McDowell and Rossby, 1978; Armi and Zenk, 1984; Elliot and Sanford, 1985).

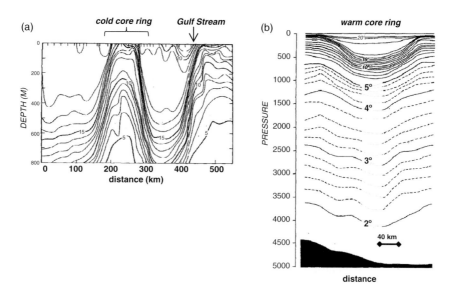

Figure 7.5 Temperature cross section through approximate center of Gulf Stream rings for (a) cold core ring Bob (from Richardson, 1980b, Figure 1b; see Figure 7.3) and (b) warm core ring 81D (from Joyce 1984, Figure 4). ©American Meteorological Society. Used with permission. See also Exercise 2.12.

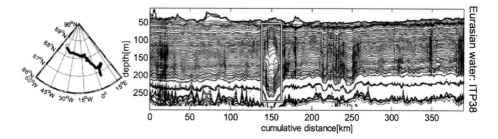

Figure 7.6 Anticyclonic eddy in the central Arctic ocean (rectangle). The six-hourly measurements were made from an ice-tethered profiler drifting for three months southeast from near the North Pole (the open circle marks the starting location in June 2010 and the mean speed was 8 km/d). The lines in the main panel are isopycnals, which are separated by 4 m on the initial profile. From Zhao et al. (2014), Figure 5. ©2014 American Geophysical Union. All rights reserved. Used with permission.

The salt lenses, called **meddies** (for "Mediterranean eddy"), are surrounded by a clockwise circulation with little or no surface expression.

Eddies are common in the halocline of the Arctic Ocean and they are somewhat unique (Manley and Hunkins, 1985). For example, Figure 7.6 shows an anticyclonic eddy near the North Pole. The pronounced area of low stratification near 150 m depth is associated with low potential vorticity and anticyclonic relative vorticity. The measurements are made from an ice-tethered profiler, a profiling CTD instrument which drifts with the sea ice. The ice-tethered profiler in Figure 7.6 traveled with the transpolar drift and encountered the anticyclone in the Eurasian Basin. Arctic eddies typically have diameters of 10–20 km (which is close to the internal deformation radius, see Subsection 1.2.1) and speeds of 5–40 cm/s (Zhao et al., 2014). They are observed throughout the Arctic Ocean, and although the sampling is inhomogeneous, there appear to be some regions with many eddies, such as near the perimeter of the Canadian Basin, and others with few of them. Instability of the boundary currents is one possible source for the eddies, as is instability of surface fronts. Interestingly, around 95% of observed Arctic eddies are anticyclonic, with cold cores. Why cyclonic Arctic eddies are so rare is unclear. Nevertheless, it seems likely that Arctic eddies play a fundamental role in the upper-ocean circulation in the Arctic, particularly for the Beaufort gyre. Subsections 10.2.2 and 10.3.3 present a model for the Beaufort gyre circulation that balances Ekman pumping from the wind with the integrated effect of eddies.

There is a class of somewhat smaller eddies, less than about 10 km width and surface intensified with strong vertical currents, sometimes called **submesoscale eddies** (McWilliams, 1985). For instance, surface submesoscale eddies occur off the northern Alaska coast in the Beaufort Sea (see D'Asaro, 1988, for a review). Coastal SST along eastern boundary currents such as the California Current, Peru Current, and Benguela Current shows waves and filaments associated with submesoscale eddies spawned by the current (Hill et al., 1998; Capet et al., 2008).

Eddy Statistics

The word **eddy** has two related meanings. One is a compact and roughly circular current system or vortex that may travel and evolve. Another is any departure from the average (usually referring to a time mean, but sometimes to the zonal mean). The two are related, because eddies (by the first definition) account for much of the ocean's eddy activity (using the second definition). We can measure the global distribution of eddy activity by using satellite altimetric measurements of surface geostrophic velocity (Ducet et al., 2000). If we divide velocity into time-mean and eddy components $\bar{\mathbf{u}}$ and $\mathbf{u}'$ so that $\mathbf{u} = \bar{\mathbf{u}} + \mathbf{u}'$, then we can define a mean and **eddy kinetic energy (EKE)** per unit mass of $\frac{1}{2}\bar{\mathbf{u}} \cdot \bar{\mathbf{u}}$ and $\frac{1}{2}\mathbf{u}' \cdot \mathbf{u}'$, respectively. Surface EKE is high where the mean kinetic energy is also high: near the equator, along western boundaries, and in the ACC (Figure 7.7). The regions where the Kuroshio and Gulf Stream leave the coast also have large EKE, as expected from the ring generation there (Figure 7.3). A somewhat puzzling area of high EKE extends across the Pacific at about 20°N. Eddy velocities used to create Figure 7.7 include the seasonal cycle, which is an important factor in the tropics.

EKE of 100 to 300 cm^2s^{-2} corresponds to an average $|\mathbf{u}'|$ of about 10–25 cm/s. Apart from isolated jets, typical time-mean velocities in the ocean are a few times 1 cm s^{-1} (Figure 3.4). Thus according to Figure 7.7, large swathes of the ocean have speeds associated with eddies that are bigger than the mean! Peak EKE (not visible in the figure) in the extensions of the Gulf Stream and Kuroshio extensions and in the Agulhas region are around 4000 cm^2s^{-2}, corresponding to a speed of 89 cm/s. The most active eddy regions are not more energetic than the strongest mean currents, but the eddies exported into slower waters dominate the kinetic energy there.

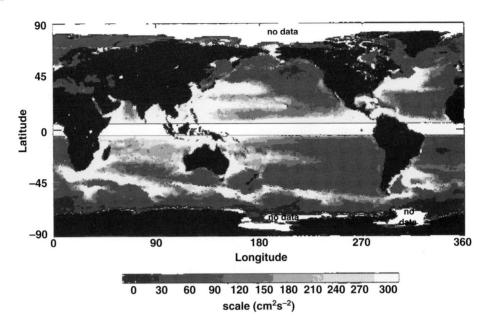

7.1.2 Small-Scale Mixing

Molecular motion causes heat and salt to diffuse through seawater with values of kinematic diffusivity of about 0.0014×10^{-4} m^2 s^{-1} for heat, about 100 times smaller for salt, and about 0.01×10^{-4} m^2 s^{-1} for momentum (Gill, 1982, pp. 71, 68, and 75, respectively; Subsection 1.3.1). We list values in terms of 10^{-4} m^2 s^{-1} because it is equal to 1 cm^2 s^{-1} and provides a convenient reference point to compare to eddy diffusivity of turbulence. Molecular values are so small that they only dominate at very short length scales compared to the scales of interest here (see Figure 1.10). The molecular diffusion is irreversible and called **mixing**. Of equal importance are other processes that **stir** the water, thus transferring large scale contrasts in temperature, salinity, and other properties to the dissipation scale, where mixing can remove them. These stirring processes are also mainly at scales shorter than our main focus, but we briefly discuss them now (see also Thorpe, 2005). We refer to stirring and mixing collectively as **small-scale mixing**; the distinction between them is made in Subsection 7.2.2.

Generally speaking, small-scale mixing involves some type of **turbulence**. It occurs on a broad spectrum of temporal and spatial scales and creates effectively random three-dimensional velocity distributions. The turbulence stirs the sea water allowing the property contrasts to then be mixed away by diffusion. Small-scale mixing is especially important for transferring properties across isopycnals. It is discussed as a small effect in wind-driven flows in the pycnocline (Subsection 5.3.1), but an important contributor to the deep meridional overturning circulation (Subsection 8.2.1 and Subsection 8.3.1).

Turbulence is generated from larger scale motion by a cascade starting with a **hydrodynamic instability** such as **Kelvin–Helmholtz instability** which develops further instabilities and smaller-scale structures as the amplitude of the disturbance grows. In the mixed layer, vertical shear in horizontal currents driven by the surface wind stress can develop instabilities. Most currents below the mixed layer have vertical shears that are too gentle to develop such small-scale instabilities, but the Equatorial Undercurrent (Subsection 6.1.1) is an exception. Throughout most of the water column, the key generator of turbulence is thought to be **internal waves** (Subsection 1.2.1; MacKinnon et al., 2017), which are undulations of isopycnals analogous to the familiar waves seen at the surface of the ocean. Like surface waves on a beach, internal waves can **break** within the ocean (Figure 7.8a), and in doing so can generate turbulence. Internal waves are generated by several processes, but perhaps the most important one is forcing by the **tides**.

Tidal forces come from gravitational interactions between the Earth and the Moon as well as between the Earth and the Sun. The primary effects of the tides are to make the sea surface oscillate vertically and to drive associated horizontal currents. The tides take the form of surface gravity waves with wavelengths on the order of the basin size of the oceans, periods of about 12 and 24 hours, and amplitudes varying from centimeters to a few meters. When tidal currents encounter topography such as sills, sea mounts, and continental slopes, water within the ocean is driven vertically by as much as tens of meters; when the water is stratified these oscillations cause internal waves. Packets of such waves have been observed near Stellwagen Bank, Massachusetts, the Strait of Gibraltar (Figure 7.8b), and elsewhere. They can be detected in subsurface temperature and velocity measurements as well as by their effects on surface waves which can then be sensed remotely.

Internal waves are also generated from other sources such as wind-forcing, surface gravity waves, and other departures from geostrophy (see Wunsch and Ferrari, 2004). Internal waves can propagate far (both horizontally and vertically) from their source, so that the internal wave field has a similar spectrum and magnitude over much of the ocean (Garrett and Munk, 1979). Internal wave frequencies range from the Coriolis parameter f to the buoyancy frequency N (Figure 1.10), with the greatest energy occurring at values close to f. While the internal wave field is more uniform than the surface wave field, it does have significant spatial variations and the resulting turbulence has even greater variation.

If we measure the amount of turbulence by an **eddy diffusivity** κ (Subsection 7.2.3), we find that κ is generally larger for the abyssal ocean than for the pycnocline. This is because stratification inhibits vertical motion, and the pycnocline is more stratified than the abyss. Typical κ values are around 0.1×10^{-4} m^{2}s^{-1} in the pycnocline and 10^{-4} m^{2}s^{-1} below the pycnocline, and much higher in the mixed layer. Thus over most of the ocean, eddy diffusivities are factors of hundreds or more greater than molecular values. Because salt and heat, and arguably momentum, are all carried in the same way by turbulent eddies, it is plausible that at a given time and place the diffusivities are the same for all three quantities. Still, some circulation models specify different values for vertical eddy diffusivities, to represent, among other processes, double diffusion (Canuto et al., 2010).

Large geographic variations exist in values of eddy diffusivities. For instance, in a section of measurements across the western South Atlantic Ocean (Figure 7.8c), κ varies from less than 0.1×10^{-4} m^{2}s^{-1} over the Brazil Abyssal Plain to over 5×10^{-4} m^{2}s^{-1} over parts of the Mid-Atlantic Ridge (Figure 7.8d). Polzin et al. (1997) attribute the larger values to the rougher sea floor topography of the Mid-Atlantic Ridge, where tidal flows over small seamounts can create stronger internal waves. Turbulence measurements across the oceans are needed in order to make more comprehensive maps of global distributions of turbulent diffusivity.

A different mechanism than turbulence may also be important for mixing. The large difference in molecular diffusivity for heat and salt (see beginning of section) can create strong mixing in otherwise quiescent fluid. One mode of mixing occurs when warm, salty water sits above cold fresh water, as occurs in the bottom of the Mediterranean Outflow in the Atlantic (see Subsection 8.1.4). If neighboring parcels of the deeper and shallower water exchange depths, they will lose heat to neighboring water faster than they lose salt,

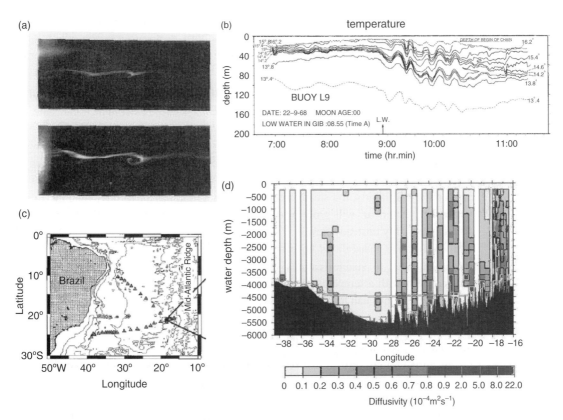

Figure 7.8 (a) Layer of fluorescent dye being deformed by an internal wave in the Mediterranean Sea, from Woods (1968), Figure 14, as shown by Turner (1973) Figure 4.20. (b) Time evolution of vertical temperature profile near the Strait of Gibraltar, from Ziegenbein (1970), Figure 4c. (c) Section map and (d) section data for eddy diffusivity as a function of depth and longitude for two tracks (combined without regard to latitude) shown in (c). Republished with permission of American Association for the Advancement of Science from Figures 1 and 2 of Spatial variability of turbulent mixing in the abyssal ocean, by K. L. Polzin et al., 276, 1997 (Polzin et al., 1997); permission conveyed through Copyright Clearance Center, Inc.

and so the fresh water continues to rise (and the salty water continues to sink) in **salt fingers**, columns of water with horizontal scales of order 1 cm and vertical scales of order 1 m. The opposite arrangement – cold, fresh water above warm and salty, as occurs at high latitudes – also undergoes a process of enhanced mixing known as **double diffusion** (Subsection 10.1.2, Schmitt, 1994, Exercise 5.3).

7.2 Concepts

7.2.1 Interaction between Eddies and Large-Scale Circulation

How does the general circulation influence where eddies are created and what properties they have? To what extent and in what ways do they affect the general circulation?

Fluctuating Circulation from Steady Forcing

Most ocean eddies are thought to be created by **hydrodynamic instabilities**. An **instability** is a process that grows spontaneously. For a physical system to exist in a certain state, the state must not only be a solution to the laws of motion, but it also must be stable. For instance, there is some position in which a bowling ball could balance on the top of a cone with a sharp point, but this state is not stable. If the ball is placed an infinitesimal distance from the balance point, that distance will grow (and accelerate) until the ball falls off the cone entirely. Similarly, if a steady circulation $\overline{\mathbf{u}}$ in a fluid has a hydrodynamic instability, a new flow pattern $\mathbf{u}'$ will grow in strength and replace the original circulation with a different, possibly unsteady flow.

Hydrodynamic instabilities have an uncanny ability to produce flow structures that appear unrelated to visible features such as basin topography. For instance, a rotating annular basin with a temperature gradient maintained between the internal and external walls of the annulus can have a zonally-symmetric flow around the annulus (Figure 7.9a).

(a) symmetric   (b) steady waves   (c) irregular

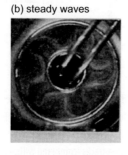

Figure 7.9   Annulus experiments showing surface circulation of water in a rotating tank traced by time-exposure photos of grains of powder, with (a) zonally symmetric flow, (b) steady meanders, and (c) irregular flow. From Fowlis and Hide (1965), Figure 2, ©American Meteorological Society. Used with permission.

For some values of the basin parameters, steady meanders, such as those in Figure 7.9b, spontaneously form on the jet. Other parameter values allow the waves to propagate, grow, and break in an irregular pattern (Figure 7.9c). This experiment is a crude analogue of the mid-latitude atmospheric circulation, with the inner cylinder representing the cold polar region and the outer cylinder representing the warm tropics. The circular flow is the analogue of the zonally flowing jet stream and the irregular eddies are analogous to high and low pressure systems that cause much of the world's mid-latitude weather.

Energy Conversion

Hydrodynamic instabilities are marked by a conversion of energy associated with the steady flow to energy associated with the growing perturbation. The horizontal temperature gradient in the steady flow in Figure 7.9a contains **available potential energy** (Subsection 8.3.1) which would be released if the lightest water migrated to the top of the water column and the densest water migrated to the bottom. This is the energy source for the perturbation in Figure 7.9b,c, which is called a **baroclinic instability** because the horizontal temperature gradient is connected via thermal wind to the vertically sheared baroclinic current (Subsection 2.1.2).

The numerical experiments of Visbeck et al. (1997), with a three-layer fluid in a periodic channel illustrate the release of available potential energy (Figure 7.10). The initial state of the model has the interface between layers 1 and 2 slanting in the cross-front direction (that is, across the channel, Figure 7.10, top panel), with along-front geostrophic shear flow

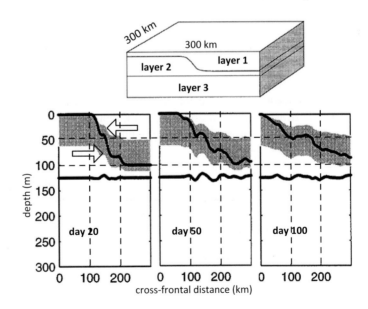

Figure 7.10 Top: perspective sketch of configuration of three-layer numerical model showing instability of a density front. Bottom: along-frontal average of layer interfaces (black curves) at days 20 (left), 50 (center), and 100 (right) of experiment. In the left panel, arrows represent direction of along-frontal average motion in upper two layers. From Visbeck et al. (1997), Figures 1 and 8, ©American Meteorological Society. Used with permission. See Exercise 7.3.

allowing the Coriolis force to prevent the fluid from falling into a uniformly stratified state. The front develops a baroclinic instability which produces eddies qualitatively similar to the ones visible in Figure 7.9c. The instability alters the along-front average interface depth (Figure 7.10, bottom panels). From days 20 to 100, the denser fluid on the left slumps downward while the lighter fluid on the right rises, as expected from the effects of gravity. Rotation allows the system to defy gravity with denser water standing higher than less dense water, but baroclinic instability provides a mechanism for gravity to partly defeat rotation and bring dense fluid to a lower level.

Most mesoscale eddies in the ocean are believed to be analogues of atmospheric weather systems. Both oceanic and atmospheric eddies are believed to be generated by baroclinic instability. Other sources of energy can drive instability: the kinetic energy in a narrow jet of fluid creates what is called a **barotropic instability** (see, for example, Vallis, 2006). In the ocean, barotropic instability of jets may also be important. The horizontal scale of the eddies is roughly proportional to the internal deformation radius (see (8.2)), which is around 30 km in the mid-latitude ocean and 500 km in the mid-latitude atmosphere. Thus mesoscale eddies in the ocean are smaller, O(100) km compared to O(1000) km for atmospheric eddies. The nomenclature is somewhat different between oceanographers and atmospheric scientists because in the atmosphere "mesoscale" also refers to O(100) km structures and hence is small compared to the atmospheric Rossby radius. The weather systems that correspond to ocean mesoscale eddies are often called **synoptic eddies**.

In the atmosphere, available potential energy grows because of the temperature gradient generated by the equator-to-pole gradient in solar radiation. At mid-latitudes, baroclinic instability is virtually the only mechanism for releasing this available potential energy. For that reason, eddies play a major role in determining the time-average state of the atmosphere. In the ocean, much of the available potential energy is generated by pycnocline excursions associated with the wind-driven gyres (Chapter 5). The steady meridional overturning circulation (Chapter 8) provides a way to release this available potential energy that does not involve mesoscale eddies. On the other hand, the oceanic pycnocline excursions associated with the wind-driven gyres (Chapter 5) provides another source of available potential energy missing in the temperature-driven atmospheric circulation. Thus it is not clear from first principles how important eddies should be in the ocean circulation (see Exercise 7.1). The fact that much of the large-scale structure of the gyres can be predicted without eddies suggests that the eddies do not contain the most important dynamics. However, eddies do play an important role in the Rhines and Young (1982) mechanism for driving the unventilated parts of the subtropical gyre (Subsection 5.2.3). Eddies may also help determine the characteristics of the recirculation gyres near the western boundary current extensions (Subsections 3.1.4, 3.2.5, and 3.3.4).

The most significant influence of mesoscale eddies in the ocean circulation probably occurs in the Southern Ocean (Chapter 9). The zonal average meridional geostrophic velocity is zero for any depth and latitude in which there are no eastern or western boundaries, such as the upper 1.5 km or so of the Drake Passage latitudes (Figure 9.1). Thus a key mechanism for the mean flow to transport heat, density, and other properties meridionally is missing. The absence of boundaries also removes the gyre circulations. The steady circulation is thus less effective at carrying the meridional transport than it is to the north, and the Southern Ocean maintains large meridional gradients in properties (Figure 8.1). Large gradients provide available potential energy, which is converted to kinetic energy by eddies and allow eddies to play an important role in property transport and in helping to determine the velocity and property distributions in the Southern Ocean (see Subsection 9.3.1 and Subsection 9.3.2).

7.2.2 Eddy Transport

The large speeds of eddy flow relative to the time-mean may appear to call into question the whole notion of steady circulation as an approximation to ocean currents. One must indeed pay careful attention to the detailed, time-varying flow to understand the path of a patch of plankton, an oil spill, or a drifting lifeboat over the course of days or weeks. Over longer periods, the time-mean flow may be quite different from the time-varying eddy velocity at any given time. For transport of heat, salt, and trace chemicals in the ocean and their influence on climate and on property distributions in the ocean, it is usually the long-term behavior that is most important.

Consider any tracer χ such as temperature, salinity, or density (we assume diffusion is negligible). The tracer flux is (see Subsection 1.3.1, Subsection 6.3.3)

$$Q_\chi = \int_0^{2L} \int_{\eta_B}^{\eta_0} v\chi \, dz \, dx, \qquad (7.1)$$

where η_B and η_0 are the heights of the bottom and top, respectively, of the ocean. The distance $2L$ is the width of the basin or, for a fluid state that is periodic in x, the wavelength. A correlation between v and χ causes a meridional transport (non-zero Q_χ).

An idealized example shows how a mesoscale eddy can transport a property down its gradient. Consider a vortex circulation that suddenly arises in a field representing a gradient of SST or some other property, with high values to the south (Figure 7.11a). Swirl velocity as a function of radius r is given by $v = a(r/R)^2(1-r/R)$, where a is an amplitude constant and R is the vortex radius. After a short time, the circulation has moved some of the water with high values of χ to the north and low values to the south (Figure 7.11b). Thus the eddy is equalizing the temperature. Later the tongues of high and low values of χ wrap further around the eddy (Figure 7.11c), though an individual stationary eddy of radius R cannot transport properties further than $O(R)$. Technically, at this point the eddy has stirred, but not mixed the fluid: individual water parcels have the same value of χ as they had at the beginning, but regions of high and low values are now more interspersed than before. This process continues. Because different radii within the eddy have different swirl velocities, water parcels that were close together in the beginning, and hence had similar values of χ, are now far apart, so that the original property gradient is stretched into long tendrils of different property values spiraling around the eddy (Figure 7.11d). In the real world the process may be continued by turbulent eddies on smaller scales. Eventually χ changes by a significant amount on small enough spatial scales that molecular diffusion can homogenize χ between neighboring parcels. At that point the eddy becomes well-mixed.

The real world has a more complicated evolution because water is influenced by more than one eddy, not all eddies are alike, the eddies evolve and propagate over time, and

initial distribution

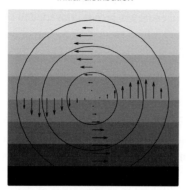

max of 1/8 rotation

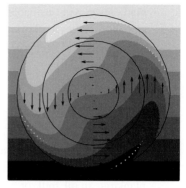

max of 1/2 rotation

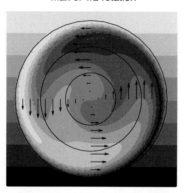

max of 1 rotation

Figure 7.11 Swirl velocity (arrows) and property χ illustrating idealized case of stirring of a uniform property gradient by a vortex at four different times. Panels are labelled in terms of angular distance traversed by fluid at radius of maximum speed. Figures created with a grid of 150×150 particles, each of which is assigned χ values at initial time and then advected an appropriate distance (based on the velocity field) at each time. See also Exercise 11.7.

properties are affected by other processes such as large-scale advection. Moreover, the relationship between the various eddy phenomena described in Subsection 7.1.1 and Subsection 7.1.2 and their property transport, stirring, and mixing is obscure. For instance, the disorganized motions between coherent mesoscale vortices may be more important than the vortices themselves. Some coherent eddies obstruct stirring via so-called **transport barriers** because they isolate a lens of water for long periods from the ambient fluid.

The example in Figure 7.11 is a two-dimensional circulation in which velocity is not parallel to isolines of χ in much of the region. This has clear resemblance to surface advection, but it is less clear how such an example applies to the ocean below the mixed layer, where we idealize the ocean as a stack of layers each with homogeneous density (and perhaps other properties). As discussed in Subsection 7.2.3, there are reasons to model eddy motion as occurring along an isopycnal, so that by definition the water flowing northward and southward within the isoycnal has the same density.

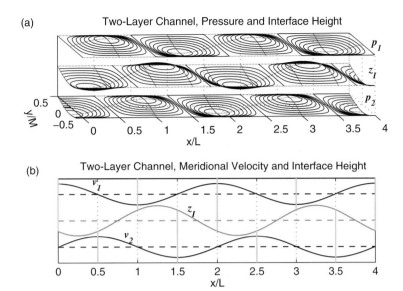

(a) Two-Layer Channel, Pressure and Interface Height

(b) Two-Layer Channel, Meridional Velocity and Interface Height

Figure 7.12 Characteristics of idealized two-layer eddy flow in a zonal channel. (a) Perspective view of p_1 (top), z_I (middle), and p_2 (bottom) as function of (x, y). (b) At $y = 0$, interface height $z_I(x)$ (gray), v_1 and v_2 (black; offset so that each black dashed line represents $v = 0$). The vertical bars mark the layer thickness of layer n when v_n is a maximum or minimum.

Another idealized example shows that eddies can make an important contribution to transport in this case as well. A two-layer fluid sits in a zonal channel which is zonally periodic. The fluid has a set of low and high pressure regions – the eddies—with the upper-layer ones offset in the x direction from the lower-layer ones (Figure 7.12a). To understand the transport property, we need to look carefully at the velocity and layer thickness field due to the eddies.

Given pressure distributions in the top and bottom layer, in Subsection 7.3.1 we calculate meridional velocity v_n ($n = 1, 2$), the depth anomaly z_I (relative to the channel mean) of the interface between the layers, and the property transport. Figure 7.12a shows that the hills and bowls in the interface are between the upper layer and lower layer pressure anomalies. Figure 7.12b shows the x dependence of v_1, v_2, and z_I. L is the zonal length of the eddies, and $0 < B < L$ is the offset distance between eddies in the top and bottom layers. Assuming that the channel length is an integer multiple of $2L$, the zonal integral of v is zero in both levels. However, the meridional transport of χ is *not* zero.

Consider a property, such as density, whose top-layer value χ_1 is less than its layer 2 value χ_2 (χ_1 and χ_2 are constants, not spatially varying in this example). As shown in Subsection 7.3.1, for $0 < B/L < 1/2$, density transport is negative: there is a net northward volume transport in the light upper layer and a net southward volume transport in the dense lower layer. We can see this in Figure 7.12b, where we can interpret the distance between the z_I curve and the top and bottom of the graph as h_1 and h_2, respectively. At the value of x where v_1 is its maximum *southward value*, h_1 (indicated by a vertical bar) is *smaller* than average, whereas at x where v_2 takes its maximum *northward value*, h_1 is *bigger* than average. Hence the net flow in the top layer is northward. By a similar argument, we can see that the net flow in the bottom layer is southward.

Because of the thermal wind relation, the correlation between velocity and thickness is associated with a tilt in the location of the eddy with depth. If the shallow velocity is a bit to the west of the deeper velocity, the net volume transport is northward in the upper layer and southward in the lower layer. Heat is transported northward and density is transported southward. In the northern hemisphere, where the south tends to be warmer and lighter than the north, this is a **down-gradient** transport, meaning that a quantity is flowing from a region with high values to one with low values, as in diffusion. The meridional density gradient can be represented by a layer model such as in Figure 7.10, with layer 1 playing the role of the light surface layer in the south. In that case the density interface slopes upward to the north. Baroclinic instability *must* generate eddies that are further to the west as we go higher in the water column, so that the mean meridional flow acts to reduce the meridional slope of the density interface and release available potential energy. This westward-tilt-with-height also occurs in mid-latitude weather systems (Vallis, 2006).

7.2.3 Parameterization of Eddy Transport

Eddies pose difficult conceptual and practical challenges to theory and models. To the extent that eddies are important for the general circulation, they challenge the notion that the circulation can be explained in terms of steady-state theories in which horizontal scales range from about 100 km for western boundary currents up to thousands of kilometers. Even with numerical models (Subsection 1.2.2), limits on computer resources make it cumbersome to adequately resolve O(100) km wide eddies in experiments that run for centuries to a quasi-steady equilibrium. Numerical models must have gridspacing of less than about 30 km in order to generate mid-latitude mesoscale eddies and 10 km or less to accurately reproduce eddy characteristics such as swirl speeds or isotherm excursions. While global-domain eddy-resolving simulations have become increasingly common since about the year 2000, it would still be useful if the effects of eddies on large-scale steady flow could be replicated without reproducing the high-frequency and small-scale evolution of individual eddies.

Numerical models which do not reproduce eddies – generally those with grid spacing of about 1° or more – parameterize the effects of eddies. Typically, parameterization of any unresolved process in a model involves adding a term to each equation of motion which represents the net effect of eddies as a function of the relatively broad flows resolved by the model (Subsection 1.2.2).

Diffusive Eddy Parameterization

The oldest and simplest such parameterization for eddy transport of any property χ (which can represent temperature, salinity, momentum, or other variables), is horizontal diffusion, $\kappa \nabla^2 \chi$ with uniform eddy diffusivity κ. Diffusion transports properties from high to low values at a rate proportional to the property gradient. Thus for a given value of κ, regions with bigger $\nabla \chi$ have larger transports which reduce the gradient. The diffusivity constant κ represents the strength of the eddy transport processes. The value of κ in the ocean is an important issue (see Subsection 7.1.2, Subsection 9.2.4). The scaling for κ in terms of, for example, the average current speed, is also important (see Subsection 7.3.1, Subsection 7.3.3).

Using horizontal diffusion to parameterize eddy transport of temperature and salinity can cause unrealistic effects in regions such as a western boundary current where isotherms are strongly tilted. In such regions, horizontal diffusion across the strong horizontal gradients in the vicinity of the tilting isotherms drives a heat flux that is believed to be unrealistically large. The excess heat transported westward across the western boundary current is balanced by cooling from spurious upwelling in the western boundary current (the **Veronis effect**, Veronis, 1975; Böning et al., 1995).

One way to eliminate these unrealistic processes is to use **isopycnal diffusion**. Instead of diffusion transporting heat or salt in the purely horizontal direction, isopycnal diffusion transports them in the quasi-horizontal direction along an isopycnal. Subsection 7.3.2 demonstrates the mathematical expression for such diffusion. Isopycnal diffusion eliminates the spuriously large mixing in simulations (Böning et al., 1995), but it introduces the opposite problem: too little mixing. This is evident because isopycnal diffusion of density cannot change the density field!

Advective Eddy Parameterization

The analysis in Subsection 7.2.2 of how eddies transport density across a tilting isopycnal suggests a solution to the problem. In the example demonstrated in Figure 7.12, eddies transport fluid in each layer from a region where the layer is thick to the region where it is thin. This tends to equalize the layer thickness or (to put it another way) to reduce the horizontal thickness gradient. Thus eddies reduce the slope of tilting isopycnals (though other processes such as wind forcing may work to simultaneously increase the slope). As discussed in the previous section, "untilting" of isopycnals releases available potential energy which baroclinic instability converts to eddy kinetic energy. We can model this transport within the layer using *advection of layer thickness* (Gent and McWilliams, 1990). Advection redistributes layer thickness as if by stirring, rather than by diffusive mixing. This parameterization is commonly referred to as **Gent–McWilliams** or simply GM.

To apply the idea of layer thickness advection to numerical models, in which density is usually defined on a grid that is fixed in space (as in Figure 1.11) rather than on a stack of uniform-density layers, we need an analogue of the layer thickness. The analogue of a layer is the water between two given isopycnals ρ_1 and $\rho_1 + \Delta\rho$, so the analogue of layer thickness is the vertical distance between the isopycnals. For a given horizontal location, we have $\rho = \rho(z)$ which we can invert (numerically if not analytically) to get $z = h(\rho)$, where h is the height of the isopycnal with density ρ. The Taylor expansion for $h(\rho)$ is

$$h(\rho_1 + \Delta\rho) = h(\rho_1) + \frac{\partial h}{\partial \rho}\Delta\rho + \cdots, \tag{7.2}$$

or

$$\Delta h \equiv h(\rho_1 + \Delta\rho) - h(\rho_1) \approx \frac{\partial h}{\partial \rho} \Delta\rho. \qquad (7.3)$$

The quantity $\Delta\rho$ depends on how finely we subdivide the vertical structure and hence is somewhat arbitrary. In any case, once we have picked the bounding isopycnals of our layer, $\Delta\rho$ does not vary within the layer. For the most convenient analogue of layer thickness, we leave out the factor of $\Delta\rho$ and use $\partial h/\partial\rho$.

The process of redistributing $\partial h/\partial\rho$ changes ρ at a given (x, y, z), unlike isopycnal diffusion of T and S. The evolution of density can be seen in a simulation with realistic dynamics and continuous stratification for the idealized case of shear flow u, temperature T and salinity S which depend on (y, z) but not x (Figure 7.13, from Gent et al., 1995). The GM parameterization flattens and thickens the thermocline from its initial state (Figure 7.13a) to a later state (Figure 7.13b). In this experiment, density (not shown) undergoes a similar evolution.

The layer thickness advection is quantified using a **bolus velocity** (Rhines, 1982; McDougall, 1991), also called an **eddy-induced transport velocity** (Gent et al., 1995; see also Plumb and Mahlman, 1987). The bolus velocity usually involves sinking of shallower isotherms and rising of deeper isotherms, as shown by the streamfunction in Figure 7.13. The mathematical formulation of the bolus velocity appears in Subsection 7.3.2.

The GM parameterization helps realistic ocean models too. It reduces mixing in western boundary currents (compared to horizontal mixing with the same diffusivity) and improves model realism by sharpening the permanent thermocline, cooling the deep ocean, and shrinking the regions where deep convection occurs (Danabasoglu and McWilliams, 1995).

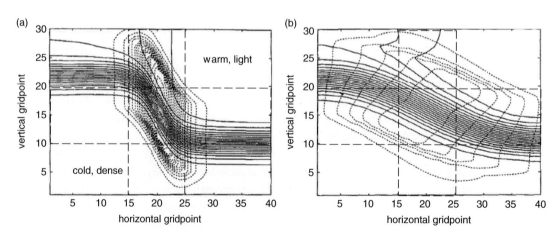

Figure 7.13 Temperature (solid contours) as a function of horizontal (y) and vertical (z) gridpoint index for (a) initial and (b) later time. Dotted contours are streamlines of bolus velocity (Subsection 7.3.3) showing counterclockwise circulation and hence slumping of the density front. From Gent et al. (1995), Figures 3 and 4, ©American Meteorological Society. Used with permission.

Conceptually, we are interested in the velocity field in order to understand the interaction between circulation and density. For instance, in Chapter 5 we assume that the flow is constrained to be along isopycnals. In Chapter 8 we postulate a balance between advective changes to density and diffusive changes. If the bolus velocity is not negligible compared to the traditional **Eulerian-mean velocity** (sometimes referred to as **advective velocity**), then we may prefer to examine the total velocity (the sum of bolus and advective components). In keeping with the idea that the Southern Ocean is a region with some of the most prominent eddy effects, bolus velocities are especially large there. Indeed, the meridional overturning streamfunction looks quite different in the southern hemisphere depending on how we compute it (Figures 9.10, 9.11).

Testing Eddy Parameterizations

While the GM parameterization follows from plausible assumptions about the nature of eddy transports, it is not rigorously derived from the equations of motion and hence is not necessarily true. We can verify the accuracy of a parameterization by comparing the evolution of an eddy-suppressing model including the parameterization to an eddy-resolving model without the parameterization. For instance, Visbeck et al. (1997) tested three flow configurations: flow around a region of convection into a stratified domain, relaxation of a front in the top 300 m of the domain, and wind-driven flow in a channel.

In the convection case (Figure 7.14a), surface cooling occurs over a disk 8 km in radius, creating a **chimney** of dense, weakly stratified fluid (Subsection 8.1.4). The density difference between the chimney and the surrounding fluid generates a baroclinic geostrophic current around the chimney. Baroclinic instability then causes this current to break into eddies, which evolve and drift over time t (Figure 7.14b). Thus the evolution of the azimuthally averaged density ρ as a function of radial distance r and vertical position z (Figure 7.14c) depends on cooling from above, geostrophic adjustment, and eddy mixing. The evolution of $\rho(r, z, t)$ is also calculated with a two-dimensional $(r - z)$ model that does not generate eddies and uses either no diffusion (MIX), horizontal diffusion (HD) or GM. The GM parameterization is implemented with uniform diffusivity (GM in figure) and with two choices of diffusivity as functions of large-scale stratification (GS and NEW). The initial stratification is itself weaker than is typical for the mid-latitude ocean, so that the internal deformation radius is only 4 km (rather than about 30 km) and the space and timescales for the eddies are correspondingly small.

The eddy-resolving experiment shows an initial cooling under the disk (Figure 7.14c, column 1, top row) followed by a slumping of the chimney (Figure 7.14c, column 1, bottom row) as eddies reduce the steep isopycnal slope by carrying denser water down and outward and lighter water up and inward. When eddies are not parameterized, the cooling in the no-eddy model is almost completely confined to the area under the cooling disk (Figure 7.14c, column 2). Horizontal diffusivity spreads the cooling out, but does not capture the slumping of isopycnals (Figure 7.14c, column 3). The GM cases reproduce the slumping of isopycnals and produces a $\rho(r, z)$ at 6 days that is quantitatively similar to the eddy-resolving case. The differences between the three GM cases are discussed in Subsections 7.3.1 and 7.3.3.

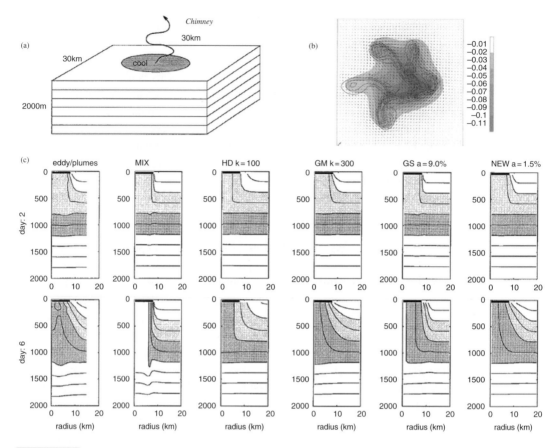

Figure 7.14   Eddy parameterization experiments. (a) Model domain showing surface cooling disk and initial linear vertical stratification. (b) Surface temperature (shading), velocity (arrows) and potential vorticity (black contours) on day 6 of eddy-resolving experiment (see Figure 7.9c). (c) Azimuthally averaged density as a function of depth and radius from cooling center, days 2 and 6, for (left to right) eddy-resolving model and depth-radius parameterized models for no lateral mixing, horizontal diffusivity, GM parameterization with uniform diffusivity, Green–Stone (GS) parameterization for diffusivity (see page 268), and NEW formula for diffusivity (see page 268). This example shows a deep convective chimney, but baroclinic instability of mixed layer fronts also occurs for shallow mixed layers (Haine and Marshall, 1998), which is the main source of submesoscale eddies. From Visbeck et al. (1997), Figures 1, 2, and 3, ©American Meteorological Society. Used with permission.

7.3 Theory

7.3.1 Flux Due to Eddy Correlation

Reynolds Stress

We want to average quantities to represent a view of the circulation in which eddies have been removed or filtered by the averaging process. The average could be in time, in a spatial dimension, or in some combination. Using an overbar to represent the average, we can write any quantity χ in terms of mean and residual (or eddy) quantities

$$\chi = \overline{\chi} + \chi'. \tag{7.4}$$

Taking the average of (7.4) shows that $\overline{\chi'} = 0$. Be aware that $\overline{\chi}$ is not a constant but can vary as a function of space and time on temporal and/or spatial scales that are large compared to the filter, and hence those associated with eddies. The decomposition (7.4) allows us to see how the evolution of average quantities depends on both average and eddy quantities. For instance the zonal component of the momentum equation (in Cartesian coordinates, from (1.16)) can be written

$$\overline{\frac{\partial}{\partial t}(\bar{u} + u')} + \overline{\nabla \cdot [(\bar{\mathbf{u}} + \mathbf{u}')(\bar{u} + u')]} = -\overline{\frac{1}{\rho_0}\frac{\partial}{\partial x}(\bar{p} + p')} + \overline{f(\bar{v} + v')} + \overline{\nu\nabla^2(\bar{u} + u')}. \tag{7.5}$$

Taking an average is a linear process which commutes with differentiation (also a linear operation), and here we assume that ν is the molecular value and is a constant. As the average of each eddy quantity is zero, the momentum equation can be rewritten

$$\frac{\partial \bar{u}}{\partial t} + \nabla \cdot \overline{[(\bar{\mathbf{u}} + \mathbf{u}')(\bar{u} + u')]} = -\frac{1}{\rho_0}\frac{\partial \bar{p}}{\partial x} + f\bar{v} + \nu\nabla^2\bar{u}. \tag{7.6}$$

All the terms of the averaged momentum equation involve only average quantities – except the advection term. Expanding the term, it becomes

$$\nabla \cdot \overline{[(\bar{\mathbf{u}} + \mathbf{u}')(\bar{u} + u')]} = \nabla \cdot \left(\bar{\mathbf{u}}\bar{u} + \overline{\bar{\mathbf{u}}u'} + \overline{\mathbf{u}'\bar{u}} + \overline{\mathbf{u}'u'}\right) = \nabla \cdot \left(\bar{\mathbf{u}}\bar{u} + \overline{\mathbf{u}'u'}\right). \tag{7.7}$$

The term $\overline{\mathbf{u}'u'}$ is called the **Reynolds Stress** and represents a flux of momentum associated with eddies.

More generally, the advection terms for any variable χ, including a momentum component, potential temperature, salinity, etc., can be written in terms of mean and residuals and the appropriate conservation equation becomes

$$\frac{\partial \bar{\chi}}{\partial t} + \nabla \cdot (\bar{\mathbf{u}}\bar{\chi}) = \bar{F} - \nabla \cdot (\overline{\mathbf{u}'\chi'}), \tag{7.8}$$

where $\bar{F}$ represents any other terms in the equation that depend on the averaged state, such as the Coriolis terms if χ is a component of horizontal momentum, or a radiation term if χ is potential temperature. Note that $\overline{\mathbf{u}'\chi'}$ is a vector, so it specifies eddy flux along each of the three coordinate axes. For a given direction, the tendency for a certain relationship between the sign of the velocity and the sign of χ' (i.e., the correlation) – either both tending to have the same sign or both having opposite signs – results in a nonzero flux.

Mixing Length Estimate of Diffusivity

The assumption that eddies act as diffusion can be written

$$-\overline{\mathbf{u}'\chi'} = \kappa \nabla \bar{\chi}. \tag{7.9}$$

The problem of calculating a κ based on the mean circulation state is unresolved in detail. Nevertheless, we can use scaling analysis to give some guidance about the magnitude. If we think that eddy transport is occurring via the process shown in Figure 7.11, then we can guess a relationship between eddy characteristics and the transport. For typical swirl speed U and typical eddy radius R, it is plausible that, at a given phase in the cycle shown in the Figure 7.11, χ' would have a magnitude that is proportional to $R|\nabla\bar{\chi}|$. Averaging over a long enough time and/or large enough region would plausibly retain this proportionality. We also write $\overline{\mathbf{u}'\chi'} = c\sqrt{\overline{\mathbf{u}'^2\chi'^2}}$, which may serve as a definition of c as the correlation between $\mathbf{u}'$ and χ', but which we hope can be approximated by some constant $0 < c < 1$. A scale relation based on (7.9) gives

$$\kappa \sim \frac{\overline{\mathbf{u}'\chi'}}{\nabla\bar{\chi}} \sim cUR. \tag{7.10}$$

This is often called a **mixing length** estimate because it is based on the idea that mixing occurs on length scale R at a rate given by speed U. This is (essentially) the GS parameterization of Green (1970) and Stone (1972) (Figure 7.14).

Near strong, mid-latitude currents such as the Gulf Stream or ACC, where $U = 0.5$ m/s and $R = 100$ km are reasonable values, the mixing length estimate is $\kappa = (5 \times 10^4 \text{m}^2/\text{s})c$. Averaging over the spatial extent and time history of an eddy, and accounting for the fact that a mesoscale eddy may be passing through a given location less than half the time, $c \sim 0.1$ is a reasonable guess which gives us $\kappa \sim 10^3$–10^4 m^2/s. We may expect κ to have smaller values away from strong currents. Note that we have not derived U from first principles but use observed values that are similar to the mean current speed of the jets from which the strongest eddies spawn.

Property Transport from Tilted Eddies in Two-Layer System

In Subsection 7.2.2, we asserted that geostrophic eddies in a two-layer fluid (like the perturbations in the channel shown in Figure 7.12) generate a meridional property transport, even if the zonal average velocity is zero. We derive the expression for transport as follows. Start with dynamic pressure, which we set to be

$$\frac{1}{\rho_0}p_1 = A\sin(\pi x/L)\cos(\pi y/M), \tag{7.11}$$

$$\frac{1}{\rho_0}p_2 = A\sin(\pi[x-B]/L)\cos(\pi y/M), \tag{7.12}$$

where M is the meridional channel width, L is the zonal length of the eddies (they are periodic over $2L$, including cyclones and anti-cyclones), and $0 < B < L$ is the offset distance. The sea surface height is proportional to p_1, and given the pressures and the hydrostatic relation we can calculate z_I, the depth anomaly (relative to the channel mean) of the interface between the layers in terms of the above parameters and the reduced gravity $\gamma_1 = g(\rho_2 - \rho_1)/\rho_0$, where ρ_0 is the characteristic density of the system.

The thermal wind gives us the geostrophic meridional velocity v_n ($n = 1, 2$). Figure 7.12b shows the x dependence of v_1, v_2, and z_I. Assuming that the channel length is an integer multiple of $2L$, the zonal integral of v is zero in both levels. However, the meridional flux of χ is *not* zero (in general). H is the average total thickness of the fluid and $\eta(x, y, t)$ contains the small departures from the average sea surface height associated with p_1. If layer n has a layer thickness $h_n(x, y, t)$ and a constant $\chi = \chi_n$, then the property flux (7.1) is

$$Q_\chi = \chi_1 \int_0^{2L} h_1 v_1 \, dx + \chi_2 \int_0^{2L} h_2 v_2 \, dx. \tag{7.13}$$

The channel-average thickness of layer n is H_n, so $h_1 = H_1 + \eta - z_I$ and $h_2 = H_2 + z_I$. Thus the property transport is

$$Q_\chi = \chi_1 \int_0^{2L} (H_1 + \eta - z_I) v_1 \, dx + \chi_2 \int_0^{2L} (H_2 + z_I) v_2 \, dx. \tag{7.14}$$

Because v_n are sinusoids in x, the zonal periodicity of the channel makes the H_n terms integrate to zero. If we define $\Delta\chi \equiv \chi_2 - \chi_1$, we can rewrite the remaining terms in the expression as

$$Q_\chi = \chi_1 \int_0^{2L} [z_I(v_2 - v_1) + \eta v_1] \, dx + \Delta\chi \int_0^{2L} z_I v_2 \, dx. \tag{7.15}$$

Geostrophy, the hydrostatic relation, and the zonal periodicity make the first integral equal to zero and allow us to solve the second integral to get

$$Q_\chi = -\frac{\pi A^2}{f\gamma_1} \Delta\chi \cos^2(\pi y/M) \sin(\pi B/L). \tag{7.16}$$

Thus $Q_\Psi \neq 0$ as long as $0 < B < L$ (see Exercise 7.4).

7.3.2 Isopycnic and Diapycnic Tracer Diffusion

The first step in the development of a theory for isopycnic diffusion of conserved quantities is to write down the mixing operator in a convenient form. This step was taken by Redi (1982), and is explained here.

In Chapter 1, we write the heat conservation equation (for example) as:

$$\frac{D\theta}{Dt} = -\nabla \cdot \mathbf{D}_\theta(\theta) + \mathcal{F}_\theta, \tag{7.17}$$

where the dissipative flux is $\mathbf{D}_\theta(\theta)$ and its divergence is $\nabla \cdot \mathbf{D}_\theta(\theta)$ (see (1.9)). The simplest form for this divergence term is (1.11)

$$\nabla \cdot \mathbf{D}_\theta(\theta) = -\kappa \nabla^2 \theta, \tag{7.18}$$

where κ is the constant, scalar diffusivity coefficient. This form accurately describes molecular diffusion but it is an inadequate description of unresolved turbulent motion, such as mesoscale eddies. We cannot simply replace κ in (7.18) with an enhanced eddy diffusivity. The reason is that, for the length and timescales of interest in this book, it has long been recognized that the diffusivity along isopycnals is greater than the diffusivity across them (Iselin, 1939; Montgomery, 1940). For example, Sarmiento and Bryan (1982) estimate that $\kappa_I/\kappa_\rho \approx 10^7$, where κ_I is the **isopycnic diffusivity** (along isopycnals, nominally caused by mesoscale eddies; Subsection 7.1.1) and κ_ρ is the **diapycnic diffusivity** (across them, caused by small-scale mixing; Subsection 7.1.2). Early attempts to account for this very strong anisotropy led to the next simplest form for the dissipative flux divergence,

$$-\nabla \cdot \mathbf{D}_\theta(\theta) = \kappa_h \nabla_h^2 \theta + \kappa_v \frac{\partial^2 \theta}{\partial z^2}. \tag{7.19}$$

Here, κ_h is the *horizontal* diffusivity, and κ_v is the *vertical* diffusivity (∇_h is the horizontal gradient). Again, κ_h and κ_v are constant scalar eddy diffusivities in (7.19).

In general, isopycnals are not horizontal, however, so the vertical direction is different to the diapycnic direction. One consequence of implementing (7.19) when the isopycnals are steeply tilted is the Veronis effect (Veronis, 1975) described in Subsection 7.2.3.

Redi (1982) explained how to handle $\mathbf{D}_\theta(\theta)$ for tilted isopycnals. Her focus was on the case where the isopycnic diffusivity κ_I greatly exceeds the diapycnic diffusivity κ_ρ. The abstract form is easy to state:

$$\nabla \cdot \mathbf{D}_\theta(\theta) = -\nabla \cdot \boldsymbol{\kappa} \nabla \theta. \tag{7.20}$$

Now $\boldsymbol{\kappa}$ is a *tensor* diffusivity that can (in principle) vary in space and time (and is of second order and symmetric). The tensor diffusivity ensures that the dissipation always respects the difference between the high isopycnic and low diapycnic mixing.

To be specific, consider the case where the heat conservation equation is computed in a Cartesian coordinate system with z^I standing for the unit diapycnic vector perpendicular to the isopycnal (pointing towards lighter water; see Figure 7.15). Let the direction of greatest downward tilt of the isopycnal be x^I, and hence y^I is in the horizontal direction forming a right-handed perpendicular set of axes (x^I, y^I, z^I). In this coordinate system, the tensor diffusivity is:

$$\boldsymbol{\kappa}^I = \begin{pmatrix} \kappa_I & 0 & 0 \\ 0 & \kappa_I & 0 \\ 0 & 0 & \kappa_\rho \end{pmatrix}. \tag{7.21}$$

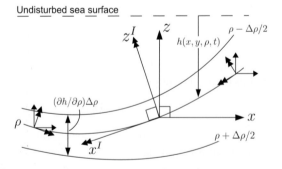

Figure 7.15 Comparison of different coordinate systems. Curves represent isopycnic surfaces seen from the side. Double-headed arrows: isopycnic coordinates, an orthonormal set (x^I, y^I, z^I) with components z^I perpendicular to, and (x^I, y^I) in plane tangent to an isopycnic surface. Single-headed arrows: geodetic coordinates (x, y, z) with vertical component z and horizontal components (x, y). $h(x, y, \rho, t)$ is the vertical distance between sea level and the isopycnal with density ρ.

In other words, unresolved processes mix heat along the isopycnal (x^I, y^I) with diffusivity κ_I, and across it in direction z^I with (much smaller) diffusivity κ_ρ (meaning the thermal diffusivity at constant density, not the diffusivity of density). To compute the final dissipative flux divergence we take the temperature gradient in the isopycnic (x^I, y^I, z^I) coordinate system, operate on it with κ – which rotates the temperature gradient vector – and then take the divergence. When κ_I and κ_ρ are constant, and the isopycnal is locally flat, we have:

$$-\nabla \cdot \mathbf{D}_\theta(\theta) = \kappa_I \frac{\partial^2 \theta}{\partial(x^I)^2} + \kappa_I \frac{\partial^2 \theta}{\partial(y^I)^2} + \kappa_\rho \frac{\partial^2 \theta}{\partial(z^I)^2}. \tag{7.22}$$

For several reasons, we need to also consider the case where the heat conservation equation is computed in a traditional (x, y, z) Cartesian coordinate system with height z as the vertical coordinate (Figure 7.15). For example, the (x^I, y^I, z^I) unit vectors move in space and time with the isopycnals themselves. This variation of the coordinate vectors introduces more terms in (7.17) and (7.22) when they are implemented in practice (which is the reason the isopycnal needs to be flat for (7.22) to hold). In the simpler (x, y, z) coordinates, (7.20) still applies. The tensor diffusivity κ is more complicated than κ^I, however. In particular, we rotate κ^I so that it acts in the (x, y, z) system. This tensor in the (x, y, z) system is called κ^g (the superscript g refers to the geodesic (x, y, z) coordinates). Redi (1982) gives full details, but for density derivatives (ρ_x, ρ_y, ρ_z) in the (x, y, z) directions, the final result is:

$$\kappa^g = \frac{1}{\rho_x^2 + \rho_y^2 + \rho_z^2} \begin{pmatrix} \kappa_\rho \rho_x^2 + (\rho_y^2 + \rho_z^2)\kappa_I & (\kappa_\rho - \kappa_I)\rho_x\rho_y & (\kappa_\rho - \kappa_I)\rho_x\rho_z \\ (\kappa_\rho - \kappa_I)\rho_y\rho_x & \kappa_\rho \rho_y^2 + (\rho_x^2 + \rho_z^2)\kappa_I & (\kappa_\rho - \kappa_I)\rho_y\rho_z \\ (\kappa_\rho - \kappa_I)\rho_z\rho_x & (\kappa_\rho - \kappa_I)\rho_z\rho_y & (\rho_x^2 + \rho_y^2)\kappa_I + \kappa_\rho \rho_z^2 \end{pmatrix}. \tag{7.23}$$

To understand the meaning of (7.23), think of the case where the isopycnal slope is in the x direction, so $\rho_y = 0$. (There is no loss of generality: we can always rotate the local (x, y, z) axes about z so that $\rho_y = 0$.) Then, (7.23) reads

$$\kappa^g = \frac{1}{\rho_x^2 + \rho_z^2} \begin{pmatrix} \rho_x^2\kappa_\rho + \rho_z^2\kappa_I & 0 & -(\kappa_I - \kappa_\rho)\rho_x\rho_z \\ 0 & (\rho_x^2 + \rho_z^2)\kappa_I & 0 \\ -(\kappa_I - \kappa_\rho)\rho_z\rho_x & 0 & \rho_x^2\kappa_\rho + \rho_z^2\kappa_I \end{pmatrix}. \tag{7.24}$$

Or, defining the slope of the isopycnal as

$$\gamma = \frac{\rho_x}{\rho_z} \tag{7.25}$$

(not the reduced gravity),

$$\kappa^g = \frac{1}{1 + \gamma^2} \begin{pmatrix} \kappa_I + \gamma^2\kappa_\rho & 0 & -(\kappa_I - \kappa_\rho)\gamma \\ 0 & (1 + \gamma^2)\kappa_I & 0 \\ -(\kappa_I - \kappa_\rho)\gamma & 0 & \kappa_\rho + \gamma^2\kappa_I \end{pmatrix}. \tag{7.26}$$

For the special case where κ_I, κ_ρ and γ are uniform (and hence κ^g is a constant), the diffusive flux divergence

$$-\nabla \cdot \mathbf{D}_\theta(\theta) = \nabla \cdot \kappa^g \nabla \theta \qquad (7.27)$$

can be written

$$-\nabla \cdot \mathbf{D}_\theta(\theta) = \left(\frac{\kappa_I + \gamma^2 \kappa_\rho}{1 + \gamma^2}\right) \frac{\partial^2 \theta}{\partial x^2} + \kappa_I \frac{\partial^2 \theta}{\partial y^2} - 2\gamma \left(\frac{\kappa_I - \kappa_\rho}{1 + \gamma^2}\right) \frac{\partial^2 \theta}{\partial x \partial z} + \left(\frac{\kappa_\rho + \gamma^2 \kappa_I}{1 + \gamma^2}\right) \frac{\partial^2 \theta}{\partial z^2}.$$
$$(7.28)$$

Or

$$-\nabla \cdot \mathbf{D}_\theta(\theta) = \kappa_I \left(\theta_{yy} + \frac{\theta_{xx} - 2\gamma \theta_{xz} + \gamma^2 \theta_{zz}}{1 + \gamma^2}\right) + \kappa_\rho \frac{\theta_{zz} + 2\gamma \theta_{xz} + \gamma^2 \theta_{xx}}{1 + \gamma^2}, \qquad (7.29)$$

where subscripts of x, y, and z represent partial derivatives.

Think of some concrete examples of these two expressions, (7.26) and (7.28): First, note that for horizontal isopycnals $\gamma = 0$. Hence, (7.28) collapses to the original form (7.19) with $\kappa_h = \kappa_I$ and $\kappa_v = \kappa_\rho$. Otherwise, they are different. Second, consider the case where the density is a function of temperature only, $\rho = \rho(\theta)$, and salinity changes are unimportant for density (this approximation applies to a large fraction of the world ocean; see Subsection 1.1.5). Then, $\gamma = \theta_x/\theta_z$, implying that $\kappa^g \nabla \theta = \kappa_\rho \nabla \theta$ (from (7.26); recall that $\partial \theta / \partial y = 0$ for $\rho = \rho(\theta)$), and we find that

$$-\nabla \cdot \mathbf{D}_\theta(\theta) = \nabla \cdot \kappa_\rho \nabla \theta. \qquad (7.30)$$

So for constant κ_ρ,

$$-\nabla \cdot \mathbf{D}_\theta(\theta) = \kappa_\rho \left(\frac{\partial^2 \theta}{\partial x^2} + \frac{\partial^2 \theta}{\partial z^2}\right). \qquad (7.31)$$

In other words, the diffusive heat flux divergence is simply the divergence of κ_ρ times the temperature gradient $\nabla \theta$. This makes sense because the temperature gradient is entirely in the diapycnic direction in this case: there is zero temperature gradient in the isopycnic directions because $\rho = \rho(\theta)$. Obviously, using (7.19) instead of (7.30) leads to big errors in this case.

Finally, consider the last term in (7.28). The isopycnic slope is rarely steeper than that of the western boundary currents, where the isopycnal descends about a kilometer in a horizontal distance of about 100 km. Therefore we find that the effective vertical diffusivity is very nearly $\kappa_\rho + \gamma^2 \kappa_I$ (from (7.28)). Thus the *isopycnal* diffusivity contributes to *vertical* diffusion, due to the nonzero isopycnal slope. The isopycnal contribution to the vertical diffusion, $\gamma^2 \kappa_I$ will be as big as the diapycnal contribution if γ^2 is at least as big as κ_ρ/κ_I, which for plausible thermocline values of $\kappa_\rho = 0.1 \times 10^{-4}$ m^2/s and $\kappa_I = 10^3$ m^2/s gives $\gamma = 10^{-4}$ (recall that κ_ρ is much larger in surface and bottom mixed layers). As shown in Figure 7.16, isopycnal slopes in the western boundary currents and at subpolar latitudes are steep enough to make a large κ_I contribution to vertical diffusion (see also Figure 5.2). Isopycnal slopes, γ, are typically smaller than 10^{-3}. They peak in the Gulf Stream, subpolar North Atlantic, Kuroshio, and the Antarctic Circumpolar Current.

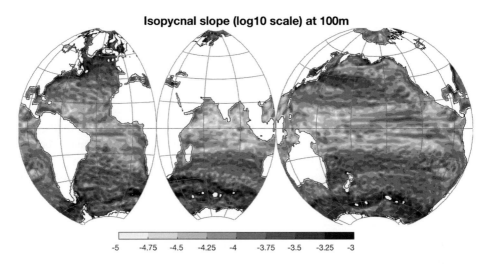

Isopycnal slope (log10 scale) at 100m

-5   -4.75   -4.5   -4.25   -4   -3.75   -3.5   -3.25   -3

Figure 7.16 Isopycnic slope γ at 100 m on a log 10 scale. The data are from the CARS climatology and are smoothed with a 2° box-car filter.

Another way to simplify the diffusivity tensor is to use the small slope approximation ($\rho_z^2 \gg \rho_x^2, \rho_y^2$):

$$\kappa^g = \frac{1}{\rho_z^2} \begin{pmatrix} \rho_z^2 \kappa_I & (\kappa_\rho - \kappa_I)\rho_x\rho_y & (\kappa_\rho - \kappa_I)\rho_x\rho_z \\ (\kappa_\rho - \kappa_I)\rho_y\rho_x & \rho_z^2\kappa_I & (\kappa_\rho - \kappa_I)\rho_y\rho_z \\ (\kappa_\rho - \kappa_I)\rho_z\rho_x & (\kappa_\rho - \kappa_I)\rho_z\rho_y & (\rho_x^2 + \rho_y^2)\kappa_I + \kappa_\rho\rho_z^2 \end{pmatrix}. \tag{7.32}$$

Using slopes $\gamma_x = \rho_x/\rho_z$ and $\gamma_y = \rho_y/\rho_z$, we can rewrite this as

$$\kappa^g = \begin{pmatrix} \kappa_I & \gamma_x\gamma_y(\kappa_\rho - \kappa_I) & \gamma_x(\kappa_\rho - \kappa_I) \\ \gamma_x\gamma_y(\kappa_\rho - \kappa_I) & \kappa_I & \gamma_y(\kappa_\rho - \kappa_I) \\ \gamma_x(\kappa_\rho - \kappa_I) & \gamma_y(\kappa_\rho - \kappa_I) & (\gamma_x^2 + \gamma_y^2)\kappa_I + \kappa_\rho \end{pmatrix}. \tag{7.33}$$

This expression is used in Subsection 7.3.3.

7.3.3 Isopycnic Layer Thickness Advection

The tensor diffusivity written by Redi (1982) (Subsection 7.3.2) shows how to handle strong isopycnic diffusivity and weak diapycnic diffusivity in the realistic case where the isopycnal depths vary in space and time. It can be applied to the diffusive mixing terms in the equations for conservation of heat, salt, and tracer (1.15) and (1.16). It does not propose how to specify the eddy diffusivity parameters (κ_I, κ_ρ), however. Nor does it propose how to handle the diffusive term in the momentum equations.

The focus here is on the case of vanishing diapycnic mixing ($\kappa_\rho = 0$), the adiabatic limit (Subsection 1.3.3). Diapycnic mixing is dominated by the small-scale mixing processes described in Subsection 7.1.2. These are relatively weak in the interior compared to in the surface (and bottom) mixed layer (Chapter 4). Nevertheless, it is not obvious that weak diapycnic mixing is less important for the general circulation than isopycnic mixing by eddies. This question remains under investigation: see Subsection 8.2.3 and Subsection 8.3.2 for arguments that diapycnic mixing controls the circulation, for example.

Eddy Parameterization for Continuity Equation

In the adiabatic limit isopycnic mixing due to unresolved processes, like mesoscale eddies, has *no effect on the density field.* The Redi (1982) dissipation simply mixes density along density surfaces. This fact is seen in (7.30) and (7.31), for example, because when $\kappa_\rho = 0$, $\nabla \cdot \mathbf{D}_\theta(\theta) = 0$. This is strange and unexpected behavior: even in the adiabatic limit, we expect the unresolved eddies to have some influence on the large-scale density field. For example, it makes intuitive sense that the eddies themselves exist because the large-scale fields (of density and circulation) are unstable. By this reasoning, the eddies should be draining energy from the large-scales, and therefore must have a nonzero impact on them. Thus, it became apparent that some essential physics was missing from the Redi (1982) proposal to account for the effects of unresolved eddies in the adiabatic limit.

Generally speaking, this is still an open research topic. Probably the most important progress appeared in the short article by Gent and McWilliams (1990) (GM), which inspired many other papers. Their focus was on parametrizing the strong along-isopycnal effects of unresolved mesoscale eddies, not the weak diapycnic part. They proposed that unresolved eddies rearrange isopycnic thickness along isopycnals. The subsequent paper by Gent et al. (1995) showed that this is mathematically equivalent to an extra *advection* due to the eddies. Their ideas are explained here.

The GM parameterization begins with the fundamental equations that satisfy the ocean circulation, which include the mesoscale eddy field. Specifically, consider the shallow water equations (Subsection 2.2.4): they describe the dynamics of the isopycnic layer thickness and the transport of tracer along isopycnals. Then one writes equations for the evolution of the large-scale thickness and tracer fields.

The shallow water continuity equation (2.80) states that, if there is no diapycnic flow, the evolution of layer thickness Δh_n is given by

$$\frac{\partial (\Delta h_n)}{\partial t} + \nabla \cdot (\Delta h_n \mathbf{u}_n) = 0. \tag{7.34}$$

As Subsection 7.2.3 shows, the layer can be interpreted as the water between two surfaces $h(x, y, \rho, t)$ defined by two different values of (potential) density ρ. Using (7.3) for Δh_n, we can define **thickness** $\partial h / \partial \rho$ for a continuous fluid and write continuity as

$$\frac{\partial}{\partial t} \left(\frac{\partial h}{\partial \rho} \right) + \nabla_\rho \cdot \left(\frac{\partial h}{\partial \rho} \mathbf{u} \right) = 0, \tag{7.35}$$

where the horizontal gradient ∇_ρ is taken at constant density (so it follows the undulating isopycnal rather than the nth layer; see Figure 7.15), and $\mathbf{u}$ is the horizontal current which varies continuously with density.

Using the division (7.4) between mean and residual quantities and writing h_ρ as shorthand for $\partial h / \partial \rho$, we can follow (7.8) to write continuity as

$$\frac{\partial \overline{h_\rho}}{\partial t} + \nabla_\rho \cdot \left(\overline{h_\rho} \overline{\mathbf{u}} + \overline{h'_\rho \mathbf{u}'} \right) = 0, \tag{7.36}$$

which highlights the separate contribution of resolved flow and eddies to the thickness transport. We can make (7.36) look more like the earlier version of the continuity equation by defining

$$\mathbf{u}_{bolus} = \frac{\overline{h'_\rho \mathbf{u}'}}{\overline{h_\rho}},$$ (7.37)

and

$$\mathbf{U} = \overline{\mathbf{u}} + \mathbf{u}_{bolus},$$ (7.38)

so that (7.36) beomes

$$\frac{\partial \overline{h_\rho}}{\partial t} + \nabla_\rho \cdot \left(\mathbf{U} \overline{h_\rho} \right) = 0.$$ (7.39)

The extra velocity term $\mathbf{u}_{bolus}$ is called the **bolus velocity** or the **eddy-induced transport velocity**. Sometimes the resolved flow $\overline{\mathbf{u}}$ is referred to as the advective velocity and $\mathbf{U}$ is referred to as the total velocity or **residual velocity**. A key insight from the GM parameterization is that the unresolved isopycnic mixing is equivalent to an extra *advection* along isopycnals due to this term (Subsection 7.2.3). (This effect was known already in the atmospheric context; see Plumb and Mahlman, 1987 and Andrews et al., 1987, for example.)

Equation 7.39 is not by itself a parameterization but a renaming of the unknown eddy quantity $\overline{h'_\rho \mathbf{u}'}$. For a parameterization, we need to express eddy thickness transport in terms of the large-scale fields. GM propose the following closure:

$$\overline{h'_\rho \mathbf{u}'} = -\frac{\partial}{\partial \rho} \left(\kappa \nabla_\rho \overline{h} \right),$$ (7.40)

where κ is a diffusivity (called the quasi-Stokes diffusivity, Kuhlbrodt et al., 2012; McDougall and McIntosh, 2001) and $\overline{h'_\rho \mathbf{u}'} = 0$ applies on the boundaries. From (7.36), the parameterization can be written in terms of the bolus velocity

$$\mathbf{u}_{bolus} = -\frac{1}{\overline{h_\rho}} \frac{\partial}{\partial \rho} \left(\kappa \nabla_\rho \overline{h} \right),$$ (7.41)

or

$$\mathbf{u}_{bolus} = -\frac{\kappa}{\overline{h_\rho}} \nabla_\rho \overline{h_\rho} - \frac{\partial \kappa}{\partial \rho} \frac{1}{\overline{h_\rho}} \nabla_\rho \overline{h}.$$ (7.42)

The simplest case concerns κ with no ρ-dependence, for which the last term in (7.42) is zero and the bolus velocity is proportional to the down-gradient flux of isopycnal layer thickness (though in general, the last term may be important). For κ constant, $\mathbf{u}_{bolus}$ is in the opposite direction to $\nabla_\rho \overline{h_\rho}$ so that the bolus flow reduces layer thickness by transporting water from the thick region of the layer to the thin region.

More generally, κ can vary in space and time, depending on the large-scale flow itself. Visbeck et al. (1997) proposed a closure of this type, for example (Subsection 7.2.1, Subsection 7.2.3). Applying baroclinic instability theory (Green, 1970; Stone, 1972), they argue that κ should be

$$\kappa = \alpha \frac{f}{\sqrt{Ri}} \ell^2 = \alpha \frac{M^2}{N} \ell^2.$$ (7.43)

In this formula f is the Coriolis parameter (2.1) and

$$Ri = \frac{f^2 N^2}{M^4},$$ (7.44)

is the large-scale Richardson number based on vertical stratification N^2 (1.1) and horizontal stratification,

$$M^2 = \frac{g}{\rho_0} |\nabla_h \rho|.$$ (7.45)

In general, the **Richardson number** equals $N^2 / (\partial |\mathbf{u}_h| / \partial z)^2$, the ratio of the vertical stratification to the square of the vertical shear of the horizontal current $\mathbf{u}_h$. Application of the thermal wind equation (2.9) yields (7.44). Finally, $\alpha \approx 0.015$ is a constant (not the thermal expansion coefficient) and ℓ is a horizontal lengthscale that characterizes the width of the eddying region, sometimes called the **baroclinic zone**. This closure, (7.43), applied as a *diffusive* eddy parameterization is what Visbeck et al. (1997) called "GS" (Figure 7.14) after the Green (1970) and Stone (1972) papers (see also Subsection 9.3.2). The Visbeck et al. (1997) "NEW" parameterization (Figure 7.14) unites the GS formula for κ, eq. (7.43), with the advective GM framework, (7.41).

Different choices exist for ℓ, such as the lengthscale set by bathymetry, or by atmospheric forcing, or by the **Rhines scale** L_I,

$$L_I = \sqrt{\frac{|\mathbf{u}_h|}{\beta}},$$ (7.46)

(also called the inertial length scale, see (3.13)). This scale separates the regime of geostrophic turbulence involving mesoscale eddies from larger Rossby waves (Subsection 1.2.1), which depend on the beta effect (Rhines, 1975, 1977). In general, however, uncertainty exists about the best choice for ℓ. Ocean GCMs typically use an ad hoc scheme to diagnose ℓ, for example, using a threshold on the magnitude of the isopycnal slope $\gamma = \rho_x / \rho_z = M^2 / N^2$ to delimit the baroclinic zone where eddies are active (Pacanowski and Griffies, 1999; γ is not the reduced gravity). Because κ depends on ℓ squared in (7.43) the eddy-induced circulation is sensitive to these somewhat arbitrary criteria, and another perspective on ℓ is that it can be tuned, within limits, to give satisfactory results.

Even more generally, κ in (7.40) can be replaced with a second-order tensor, as Griffies (2004) discusses. Understanding exactly how to compute $\overline{h'_\rho \mathbf{u}'}$ from the large-scale fields remains an important open research question.

Eddy Parameterization for Tracer Equation

In addition to generating a flux of layer thickness, eddies also transport passive tracers. The layer-weighted passive tracer concentration changes due to convergence/divergence of the advective flux of tracer. For a given shallow water layer,

$$\frac{\partial}{\partial t}(h_n \chi_n) + \nabla \cdot (h_n \mathbf{u}_n \chi_n) = 0.$$ (7.47)

Here we neglect molecular mixing; such mixing can be included, but we anticipate that for the large-scale fields it will be overwhelmed by the action of the eddies. In the continuously stratified case, the passive tracer equation becomes

$$\frac{\partial}{\partial t}\left(h_\rho \chi\right) + \nabla_\rho \cdot \left(h_\rho \mathbf{u}\chi\right) = 0, \tag{7.48}$$

which makes sense by analogy with (7.35).

Partitioning variables between resolved and eddy values, (7.48) becomes

$$\frac{\partial}{\partial t}\left(\overline{h_\rho}\,\overline{\chi} + \overline{h'_\rho \chi'}\right) + \nabla_\rho \cdot \left[\overline{h_\rho}\,\overline{\mathbf{u}}\,\overline{\chi} + \overline{h'_\rho \mathbf{u}'}\,\overline{\chi} + \overline{(h_\rho \mathbf{u})'\chi'}\right] = 0, \tag{7.49}$$

where $(h_\rho \mathbf{u})' = \overline{h_\rho}\mathbf{u}' + h'_\rho \overline{\mathbf{u}} + h'_\rho \mathbf{u}'$. Now, write the first term in (7.49) as $\partial(\overline{h_\rho}\,\overline{\chi})/\partial t = \overline{h_\rho}\partial \overline{\chi}/\partial t + \overline{\chi}\partial \overline{h_\rho}/\partial t$ and substitute from (7.36). The first two terms in the divergence part of (7.49) simplify and the result is:

$$\overline{h_\rho}\frac{\partial \overline{\chi}}{\partial t} + \frac{\partial}{\partial t}\overline{h'_\rho \chi'} + \left(\overline{h_\rho}\,\overline{\mathbf{u}} + \overline{h'_\rho \mathbf{u}'}\right) \cdot \nabla_\rho \overline{\chi} + \nabla_\rho \cdot \left[\overline{(h_\rho \mathbf{u})'\chi'}\right] = 0. \tag{7.50}$$

At this stage, (7.50) is exact. Now we argue that it can be simplified with two reasonable assumptions. First, thickness and tracer concentration are not likely correlated, so $\overline{h'_\rho \chi'}$ can be neglected. Second, we assume that the final term in (7.50) acts as an isotropic diffusion along isopycnals with diffusivity κ_I (other choices are possible). The form of this diffusive closure assumption is:

$$\nabla_\rho \cdot \left(\overline{(h_\rho \mathbf{u})'\chi'}\right) = -\nabla_\rho \cdot \left(\kappa_I \overline{h_\rho}\mathbf{J}\nabla_\rho \overline{\chi}\right), \tag{7.51}$$

and

$$\mathbf{J} = \frac{1}{1 + \overline{h_x}^2 + \overline{h_y}^2}\left(\begin{array}{cc} 1 + \overline{h_y}^2 & -\overline{h_x}\,\overline{h_y} \\ -\overline{h_x}\,\overline{h_y} & 1 + \overline{h_x}^2 \end{array}\right). \tag{7.52}$$

Here, $\kappa_I \mathbf{J}$ is a diffusion tensor that takes account of the somewhat unusual coordinate system used for the isopycnic equations. (The (x, y, ρ) coordinate system is called the projected nonorthogonal coordinate system. $\nabla_\rho \chi$ is a two-dimensional vector in the (x, y) horizontal plane, but the value of ρ is held constant in computing the gradient. See Figure 7.15 and Griffies (2004) sections 6.5 and 9.4.3 for details.) It corresponds to the isopycnic coordinate version of the Redi (1982) diffusion tensor from Subsection 7.3.2 (for example, (7.23)) where the diapycnal diffusivity κ_ρ vanishes. Gathering these pieces together, we have,

$$\frac{\partial \overline{\chi}}{\partial t} + \mathbf{U} \cdot \nabla_\rho \overline{\chi} = \nabla_\rho \cdot \left(\kappa_I \overline{h_\rho}\mathbf{J}\nabla_\rho \overline{\chi}\right)/\overline{h_\rho}. \tag{7.53}$$

The GM parameterization guarantees some desirable properties of the large-scale circulation (see Exercise 7.8). These are: first, the volume of fluid within the layer between two isopycnals is conserved. Second, the average value of tracer χ in the layer between two isopycnals is constant, and all the higher moments decrease as the χ gradients in the layer are mixed away (for $\kappa_I > 0$ in (7.53)). Third, as ρ satisfies (7.53) exactly (the diffusive right-hand side term vanishes for $\chi = \rho$), all domain-averaged moments of ρ are constant. Finally, for small Rossby numbers, and flow that is dominated by geostrophy, the GM closure (7.40) implies down-gradient vertical diffusion of large-scale (horizontal) momentum (for examples see 5.2.3 and 9.2.3). That means that the isopycnals (in Figure 7.13, for example) flatten because vertical diffusion of horizontal momentum implies less steep isopycnals from the thermal wind relation (2.9). For this reason, and because it reduces the gradients in layer thickness, the GM closure converts mean potential energy to eddy potential energy. This draining of energy from the large-scale stratification into the small-scale eddies makes physical sense, as discussed above.

Thickness Advection in Geodesic Coordinates

The preceding discussion concerns the effect of the unresolved eddies on the large-scales in (x, y, ρ) isopycnic coordinates, which most clearly expose the physics. How does it work in the more familiar (x, y, z) geodesic coordinates? There are two equations to consider: the equation for evolution of density, which replaces the thickness equation (7.36), and the equation for evolution of tracer. The large-scale density equation in geodesic coordinates reads

$$\frac{\partial \overline{\rho}}{\partial t} + \mathbf{U} \cdot \nabla_h \overline{\rho} + W \frac{\partial \overline{\rho}}{\partial z} = 0. \tag{7.54}$$

Here $\mathbf{U}$ is the same as that used in (7.39), the sum of the large-scale horizontal velocity and the horizontal bolus velocity, and

$$W = \overline{w} + w_{bolus}. \tag{7.55}$$

In this expression $\overline{w}$ is the large-scale Eulerian vertical velocity and w_{bolus} is a vertical bolus velocity.

Similarly, the large-scale tracer equation in geodesic coordinates reads

$$\frac{\partial \overline{\chi}}{\partial t} + \mathbf{U} \cdot \nabla_h \overline{\chi} + W \frac{\partial \overline{\chi}}{\partial z} = \nabla \cdot \kappa^g \nabla \overline{\chi}, \tag{7.56}$$

where κ^g is the Redi diffusivity tensor corresponding to Laplacian isopycnic mixing. Specifically, from (7.23) with $\kappa_\rho = 0$

$$\kappa^g = \frac{\kappa_I}{\rho_x^2 + \rho_y^2 + \rho_z^2} \begin{pmatrix} \rho_y^2 + \rho_z^2 & -\rho_x \rho_y & -\rho_x \rho_z \\ -\rho_y \rho_x & \rho_x^2 + \rho_z^2 & -\rho_y \rho_z \\ -\rho_z \rho_x & -\rho_z \rho_y & \rho_x^2 + \rho_y^2 \end{pmatrix}, \tag{7.57}$$

and for small isopycnic slopes (see Subsection 7.3.2 and Figure 7.16), which are typical, the diffusivity tensor is

$$\kappa^g \approx \frac{\kappa_I}{\rho_z^2} \begin{pmatrix} \rho_z^2 & 0 & -\rho_x \rho_z \\ 0 & \rho_z^2 & -\rho_y \rho_z \\ -\rho_z \rho_x & -\rho_z \rho_y & \rho_x^2 + \rho_y^2 \end{pmatrix}. \tag{7.58}$$

The new vertical bolus velocity w_{bolus} appears in the density and tracer equations written in geodesic coordinates for an interesting reason: Think of the equation for large-scale layer thickness (7.39) for a piece of ocean as in Figure 7.17. The isopycnic thickness $\overline{h}_\rho$ increases smoothly with depth apart from the bump in the middle (a large-scale eddy). By the GM closure (7.40), the horizontal bolus velocities $\mathbf{u}_{bolus}$ are from areas of large

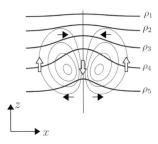

Schematic example of density distribution (thick curves) and associated bolus currents which in geodesic (x, y, z) coordinates have a horizontal component $\mathbf{u}_{bolus}$ (solid arrows) and vertical component w_{bolus} (unfilled arrows). For the case in which $\rho_y = 0$, the bolus velocity is represented by a streamfunction (thin curves).

thickness (weak stratification) to small thickness (strong stratification), and are indicated schematically with black arrows. Between ρ_2 and ρ_3, for example, the horizontal bolus velocity is convergent, so by (7.48) $\overline{h_\rho}$ is increasing. Below ρ_5 the horizontal bolus velocity is divergent, so $\overline{h_\rho}$ is decreasing. Some time later the bump has disappeared and $\overline{h_\rho}$ is constant everywhere. In the isopycnic framework no more information is needed because the thickness variable $\overline{h_\rho}$ completely defines the distribution of density. The isopycnal heights $\overline{h}(x, y, \rho, t)$ can be diagnosed by integrating the thickness $\overline{h_\rho}$ with respect to ρ. One finds that at the bump the $\overline{h}$ is now lower than before.

The geodesic framework portrays the processes above differently. There, the density is decreasing at the location of the bump corresponding to a vertical bolus velocity w_{bolus} advecting the ρ_4 contour down. For these reasons, the total three-dimensional bolus velocity $(\mathbf{u}_{bolus}, w_{bolus})$ is non-divergent;

$$\nabla_h \cdot \mathbf{u}_{bolus} + \frac{\partial}{\partial z} w_{bolus} = 0, \tag{7.59}$$

just as for the resolved velocity.

Finally, what is the GM closure proposal for $(\mathbf{u}_{bolus}, w_{bolus})$ in geodesic coordinates? Recall that GM says (7.40)

$$\overline{h_\rho} \mathbf{u}_{bolus} = -\frac{\partial}{\partial \rho} \left(\kappa \nabla_\rho \overline{h} \right), \tag{7.60}$$

with $\mathbf{u}_{bolus} = 0$ on the boundaries. In geodesic coordinates,

$$\partial \overline{h}/\partial \rho = -\frac{1}{\partial \rho / \partial z}, \tag{7.61}$$

and

$$\nabla_\rho \overline{h} = (1/\rho_z) \nabla_h \rho. \tag{7.62}$$

Hence, converting the $\partial/\partial \rho$ to $\partial/\partial z$,

$$\mathbf{u}_{bolus} = \frac{\partial}{\partial z} \left(\frac{\kappa}{\rho_z} \nabla_h \rho \right), \tag{7.63}$$

and, from (7.59),

$$w_{bolus} = -\nabla_h \cdot \left(\frac{\kappa}{\rho_z} \nabla_h \rho \right). \tag{7.64}$$

These formulae have a useful interpretation worth remembering (and used to quantify the Deacon Cell in Chapter 9 and in the Beaufort gyre model of Subsection 10.3.3): Think of the case where the horizontal density gradient vanishes $\rho_y = 0$ (as we assume in Figures 7.15 and 7.17), then

$$(u_{bolus}, w_{bolus}) = \left(-\frac{\partial \Psi}{\partial z}, \frac{\partial \Psi}{\partial x} \right). \tag{7.65}$$

Here the eddy-induced overturning streamfunction (Subsection 2.1.4, Subsection 2.2.3) in the (x, z) plane is

$$\Psi = \frac{\kappa \rho_x}{\rho_z}, \tag{7.66}$$

and is sometimes called the quasi-Stokes streamfunction (McDougall and McIntosh, 2001; Gent, 2011; Kuhlbrodt et al., 2012).

This formula means that the eddy-induced overturning streamfunction is proportional to the isopycnic slope ρ_x/ρ_z, as in Figure 7.17. An isolated (large-scale) eddy displaces the isopycnals upwards. The horizontal bolus velocity is from large-to-small thickness, as seen with the black arrows. Convergence and divergence of u_{bolus} drives a vertical bolus velocity w_{bolus}. The net effect is a dipolar overturning circulation to relax the isopycnic displacements, seen with the thin lines which are proportional to Ψ in (7.66). The effect of the eddy-induced flow is to flatten isopycnals reducing potential energy and, for a geostrophic flow, to reduce vertical shear in the large-scale current. In this case, the bolus velocity (u_{bolus}, w_{bolus}) vanishes at large distances from the eddy. As $\rho_x/\rho_z \neq 0$ at the sea surface and sea floor in general, κ in (7.66) must be tapered to zero approaching boundaries to ensure no normal flow (constant Ψ) due to the eddies.

The κ in (7.66) is a constant according to the GM closure. Alternatively, Visbeck et al. (1997) propose that it depends on the large-scale density field itself, (7.43). Using their closure we find

$$\Psi = \alpha N \ell^2 \left| \frac{\rho_x}{\rho_z} \right| \frac{\rho_x}{\rho_z}, \tag{7.67}$$

instead of (7.66). The pre-factor $\alpha N \ell^2$ in this expression varies between different problem configurations, but is (more or less) constant for a given problem with characteristic N and ℓ. Recalling that $\rho_x/\rho_z = \gamma$ is the slope of the isopycnic surfaces, we see that both the GM and Visbeck et al. (1997) parameterizations are of the general form

$$\Psi = k|\gamma^{n-1}|\gamma, \tag{7.68}$$

for **eddy efficiency** k (Manucharyan et al., 2016), where $n = 1$ and k is a constant for GM, and $n = 2$ and $k = \alpha N \ell^2$ for Visbeck et al. (1997). In other words, the GM closure proposes that the eddy-induced overturning streamfunction scales linearly as the isopycnal slope and the Visbeck et al. (1997) closure proposes it scales quadratically. This form of eddy parameterization is used in Subsection 10.3.3.

Exercises

7.1 Select a region rich in mesoscale eddies (see the figures in Section 7.1). Using sea-surface height (SSH) and sea-surface temperature (SST) data (see Appendix A), answer the following questions:

(a) Make SSH and SST maps of successive snapshots of eddies. Identify specific eddies and track them over time. Include cyclonic and anticyclonic features.

(b) Estimate the formation rate, propagation speed, and propagation direction of eddies in your data.

(c) Hence make a simple estimate of the transport of volume and heat associated with your eddies.

(d) Comment on the magnitude of these eddy fluxes compared to corresponding fluxes from the time-mean flow.

Hints: Consult Chapter 3 and Chapter 11, for example, for guidance on typical time-mean fluxes. Exercise 3.2 is also helpful.

7.2 Use a gridded satellite altimetry product (see Appendix A) for this question. Select an ocean basin (as in Exercise 3.2, for instance; coordinate with your classmates to avoid replicates). Then:

(a) Compute (or download) the zonal and meridional components of surface geostrophic velocity anomaly (u'_G, v'_G) for many different times (the anomaly is from the long time average). Make maps of a few representative snapshots.

(b) Make maps of u'_G and v'_G averaged over periods of 32, 128, 512, and 2048 days.

(c) Describe and comment on the coherence of the surface geostrophic flow on these timescales.

Hint: See Maximenko et al. (2005) and Exercise 3.2.

7.3 Potential energy, E_P, is defined by (8.5) (see Subsection 8.2.2; (8.6) is also relevant). The available potential energy is the part of E_P that can be converted to another form by an adiabatic rearrangement of the fluid (Lorenz, 1955; Vallis, 2006).

(a) Use the E_P formula to explain why there is potential energy available in the three-layer idealized model of Visbeck et al. (1997), shown in Figure 7.10.

(b) Writing the interface height between layers 1 and 2 as η_1, and so on, write an expression for E_P. Assume that layer 3 is motionless. Define all variables and make suitable assumptions.

(c) Compute the geostrophic current in the fluid and hence the total kinetic energy. Define all variables and make suitable assumptions.

(d) Hence, compute the ratio of available potential energy to kinetic energy in the idealized model of Figure 7.10 for the three times shown. Comment on your result and speculate on the ratio of available potential energy to kinetic energy for the global ocean.

7.4   Consider the two-layer system in Figure 7.12, Subsection 7.2.2, and Subsection 7.3.1.

(a) Prove (7.16). For what values of B is the tracer flux a maximum and a minimum? Which is more likely in practice for ocean mesoscale eddies and why?

(b) Write a formula for the bolus velocity defined analogously to (7.37).

(c) Now assume that the eddies in Figure 7.12 derive from baroclinic instability of a mean geostrophic zonal flow with a characteristic speed that equals the characteristic eddy speed. Hence write a scaling formula for the timescale for spin-down of the density front (as in Figure 7.13) associated with the mean zonal flow. Make suitable assumptions that you clearly explain.

(d) Using reasonable parameters for an eddying current (see Figure 7.2, for example), estimate the size of the bolus velocity and the spin-down timescale from part (c).

7.5   Make annotated sketches with brief explanations to show:

(a) Horizontal stirring of a buoyant passive tracer released at the surface of the ocean.

(b) Horizontal mixing of a buoyant passive tracer released at the surface of the ocean.

(c) Mixing of a passive tracer released in the ocean interior with isotropic diffusion.

(d) Mixing of a passive tracer released in the ocean interior with diffusion that is much greater in the horizontal than the vertical direction.

(e) Mixing of a passive tracer released in the ocean interior with diffusion that is much greater along isopycnals than across them.

7.6   Explain the following parameterization ideas, giving examples from the book (or otherwise):

(a) Heuristic scaling law.

(b) Parcel theory.

(c) Mixing length theory.

(d) Parameterization based on recovery of marginal stability.

(e) Parameterization based on estimating the flux of a conserved quantity.

(f) Parameterization based on a budget of a conserved quantity.

7.7   Think about the distribution of isopycnals shown in Figure 7.18. Sketch the eddy-induced overturning streamfunction in each case according to the GM parametrization (7.66) with constant κ and indicate the direction of flow. Assume that the contour spacing is the same everywhere and there is no isopycnic slope normal to the plane of the page ($\rho_y = 0$). The sea surface and sea floor are at a large distance from the isopycnals shown.

7.8   Consider a large scale ocean front, such as the one associated with the Antarctic Circumpolar Current (see Figure 8.1). Use Figures 7.10 and 7.13 to conceptualize the density field. Assume that there are no variations in any property along the front and

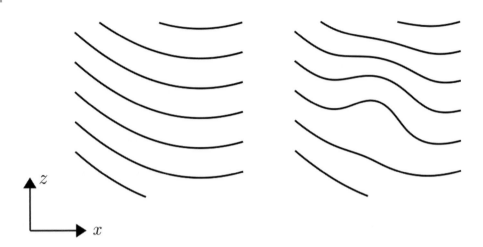

Figure 7.18   Arrangement of isopycnals at a large-scale front (left) and at a large-scale front with an eddy (right). For Exercise 7.7.

ignore salinity variations. Write a numerical code that applies the eddy parameter-ization (7.68) to relax the front. Clearly state the mathematical problem you solve. Using realistic parameter values compare and contrast the GM and Visbeck et al. (1997) eddy parameterizations. Present and discuss results for:

(a) The time-evolution of the density and geostrophic velocity field.
(b) The time-evolution of the water mass distribution over density classes.
(c) The time-evolution of the available potential energy, mean kinetic energy, and total mechanical energy. Where does the lost energy go?

Hence, demonstrate the desirable properties of the GM parameterization mentioned on page 269. Are these properties also true for the Visbeck et al. (1997) parameteri-zation?

Hints: Exercise 2.10 is useful. Taper the eddy diffusivity approaching the ocean sur-face and floor, so that the eddy-induced overturning streamfunction vanishes there.

Deep Meridional Overturning

The ocean circulates in a single connected global loop. It involves flow at all depths and in all basins. To understand these planetary currents we discuss the patterns of surface water sinking into the abyss, flowing over global scales, upwelling, and returning in the pycnocline. We discuss how this flow alters the gyre circulations and influences the pycnocline and abyssal stratification. We present evidence for deep meridional overturning cells, their shallow upstream sources, and their western boundary currents. The roles of interior mixing and surface density forcing as drivers of overturning are highlighted. The importance of topography is discussed, especially sills and marginal seas; and the horizontal circulations that occupy the deep basins.

8.1 Observations

8.1.1 Evidence of Deep Circulations

Seawater is cold (see Exercise 1.12). Even though surface water in the tropics is mostly warmer than 25°C (Figure 4.1), comfortable for a human to swim, the water below 1 km is universally less than about 5°C (Figure 8.1). Why is the deep ocean so cold?

Tropical surface water warms by interaction with the sun and the atmosphere. This warmth should diffuse downward due to turbulence. If the deep water were stagnant, such diffusion would eventually warm deep water to surface temperatures after a sufficiently long time (and the ocean has existed for very long). There are no processes which remove heat from the deep ocean. As the Earth is generating geothermal heat through radioactive decay, any transfer of heat through the sea floor warms the deep ocean rather than cools it (see Exercise 8.4).

To answer our question: deep water is cold because it travels from the surface in high latitudes where the equilibrium surface temperature is cold. The cold water sinks from the surface and then spreads laterally to fill the deep ocean. In a closed basin, conservation of mass dictates that if there is a net flow of water down and away from the polar sinking region, then it must flow upwards again and return in a shallower depth range to the sinking region. This is the essence of deep meridional overturning.

Evidence for deep pathways can be seen in various **tracer** fields. We refer to any substance that is carried by sea water as a tracer (Subsection 1.3.1). At the surface, the salinity

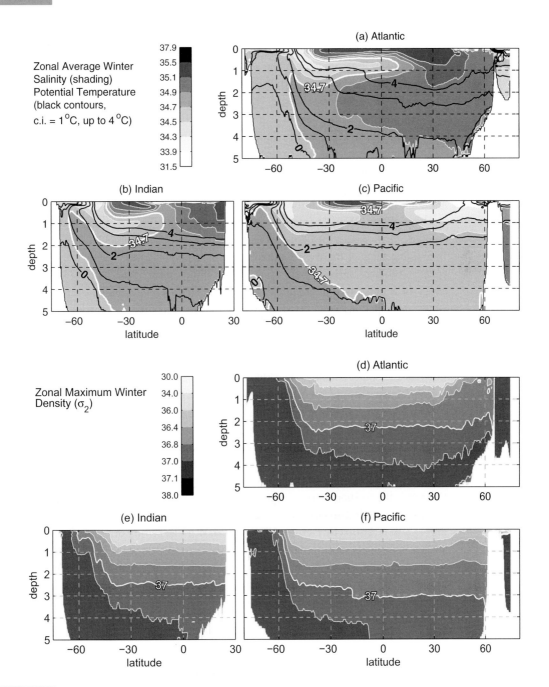

Zonal Average Winter
Salinity (shading)
Potential Temperature
(black contours,
c.i. = $1\,^{\circ}$C, up to $4\,^{\circ}$C)

Zonal Maximum Winter
Density (σ_2)

Figure 8.1   Latitude–depth cross section of zonal average salinity (shading) and potential temperature (black contours) for
(a) Atlantic, (b) Indian, and (c) Pacific Oceans, and zonal maximum density (σ_2) for (d) Atlantic, (e) Indian, and (f)
Pacific. Data from WOA 2013 (see Appendix A). Contour interval is $1\,^{\circ}$C for temperature and is given by color bars for
salinity and density. A black-and-white version of this figure appears in some formats. For the color version, please
refer to the plate section.

of a water parcel can undergo fast changes due to evaporation, precipitation, and runoff from the land. (By "fast" we mean fast enough to induce changes in the salinity of a water parcel of O(1) g/kg, which is comparable to the salinity range over much of the ocean). Once the water leaves the mixed layer, it changes at a much slower rate due to mixing with neighboring water of different salinity (Subsection 7.3.2). Just as a dye poured into the top of a laboratory tank traces the flow of water in an experiment, salinity and the concentration of other tracers set at the surface indicate the subsurface ocean flow.

The Atlantic has a salinity pattern that is asymmetric about the equator (Figure 8.1a), in contrast to potential density, which is more symmetric (Figure 4.6). The surface salinity is high in the subtropics and low poleward of about 45°S and 50°N. Moreover maximum and minimum surface salinities are higher in the northern hemisphere than in the southern hemisphere. The saltier northern subpolar water (with $S > 34.8$ g/kg or so) forms a "tongue" that crosses into the southern hemisphere between about 1200 m and 3600 m. Below this tongue, the fresher southern subpolar water extends into the northern hemisphere. Another source of fresh water ($S < 34.6$ g/kg) extends from the surface south of about 45°S down to more than 800 m depth where it crosses the equator. This tongue of water stands in contrast to the slightly saltier northern-source water below it and the much saltier subtropical water above about 500 m.

Temperature decreases monotonically with depth nearly everywhere (Subsection 1.1.5), but it also varies with latitude (Figure 8.1a) and longitude. Generally speaking, the fresher water in the south is colder than saltier water to the north at the same depth. Temperature differences at a given depth are as great as 4°C, which is comparable to the vertical temperature range below about 1 km.

A **water mass** consists of water falling into a given temperature and salinity range and is believed to be "formed" (that is, have its temperature, salinity, and other tracer properties set at the surface; see Chapter 4) in a single region. Thus the cold, salty water mass crossing from the North Atlantic to the South Atlantic is called **North Atlantic Deep Water (NADW)**, the fresher, warmer, lighter, and hence shallower water mass coming from the south is **Antarctic Intermediate Water (AAIW)**, and the water flowing northward under the NADW is **Antarctic Bottom Water (AABW)**. The setting of properties and sinking of a water mass to about a kilometer or deeper is often called **deep water formation** (see also Subsection 4.2.2). NADW gets fresher flowing southward, while the AAIW and AABW get saltier, consistent with the idea that there is some mixing between the water masses, albeit weak, as they flow.

Generally, the water column below the pycnocline is divided into intermediate, deep, and bottom water masses although the distinctions are somewhat ambiguous. For example, the boundaries of the pycnocline itself are not precisely defined, so one can refer to intermediate water as at least partly including lower pycnocline water. The idea of using three broad classes of water mass below the pycnocline is justified by the existence of up to three distinct salinity extrema and flow directions in a zonally averaged water column.

The structure of the southern hemisphere Indian Ocean resembles the Atlantic (Figure 8.1b). However, there is no sign of deep or intermediate fresh water crossing the equator, and there is no surface source of water with the same salinity (about 34.7 g/kg) as most of the deep water. The Pacific has a fresh tongue descending to about 1 km and flowing equatorward from latitudes 40–50° in each hemisphere (Figure 8.1c). Thus the top kilometer is more symmetric about the equator, though the deep water has a gradient from cold, salty water in the south. Generally speaking, the deep salinity is highest in the North Atlantic (34.9–35 g/kg) and decreases southward in the Atlantic (to 34.7 g/kg) and northward in the Pacific (34.6–34.7 g/kg). The Indian Ocean deep water is of intermediate salinity (about 34.7 g/kg). Dissolved oxygen (O_2) is another tracer of the flow field. A water parcel's O_2 content is highest when it is in contact with the atmosphere and decreases over time due to respiration. High values in the deep North Atlantic and low values in the deep North Pacific indicate sinking in the former but not in the latter (see Exercise 8.1).

The density field also indicates possible connections between surface and deep water. Because near-surface density varies with longitude and season, we look at late-winter properties (March in northern hemisphere and September in southern; see Figure 4.6 and Subsection 5.2.4) and at the maximum value at each latitude and depth (Figure 8.1d,e,f). As with salinity, density has vertically near-homogeneous properties in the Southern Ocean (especially the Atlantic sector). The Nordic Seas (Figure 8.1d; see Subsection 8.1.4 for geographic details of northern Atlantic) are also dense enough (actually denser) to act as a deep water source. The subpolar Atlantic is slightly less dense (about 0.1 kg/m^3) then the deep water, mostly because of temperature (Figure 8.1d). The subpolar northern Pacific is cold, but also fresh (Figure 8.1c), with a vertical density difference of about 0.4 kg/m^3 forming a barrier between the surface and deep water (Figure 8.1f).

To summarize, salinity, density, and oxygen data all point to deep water formation in the Southern Ocean and in the northern North Atlantic (including perhaps the Nordic Seas) but not in the North Pacific or Indian Oceans.

8.1.2 Horizontal Deep Flow

If the southward flow of NADW were distributed uniformly across the roughly 2 km deep layer and across the entire width of the Atlantic (around 6000 km), the current speed would be on the order of 0.1 cm/s, assuming the volume transport is no larger than the subtropical gyre volume transport (we will show below that it is somewhat less). This is a little small to observe directly. However, Stommel (1957) speculated from theoretical considerations that there should be western boundary currents associated with the NADW flow (see also Chapter 5). Such currents should concentrate the transport in a band only about 100 km wide. Floats dropped to about 2.5 km depth at 33°N off the east coast of North America drifted southwards at speeds of up to 17 cm/s (Swallow and Worthington, 1957). This is the most famous example of a rare occurrence: an ocean feature predicted theoretically before it was observed in the sea.

Oceanographers have conducted more comprehensive measurements of the North Atlantic's deep western boundary currents. For example, observations near Cape Hatteras show a southward current of around 10 cm/s which is under the Gulf Stream, is over 100 km wide, and clings to the continental slope in a layer extending from 1 to 4 km depth (Figure 8.2a,b, from Pickart and Smethie, 1993). The velocity profile was obtained by combining hydrographic data for the vertical shear of geostrophic velocity with data from "Pogo floats," which measure depth-average velocity by being dropped off a ship,

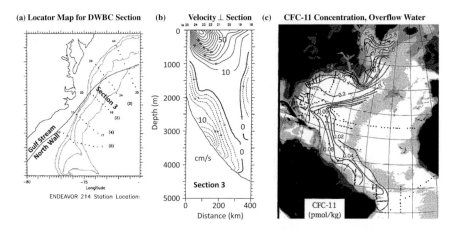

(a) Locator Map for DWBC Section   (b) Velocity ⊥ Section   (c) CFC-11 Concentration, Overflow Water

Figure 8.2   Portrait of a deep western boundary current. (a) location of section. (b) Geostrophic velocity perpendicular to the section (roughly northeastward) as a function of depth and distance along the section. From Pickart and Smethie (1993), Figures 1 and 3, ©American Meteorological Society. Used with permission. (c) Time-adjusted CFC-11 concentration for depth and density corresponding to Overflow Water From Smethie et al. (2000), Figure 5b, ©2000 American Geophysical Union. All rights reserved. Used with permission.

sinking to a specified depth, and returning to the surface a measureable distance from the drop location.

The large-scale extent of the deep western boundary currents can be seen in maps of chlorofluorocarbon (CFC) concentrations. CFCs are human-made chemicals with no known natural sources or sinks in the ocean. They started being manufactured in the early twentieth century for refrigeration and air conditioning. Trace quantities of CFCs find their way into the atmosphere and from there into the surface ocean. In the ocean, they are dissolved in minute concentrations (measured in pmole/kg). Their presence in the deep ocean is a sign that water has travelled from the surface. The distribution of CFC-11 in the Atlantic (Figure 8.2c from Smethie et al., 2000) confirms that water sinks into the deep layers in the northern North Atlantic and flows southward off the western coast. CFC concentration decreases from north to south, which is explained by mixing with neighboring low-CFC water and also because water that is further south is "older" (meaning having spent longer since surface exposure) and was last influenced by the atmosphere at a time when surface CFC concentrations were smaller. The CFC plume is wider than the boundary current in (Figure 8.2b) because eddies and slow permanent currents exchange some deep western boundary current flow with the interior. (The inference about currents and mixing from tracer data, like CFCs, is complicated in general; see Haine and Hall, 2002 and Exercise 8.2.)

In the southern hemisphere, the plume of higher-salinity water associated with NADW (Figure 8.1a) has highest salinities in the west for a wide range of latitudes (Figure 8.3a,b). A latitude–longitude view of the deep salinity maximum (excluding the top kilometer of the ocean) shows a tongue of high salinity water extending southward along the western boundary (Figure 8.3c). This distribution implies that the western boundary current

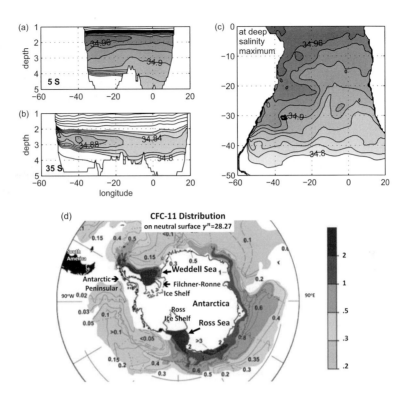

Figure 8.3   South Atlantic annual average salinity fields (from CARS; see Appendix A) for (a, b) longitude–depth slices at latitudes 5°S, and 35°S, and (c) longitude–latitude plot of maximum salinity below 1000 m. Contour interval is 0.02 g/kg ($S > 34.8$ g/kg) and 0.1 g/kg ($S < 34.8$ g/kg). (d) CFC-11 concentration (pmole/kg) depth-averaged over the AABW layer, from Sen Gupta and England 2004, Figure 12, ©American Meteorological Society. Used with permission.

continues further southward, at least to about 35°S. Current meters moored near the western boundary at 26.5°N (Bryden et al., 2005), ADCP sections at 5°S and 11°S (Schott et al., 2005), and hydrography at 19°S and 30°S (Zangenberg and Siedler, 1998) all show southward-flowing deep western boundary currents, although direct current measurements detected only a trace of southern flow at 28°S (Müller et al., 1998). At 11°S the deep western boundary current appears to be breaking into eddies (Dengler et al., 2004) and at 20°S much of the transport seems to turn eastward at the Vitoria-Trindade Ridge which acts as a barrier to deep southward flow within about 500 km of the shelfbreak (Zangenberg and Siedler, 1998).

In contrast to the wind-driven gyres, which have different circulation directions in meridional bands, the maps of CFC and salinity suggest a single current stretching from Greenland to south of the equator. This agrees with the basin-scale, cross-equatorial flow suggested by the salinity distribution (Figure 8.1a).

High CFC-11 values indicate sources of bottom water around Antarctica, especially in the **Weddell Sea** and **Ross Sea** (Figure 8.3d). The Weddell CFC plume extends along the Antarctic Peninsula and is carried eastward by the Antarctic Circumpolar Current (ACC). Measurements with various methods find that volume transports of AABW

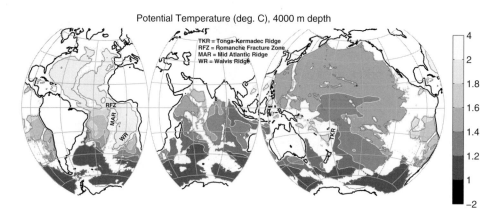

Potential Temperature (deg. C), 4000 m depth

TKR = Tonga-Kermadec Ridge
RFZ = Romanche Fracture Zone
MAR = Mid Atlantic Ridge
WR = Walvis Ridge

Figure 8.4 Potential temperature of water at 4000 m depth, from CARS data (see Appendix A).

or its precursors along the Antarctic Peninsula are in the range of 2–5 Sv. The Ross Sea CFC plume stretches westward, carried by currents near Antarctica which flow counterclockwise around the continent. Some CFC that extends further offshore near the Ross Sea is also carried by the ACC. Chapter 9 discusses Southern Ocean circulation further.

The deepest water is strongly influenced by bathymetry, as we can see in the distribution of temperature at 4000 m depth (Figure 8.4), which traces the flow out of the Southern Ocean of the coldest bottom water (see also Johnson 2008, who uses multiple tracers to map AABW fraction). In the Atlantic, high concentrations of AABW water reach further north to the west of the Mid-Atlantic Ridge than in the East, where the **Walvis Ridge** stretches southeastward from about 20°S on the African coast and blocks northward flow (see Figure 1.9 for a bathymetric map). In the western basin, hydrography (not shown) reveals northward flow off the east coast of South America and there is some evidence that when it crosses the equator it switches to the western flank of the Mid-Atlantic Ridge. At the equator, some AABW leaks into the eastern basin through a gap in the the Mid-Atlantic Ridge called the **Romanche Fracture Zone**. From there it appears to spread both northward and southward into the eastern basins of the Atlantic.

AABW appears to spread into all the deep basins of the Indian Ocean (Figure 8.4). Hydrography shows sloping isopycnals marking presumed northward deep flow off the coast of Madagascar and on the eastern flank of the Ninety-East Ridge in the eastern Indian Ocean (Warren, 1981). CFCs confirm the northward abyssal flow into the southwest Indian Ocean (Haine et al., 1998).

The **Tonga-Kermadec Ridge**, which stretches northward from New Zealand, acts as a western boundary for the deep flow even though it is basically in the middle of the Pacific. A tongue of cold water reach northward along the ridge (Figure 8.4) suggests a "western boundary current." Hydrography at 28°S and later an extensive current meter array at 32.5°S found a northward flowing current below 2000 m (Whitworth III et al., 1999; Warren, 1973). Hydrography and moored current meters found northward flow continuing in and near the Samoan Passage (Roemmich et al., 1996), which is north of Samoa at about 10°S, 170°W.

Subsection 8.1.3 has a more quantitative discussion of the volume transports associated with these deep flows. Then, we look at the sources of water sinking into the deep ocean.

8.1.3 Deep Meridional Overturning Circulation

Oceanographers have found it useful to think about the deep circulation discussed in the previous section in terms of a meridional overturning. While the circulation has interesting zonal structure, including features such as western boundary currents, the concept of the **Deep Meridional Overturning Circulation** captures the core idea that water is sinking at high latitudes, flowing away from the deep water formation sites below the thermocline, rising, and eventually returning to the sinking sites. As in Chapter 6, we use a meridional overturning streamfunction to characterize the overturning, thus describing the three-dimensional flow with a single scalar function of latitude and depth (and perhaps time, if we are interested in variability in the overturning). As in the local Ekman-driven cells described in Chapter 6, the overturning is closely related to climate processes such as transport of heat, carbon, and other sea water properties (see Chapter 11).

Starting in the 1980s, many modeling and theoretical studies have examined the overturning, and so it is common to see graphs of meridional overturning streamfunction. Unfortunately, constructing the meridional overturning streamfunction for the real ocean is more difficult than measuring the relatively strong velocity in a western boundary current or using wind stress to calculate it for an Ekman layer. Calculating the meridional overturning streamfunction is hampered by the same problems hampering calculation of the gyre streamfunction (Chapter 3): the difficulty of measuring absolute velocity and the large number of observations needed to properly average out variability associated with eddies, interannual variability, etc. Nonetheless, estimates as early as that of Sverdrup et al. (1942) (their Figure 187 in Chapter XV) show a net volume transport from the subtropical gyre to the subpolar gyre in the top kilometer of the North Atlantic.

To calculate the streamfunction, we need the zonal average meridional velocity as a function of depth at different latitudes. An estimate of this quantity is based on geostrophic calculations from the density field, surface height, and estimates of the level of no motion, and on Ekman velocity calculated from wind measurements. It is important to have measurements with high enough spatial resolution to capture the 100 km wide western boundary currents. This reduces the usefulness of temperature and salinity atlases which smooth the data enough to remove signals at these scales. Thus the basin-wide velocity calculation has only been performed for a few latitudes where hydrography measurements have provided the density field for the entire width of individual basins.

Such a calculation (Talley et al., 2003) shows that, for the Atlantic, a meridional overturning streamfunction can be constructed that is consistent with the picture we inferred from property distributions discussed in Subsection 8.1.1. The streamfunction is dominated by a northward flow in the top kilometer, southward flow (of NADW) at about 1500–4000 m, and a weaker cell with AABW flowing northward at the sea bottom (Figure 8.5a). The figure shows that the NADW cell has the same strength at all the latitudes where velocity was measured, implying that downwelling must occur north of about 55°N and

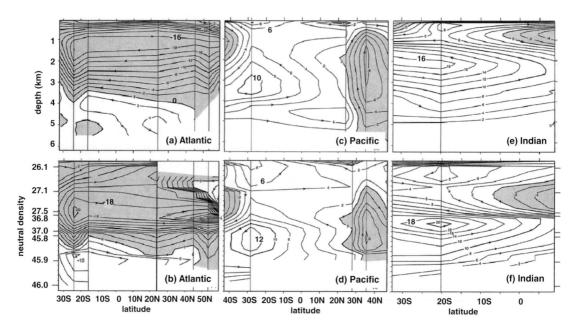

Figure 8.5 Meridional streamfunction for (top) depth-coordinate overturning and (bottom) density-coordinate overturning and for (left) Atlantic, (middle) Pacific, and (right) Indian oceans. Contour interval 2 Sv, gray denotes clockwise rotation. From Talley et al. (2003), Figure 2, ©American Meteorological Society. Used with permission.

upwelling must occur south of 24°S or in other oceans. When we calculate zonal average velocities along isopycnals rather than depth levels, the density-coordinate overturning shows a similar magnitude and shape (Figure 8.5b). In both the Pacific and Indian Oceans, the most prominent overturning cell consists of AABW flowing northward near the bottom, rising, and returning southward at 2–3 km in the Pacific (Figure 8.5c) and 1–2 km in the Indian (Figure 8.5e) Oceans. The strong overturning cell confined to midlatitudes of the northern Pacific runs counter to the picture we have from mixed layer properties, which implies no source of water dense enough to sink to the bottom, and tracer properties, which implies that the deep water in the apparent cell should not come from a local downwelling. Several (though not all) other estimates of North Pacific deep velocity do not show a sign of the cell, which may be an artifact caused by time variability (Talley et al., 2003).

Numerous estimates of Atlantic volume transport have converged on an upper cell (southward NADW flow compensated by lighter and shallower northward flow) strength over a wide range of latitudes of about 15 Sv with uncertainty of a few Sverdrups. The Atlantic AABW cell strength is 2–7 Sv. The Rapid Climate Change/Meridional Overturning Circulation/Heat flux Array was created to continuously monitor the Atlantic at 26.5°N with multiple moorings across the basin to measure currents, temperature, and salinity. That latitude has a long history of measurements of Gulf Stream flow in the 700 m deep channel between Florida and the shallow bank on which the island of Abaco in the Bahamas sits and numerous measurements east of Abaco. Based on four years of measurements, the time average strength of the NADW cell there was 18.7± 2.1 Sv. The overturning has a seasonal cycle of about 6.7 Sv between an autumn maximum and a spring minimum which appears to be associated with wind curl in the eastern basin (Kanzow et al., 2010). The large size of the seasonal cycle points to the danger of basing an estimate of climatological properties from measurements taken in a single season.

While CFC distributions suggest a relatively simple deep western boundary current connecting the Labrador Sea with the South Atlantic (Figure 8.2c), current measurements show strong northward countercurrents offshore of the deep western boundary current. Current meter measurements find 35 Sv going southward within 160 km of Abaco and about 13 Sv going northward further offshore to about 600 km (Bryden et al., 2005). Similarly, repeat ADCP sections at 11°S show large NADW recirculation and eddy influences, with about 36 ± 15 Sv flowing southward and 11 ± 16 Sv flowing northward (Schott et al., 2005, their Figure 7b). Hydrographic estimates (Zangenberg and Siedler, 1998) show southward NADW transport switching from flow west of the Mid-Atlantic Ridge at 19°S (22 Sv of 26 Sv) to flow east of the Ridge at 30°S (17 Sv of 23 Sv). These estimates include NADW transport which is compensated by northward flow both above and below.

Due to the overturning, water must cross from the subtropical to the subpolar gyre in the top kilometer or so of the North Atlantic. This looks plausible in surface geostrophic flow (Figure 3.3) and the surface circulation estimated by drifters (Figure 6.6), but among about 1000 near-surface drifters deployed in the North Atlantic, only one was observed to cross between the gyres (Brambilla and Talley, 2006; Burkholder and Lozier, 2011). Part of the reason is southward Ekman transport deflecting the drifters from the gyre boundary, but simulated drifters in a North Atlantic simulation indicate that most of the water crossing the gyre boundary travels upward along the sloping density interfaces from hundreds of meters depth (Figure 5.1a). Shallow floats are trapped in the subtropical gyre by the isopycnal bowl (Figure 5.3).

There is much greater uncertainty and disagreement in the overturning circulation of other oceans: for instance bottom flow into the Indo-Pacific has been estimated to be about 15 ± 4 Sv (Ganachaud and Wunsch, 2000), 20 Sv (Schmitz, 1995), 27 Sv (Lumpkin and Speer, 2007; Talley, 2008) and 49 Sv (Sloyan and Rintoul, 2001). Reflecting uncertainty over cold route/warm route transports (see Subsection 9.1.1), it is not clear how much of the return of deep water to the thermocline occurs in the Indo-Pacific and how much occurs in the Southern Ocean. Moored current meter measurements of deep western boundary currents (see Subsection 8.1.1) give additional evidence of significant flow into the Pacific from the Southern Ocean. Transport is estimated to be 16.0 ± 11.9 Sv along the Tonga-Kermadec Ridge at 32.5°S (Whitworth III et al., 1999) and 10.6 ± 1.7 Sv in the vicinity of the Samoa Passage at about 10°S (Roemmich et al., 1996, see Figure 8.4 for locations).

8.1.4 Deep Water Formation

Subsection 8.1.1 shows that the properties of deep water can be explained by connections to near-surface water in high-latitude regions of the North Atlantic and the Southern Ocean. Here we look at the connection between shallow and deep water in the regions of deep water formation.

Deep Convection

It is important to distinguish vertical *mixing* and vertical *mass transport*. In the northern North Atlantic during winter, strong convective mixing reaches to great depth. Convective currents carry water downward in plumes that are hundreds of meters wide while somewhat broader currents return deep water to the surface (Schott and Leaman, 1991). Such currents can be locally strong, with vertical velocities as high as 10 cm/s, and can drive strong vertical fluxes of heat, salt, and tracers. A region of convection does not necessarily transport mass however, because the downward flow in the dense plumes may be balanced by upward flow of lighter water in between. The most obvious effect of convection is to allow the surface heat loss to cool and deepen the mixed layer during the winter. Warming then restratifies the deep mixed layer during the summer (Subsection 4.2.2). In general, convection has a less direct effect on the large-scale flow than it does on the annual mixed layer evolution.

Nevertheless, deep convection provides a mechanism for the deep water properties to be influenced by surface conditions, whether or not the net downward flow is zero. In fact, net deep sinking does appear to occur in the vicinity of deep convection regions. As Subsection 8.2.1 explains, there are theoretical reasons why we expect deep convection to be related to deep sinking. For all these reasons, it is worth noting the distribution and characteristics of deep convection.

The complex geography of the northern Atlantic provides an important constraint on the behavior of the regions of deep convection (Figure 8.6). The **Greenland–Iceland–Scotland Ridge** separates the open North Atlantic from the **Nordic Seas** to the north. The

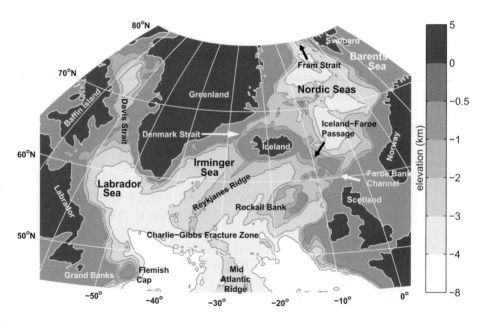

Figure 8.6   Bathymetry (km) and labels for major features in the northern North Atlantic Ocean and adjacent seas, based on ETOPO5 data (1/12° resolution, see Appendix A) with 5×5 gridpoint smoothing.

deepest passages through the Greenland–Iceland–Scotland Ridge are the **Denmark Strait** (about 600 m deep) the **Iceland–Faroe Passage** (about 500 m), and the complex undersea valley of the **Faroe Bank Channel** (about 800 m). The Nordic Seas themselves reach depths of 3 km, but their only deep passage to the Arctic Ocean is through the relatively narrow **Fram Strait** to the north; the wider boundary with the shallow **Barents Sea** to the east provides another route to the Arctic. The western subpolar Atlantic is called the **Labrador Sea**, which connects to the Arctic via the relatively shallow **Davis Strait** and even shallower passages through the **Canadian Arctic Archipelago** (see also Figure 10.1a).

The Nordic Seas are colder and denser than the subpolar Atlantic, and also have weaker stratification. There are indications that in the central Greenland Sea, convection has sometimes reached the sea floor. The strongest circulation goes around the periphery of the Nordic Seas (Mauritzen, 1996), with the deepest mixing occuring in a region of weaker horizontal flow. South of the Greenland–Iceland–Scotland Ridge, the central Labrador Sea has the deepest convection, typically 1000 m but ranging from a few hundred to two thousand meters in different years (Yashayaev, 2007). Convection in the **Irminger Sea** (Figure 8.6), also subject to great interannual variability, can be as deep as 1000 m (Pickart et al., 2003). The rest of the subpolar ocean is more stratified due to relatively warm and salty water delivered from the subtropical gyre by the North Atlantic Current.

The Mediterranean Sea also undergoes some of the deepest convection even though minimum temperatures are around $12°C$. Deep water from high latitudes has no way to penetrate the Strait of Gibraltar, which has a sill depth of 300 m. South of Marseille, France, at the northwest edge of the sea, winter convection reaches the bottom (MEDOC Group, 1970), about 2 km there. The Red Sea, another deep basin with a sill blocking deep access to the rest of the ocean, also appears to undergo some deep convection (see Cember, 1988 for a review). And deep homogenized layers have been observed in the central Weddell Sea, suggesting recent convection (see Killworth, 1983, for a review).

In contrast, the winter convection in the North Pacific Ocean generally reaches to a depth of just a few hundred meters. This fact reflects the stabilizing influence of the vertical salinity gradient in the North Pacific, compared to the North Atlantic, for example (Figure 8.1a,c).

Overflows

Plumes of dense water flowing down sloping topography provide another way for water to sink, and are a significant source of deep and bottom water. They are the ocean's version of river rapids and waterfalls, but with the stream typically tens of kilometers wide and hundreds of kilometers long. They are influenced by gravity, friction, mixing, and the Earth's rotation (Pratt and Whitehead, 2008). Because of the strong turbulence typically found in plumes flowing down a sloping bottom, the plume entrains the surrounding water (draws it into the plume). The plume can also detrain water (transfer in the opposite direction). Tracing the flow downstream, entrainment typically increases the volume transport and decreases the anomalies in T, S, and σ_0 of the plume.

Some bottom plumes originate on continental shelves. In these shallow regions, it is relatively easy for the mixed layer to reach the sea floor (typically about 200 m).During

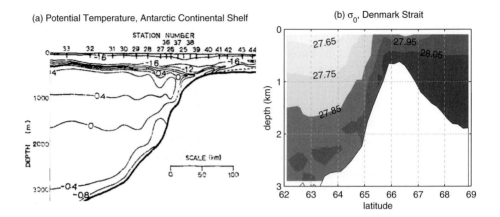

Figure 8.7 (a) Temperature cross section at the edge of the Antarctic continental shelf in the southern Weddell Sea, 40°W. Reprinted from Foster and Carmack (1976), Fig. 4a, Copyright 1976, with permission from Elsevier. (b) Maximum density (σ_0) at each latitude and depth in the vicinity of Denmark Strait, based on WOD 2013 individual profiles and CARS gridded data (see Appendix A and Exercise 8.3).

winter the shelf density increases without further mixed layer deepening. Geostrophy requires that dense water on the continental shelf associates with currents along the shelf edge, but ageostrophic effects cause some water to flow down the continental slope. The process can be seen in a cold, dense plume of water on the Antarctic continental shelf in the Weddell Sea (Figure 8.7a, from Foster and Carmack, 1976). Similar plumes of shelf water occur elsewhere around Antarctica, including the Ross Sea (Figure 8.3d), the Siberian shelves in the Arctic, and in the northern Adriatic arm of the Mediterranean.

Some of the densest water occurs in relatively small basins such as the Nordic Seas and the Mediterranean Sea, which are isolated from the rest of the ocean by shallow passages. These **marginal seas** import relatively light water near the surface and export dense water in a strong, turbulent bottom current often referred to as an **overflow** or outflow. For example, a reservoir of dense water sits north of the Greenland–Iceland–Scotland Ridge and cascades down the southern face of the ridge into the deep northern Atlantic (Figure 8.7b, which for each latitude and depth shows maximum density in a longitude range from Greenland to about 16° to the east).

Figure 8.7b suggests that the plume is a primarily two-dimensional structure, but a map of maximum density (Figure 8.8a) shows some of the structure in the horizontal plane. Tongues of dense water can be seen extending out of the Nordic Seas in the relatively deep passages of the Denmark Strait, Iceland–Faroe Passage, and Faroe Bank Channel (see Figure 8.6 for a regional map). The Denmark Strait tongue extends down the steep topography south of the Strait and then continues in a direction more parallel to the Greenland continental slope. In this region, the densest water between Greenland and the **Reykjanes Ridge** does not occur on the deepest bathymetry but further up the continental slope at 1500–2500 m. Density cross sections show that the densest water descends and clings to the Greenland continental slope for hundreds of kilometers south of the Denmark Strait

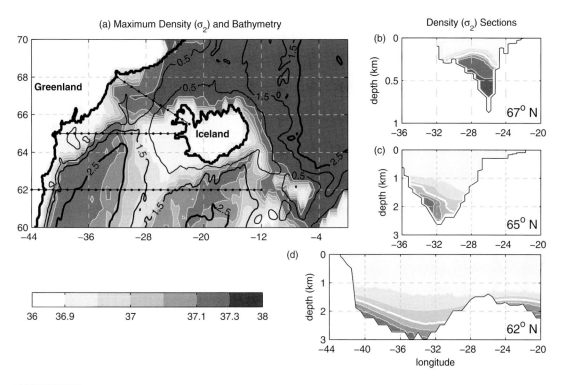

Figure 8.8 Density of Greenland–Iceland–Scotland Ridge overflows (CARS climatology; shading and white contours). (a) Maximum σ_2 and bathymetry (black contours, 0.5, 1.5, and 2.5 km, ETOPO5 with 3×3 smoothing). (b), (c), and (d) σ_2 off the east coast of Greenland along latitude–depth sections marked by black dots in (a); $\sigma_2 = 37$ marked by thick white contour. See Appendix A for data sources.

(Figure 8.8b,c,d). The section at 62°N also shows similar dense water leaning against the eastern flank of the Reykjanes Ridge (Figure 8.8d). This water comes from the more eastern passages and then flows around Reykjanes Ridge and through gaps such as the **Charlie Gibbs fracture zone** (Figure 8.6).

Measurements closer to the Denmark Strait show a strong current flowing southwestward primarily along the western flank of the deepest part of the strait. At the sills, the volume transport of water denser than $\sigma_\theta = 27.8$ is 3.4 Sv in the Denmark Strait, 1 Sv in the Iceland-Faeroes Strait, 2 Sv in the Faroe Bank Channel, and less than 0.3 Sv across the Wyville Thomson Ridge between the Faroes and the Hebrides (Hansen and Østerhus, 2000). Due to entrainment of ambient water by the turbulent plume, these sources grow to 7.3 Sv at 65°N south of the Denmark Strait and 3.5 Sv south of Iceland (Haine et al., 2008). Thus, the source for a significant fraction of NADW can be seen descending into the abyss.

A similar exchange occurs between the dense, salty Mediterranean and the Atlantic (at the Strait of Gibraltar) and between the Mediterranean and the Black Sea, which is fresher due to a large inflow of fresh water from rivers. The **Mediterranean Outflow** is a plume of dense, fast, and turbulent water flowing downward from Gibraltar and reaching

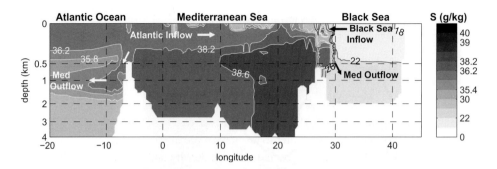

Figure 8.9 Zonal section through Atlantic-Mediterranean-Black Sea system. Salinity maximum (Atlantic and Black Sea) and minimum (Mediterranean) at each longitude and depth. Atlantic sector only includes data between 32°N and 39°N. Plot excludes Mediterranean data east of the Bosphorus Strait. The vertical scale is nonlinear to better resolve shallow depths. All data from CARS climatology annual average data (see Appendix A).

an equilibrium depth of about 1 km along the southern continental slope of Iberia. Volume transport increases by entrainment from 0.9 Sv just downstream of the sill (approximately 400 m depth) to 1.4 Sv about 100 km downstream (Baringer and Price, 1997). Interestingly, the Mediterranean water is denser than Nordic Seas water, but does not penetrate as deeply (Price and Baringer, 1994), as we discuss in Subsection 8.2.5.

A large-scale zonal section through the Mediterranean, Black Sea, and Atlantic shows the exchanges with the Mediterranean (Figure 8.9). The figure shows maximum salinity at each longitude and depth in the Atlantic and Black Sea in order to capture the salty outflows, and minimum salinity in the Mediterranean in order to capture fresh inflows. Compared to the Atlantic, the Mediterranean is salty and dense while the Black Sea is very fresh and light. The Mediterranean plume of salty water, barely resolved in the 0.5° resolution climatology used here, can be seen descending to about 1200 m in the Atlantic (note that the figure distorts the plume's flow along the Iberian continental slope) while surface tongues of fresh water can be seen entering the Mediterranean from the other basins. The Mediterranean plume in the Black Sea is not obvious in this figure but accounts for the large vertical salinity gradient there. The sill depth for the Strait of Istanbul (Bosphorus) is shallower than the Strait of Gibraltar and the flow is weaker (Iorio and Yüce, 1999).

8.2 Concepts

Oceanographers are less certain about the mechanism of the deep meridional overturning circulation than they are about the gyres. Two different mechanisms have been shown to be important in driving overturning in numerical models of the ocean, and likely the real ocean too. Here we discuss the overturning as a product of buoyancy (thermohaline) forcing and interior mixing. Subsection 9.2.2 introduces the idea that Ekman pumping from wind in the Southern Ocean can drive the deep overturning.

8.2.1 Thermohaline Overturning

Until about the year 2000, the term **thermohaline overturning** was considered syn-
onymous with the deep meridional overturning circulation (Wunsch, 2002). The term
represents a view of the driving mechanism of the overturning that we will examine here.
It refers to the role of temperature and salinity variations in causing the overturning circulation through their effect on density. We use the terms thermohaline, density-driven, or
buoyancy-driven to refer to the same circulation, because of the close links between the T
and S fields and density, and between density and the buoyancy. **Buoyancy** b is defined as

$$b = -\frac{g\rho'}{\rho_0}, \qquad (8.1)$$

where ρ' is the departure of density from a reference value ρ_0 (b is also the reduced gravity). We will explore the connection between density and overturning through a series of
fluid dynamics scenarios of increasing relevance to the real ocean. As discussed in Subsection 9.2.2, wind is also thought to be important for the deep meridional overturning
circulation, but in this section we concentrate on the thermohaline driving.

   Before beginning, consider that the very idea of speaking about internal ocean processes
driving a steady state circulation is ambiguous. By driving, we mean a one-way causal
relationship. For a hydrostatic and geostrophic fluid, there exists a diagnostic link between
density, pressure, and velocity (Subsection 2.1.2). That means that these fields mutually
influence each other to create a steady state. Nevertheless, to some extent we can isolate
some processes as particularly important for the overturning circulation. They include turbulent mixing and the dependence of surface radiative balance on latitude. These processes
are not strongly influenced by the overturning itself. Of course, the circulation is part of
the larger ocean–atmosphere–climate system, so they are all ultimately connected. Yet, it
is still useful to mentally isolate the main processes in order to estimate the circulation's
sensitivity to them.

Adjustment in a Non-rotating System

Perhaps the simplest example of a density-driven flow is a (non-rotating) laboratory basin
in which water of two different (but similar) densities are initially side-by-side (Figure 8.10a). This configuration represents the presence of cold, dense water at high latitudes
in the real ocean and warm, light water at low latitudes. The initial configuration has a
wall separating the two regions. Essentially, this experiment was conceived by Luigi Ferdi-
nando Marsili in 1679, after considering reports of fishermen that the flow in the Bosphorus
changes direction with depth (Figure 8.9).

   The bottom pressure is given by $g\rho_L h_L$ in the light fluid and $g\rho_D h_D$ in the dense fluid,
where ρ_L and ρ_D are the two known densities, and h_L and h_D are the heights of the fluid.
We assume that ρ for each parcel of water stays constant (true if we ignore mixing, which
is not terribly accurate for a table-top basin) but we allow h_L and h_D to vary in time, with
initial condition of $h_L = h_D$. When the wall is first removed, at every depth pressure is
greater in the dense fluid than in the light fluid. Indeed, the pressure gradient increases

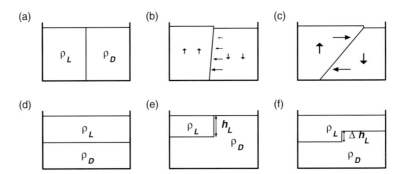

Figure 8.10   Schematic of influence of horizontal density gradient in a non-rotating fluid, showing side view of basin with two different densities ρ_L and ρ_D. (a) Initial state with two different densities next to each other. (b) Horizontal pressure gradient and vertical motion associated with initial state. (c) Flow and location of density interface after system begins to evolve. (d) Final equilibrium state. (e) Same as (a) except showing shallow light water resting above deep dense water. (f) Light layer extends across entire basin and pressure gradient is generated only by the step in upper layer thickness.

with depth. Under this force, the water accelerates to the left, faster towards the bottom (Figure 8.10b).

In addition, gravity waves radiate away from the density jump. As they pass, the heights of the fluid adjust (the light fluid rises and the dense fluid sinks) so that the pressure gradient changes sign near the surface. That accelerates the water near the top of the column to the right (Figure 8.10c). Continuity dictates that the volume flux going in each direction is approximately equal, both in the horizontal and vertical directions. This is the essence of a density-driven overturning circulation. In this thought experiment nothing maintains the horizontal density gradient, so friction eventually damps all motion and the light water floats on top of the dense water (in the absence of mixing, Figure 8.10d).

Other initial configurations behave similarly. For instance, the light water can be confined to a shallow layer of thickness $h_L < h_D$ (Figure 8.10e). In this case, the horizontal pressure gradient still extends to the bottom even though the layer of light water does not. When the wall is removed, the deep water below depth h_L flows to the left. Another example consists of a dense bottom layer and a light upper layer. In this case, both sides of the wall initially have both dense and light water, but one side has a thinner light layer and one side has a thinner dense layer (Figure 8.10f). All these models crudely represent an ocean with a surface density that varies with latitude. The last one (Figure 8.10f) shows that some of the dynamics associated with a horizontal *density* gradient can be represented by a layer *thickness* gradient (Chapter 7; Pratt and Whitehead, 2008, Section 1.2).

Adjustment in a Rotating System

Rotation causes different behavior. As an example, the configurations in Figure 8.10a,e,f can represent a channel periodic in the direction perpendicular to the page. The annulus in Figure 7.9 is an example of such a channel. In that case, when we remove the wall separating light and dense water, the Coriolis force pushes the water in and out of the plane of the page. This is the classic **Rossby adjustment problem**, in which the system reaches an approximate equilibrium which, unlike Figure 8.10d, does *not* eliminate horizontal pressure gradients. Instead of inducing meridional motion (as in Figure 8.10c), the pressure step generates a pair of *zonal* geostrophic currents along the pressure step (as well as inertia gravity waves). For the northern hemisphere, the pair consists of eastward flow in the top layer and westward flow in the bottom layer. Density-driven currents are confined to a region whose width is proportional to the **internal deformation radius**,

$$R_D = \frac{1}{f}\sqrt{\frac{gH(\rho_D - \rho_L)}{\rho}} = \frac{\sqrt{\Delta bH}}{f}, \tag{8.2}$$

where f is the Coriolis parameter, H is the thickness scale (slightly different in each example in the figure), and $\Delta b = b_L - b_D$ is the buoyancy difference. The step in density is smoothed out over this scale but does not disappear. The rotating, periodic case does not redistribute density to eliminate horizontal gradients the way the non-rotating case does. Rotation allows permanent horizontal pressure gradients (neglecting friction).

The zonally periodic channel brings to mind the Southern Ocean, but over most of the ocean there are continental boundaries. If we include (again) eastern and western boundaries to the configurations in Figure 8.10, the rotating system becomes more like the non-rotating case. If the Rossby radius is small enough compared to the basin dimensions, there can be a geostrophic flow along the initial pressure step, as in the periodic case. This is illustrated for a case similar to Figure 8.10f, with an initial step in the internal interface height as well as a small step in the top surface. A perspective view shows the initial state (Figure 8.11a) and the geostrophic current soon after the adjustment begins (Figure 8.11b).

Because the basin is closed, the geostrophic currents interact with the basin walls. The flow generates **Kelvin waves** propagating away from the initial pressure gradient along both eastern and western boundaries (Figure 8.11c). While the Kelvin waves propagate away from the initial pressure step in opposite directions (north and south), the current is northward on both walls of the top layer and southward in both sides of the bottom layer. Thus there is a meridional overturning as in the non-rotating case and a flattening of the density interface, but it is confined to the vicinity of the walls, again within one internal deformation radius (Figure 8.10c). Wajsowicz and Gill (1986) show similar behavior in numerical models of the adjustment of a smooth meridional density gradient (Figure 8.11d).

8.2.2 Thermohaline Equilibrium

The Importance of Buoyancy Forcing

In the absence of friction, each system above would continue to experience currents due to reflected gravity waves (non-rotating case), meridional pressure gradients (rotating, zonally periodic channel), or pressure gradients between the walls and a pressure maximum (or minimum) in each hemisphere (rotating, closed basin). The presence of friction would

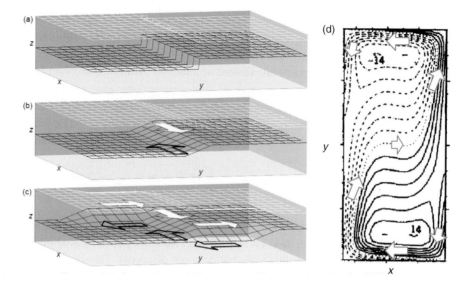

Figure 8.11 Left panels: schematic of gravitational adjustment of two-layer rotating fluid showing density interface height h (black grid), sea surface height η (white grid), and flow in upper (white arrows) and lower (black arrows) layers, for (a) initial condition, (b) early snapshot, and (c) later. (d) η (contours) and schematic surface velocity (gray/white arrows) for rotating shallow water adjustment from initial uniform meridional slope in η, based on Wajsowicz and Gill 1986, Figure 5b; ©American Meteorological Society. Used with permission.

eventually damp all motion so that the final equilibrium state consists of motionless, horizontally uniform, stably stratified fluid. In order for the equilibrium to maintain a density gradient that continues to drive circulation, there must be a mechanism for density of individual water parcels to change.

As an example of such an equilibrium, consider the enclosed (non-periodic) system with a relatively thin layer of light fluid (Figure 8.10e). We replace the region of initial low density with a region in which density is continuously lowered toward $\rho = \rho_L$, for instance by heating the fluid. At the same time, we somehow cool water of depth less than h_L in the other half of the basin so that it stays close to $\rho = \rho_D$. This maintains a pressure gradient which drives an overturning circulation (as in Figure 8.10c). Individual parcels undergo a cycle of density change, cooling as they flow horizontally near the surface from the ρ_L region to the ρ_D region, and warming as they rise into the ρ_L region from below. An example of a warming and cooling law is to represent the density evolution of the parcel by

$$\frac{d\rho}{dt} = \lambda(\rho_* - \rho), \tag{8.3}$$

where ρ_* is the target density (either ρ_L or ρ_D depending on location). The parameter λ determines how quickly the parcel's density changes and hence how close to the target density each region remains.

In the real ocean, there is no process that corresponds to (8.3) throughout the water column. On the other hand, (8.3) is a good representation of *surface* heating and cooling in the ocean by radiation and latent and sensible heat exchange with the atmosphere. Arguably, the depths to which (8.3) directly applies only includes the mixed layer, typically a few tens of meters thick in most of the ocean (Chapter 4; Figure 4.9). The thickness h_L of the light layer is important for the magnitude of the velocity, which we can see by noting that the maximum horizontal pressure difference associated with the light fluid is

$$\Delta p = g h_L (\rho_D - \rho_L), \tag{8.4}$$

and by assuming that current speed increases with pressure difference. As Figure 8.1 shows, large horizontal density gradients in the real ocean extend down to depths of hundreds of meters. If these density gradients are responsible for overturning circulation, it is important to understand the mechanism(s) by which meridional density gradients at the surface creates meridional density gradients below the mixed layer.

Joint Contribution of Buoyancy Forcing and Mixing

Small-scale mixing provides a potentially important mechanism for deepening horizontal density gradients. We can see a steady state with this mechanism in a numerical simulation of a single-hemisphere basin on a rotating sphere (Figure 8.12). The only surface forcing is by restoring the surface temperature to a meridionally varying profile. With a surface temperature range of about 30°C, a basin width of 60° longitude, and a uniform vertical diffusivity of 0.5×10^{-4} m²/s, and a linear equation of state (see (1.6)), the system develops a thermocline of about 800 m thickness and an overturning of about 10 Sv. The mixed layer deepens as one approaches the northern boundary, like the real North Atlantic (Figure 8.1 and Figure 4.9). Like the real ocean, there is a strong asymmetry between sinking and rising, with sinking confined to a narrow region near the maximum surface density and upwelling more broadly distributed.

The single-hemisphere example illustrates the two essential elements of thermohaline theories of the large-scale circulation: horizontal density gradients and downward diffusion of surface density via turbulent mixing. In the limit of no mixing (and no penetration of light into the ocean), density variations are confined to the surface and produce negligible

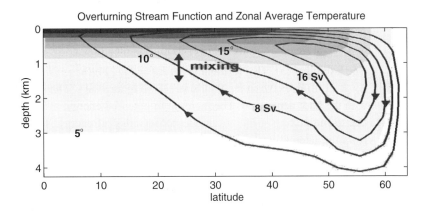

Figure 8.12 Meridional overturning streamfunction (contours) and temperature (shading) from numerical solution for a single-hemisphere basin on a rotating sphere forced by restoring surface temperature to a profile with a 30° range.

pressure gradients, velocities, and volume transports. Vertical (or more generally, cross-isopycnal) mixing is necessary to generate substantial currents. At the same time, the density variations are also necessary, because without them mixing would only produce a horizontally uniform stratification that would not generate any motion. The horizontal density gradients in the interior are required for horizontal pressure gradients, and hence for currents, that vary with depth.

Energy Source of Buoyancy-Driven Circulation

The concept of **available potential energy** gives us insight into the importance of mixing in a steady-state thermohaline circulation. The potential energy of a mass M at height z is Mgz, and we define E_P to be the potential energy integrated over the entire volume of water. The evolution shown in Figure 8.10a–d is an example of a state of initially high E_P (due to dense water having a higher center of mass) converting E_P to kinetic energy as the system evolves to a state with low E_P. Viscosity eventually removes the kinetic energy (after a long time if viscosity is small) and converts it to heat (see Exercise 8.7). For typical oceanographic applications, the heat generated in this way makes a negligible contribution to temperature and density. For the sequence in Figure 8.10, the high potential energy is given as an initial condition, but in an equilibrium state, we need a mechanism to continually resupply the E_P of the system as it is converted to other forms.

The transformation of energy is summarized by a budget equation for E_P (Huang, 1998; see Subsection 8.3.1 for a derivation). For a closed basin in equilibrium, the budget is a balance between the creation of E_P by mixing and removal by conversion to the kinetic energy of the overturning circulation. A separate budget can be written for kinetic energy, for which the conversion of potential energy to kinetic energy is balanced by dissipation due to friction.

It may seem strange that mixing (including diffusion) can create potential energy, but a simple tabletop experiment illustrates the process. Consider a tank of fresh water of height H, surface area A and density ρ_0 with a layer of salty water at the bottom that has a thickness h and density $\rho_0 + \Delta\rho$. After a long enough time, the salinity mixes to make the whole tank have a uniform density $\rho_0 + (h/H)\Delta\rho$. The potential energy is given by

$$E_P = Ag \int_{-H}^{0} \rho z \, dz \tag{8.5}$$

(see Exercise 7.3). This formula gives $E_P = (1/2)Ag(h^2 - 2hH)\Delta\rho$ in the beginning and $E_P = (1/2)AgHh\Delta\rho$ in the final state, if we ignore the unchanging component associated with the base density ρ_0. The increase in potential energy, $\Delta E_P = (1/2)(H - h)hAg\Delta\rho$, due to the mixing seems to violate conservation of energy. The potential energy comes from the internal energy of the water, which decreases by a miniscule amount when E_p increases. Loss of turbulent kinetic energy associated with the mixing can also occur.

On oceanic scales, the mixing from molecular diffusion alone is quite small; turbulence greatly strengthens the mixing. Therefore the energy for the thermohaline circulation comes from processes that drive turbulence, notably internal waves generated by wind forcing at the surface and tidal currents interacting with bottom topography (Munk and

Wunsch, 1998; Wunsch and Ferrari, 2004; see also Chapter 7). This is in contrast to the atmosphere. Like the ocean, the atmosphere has horizontal density contrasts which drive the general circulation. Unlike the ocean, the atmosphere is a heat engine (Peixoto and Oort, 1992, Chapter 14), that is, a system that converts temperature contrasts into motion. The kinetic and potential energy of the atmosphere is supplied by the radiation heating the tropics, not by turbulent mixing from other sources.

In some ways the atmosphere resembles an upside-down ocean, with warming and cooling occurring at the atmosphere bottom/ocean top (here we ignore the complex horizontal circulation of the atmosphere). Warm air rising in the tropics is analogous to cold water sinking in the polar zones, and sinking midlatitude air is analogous to broad upwelling in the ocean. The atmospheric analog to the thermocline is a layer of cold air at the bottom of the atmosphere in mid-latitudes[1]. In the enclosed basin discussed earlier in this section, the large thermocline thickness is due to mixing, which is the only process that can change the temperature or salinity of a water parcel below the mixed layer. In contrast, electromagnetic radiation can propagate much more effectively in the atmosphere than in the ocean. Thus the atmospheric analogue to heating of deep water parcels by mixing is the cooling of tropospheric air parcels by emission of radiation. A fundamental property of a heat engine is that the heating occurs at a higher pressure than the cooling; its application to the atmosphere is known as **Sandström's Theorem** (Houghton, 1986; Huang, 1999, Section 1.4). The atmosphere can satisfy this property because the net heating (averaged over the entire Earth) occurs at the surface and the net cooling occurs at higher altitudes.

8.2.3 Overturning Strength

Scaling Laws, Rotation, Sphericity

The ideas of the previous section can be used to deduce scale relations for pycnocline thickness D, vertical and horizontal components of velocity W and V, and volume transport Ψ for the thermohaline overturning circulation shown in Figure 8.12. The scaling is based on the assumptions of thermal wind, continuity, and the **vertical advective diffusive balance** which states that the decrease of a water parcel's density as it rises up through the pycnocline is caused by vertical mixing. The advective-diffusive balance is necessary for the system to be in steady state.

The complete scaling is derived in Subsection 8.3.2. Here we look at the dependence of overturning characteristics on surface density range $\Delta\rho$ and vertical diffusivity κ_v. The strength of the overturning increases with both parameters, with W proportional to $(\kappa_v^2 \Delta\rho)^{1/3}$, V proportional to $\kappa_v (\Delta\rho^2)^{1/3}$, and Ψ proportional to $(\kappa_v^2 \Delta\rho)^{1/3}$. This is consistent with the idea that both surface density gradients and vertical diffusivity control the overturning in steady state. These different measures of flow have different dependence on the parameters because V, W, and Ψ all differ by factors of D, which is proportional to $(\kappa_v / \Delta\rho)^{1/3}$. The D scaling is consistent with the idea that more mixing thickens the pycnocline.

---

[1] This argument applies to potential temperature. In situ temperature decreases with height due to pressure effects.

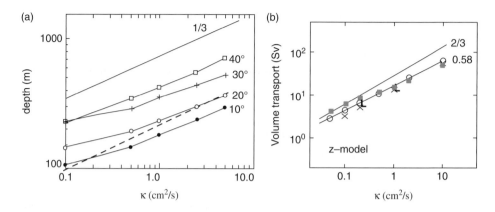

Figure 8.13 (a) Pycnocline thickness as a function of vertical diffusivity κ_v for a series of numerical experiments. From Bryan (1987), Figure 4, ©American Meteorological Society. Used with permission. (b) Peak volume transport calculated from depth-coordinate streamfunction (circle and "x") and from density-coordinate streamfunction (square), based on Park and Bryan, 2000, Figure 11a with data from Figure 7 inserted. ©American Meteorological Society. Used with permission.

A density-driven overturning in a non-rotating system can be easily produced in a small tank of warm water in which ice cubes float at one end. Equilibrium states have been studied in the laboratory (Rossby, 1965; see Hughes and Griffiths, 2008, for a review of recent work). An analogous scaling can be derived for overturning in this system too. The parameter dependence is different for the non-rotating state and is sensitive to the viscosity (unlike the scaling for a rotating system). Unlike the wind-driven gyres, the thermohaline overturning scaling for a rotating system is sensitive to f rather than to β, and numerical experiments in fact show little sensitivity to β (Winton, 1996).

Aspects of the thermohaline scaling for a basin on a rotating sphere have been tested with primitive equation numerical models in idealized configurations similar to the one in Figure 8.12. Bryan (1987) finds that D is roughly proportional to $\kappa_v^{1/3}$ (Figure 8.13a), while Klinger and Marotzke (1999) find D and W dependence on $\Delta\rho$ in accord with the scaling. Park and Bryan (2000) find that Ψ has a roughly $\kappa_v^{2/3}$ dependence when volume transport is measured as a function of density, but a bit weaker dependence when measured as a function of depth (Figure 8.13b).

Gyre and Overturning Influence on Pycnocline

Chapter 5 discusses the way in which the wind curl sets the thickness of the pycnocline "bowl" in the subtropical gyres. Thus, we have two measures of pycnocline thickness, one controlled by the wind and the other by thermohaline overturning. Both factors should operate in the real ocean, and the Bryan (1987) experiments show that D is less sensitive to κ_v as κ_v is reduced to 0.1 cm²/s, consistent with the idea that the pycnocline thickness is nonzero even in the limit of no mixing. Chapter 5 also shows that in wind-driven theories, the stratification at the eastern boundary is undetermined; the thermohaline scaling may represent one way to estimate this parameter. Furthermore, some evidence exists for *two* pycnoclines, which may be due to the ventilated mechanism of Chapter 5 and the diffusive mechanism of this chapter (Salmon, 1990; Samelson and Vallis, 1997; Vallis, 2000; see Exercise 8.11).

Comparison: Scaling, Models, Observations

When applied to observed temperature and salinity profiles and chemical tracer data in the deep Pacific (Munk, 1966) the vertical advective-diffusive balance is consistent with a vertical diffusivity of about 1 cm^2/s and upwelling of about 1 cm/day (about 10^{-7} m/s or about 4 m/yr). Integrated over the ocean excluding high latitudes, this is equivalent to about 30 Sv of upwelling. As discussed in Chapter 7, turbulence measurements and tracer release experiments indicate that κ_v has typical values of 0.1 cm^2/s in the pycnocline and 1 cm^2/s in the abyss, but that there are large geographic variations so that the large-scale average κ_v is not known. In the single-hemisphere numerical experiments referenced above, for surface density ranges similar to the real world and κ_v in the 0.1–1 cm^2/s range, pycnocline thickness is several hundred meters and overturning is around 5 to 20 Sv. Thus numerical experiments with parameters not far from real-world values give overturning strength and pycnocline thickness that are within a factor of two or three of ocean observations. In order to attain better accuracy in matching models with observations, we must include the geometrical complications of the real ocean. Because of the central importance of the Southern Ocean to the multi-basin overturning circulation, we discuss more geometrically realistic cases in Chapter 9.

Two Hemisphere Overturning

A two-hemisphere basin introduces an additional complexity: multiple regions of high surface density and potentially of deep water formation. The equatorially symmetric version of the one-hemisphere case produces a symmetric circulation, with identical sinking at the two poleward boundaries. Numerical experiments show that if the maximum surface density is much smaller in one hemisphere, the denser, "dominant" hemisphere will fill the deep water even in the other, "subordinate" hemisphere (Figure 8.14a). The densest water in the subordinate hemisphere cannot penetrate the denser water filling the bottom from the dominant hemisphere, and so any thermohaline overturning associated with sinking in the subordinate hemisphere will be confined to shallow depths. We can measure the asymmetry in surface density by $\Delta\rho_P/\Delta\rho$, where $\Delta\rho$ is the surface density range in the dominant hemisphere and $\Delta\rho_p$ is the difference in maximum surface density between the two hemispheres. Even a very small difference, such as the $\Delta\rho_P/\Delta\rho = 0.02$ in Figure 8.14b, causes the "dominant" cell to fill the deep water of much of the subordinate hemisphere. Pycnocline thickness and total upwelling are insensitive to $\Delta\rho_P/\Delta\rho$, so we can think of changes in the degree of asymmetry as changing the division of the upwelling between the dominant and subordinate cell.

8.2.4 Deep Western Boundary Currents

Previous sections focus on the magnitude of the meridional overturning circulations. Here we consider the horizontal structure of the flow. How does the meridional current depend

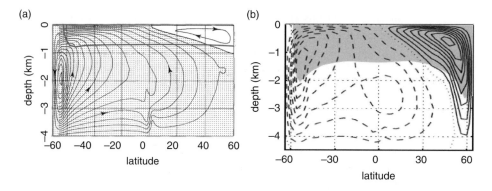

Figure 8.14 Meridional overturning streamfunction for closed basin with $\Delta \rho_P / \Delta \rho$ of (a) 0.2, and (b) 0.02. Contour interval is 2 Sv except for solid contours in (b) which are 1 Sv. Region with density no greater than maximum density of subordinate hemisphere is shown by thick contour in (a) and by shading in (b). (a) from Cox (1989) Figure 5; (b) from Klinger and Marotzke (1999) Figure 2. ©American Meteorological Society. Used with permission.

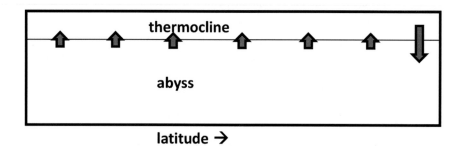

Figure 8.15 Schematic of deep flow showing uniform-density layers and imposed vertical velocity at the interface between layers.

on longitude? Is there any zonal current and, if so, what is it? The concepts discussed here are not definitive answers but give us some guideposts for understanding the real ocean.

Once again we find that Henry Stommel and his collaborators have supplied some key ideas for conceptualizing the flow. They made a simplified model in which the density-driven circulation is calculated based on a uniform-density flow field. In their simplification, the density structure of the ocean is represented by two uniform-density, approximately uniform-depth layers (Figure 8.15). Such a model is already too simplified to express the magnitude of the overturning, which depends on horizontal and vertical density gradients (Subsections 8.2.2 and 8.2.3) However, given the magnitude of a single overturning cell, this model can represent the horizontal flow patterns. We assume that the vertical velocity field at the layer interface is known. In the example shown it consists of a small region of sinking and a broad region of upwelling, which is similar to numerical simulations such as Figure 8.11 and perhaps to the real ocean.

Stommel's insight was to notice that the lower layer in this two-layer model looks similar to uniform-density models of the wind-driven circulation. In the wind-driven case, the forcing can be considered to be vertical flow at the bottom of the Ekman layer driving the geostrophic flow below it. Here the forcing is the vertical flow associated with meridional overturning. Once we make this analogy, we can remove the top layer and model the deep flow with a single uniform-density layer driven by vertical currents at the surface.

In the gyres, the Sverdrup balance determines the meridional velocity away from the western boundary current. Downwelling produces equatorward flow and upwelling produces poleward flow. For the case shown in Figure 8.15, this produces a surprising result: over most of the width of the basin, water is flowing *towards* the source of deep water even though the zonal integral of the deep flow must be *away* from the source. As in the wind-driven circulation, there is also a frictional western boundary current. This should have the correct direction and magnitude to satisfy continuity given the upwelling and downwelling imposed at the surface of the layer and the Sverdrup-like interior flow.

Stommel et al. (1958) found the flow patterns implied by this model "so bizarre and contrary to intuition that it would be reassuring if they could be partially tested by [laboratory] model experiments." The β effect associated with the sphericity of the Earth is an essential part of the Sverdrup balance, but it is extremely difficult to make a laboratory apparatus in which the ocean stays on the surface of a sphere. However, a gradient in layer thickness, for instance due to bottom topography, produces an analogue of the β-effect (called the **topographic β-effect**, see Exercise 8.5). In a rotating laboratory tank, the direction towards smaller layer thickness corresponds to north on a rotating sphere. Henceforth we will use latitude–longitude terminology for the simulated compass directions. The thickness gradients in the Stommel et al. (1958) experiments were due to the free surface sloping upward from the middle of the tank due to centrifugal force. Stommel et al. (1958) imposed vertical flow in small regions at the surface by inserting tubes with water slowly pumped in or out of the basin. A broadly distributed vertical velocity could be created by having an unbalanced source or sink. Surface upwelling, for example, is provided by an unbalanced source which causes the water surface to rise over time. Strictly speaking this is not an equilibrium state but for a slow enough filling time the circulation is close to equilibrium.

An equal source and sink placed at different latitudes has no meridional flow between them because the vertical velocity is zero everywhere except at the source and sink themselves. Yet continuity demands that there be a net flow of water from the source to the sink. The resulting circulation is the circuitous path in which water flows from the source westward to the western boundary, south along the western boundary, and finally eastward to the sink (Figure 8.16, top left). An experiment in which the injected water is dyed shows a similar circuitous path (Figure 8.16, top photos). The zonal currents are broader and more complex than indicated by the schematic but the importance of the western boundary current is evident.

The case of a source at the northern boundary (representing deep water formation) and upwelling everywhere else should produce a northward flow almost everywhere but a southward flow along the western boundary (Figure 8.16, bottom left). Again, laboratory experiments show similar behavior, with the dye first travelling southward along the western boundary current and eventually flowing northward as a slow current across the rest of the basin (Figure 8.16, bottom photos).

The "reverse" flow associated with rotation may seem less strange if we think about the horizontal divergence associated with the imposed vertical current, w. For an upwelling

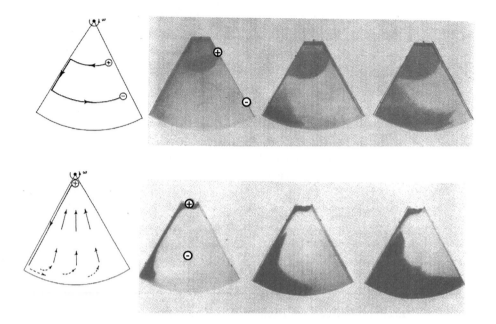

Figure 8.16 Top view of flow patterns in rotating tank with topographic β effect (north is to the top of image) and uniform density for (top) tubes acting as source and sink and for (bottom) single source and distributed sink. The "+" and "−" symbols represent locations of sources and sink, respectively. Left panels show schematic, right panels show photos of dye distributions at 20, 80, and 220 minutes after introduction of dye. From Stommel et al. (1958), Figures 2, 3, 7, and 8; used with permission of Taylor & Francis Ltd.

fluid, water must converge throughout the basin, with a net flow towards some point in the interior of the upwelling region (Figure 8.17). Because of the rotation of the basin, such converging flow circulates cyclonically (counterclockwise for the northern hemisphere, Figure 8.17). On an f plane the flow towards the source and away from the source occupy similar fractions of the basin (Winton, 1996), but the β effect causes westward intensification which breaks the symmetry between flow on the eastern side of the basin and on the western side. It is the combination of the w-driven gyre and the westward intensification that produces the unexpected flow pattern seen in Figure 8.16. The same reasoning works if we reverse the hemisphere or direction of w, as for the wind-driven gyre (Chapter 3).

For the two-layer model (Figure 8.15), the top layer should have the opposite forcing to the deep layer – upwelling stretches water columns in the deep layer but squashes them in the top layer, changing the vorticity in opposite directions. Although the Stommel abyssal circulation is often described as a model of the deep water, it implies that similar circulation patterns in the pycnocline should mirror the deep water. As the pycnocline is much thinner than the abyss in the real ocean, and volume transport is the same in both layers, horizontal velocities associated with the overturning should be larger in the pycnocline.

Numerical models reproduce some of the features seen in both observations and the idealized experiments and theory above. Simulated CFC fields show the southward path of NADW in the deep western boundary current (Sen Gupta and England, 2004) which is similar to observations (Figure 8.2c). CFC in the same model traces AABW flowing away

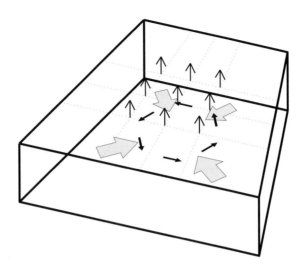

Figure 8.17 Illustration of rotational effects on horizontal upwelling-driven large-scale currents. Three-dimensional perspective sketch of f-plane basin includes vertical flow (upward-pointing arrows at surface of layer), convergent horizontal flow (gray arrows) induced by upwelling, and cyclonic flow (horizontal black arrows) due to the Coriolis effect influencing convergent flow.

from Antarctica along paths that are similar to observations (Figure 8.3d). **Tracer age** is a measure of time it takes for water to flow from AABW source region to each location on the sea floor. It reproduces key inferences from observations (Johnson, 2008), particularly flow into the Pacific east of the Tonga-Kermadec Ridge and flow into the eastern Atlantic to the Walvis Ridge (Sen Gupta and England, 2004). For more on tracer age see, for example, Haine and Hall (2002).

8.2.5 Dynamics of Sinking Regions

Deep convection constitutes a dramatic part of the ocean's overturning circulation, but in reality it plays a small role in the sinking. Most of the convection erodes the seasonal pycnocline (Chapter 4) without net downward flow. Instead, most sinking seems to occur in the outflows from marginal seas and around around their edges. Several questions arise: why is there an abrupt density difference (Figures 8.8 and 8.9) between the marginal sea and lighter water just outside? What determines the density difference? What determines the volume flux exchanging water in and out of the marginal sea? What determines how water properties change as water descends in the outflow? How is inflow water into the marginal sea transformed into outflow water? These questions have not been definitively answered but here we discuss some of the concepts that seem to be most relevant.

Exchange With Marginal Seas

The flow of dense water over a sill such as the Denmark Strait or the Strait of Gibraltar bears some resemblance to the flow of uniform-density water over an obstacle or bump in

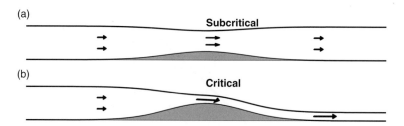

Figure 8.18   Schematic cross sections of flow over a topographic bump (gray) in a channel showing fluid velocity (arrows) and fluid surface height, η (top curve) for (a) subcritical case and (b) critical case.

a non-rotating channel (Whitehead, 1998). If the obstacle is short compared to the thickness of the fluid, then it has a relatively small effect on the flow (Figure 8.18a). Because the bump reduces the thickness of the fluid, the water flows faster at the bump than elsewhere. **Bernoulli's principle**, states that when the parcel's kinetic energy increases, its pressure decreases (neglecting friction). Thus the faster flow over the bump must have lower pressure, which is attained by lowering the height of the water surface η there. The flow downstream of the bump is identical to the flow upstream. This state is called **subcritical** (Figure 8.18a).

A tall bump can behave differently. In some cases, there is no steady flow which satisfies Bernoulli's principle and allows the volume flux, Ψ, at the bump to equal its value upstream. Because of this incompatibility, the system evolves to a different state. The inadequate Ψ at the bump causes water to "pile up" upstream of the bump, which increases the thickness of the fluid and thereby reduces the obstruction. The flow reaches the **critical state** in which η upstream is high enough for the system to reach equilibrium. The system is no longer symmetric and downstream of the bump the water continues to flow rapidly in a thin layer (Figure 8.18b, see also Exercise 8.6). This state resembles the overflows described in Subsection 8.1.4.

In the ocean, the interface in Figure 8.18 does not represent the sea surface but an interface separating denser water from lighter water. Thus the dense flow leaving the marginal sea may be compensated by light flow above it going the other way. Another important difference is the presence of rotation. Rotation causes the opposing currents to cling to opposite walls (as in Figure 8.11). In both rotating and non-rotating systems, shallow and/or narrow straits can limit the exchange flux. This is sometimes referred to as **hydraulic control** or **topographic control** (Pratt and Lundberg, 1991; Whitehead, 1998; Pratt and Whitehead, 2008).

Overflows and Water Mass Properties

When the boundary of the overflow fluid is an interface rather than the ocean surface, mixing across the interface can also affect the dynamics. The densities of the two water masses being exchanged are no longer fixed, but are partly determined by their interaction. Observations show large changes in salinity, temperature, and density in real ocean overflows (Subsection 8.2.5), indicating the importance of mixing. It is not surprising that mixing is strong, as the O(1) m/s currents within tens of meters of quiescent or oppositely flowing water is a good source of Kelvin–Helmholtz instability and turbulence (Subsection 7.1.2).

Process models give some insight into the diversity of overflows. Such models estimate how velocity, mixing, and density of the overflow plume evolve along the the plume. For example, **streamtube models** represent the outflow as a tube of fluid with properties that depend only on distance along the tube. The goal is to predict the density of water in the plume when it reaches its deepest level. Price and Baringer (1994) use a streamtube model to investigate why the Mediterranean Sea, which has *denser* source water than the Nordic Seas, produces an outflow that is *less dense* than those from the Nordic Seas.

The model is applied separately to conditions appropriate to the Mediterranean Outflow (Figure 8.19a) and Denmark Strait Outflow (Figure 8.19d). Starting from the ambient stratification (Figure 8.19b,e), Price and Baringer (1994) compute the density and other properties of the outflow at different locations along its path (Figure 8.19b,e). The model reproduces the observation that the Mediterranean Outflow starts at a high density of

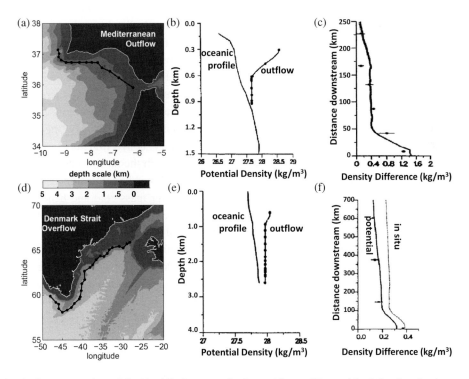

Figure 8.19 Results from streamtube models of (top) Mediterranean Outflow and (bottom) Denmark Strait Overflow showing (left) path of outflow superimposed on local bathymetry, (middle) model-derived outflow density as a function of depth (curves with dots) and density of water adjacent to the plume (plain curves), and (right) model (curves) and observed (dots with error bars) density difference between outflow and adjacent water as a function of distance from the sill. Dots in middle panels correspond to locations marked by dots in right and left panels. Reprinted from Price and Baringer (1994), from Figures 4, 6, 8, and 9, Copyright 1994, with permission from Elsevier.

$\sigma_\theta > 28.5$ and is diluted to about $\sigma_\theta \approx 27.7$, whereas the Denmark Strait outflow is mixed from $\sigma_\theta \approx 28.2$ down to $\sigma_\theta \approx 28.0$. The evolution of density difference between the plume and surrounding water (Figure 8.19f,g) demonstrates the key difference between the Mediterranean and Denmark Strait plumes. At the beginning of its descent from the sill, the Mediterranean water is over 1.4 kg/m^3 denser than ambient water, while the Nordic Seas water is less than 0.4 kg/m^3 denser. The larger density difference drives the larger dilution by stronger entrainment of ambient water. A similar result is found in high-resolution simulations of the Denmark Strait Overflow (Koszalka et al., 2013).

Sinking in Marginal Seas

The downward motion in overflow plumes is one way for water to sink in the deep meridional overturning circulation, but there are others. There seems to be sinking in the Labrador and Nordic Seas, for example. and perhaps elsewhere in the subpolar gyre. Where and how does it occur?

A series of eddy-resolving numerical experiments (Spall, 2003, 2004) show that in a marginal sea with deep convection, the sinking occurs *outside* the convection zone. The model domain (Figure 8.20a) consists of an idealized marginal sea linked to an external ocean by a strait. The overturning circulation is simulated by cooling surface water in a disk at the center of the marginal sea for two months every year, and warming subsurface water in the external ocean to maintain a thermocline. The experiment in Figure 8.20 has a continental slope around the periphery of the domain. Warm water enters the marginal sea, then flows counterclockwise around the basin and cools slightly before exiting (Figure 8.20a,c). The central basin, cooled from above, stays much colder than the periphery. Though the central basin is stratified in the annual average, convection homogenizes it during the winter. The strong vertical motions associated with the narrow convective plumes cause little net sinking because the downward plume flow is largely compensated by upward motion between plumes. They keep the central water mass cold and hence support the cyclonic rim current in thermal wind balance (Figure 8.20c). The cooling of the inflow as it circles the basin is due to lateral eddy mixing with this cold central water (Figure 8.20b). In the marginal sea, average vertical flow is dominated by motion in the boundary current (Figure 8.20d). The vertical flow exhibits both upwelling and downwelling (the flat-bottom case just has downwelling). These idealized experiments suggest that sinking in the meridional overturning circulation occurs near the edges of the marginal sea, rather than in deep convective regions further from the coast.

8.3 Theory

8.3.1 Energetics

Subsections 8.2.1 and 9.2.2 claim that various mechanisms drive the meridional overturning circulation. Confidence in the relationship between a mechanism and the

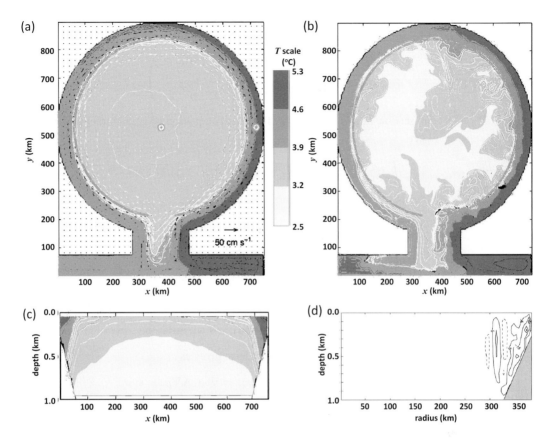

Figure 8.20 Buoyancy-driven flow in a high latitude marginal sea. Upper-level temperature (shading, see figure for scale) for (a) time mean and (b) snapshot 2 months after the end of cooling season. (c) Time mean temperature as a function of zonal distance and depth at mid latitude of the basin. (d) Azimuthally integrated time mean vertical velocity (c.i. = 10 m^2/s; solid contours downward). From Spall (2004), Figures 2a, 8, 4a, and 10; ©American Meteorological Society. Used with permission.

circulation increases by examining the energy balance. It tells us how various forms of energy are transformed and isolates essential processes.

Subsection 8.2.1 discusses the creation of potential energy through mixing for a simple case. Here we write down the potential energy budget in general. Our derivation follows that of Huang (1998). First we define the potential energy for the entire basin by integrating:

$$E_P = \int g\rho z \, dV. \tag{8.6}$$

The part of E_P that can be converted to other forms, for instance kinetic energy, is called the available potential energy (Lorenz, 1955; Huang, 1998; Vallis, 2006). To derive the budget for E_p, we start with the density equation (see Subsection 1.3.1),

$$\rho_t + \mathbf{u} \cdot \nabla \rho = \nabla \cdot (\mathbf{A}\nabla \rho) + C, \tag{8.7}$$

where $\mathbf{u}$ is velocity, C represents convection (which is a nonlinear and nonlocal function of ρ), and $\mathbf{A}$ is the diffusivity tensor. One form of $\mathbf{A}$ is

$$\mathbf{A} = \begin{pmatrix} \kappa_h & 0 & 0 \\ 0 & \kappa_h & 0 \\ 0 & 0 & \kappa_v \end{pmatrix} \tag{8.8}$$

(Subsection 7.3.2). We multiply the density equation by gz and integrate over the domain, to get

$$\int g\rho_t z \, dV = -\int (\mathbf{u} \cdot \nabla \rho) \, gz \, dV + \int \nabla \cdot (\mathbf{A}\nabla\rho) gz \, dV + \int Cgz \, dV. \tag{8.9}$$

Now consider the advection term: continuity gives

$$z\mathbf{u} \cdot \nabla \rho = z\nabla \cdot (\rho\mathbf{u}) \tag{8.10}$$

and the product rule allows us to rewrite this as

$$z\nabla \cdot (\rho\mathbf{u}) = \nabla \cdot (z\rho\mathbf{u}) - \rho\mathbf{u} \cdot \nabla z. \tag{8.11}$$

We use the fact that $\nabla z = \hat{\mathbf{z}}$ (the unit vector in the z direction) so that

$$z\nabla \cdot (\rho\mathbf{u}) = \nabla \cdot (z\rho\mathbf{u}) - \rho w. \tag{8.12}$$

A similar exercise with the diffusion term gives us

$$\nabla \cdot (\mathbf{A}\nabla\rho)z = \nabla \cdot (z\mathbf{A}\nabla\rho) - \kappa_v \rho_z, \tag{8.13}$$

using (8.8). Restrict attention to the case where κ_v is uniform in z so that

$$\int \rho_z \, dV = \iint \rho_z \, dz \, dA = \int [\rho]_B^T \, dA, \tag{8.14}$$

where T and B represent the top and bottom of the domain and $\int dA$ represents area integration over the horizontal extent of the domain. Inserting (8.12) and (8.13) into (8.9) and using the divergence theorem gives

$$\frac{dE_p}{dt} = -\int gz\rho\mathbf{u} \cdot \hat{\mathbf{n}} \, dS + \int g\rho w \, dV$$
$$+ \int gz(\mathbf{A}\nabla\rho) \cdot \hat{\mathbf{n}} \, dS + \kappa_v g \int \rho(B) - \rho(T) \, dA$$
$$+ \int Cgz \, dV, \tag{8.15}$$

where $\int dS$ represents integration over the surface of the domain and $\hat{\mathbf{n}}$ is the unit outward normal to this surface.

Each term in the potential energy budget as written in (8.15) has a physical interpretation. The $\rho\mathbf{u}$ term represents horizontal advection. For a closed basin, there is no advection through the boundaries, so the term is zero. Similarly, the $\mathbf{A}\nabla\rho$ term represents diffusive flux of mass into the basin. This too should be zero for solid walls and should integrate to zero for surface density fluxes for an equilibrium case (otherwise the mass of the ocean would change) assuming that the sea surface is a uniform value of z over the basin. It can be shown that for small variations in z in the real ocean, this term is negligible compared to the ones we discuss below.

Convection is sometimes portrayed as "driving" the meridional overturning, but the potential energy equation suggests otherwise. As $C < 0$ higher in the water column, and $C > 0$ lower in the water column, the vertical integral of Cgz in (8.15) is negative. Therefore convection actually *reduces* the potential energy. See Subsection 4.3.3 for details on this point.

The ρw term is nonzero if vertical velocity is correlated with density. It is positive if dense fluid rises and light fluid sinks, which makes intuitive sense as such a circulation lifts the center of mass of the fluid in an analogous way to the example in Subsection 8.2.1. As motion is needed to make this term positive, it can be interpreted as the conversion of kinetic energy to potential energy. The $\kappa_v \Delta\rho$ term (for $\Delta\rho = \rho(B) - \rho(T)$) represents the diffusive process discussed in Subsection 8.2.1, and so we interpret it as the conversion of energy associated with diffusion to potential energy. If κ_v represents turbulent diffusion, as in the real ocean, the term is another form of conversion from kinetic to potential energy, except this kinetic energy is associated with small-scale turbulent motions unresolved by the model or measurements. The form of the diffusive term is quite simple; for uniform κ_v it is independent of the details of the density profile. Diffusion works to remove any density difference between top and bottom of the water column, which raises the center of mass and hence the potential energy in the usual stably-stratified case $(\rho(B) > \rho(T))$.

In equilibrium, and removing the terms that were declared negligible, (8.15) becomes

$$\int g\rho w \, dV + \kappa_v g \int \rho(B) - \rho(T) \, dA + \int Cgz \, dV = 0. \tag{8.16}$$

Further consideration of the kinetic energy term shows that ρw integrated over the horizontal area of the domain is simply the net vertical mass transport. So if the net vertical mass transport is zero, as one expects at equilibrium, then there can be no net transfer of energy between potential and kinetic. Similarly, the diffusive term in the energy equation is proportional to the vertical mass transport due to diffusion. It is the pair of these terms that allows us to solve the paradox. In equilibrium, there *is* a vertical mass transport due to advection, and an opposite one due to diffusion. This corresponds to the advective diffusive balance discussed in Subsection 8.2.3. Because $\Delta\rho > 0$, diffusion adds potential energy. This can be part of an equilibrium state if $\rho w < 0$, signifying the conversion of potential energy to kinetic energy (recall that the C term in (8.16) is negative). This energy balance is another way of saying that potential energy derived from mixing provides the energy to drive the circulation.

8.3.2 Advective-Diffusive Scaling for Overturning Strength

The problem of deducing the density and velocity fields for a basin driven by surface density variations and by subsurface mixing is difficult to solve analytically. Unlike the gyre, the overturning does not have a linear solution like the Sverdrup balance which sets the magnitude of the volume transport. Unlike LPS theory (Subsection 5.2.2 and Subsection 5.3.1), the nonlinearity associated with the density equation must be coupled to a diffusive process. Though oceanographers have not found a detailed solution to the problem, it is possible (as mentioned in Subsection 8.2.3) to deduce scaling relations that predict key parameters. We consider a basin meant to represent a single ocean such as the Atlantic, with a density gradient between dense polar water and lighter equatorial water imposed at the surface. Parameters we want to know are the pycnocline thickness D, the horizontal and vertical velocity scales V and W, and the maximum volume transport Ψ. To find the three unknowns D, V, and W, we need three equations.

The first scaling relation is based on the density equation (8.7) which we write as

$$\rho_t + u\rho_x + v\rho_y + w\rho_z = (\kappa_v \rho_z)_z + \nabla_h \cdot (\kappa_h \nabla_h \rho). \tag{8.17}$$

Equation 8.17 is derived from similar equations for potential temperature θ and salinity S. It is only true to the extent that diffusivities are the same for both components. and the linear equation of state (1.6) holds.

A subtlety of the density equation is that it hides the surface density forcing. The forcing is present in the boundary condition on ρ. Most commonly the boundary condition is on the density flux Q_ρ

$$-\hat{\mathbf{n}} \cdot \nabla \rho = Q_\rho, \tag{8.18}$$

where $\hat{\mathbf{n}}$ is the unit outward normal to the boundary and Q_ρ is derived from the fluxes of freshwater and heat (Chapter 4). Density flux is zero at all the solid boundaries, and here we assume (unless otherwise noted) that at the sea surface the flux is specified by restoring to a given density profile $\rho_*(x, y)$, with strong enough restoring that the boundary condition is equivalent to approximately specifying ρ at the surface.

We characterize the flow by horizontal and vertical velocity scales (V, W) and horizontal and vertical length scales (L, D) (where L scales the zonal width of the basin). The relative size of horizontal and vertical advection terms in (8.17) is given by

$$\frac{u\rho_x}{w\rho_z} \sim \frac{VD}{WL}, \tag{8.19}$$

and the relative size of diffusion terms is given by

$$\frac{\nabla_h \cdot (\kappa_h \nabla_h \rho)}{(\kappa_v \rho_z)_z} \sim \frac{\kappa_h D^2}{\kappa_v L^2}. \tag{8.20}$$

In order to make the problem tractable, we assume that the horizontal terms are no greater than the vertical; in other words, $VD/WL \leq 1$ and $\kappa_h D^2/(\kappa_v L^2) \leq 1$. Therefore we can assume that the vertical terms are dominant, because adding the horizontal terms only doubles the size of the advective or diffusive terms (compared to vertical only), so for scaling purposes we can then ignore the horizontal terms. In steady state, the density equation becomes

$$w\rho_z = (\kappa_v \rho_z)_z, \tag{8.21}$$

and is often called the **vertical advective-diffusive balance** (Munk, 1966). It represents the fact that the density must decrease at a certain rate as water rises through the pycnocline with vertical speed w, and that the density change must be caused by mixing, which is controlled by κ_v.

The second scaling relation is continuity,

$$\bar{v}_y + \bar{w}_z = 0, \tag{8.22}$$

where the bar represents a zonal average across the basin. The final relation is the thermal wind law for meridional velocity, integrated zonally across the entire basin:

$$fL\bar{v}_z = \Delta b, \tag{8.23}$$

where buoyancy difference $\Delta b = g(\rho(0, y) - \rho(L, y))/\rho_0$, ρ_0 is a typical density, and $x = (0, L(y))$ are the western and eastern boundaries. Except for a small region near the equator, f does not vary by more than about a factor of two, and so we use its mid-basin average. For the scaling analysis, we assume that $\Delta b = g(\rho_{*D} - \rho_{*L})/\rho_0$, where (ρ_{*D}, ρ_{*L}) represent (maximum, minimum) target densities, respectively. It is easy to accept that the difference between polar and equatorial ocean density is approximately imposed by the surface forcing, but not so obvious that the zonal density difference should also be proportional to $(\rho_{*D} - \rho_{*L})$. For instance, it is conceivable that the zonal difference is some power of the meridional difference. Nevertheless, the assumption of proportionality does seem to be accurate in some numerical overturning models (Klinger and Marotzke, 1999).

Equations 8.21, 8.22, and 8.23 imply the scale relations

$$W = \kappa_v/D, \tag{8.24}$$

$$V/L = W/D, \tag{8.25}$$

$$fV/D = \Delta b/L. \tag{8.26}$$

We can solve this system of equations for the unknowns to get (see also Subsection 8.2.3)

$$D = \left(\frac{\kappa_v L^2 f}{\Delta b}\right)^{1/3}, \tag{8.27}$$

$$W = \left(\frac{\kappa_v^2 \Delta b}{L^2 f}\right)^{1/3}, \tag{8.28}$$

$$V = \left(\frac{\kappa_v \Delta b^2}{L f^2}\right)^{1/3}, \tag{8.29}$$

$$\Psi = \left(\frac{\kappa_v^2 \Delta b L^4}{f}\right)^{1/3} = LDV. \tag{8.30}$$

Notice how the small-scale mixing controls the strength of the circulation, via diffusivity κ, and the wind strength is excluded. This result is qualitatively different from the adiabatic (and inviscid) circulation theories in Chapters 3, 5, and 6 (see also Exercise 8.11). For an overturning theory that includes diffusivity and the wind strength, see Subsection 9.3.1.

8.3.3 Horizontal Flow Patterns

The behavior of deep western boundary currents discussed in Subsections 8.1.1 and 8.2.4 can be predicted from theoretical considerations. Following Stommel et al. (1958), we consider the equations of motion for a uniform-density layer of a rotating fluid in which friction can be neglected. We believe this approximation applies to what we called the "Sverdrup interior" in Chapter 3; in other words, outside of the western boundary layer. The derivation is a variant of the Ekman-pumping version of the Sverdrup balance (Subsection 3.2.4, Subsection 3.3.1).

Start with the formula for geostrophic divergence (2.62). If the top and bottom of the layer are at height $h_1(x, y)$ and $h_2(x, y)$, respectively, so that the thickness is given by $H = h_2 - h_1$, the vertical integral of this equation gives

$$\beta v H = f \left[w(x, y, h_1) - w(x, y, h_2) \right]. \tag{8.31}$$

In Chapter 3, we look at the simple case where h_1 and h_2 are independent of (x, y). Here we keep the more general case, though again we assume no flow through the bottom of the layer:

$$w(x, y, h_1) = w_* + \mathbf{u}_h \cdot \nabla h_1, \tag{8.32}$$

$$w(x, y, h_2) = \mathbf{u}_h \cdot \nabla h_2, \tag{8.33}$$

where w_* is the vertical velocity through the layer interface and $\mathbf{u}_h$ is the horizontal velocity. Hence,

$$\beta v H = f \left(w_* + \mathbf{u}_h \cdot \nabla H \right). \tag{8.34}$$

We start by looking at two special cases. For a rotating sphere with a layer of uniform depth, we have

$$\beta v H = f w_*, \tag{8.35}$$

which is just the Sverdrup balance with Ekman pumping w_E replaced by cross-isopycnal vertical velocity w_*. For a rotating laboratory tank ($\beta = 0$) with a bottom that slopes so that layer thickness decreases in the y direction

$$H(x, y) = H_0 - sy \tag{8.36}$$

(H_0 and s are constants representing typical thickness and slope of thickness), the vorticity equation (8.34) becomes

$$sv = w_*. \tag{8.37}$$

The sloping-bottom f-plane solution (8.37) is analogous to the flat-bottom, β-plane solution (8.35), with s/H equivalent to β/f. This is why the Stommel et al. (1958) laboratory experiment is relevant to flow on the spherical Earth, as discussed in Subsection 8.2.4. Another difference from the real world is that in Stommel et al. (1958), $w_* = 0$, but as the total amount of water in the tank can change, w_* is replaced by H_t which is given if we know the net flow rate into the basin. Stommel et al. (1958) formulated a slightly more complicated version of our case, using polar coordinates and a parabolic H variation, but the behavior is similar.

To calculate the zonal velocity, return to the vertical integral of the continuity equation, integrate vertically, and insert expressions for w at top and bottom of layers to get

$$H\nabla \cdot \mathbf{u}_h + w_* + \mathbf{u}_h \cdot \nabla H = 0, \tag{8.38}$$

or in Cartesian coordinates,

$$(uH)_x + (vH)_y + w_* = 0. \tag{8.39}$$

For the two special cases considered above, we can easily find v, and because we assume that we are given w_*, we can solve (8.39) for u as

$$u(x, y)H(x, y) - u(L, y)H(L, y) = \int_x^L \left[w_* + (vH)_y \right] dx, \tag{8.40}$$

where $x = L(y)$ is the location of the eastern boundary. The more general case, with nonzero β and with H depending on both x and y, is harder to solve. Equation 8.34 allows us to write u in terms of v (or vice versa), and then (8.39) becomes a single first-order partial differential equation in one variable.

Remember that these equations are not a complete solution for the flow. There must be a frictional (and perhaps nonlinear) western boundary current that connects to the "interior" circulation discussed so far. Chapter 3 discusses the detailed solution for the wind-driven case, but here we are only interested in the volume transport of the western boundary current. This can be deduced from continuity. Integrating (8.39) over the entire width of the basin (from $x = 0$ to $x = L$) and from the northern boundary $y = M$ to some meridional location y, applying the divergence theorem, and assuming no flow through northern, eastern, or western boundaries, we get

$$-\int_0^L Hv \, dx + \int_y^M \int_0^L w_* \, dx \, dy = 0. \tag{8.41}$$

The negative sign is because the outward direction for the southern boundary of the integration volume is southward (negative). If we refer to this as the inviscid solution, $\mathbf{u}_I$, then the western boundary current transport Ψ_{WBC} is given by the expression

$$\Psi_{WBC} + \int_0^L Hv_I \, dx = \int_y^M \int_0^L w_* \, dx \, dy. \tag{8.42}$$

For a northern hemisphere basin with a source in the north and upwelling everywhere else (Figure 8.16, bottom panels), the southward western boundary current transport is *greater* than the volume transport sinking in the north, in order to compensate for the northward interior flow. As one travels southward to meridional position y in this basin, an increasing fraction of the volume transport upwells out of the lower layer between y and the source, so that western boundary current transport decreases. Figure 8.21a shows the velocity field for an equatorial β-plane with downwelling in a small region near the northern boundary and uniform upwelling everywhere else.

At the equator, (8.35) shows us that $v = 0$ for the interior flow as $f = 0$ makes the right-hand side vanish (Figure 8.21b). All flow between the hemispheres must occur in

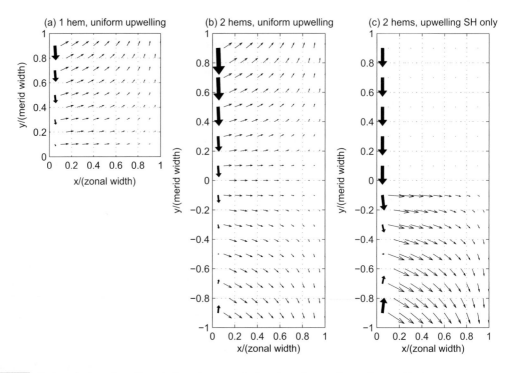

Horizontal velocity for uniform-thickness layer on a β-plane driven by upwelling at the top of the layer (Equations 8.35, 8.40, and 8.42). Western boundary currents (thick arrows) are reduced in length by a factor of 10 relative to the interior flow. Flows are shown for case of (a) single hemisphere, uniform upwelling; (b) two hemispheres, uniform upwelling; and (c) two hemispheres with same overturning strength but with no upwelling in northern hemisphere and w_* uniform in southern hemisphere.

the western boundary current. Continuing with the example of a northern source of deep water such as NADW, the interior flow switches sign in the southern hemisphere (because f switches sign). For uniform w_*, there can be a latitude where the interior volume transport has the same sign and magnitude as the total meridional volume transport in the layer; south of this point the western boundary current switches direction and flows northward rather than southward (Figure 8.21b).

In the real world, one can imagine different distributions of w_* depending on the relative strengths of mixing-driven and wind-driven upwelling, and also depending on spatial variations in mixing strength. To give an illustration of how the flow can change for a different meridional distribution of upwelling, Figure 8.21c shows an example with the same overturning strength as the flow in Figure 8.21b, but with uniform upwelling in the southern hemisphere only. Thus in the northern hemisphere, there is no interior flow, so all meridional transport in the layer is carried by a western boundary current that does not change strength with latitude. The southern hemisphere flow is similar to that in Figure 8.21b, but with stronger w_* and horizontal flow. Zonal variations of mixing, perhaps associated with differences in tidal mixing near the basin edges or over rough topography, could make other changes in the interior flow patterns as well.

In a closed basin with vertical walls, as in Figure 8.21, the horizontal velocity is zero at the walls. The interior solution does not satisfy this condition at the polar boundaries. When friction terms are included in the solution, boundary layers occur in which the meridional velocity goes to zero near the boundary and transport from the interior flow as well as from deep water formation is carried westward to the western boundary current.

8.4 Excursions: Oceanography through the Ages

Aristotle's student Theophrastus (371–287 BCE) may have taken the first known current measurement. He threw a bottle into the Mediterranean with a message asking that it be returned to him (Pollard and Reid, 2007).

Strong surface currents were recognized by sailors long before oceanographers studied them. Al-Mas'udi of Baghdad (896–956) discussed Indian Ocean currents reversing with the monsoon (Deacon, 1997). In 1513, when Ponce de Leon was pushed northward while trying to sail southward off the east coast of Florida, he reported that the current "was more powerful than the wind" (Kelley Jr., 1991): the Gulf Stream. Benjamin Franklin commissioned a chart of it in 1769 (Richardson, 1980a) based on advice from a ship captain. In 1785, Shihei-Hayashi wrote about the Kuroshio as "a very rapid current, called Kouro se gawa, or the 'black gulf current'," and older Chinese stories of a strong current carrying ships eastward into an abyss may also refer to the Kuroshio (Jones and Jones, 2002).

Subsurface currents were another matter. In the seventeenth century, several scientists recognized that continuity demands that the surface inflow from the Atlantic and from rivers into the Mediterranean must exit somewhere. Edmond Halley (1646–1742) and others believed this occurred through evaporation, which is now known to export a small part of the inflow. Some argued that there was a hidden subsurface Mediterranean outflow at the Strait of Gibraltar, but they did not know a mechanism that would cause it.

Meanwhile, at the Bosphorus strait at the other end of the Mediterranean, local fishermen knew that the flow of fresh water from the Black Sea was countered by a Mediterranean outflow beneath. Count Luigi Marsigli (1658–1730) lowered a brightly colored weight and observing its drift changing direction with depth; the outflow is closer to the surface than at Gibraltar because the Bosphorus is shallower. Marsigli illustrated the dynamics in 1681, six years before Newton's *Principia*, with a simple experiment analogous to the configuration shown in Figure 8.10. Nevertheless, the similar outflow at Gibraltar was not confirmed until William Carpenter (1813–1885) lowered a drogue and saw it dragging a boat towards the Atlantic.

The equatorial undercurrent was discovered, forgotten, and discovered again. The first time (as far as we know) occurred in the Atlantic in 1886, when John Buchanan (1844–1925) saw that a surface buoy attached to a drogue at 55 m depth was dragged rapidly eastward, against the surface current. But by the time he died, his discovery of the undercurrent was forgotten – not even mentioned in a memorial by his colleague (McPhaden, 1986). Almost 70 years later, Cromwell et al. (1954) measured a similar Pacific equatorial undercurrent. The second discovery made a much greater impact on the field, perhaps because dynamics had evolved to the point where physical oceanographers could hope to understand how the current was driven (McPhaden, 1986).

Exercises

8.1 Review Figure 8.1a–c then:

(a) Make the corresponding plots of meridional dissolved oxygen concentration, and the percent oxygen concentration. Comment on your results and explain your interpretation in terms of water masses.

(b) Repeat part (a) for dissolved silicate, SiO_4 (which exists primarily as silicic acid, $Si(OH)_4$).

Hints: You may find it helpful to also plot the global surface silicate distribution. Review Subsection 6.1.2 and consult Sarmiento and Gruber (2006) as needed.

8.2 (a) Search for CFC data on long zonal sections, preferably from coast to coast, with repeat occupations (i.e., data from different years; see Appendix A). Select sections in either the Atlantic, Pacific, or Indian Oceans. Try to cover the full range of latitudes for your Ocean, with several repeats.

(b) Make plots of the CFC concentration (CFC-11 or CFC-12) on your sections, similar to those you made in Exercise 2.11.

(c) Download and plot the time history of atmospheric concentration of CFCs, for example from www.nodc.noaa.gov/ocads/oceans/.

(d) Hence, make inferences about the timescales and pathways of the deep meridional overturning circulation in your Ocean. You may assume that the surface ocean CFC concentration is proportional to the atmospheric concentration. State and discuss any other assumptions.

8.3 Consider the portrait of the Denmark Strait overflow in Figure 8.7 and comment on its realism. Using hydrographic profiles from WOD 2013, CARS, Argo, and climatologies as necessary, make your own section of temperature, salinity, and density for Denmark Strait. Try to show the overflow more realistically and remark on the challenge of observing it.

8.4 A simple model for a stagnant, diffusive ocean says that the temperature profile $T(z, t)$ satisfies:

$$\rho c_p \frac{\partial T}{\partial t} = \kappa \nabla^2 T. \tag{8.43}$$

(a) Explain the origin of (8.43), and define each term.

(b) Find steady solutions of (8.43) for an ocean of depth H with surface temperature T_s, set by the atmosphere, and geothermal heat flux $\mathcal{F}$ (typically, 8×10^{-2} W m^{-2}). Plot the $T(z)$ profiles for reasonable choices of T_s and κ. Comment on the realism of your solution.

(c) Using a scaling argument (or otherwise), write a formula for the time taken to establish the steady solution from part (b). Estimate this timescale for reasonable ocean parameters and comment. (See also Exercise 3.6.)

8.5 By considering the shallow water potential vorticity of a uniform density fluid (or otherwise):

(a) Show that a bottom slope is dynamically equivalent to the planetary β-effect.
(b) Derive a formula that quantifies this topographic β-effect.
(c) Hence show that an undamped, unforced, linear flow circulates along contours of constant f/h, where f is the Coriolis parameter and h is the bottom depth.
(d) Discuss whether changes in f or h dominate changes in potential vorticity in the real ocean circulation.

(For more on the topographic β-effect, consult Vallis, 2006 or Haine and Fuller, 2016.)

8.6 **Bernoulli's principle** for steady frictionless incompressible flow says that the quantity (the Bernoulli function)

$$\frac{v^2}{2} + gz + \frac{p}{\rho} \tag{8.44}$$

is constant along a streamline.

(a) Define all the variables involved.
(b) Derive Bernoulli's principle from the governing equations (1.16) (or (2.83)–(2.84)) under suitable assumptions that you state.

Hint: Write the momentum equation as the gradient of a potential and show that this gradient vanishes.

(c) Consider flow over a topographic bump, as in Figure 8.18. Show that for sufficiently tall bumps, Bernoulli's principle cannot be satisfied. Derive a formula for the critical bump height.
(d) Hence show that there exists a maximum steady volume flux that can cross the bump and derive an expression for it. Explain any additional assumptions.

Hints: Assume, if you wish, that the width of the flow is uniform. Pratt and Whitehead (2008) Section 1.4 is also helpful.

8.7 A box of water of height H, area A, and density ρ_0 has a layer at the bottom of thickness h cooled so that it becomes denser by an amount $\Delta\rho$. Diffusion then spreads this density anomaly so that the density of the entire box becomes constant. No heat enters or leaves the box at any time.

(a) What is the change in potential energy of the box?
(b) If all the potential energy becomes kinetic energy associated with currents in the box, what is the average water speed?
(c) If all this kinetic energy becomes heat energy, what is the change in water temperature in the box?

First answer these questions with formulae, then compute numbers for the case of $H = 1000$ m, $h = 100$ m, $A = 10^4$ m^2, $\rho = 1000$ kg/m^3, $\Delta\rho = 5$ kg/m^3, and any other physical constants that are necessary.

8.8 Consider the advective-diffusive scaling for thermocline thickness D, upwelling speed W, and horizontal velocity V.

(a) Show that these three scalings follow from the scaling equations ($W = \kappa/D$ etc.) in Subsection 8.3.2.

(b) We assumed that the surface buoyancy range Δb is known. Suppose that we know the midlatitude heat transport Q instead, which is approximately equivalent to knowing the buoyancy transport $B = \Psi\Delta b$. In other words, we have specified the surface heat flux leaving/entering the ocean. How do D, W, and Ψ depend on the parameters L, M, f, κ, and B? A good way to check your answer is to make sure the dimensions of each answer are correct.

8.9 For the advective-diffusive scaling in Subsection 8.2.3 and Subsection 8.3.2, give a rough estimate of the strength of conversion of potential energy to kinetic energy as a function of parameters $L, M, f, \Delta b, \kappa$, and ρ_0. What is the numerical value for $L = M = 6000$ km, Δb based on $\Delta\rho = 5$ kg/m^3, and $\kappa = 1$ cm^2/s?

8.10 Consider a Stommel abyssal flow in a rectangular basin bounded by $y = \pm M$ and by $x = 0$ and $x = L$. In spherical coordinates, note that latitude $\theta = y/R_E$ and longitude $\phi = x/(R_E \cos\theta)$, where R_E is the radius of the Earth. Next to the northern boundary of the basin, there is a source of water into the basin with a volume transport Ψ_S. Some fraction α of this water upwells uniformly throughout the basin; the rest leaves the basin at the southern boundary. The layer of water has a uniform thickness H.

(a) Calculate the meridional velocity in the "interior" as well as the zonal average value in the western boundary current.

(b) In the northern hemisphere of this basin, there is always a southward flow in the western boundary current. Derive an equation for the latitude θ where the western boundary current reverses direction (in the southern hemisphere). For what values of α is there no flow reversal, both in general and for the special case $M = R_E$? If $\alpha = 1$, does the western boundary current reverse direction? If so, what is the latitude (given $M = R_E$)?

(c) Using continuity and v_I derived above, derive the zonal interior velocity component.

(d) Create a vector plot showing the interior velocity field for the basin for the special case $M = L = R_E$. Does the flow pattern depend on α?

8.11 This question is based on the **diffusive pycnocline** ideas of Vallis (2000) and Samelson and Vallis (1997) (see also Welander, 1971). It addresses the interface between the wind-driven ventilated pycnocline (Chapter 5) and the buoyancy-driven circulation below it (Chapter 8). It unites several concepts from Chapters 5 and 8.

(a) Referring to Figure 1.6, make a plot showing the vertical profiles of $\partial T/\partial z$, $\partial S/\partial z$, and $\partial \rho/\partial z$ for the subtropical North Atlantic Ocean. Identify the mode water layer, and hence the shallow pycnocline and the deep pycnocline.

(b) By plotting other stations in the subtropical gyre, roughly map the spatial extent of the (permanent) shallow and deep pycnoclines.

Hint: You may find this task easier by using a gridded climatology rather than individual hydrographic stations.

(c) Refer to the scaling arguments in Subsection 5.3.1 and check that the shallow pycnocline depth is consistent with the neglect of vertical diabatic mixing. Comment on the validity of the argument for the deep pycnocline.

Hint: You may assume that the horizontal distance L in the scaling equals the meridional distance between the station and the winter-time outcrop of the density surface in question (see Figure 4.6 for guidance).

(d) Referring to the Sverdrup balance for a $1\frac{1}{2}$-layer wind-driven model, (5.24), write a scaling for the depth of the wind-driven pycnocline. Estimate this depth for the subtropical North Atlantic Ocean then compare to the depths of the shallow and deep pycnoclines you see in the hydrographic profiles from part (a) and comment.

(e) Refer to the scaling arguments in Subsection 8.3.2 and investigate the relevance of the vertical advective-diffusive balance for the two pycnoclines. Estimate the depth from the vertical advective-diffusive balance then compare to the depths of the shallow and deep pycnoclines and comment.

(f) Hence discuss the proposition that the shallow pycnocline derives from the wind-driven ventilated mechanism of Chapter 5 and the deep pycnocline derives from the vertical advective-diffusive balance of Chapter 8. How would you test the proposition?

9

The Southern Ocean Nexus

The **Southern Ocean** is a nexus that draws together water from all the major oceans. In thinking about the Southern Ocean it is difficult to separate any of the themes in this book. The Antarctic Circumpolar Current (ACC) is the principal circulation feature. It is a wind-driven current (Chapter 3), dominated by eddies (Chapter 7), and connects both the local meridional Ekman cell (Chapter 6) and the global deep meridional overturning circulation (Chapter 8). These factors, along with difficulties in understanding the vertical structure and interactions with topography, make the Southern Ocean one of the least understood regions of the ocean. In addition, its location far from most oceanographers and its harsh subpolar conditions make it less well observed. Therefore, this chapter is more tentative and open-ended than others. Like Chapter 8, our understanding relies more on numerical model experiments than the topics earlier in the book. Here we present some of the main elements that govern our understanding of the Southern Ocean circulation.

9.1 Observations

9.1.1 Global Overturning

Southern Ocean Topography

One of the most distinctive features of the Southern Ocean is its zonal periodicity. At the latitude of the **Drake Passage**, which is a strait between South America and the Antarctic Peninsula at, nominally, 55–60°S, one can travel eastward or westward without ever encountering land (Figure 9.1a). Besides bordering (and connecting) all the other oceans, this fact has important dynamical consequences, which we discuss in Subsection 9.2.1.

The periodicity of the Drake Passage latitudes is somewhat complicated. Several undersea ridges block the periodicity (Figure 9.1b). The **Kerguelen Plateau** rises to depths of less than 1 km in the Indian Ocean sector, (Figure 9.1a,b,c), and the East Pacific Rise arcs into high latitudes in two locations (Figure 9.1a) but does not rise as high (Figure 9.1b). The Atlantic sector topography is particularly complex, with island arcs stretching eastward from South America and Antarctica around the **Scotia Sea** (Figure 9.1d). The combined

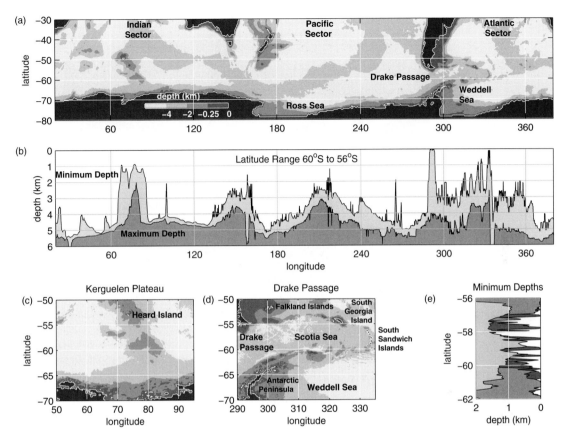

Figure 9.1 Southern Ocean bathymetry, as seen by (a) bathymetric shading in longitude–latitude plane, (b) maximum (dark) and minimum (light) depth in latitude range 56–60°S at each longitude, high-resolution charts of (c) Kerguelen Plateau, and (d) Drake Passage, and (e) minimum depths at each latitude for: "Kerguelen" (light shading), "South Sandwich Island" (dark shading), and "Drake Passage" (medium shading) longitudes. Panels (a) and (b) use ETOPO5 data averaged on 0.5° squares, and the rest use ETOPO1 data (see Appendix A).

effect of all these features is to subdivide the Drake Passage latitudes into a series of gaps (Figure 9.1e), each less than 100 km wide and up to about 1.5 km deep. It is not clear to what extent this topographic "comb" or sawtooth is similar to a single opening spanning the Drake Passage. Moreover, the ACC is not completely zonal (see Figures 9.4 and 9.6), and avoids some of these obstructions by shifting latitude (see also Exercise 9.3).

Global Zonal-Average Overturning

The meridional property sections in Figure 8.1 show that in all three sectors of the Southern Ocean (Atlantic, Indian, and Pacific), the coldest and densest water outside the Arctic and some marginal seas occurs near Antarctica. There are indications of salty water rising to

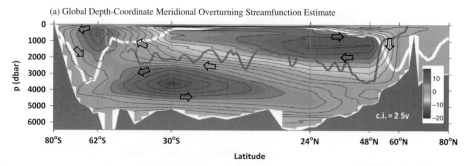

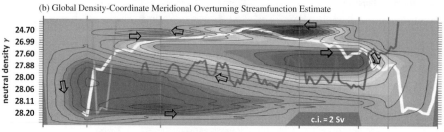

Figure 9.2   Global meridional streamfunction in (a) depth-coordinates and (b) density coordinates based on observations. The contour interval is 2 Sv. The white line shows typical winter mixed layer depths and densities. The dark gray line shows the mean depth of the ocean ridge crests and the light gray line shows the depth of the Scotia Arc east of Drake Passage. From Lumpkin and Speer (2007) Figure 2. ©American Meteorological Society. Used with permission.

the surface from great depth in the general vicinity of 60°S, and fresher water sinking to about 1 km near 45°S.

The meridional streamfunction is not defined for individual basins at latitudes where they connect to other basins. Thus the Indonesian passages make interpretation of the Pacific and Indian streamfunctions ambiguous, and in the Southern Ocean we can only define a single streamfunction based on velocity from all longitudes. Such a global meridional overturning streamfunction (Lumpkin and Speer, 2007) shows both the overturning cell, in the top couple of kilometers, associated with North Atlantic Deep Water (NADW) formation, and a bottom cell associated with Antarctic Bottom Water (AABW) formation (Figure 9.2). The NADW and AABW water masses are introduced in Subsection 8.1.1.

As with individual basins, there are strong similarities between the depth-coordinate (Figure 9.2a) and density-coordinate (Figure 9.2b) meridional overturning streamfunction. (The difference between the two streamfunctions is discussed in Subsection 6.3.3). The Southern Ocean shows one major difference: in depth-coordinate overturning, much of the zonal average flow sinking south of 62°S rises nearby and returns southward at the surface, while in the density-coordinate overturning, all of the sinking near Antarctica flows far to the north in the AABW cell before it rises and returns. Note that the zonal variation in currents means that actual water parcels do not necessarily flow along the paths shown by the streamfunction, but the difference between the two varieties of streamfunction is significant. It should also be noted that model results (see Figure 9.10) indicate that there is important structure (such as northward Ekman transport between 62°S and 32°S; see also Figure 6.4) that is not resolved by this figure.

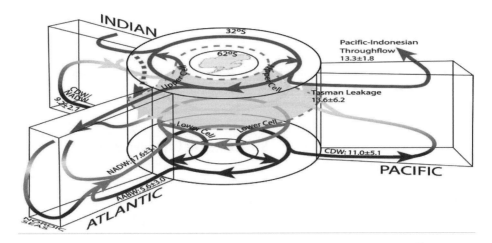

Figure 9.3 Schematic cartoon showing elements of global conveyor belt circulation, showing volume transport (Sv) of major branches of overturning. From Lumpkin and Speer (2007) Figure 4. ©American Meteorological Society. Used with permission.

Inter-Basin Overturning

Tracer measurements and numerical experiments indicate that the overturning cells in different basins interact, so it is helpful to illustrate the global nature of the overturning circulation. This is most often done with a circulation cartoon such as Figure 9.3 by Lumpkin and Speer (2007) (see also Schmitz, 1995). The cartoon does not include the gyre circulations. According to the Lumpkin and Speer (2007) cartoon, deep flow leaving the Atlantic enters the ACC and feeds bottom flow into the Indian and Pacific, bottom water formation in the Weddell Sea, and upper layer water in the Pacific (Figure 9.3). The Weddell water feeds the AABW returning into the Atlantic.

Much of the bottom water that flows into the Indo-Pacific returns to the ACC as deep water, but some of this deep water rises in the Indian Ocean to the top layer there. This top layer, fed by Indian and top layer Pacific water, exports some water from the tip of the Agulhas Current off South Africa to the Atlantic and returns some to the top layer of the ACC. Some water from the ACC flows into the Atlantic after passing through the Drake Passage. These two pathways then feed the top layer inflow into the Atlantic. The South African and ACC paths are called the **warm route** and **cold route**, respectively, because of the relative temperature of each flow. The warm flow from near South Africa is sometimes called the **Agulhas Leakage**, referring to the relatively small volume transport of water that does not return to the Indian Ocean in the Agulhas Retroflection. Much of the Agulhas Leakage is due not to steady flow but to the formation and westward propagation of warm Agulhas rings (eddies; see Subsection 7.1.1). It is debated about what fraction of the Atlantic inflow is due to each route, with different papers arguing that one or the other route dominates (see de Ruijter et al., 1999 for a review and Talley, 2013).

The large-scale pattern of water sinking in the Atlantic and rising elsewhere is sometimes called **the conveyor belt** after the label of an illustration (Broecker, 1987) based on a flow schematic emphasizing the warm route of the return flow (Gordon, 1986).

The Atlantic NADW and AABW cells and the Indo-Pacific AABW cell are well established in observations, but many of the details displayed by Figure 9.3 are uncertain due to data limitations. Studies with inverse methods (such as, Macdonald, 1998; Ganachaud and Wunsch, 2000; Sloyan and Rintoul, 2001, and Lumpkin and Speer, 2007) and data assimilation (such as, Forget et al., 2015) quantify this uncertainty, but do not eliminate it.

9.1.2 Antarctic Circumpolar Current

As we saw in Chapter 3, all longitudes in the vicinity of the Southern Ocean are characterized by strongly zonal westerly winds (Figure 3.1a). Latitudes of about 45–60°S have the greatest values of zonally and annually averaged stress (Figure 3.2).

The Southern Ocean circulation is dominated by the **Antarctic Circumpolar Current** (ACC), a roughly 1000 km-wide band of eastward flow around Antarctica (Figure 9.4). The center of the current meanders from around 58°S in the Drake Passage to about 45°S

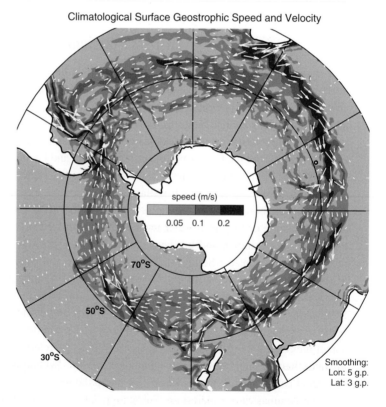

Figure 9.4   Geostrophic surface velocity (vectors) and speed (shading), based on CMEMS climatology of sea surface height (see Appendix A). Velocities are smoothed with five-point averages in both longitude and latitude. Arrows show a small subset of climatology which has resolution of 1/4°. See also Figure 7.2.

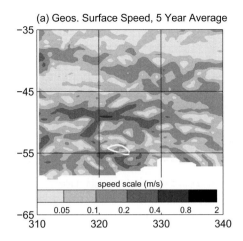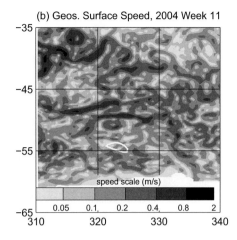

Figure 9.5 Surface geostrophic speed in the vicinity of Drake Passage for (a) five-year average and (b) week 11 of 2014, from CMEMS data (see Appendix A).

in parts of the Atlantic and Indian sectors. The maximum speed south of 30°S at each longitude, when averaged zonally around the Southern Ocean, is about 30 cm/s. Even in the multi-year average (Figure 9.4), the ACC has a complex structure that includes filaments, meanders, and significant variations in current speed along its path. There are at least two parallel streams south of South Africa and three south of Australia. However, the time-averaging smooths out much of the complexity of the flow, as shown by comparing charts in the vicinity of the Drake Passage for the time-average (Figure 9.5a) and a single-week snapshot (Figure 9.5b) of speed. For the entire ACC, this snapshot has a zonal average maximum of approximately 60 cm/s.

High speeds correspond to high gradients in sea surface height (SSH), and so the jets visible in Figure 9.5b correspond to fronts: regions of enhanced property gradients. At any time the SSH field is complex. The fronts span a restricted longitude range, and they move north and south over time. However, by calculating how frequently each Southern Ocean grid point in the altimetry data set had an enhanced SSH gradient, Sokolov and Rintoul (2009) showed that the flow jumps between several quasi-stable front locations that form continuous paths around the entire Southern Ocean (Figure 9.6). These SSH fronts correspond to long-recognized temperature fronts (Deacon, 1937; Orsi et al., 1995). The northernmost one is the Subantarctic Front, south of which is the Polar Front, south of which is the Southern ACC Front.

The volume transport of the ACC has been measured most often in the Drake Passage, where the geographical constriction facilitates capturing the entire width of the current with hydrography and moored current meters. Like the gyres, the ACC flow is fastest near the surface and slowest near the bottom, with typical bottom velocities in the Drake Passage about 10% or less of surface values (Figure 9.7). Early measurements indicated volume transports in the range of 120–140 Sv (Whitworth III et al., 1982; Cunningham et al., 2003), but recent, more extensive, measurements show the volume transport to be 173 ± 11 Sv

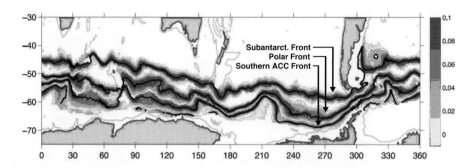

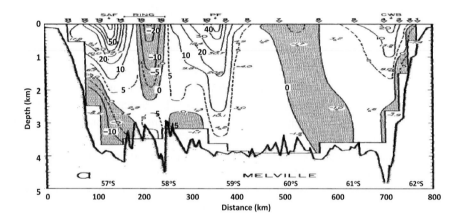

(Donohue et al., 2016; the increase is attributed to better data, not secular change). If we split the ACC volume transport between the contribution of $u_B H$ (for ocean depth H and bottom zonal current speed u_B) and the contribution associated with vertical shear, the Donohue et al. (2016) observations indicate that about a quarter of the transport is due to the extended bottom velocity.

Naveira Garabato et al. (2014) used an inverse model to estimate the ACC volume transport, in different density classes, at the longitudes of Drake Passage, South Africa, and Tasmania. They found a little less than 40 Sv each in Intermediate (neutral density 27.6–27.10), Upper Circumpolar Deep (27.98–27.60), and Lower Circumpolar Deep (28.18–27.98) waters. Almost 20 Sv consisted of surface and mode waters and 9 Sv were Antarctic Bottom Water. Cunningham et al. (2003) found broadly similar water mass divisions of the flow.

9.2 Concepts

9.2.1 Global Ocean Meridional Overturning

Whereas the thermohaline mechanism described in Subsection 8.2.3 explains the existence of a pycnocline and deep overturning, single-basin models are too simple to reproduce the complexities of the real ocean. The fact that water masses in one location can be influenced by remote surface conditions raises the possibility of interactions between different oceans. Moreover, the Atlantic, Pacific, and Indian basins all abut the Southern Ocean, whose unique geometry has a global influence on the overturning.

Zonal Periodicity

As Subsection 9.1.1 and Figure 9.1 show, a key feature of the Southern Ocean is its zonal periodicity at the latitude range of the Drake Passage. Zonal periodicity is an important constraint on overturning because of its effect on meridional flow. Consider the geostrophic meridional speed equation (2.3)

$$f v_G = p_x / \rho_0, \tag{9.1}$$

and integrate in x over a zonally periodic basin of zonal length L. Then (9.1) becomes

$$f L \overline{v}_G = \Delta p / \rho_0, \tag{9.2}$$

where overbar represents zonal averaging and Δ represents the difference between eastern and western boundaries. Because of the zonal periodicity, $p(x, y, z) = p(x + L, y, z)$, so $\Delta p = 0$ and there is no zonal average geostrophic meridional flow, $\overline{v}_G$ equals zero. There can be meridional flows that depend on longitude as well, but at each depth, the integrated northward velocity must be balanced by equal integrated southward velocity. We expect that imposed meridional density gradients creates a roughly zonal flow rather than meridional overturning.

The Southern Ocean is crossed by several ridges that rise to shallower than 2000 m depth, however (Figure 9.1). For depths that are blocked by a ridge, the constraint that $\overline{v}_G = 0$ does not apply. There can be a pressure difference between the eastern flank of the ridge and the western flank and hence a nonzero meridional current.

The prohibition on net geostrophic $\overline{v}_G$ in a zonally periodic channel such as the Southern Ocean separates the flow to the south of the channel from flow to the north (Gill and Bryan, 1971; Cox, 1989). This is dramatically illustrated in a numerical model of a single basin with the highest surface density near the southern boundary. For a closed basin, the deep water formation in this experiment occurs near the southern boundary and connects to a dominant overturning cell as discussed in Chapter 8 (see Figure 8.14a). For a similar basin in which a periodic channel is imposed above 2545 m from 48°S to 64°S (Figure 9.8a), the deep cell with water sinking in the south becomes more confined to the bottom, and the subordinate northern-sinking cell occupies much of the water column and extends across

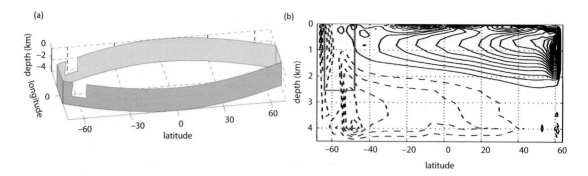

Figure 9.8 (a) Perspective view of basin geometry with zonally periodic region near 60°S. (b) Overturning in geometry shown in (a), with similar forcing to in Figure 8.14a. Contour interval is 1 Sv. (b) is from Klinger et al. (2003) Figure 4. ©American Meteorological Society. Used with permission.

the basin to the channel. The deep stratification also increases substantially. In this way, the zonal channel qualitatively affects the circulation.

Multiple Basins

In the real ocean, the deep overturning has a unique pattern in each basin and interactions among basins (Figure 9.3). Therefore, to understand the behavior of the global overturning, our conceptual and numerical models must include elements of the ocean's geography missing from the single-hemisphere models.

Even a coarse-resolution model with idealized topography (Figure 9.9a) captures essential features of the overturning (Cox, 1989). Among the real-world geographic features of this model are three sites in cold, high-latitude regions with potential to form deep water: the North Atlantic, North Pacific, and coastal Antarctica. These *potential* sites are *actual* sources of deep water in the model North Atlantic and Antarctic, but not in the North Pacific, where low surface salinity (imposed as a boundary condition in the model) gives it a low surface density. Subsection 8.1.1 discusses a similar pattern in the real world, and Subsection 11.2.4 discusses the reasons for the low salinity.

When sea surface temperature and salinity are given broadly realistic latitude distributions, the model Atlantic shows a salinity structure (Figure 9.9b) reminiscent of the real ocean (Figure 8.1a,c). Like the real ocean, the salinity tongues in the model Atlantic suggest northward flow near the bottom (Antarctic Bottom Water, AABW), southward flow centered at about 2 km depth (North Atlantic Deep Water, NADW), and northward flow in the top kilometer (Antarctic Intermediate Water, AAIW). The model Pacific, like the real Pacific, has approximately symmetric tongues of relatively fresh intermediate water (Figure 9.9c) flowing equatorward at about one kilometer depth. The South Indian (not shown) also has a plume of intermediate water stretching towards the equator. In all basins, a column of the coldest water sits near the southern boundary, similar to both a single-basin model with a channel and to the real world (Figures 8.1a,c and 9.8b).

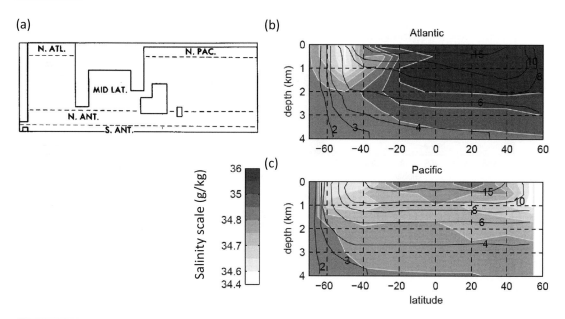

Figure 9.9 Coarse resolution GCM simulation of near-global ocean. (a) Basin geometry. Zonal average salinity (shading) and temperature (contours) for (b) Atlantic and (c) Pacific basins, based on Cox (1989), Figures 7, 9, and 10. ©American Meteorological Society. Used with permission.

Models allow us to visualize meridional overturning streamfunctions that in the real ocean, must be inferred from property distributions (Figure 8.1) or sparse and uncertain measurements (Figures 8.5 and 9.2). Hirst et al. (1996) show good examples of the meridional streamfunction in an eddy-suppressing model with realistic topography and surface forcing. The global-average flow (that is, including all ocean basins at any given latitude) is dominated by two major cells (Figure 9.10a). A deep cell carries far-northern water down to 2 km and then southward to the Southern Ocean. Another cell, with sinking near Antarctica, sits on the bottom, mostly in the southern hemisphere. Circulation in the top kilometer includes shallow subtropical cells (Subsection 6.2.2) straddling the equator.

The global-average circulation pattern is strikingly similar to that of the single basin with a periodic channel (Figure 9.8). The two major overturning cells correspond to the two areas of high surface density: northern and southern subpolar regions. In contrast to a single closed basin, in which one cell dominates the entire water column (Figure 8.14), the existence of an unblocked-blocked channel deepens the return path of the densest water in a domain that consists of either a single basin (Figure 9.8) or multiple basins (Figure 9.10). As Figure 9.2a shows, the real world appears to have a similar two-cell structure in the global-average streamfunction.

The deep northern-sinking cell seen in the global overturning is almost entirely due to the NADW cell in the Atlantic (Figure 9.10b). The bottom AABW cell outflow from the Southern Ocean (Figure 9.10a) is primarily in the Indo-Pacific (Figure 9.10d), but the return flow superposes circulation in all three basins. Although the bottom cell in the global streamfunction occupies a substantial fraction of the water column, its density range is much

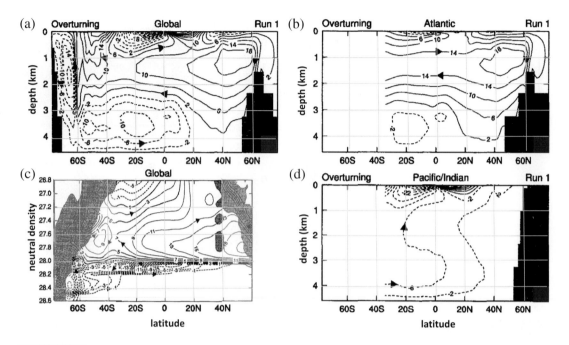

Figure 9.10 Meridional overturning streamfunction (Sv) in an eddy-suppressing model for (left panels) global zonal integral, (b) Atlantic, and (d) Indo-Pacific. All panels are depth-coordinate streamfunctions except (c) in which the zonal integral is along neutral density surfaces and shading represents densities at the surface and in the model Mediterranean Outflow. In all panels, the contour interval is 4 Sv and solid contours show clockwise circulation. The model represents eddies using horizontal diffusion (compare to Figure 9.12). From Hirst et al. (1996), Figures 2 and 5. ©American Meteorological Society. Used with permission.

smaller than the deep cell at most latitudes (Figure 9.10c). The individual basins connected via the Southern Ocean (not to mention the Indonesian passages and a weaker connection through the Arctic Ocean) improve the realism of the single-basin model. Nevertheless, the single-basin model provides insight into Atlantic–Pacific interaction. The domain in Figure 9.9a can be thought of as a single basin in which the North Atlantic and the North Pacific are at opposite ends. The much shallower penetration of North Pacific water above a barrier of denser water is analogous to the subordinate cell in Figure 8.14a.

The surface density difference between high latitude regions is fairly small compared to the tropical-subpolar difference (see Exercise 5.1). Therefore, small changes in high latitude temperature or salinity can make a big difference in the global overturning. This issue is germane to climate dynamics and especially to paleoclimate (see Subsection 11.1.5 and Subsection 11.2.4). The tendency for strengthening of one cell to be associated with weakening of another is often referred to as a **seesaw** between northern and southern hemispheres (Stocker, 1998) or between North Atlantic and North Pacific (Hughes and Weaver, 1994; Saenko et al., 2004). Numerical experiments in which forcing, ocean geometry, and other parameters are varied have wide ranges in strength of the AABW cell, NADW cell, or both (England, 1993). For instance, adding an annual cycle to annual average sea surface temperature and salinity strengthens the NADW cell and weakens the AABW cell. Changing the vertical diffusivity in deep water also affects the relative strength of the AABW and NADW cells (Kamenkovich and Goodman, 2000).

As Subsection 8.2.3 discusses, vertical mixing is crucial for the deep meridional overturning. Lateral mixing by mesoscale eddies is also important. The horizontal diffusion used in the numerical experiments discussed above is only a crude approximation to eddy mixing (Chapter 7). Deep water formation regions are unrealistically large due to the spurious mixing, because the densest mixed layer water is efficiently spread over a relatively large area. Eddy effects are represented better by redistributing temperature, salinity and layer thickness (Gent and McWilliams, 1990) along isopycnals. Implementing this in a general circulation model reduces spurious mixing so that regions of deep water formation shrink to more realistic sizes (Danabasoglu and McWilliams, 1995). It is in the Southern Ocean that eddies play their largest role, as we discuss below and in Section 9.3.

The Deacon Cell

The global overturning streamfunction in numerical models (for example, Figure 9.10) includes signatures of the subtropical cells as well as a southern hemisphere subpolar cell with equatorward Ekman transport. This upper Southern Ocean overturning cell is often referred to as the **Deacon Cell** after British oceanographer George Deacon. The great depth of the Deacon Cell presents something of a paradox because the downwelling at around 40°S penetrates more than one kilometer through significant stratification. This suggests a degree of density change that is much greater than is thought to occur through vertical mixing. Different models exhibit different ways to resolve the paradox, as we now discuss.

In the Fine Resolution Antarctic Model, an early eddy-permitting model of the southern hemisphere, some of the northward Ekman transport between 40°S and 60°S feeds the Deacon Cell. Some of the Deacon Cell is part of the NADW overturning cell (schematic

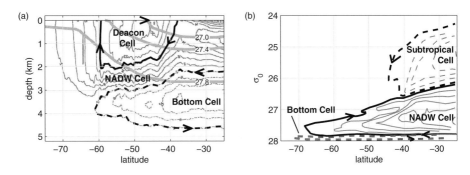

Figure 9.11 Southern hemisphere meridional overturning streamfunction (a) Ψ_z and (b) Ψ_ρ, from the Fine Resolution Antarctic Model, 2 Sv contour interval, thick black curves for bounding streamlines for clockwise (solid) and counterclockwise (dashed) flow. In (a), thick gray contours show selected zonal-average σ_0. Based on Döös and Webb (1994) Figures 4a and 5a and Döös (1994) Figure 3. ©American Meteorological Society. Used with permission.

based on actual streamfunction shown in Figure 9.11a), which Döös and Webb (1994) call the "Sub-polar Cell." The Deacon Cell shown in Figure 9.11a appears in depth-coordinate overturning, Ψ_z. In density-coordinate overturning Ψ_ρ the picture is different: then (Figure 9.11b) *all* of the northward transport of relatively light water is carried by the NADW cell. To the extent that there is increase in density of water corresponding to the downward limb of the Deacon Cell, it occurs in *southward* flow at the surface of the subtropical cell, which in Ψ_ρ extends past 40°S. We expect this southward flow to involve density increases as the water travels polewards, principally in western boundary currents. This is one of the most dramatic examples of differences between Ψ_ρ and the depth-coordinate overturning Ψ_z. In this case, the Deacon Cell in Ψ_z disappears in Ψ_ρ (see also Zika et al., 2013).

In Chapter 7 we introduce the concept of a transport-velocity or eddy-velocity, $\mathbf{u}_E$, associated with eddy transport of tracers, in addition to the $\mathbf{u}_A$, the Eulerian average velocity. The velocity that is relevant for the tracer budget is the total velocity $\mathbf{u} = \mathbf{u}_A + \mathbf{u}_E$. The Deacon Cell, as measured by $\mathbf{u}_A$, reaches a greater depth in the Gent and McWilliams (1990) experiment (Figure 9.12a) than in the horizontal diffusion experiment (Figure 9.10a), virtually isolating the AABW cell from the sea adjacent to Antarctica. However, there is also a strong eddy transport overturning (Figure 9.12c) flowing in the opposite direction. The counterclockwise $\mathbf{u}_E$ (sinking occurring in the south) represents the tendency of eddies to reduce the steep isopycnal slopes in the Southern Ocean (Figure 8.1). A similar tendency occurs at high latitudes in the northern hemisphere as well, but it is weaker because much of the northern latitude band is occupied by land.

In Ψ_ρ, unlike the Fine Resolution Antarctic Model (Döös and Webb, 1994), in the experiment of Hirst and McDougall (1998), the Deacon Cell is somewhat weaker in density- than depth-coordinate overturning, but it does not disappear (Figure 9.12b). The eddy transport cell also persists in the depth-coordinate streamfunction (Figure 9.12d). In both versions of the streamfunction, the eddy overturning and the advective overturning cancel each other, so the total velocity field (Figure 9.12e,f) has virtually no Deacon Cell and the AABW cell is even more directly connected to the surface than in the horizontal diffusion experiment (Figure 9.10).

9.2.2 Remote Wind-Driven Overturning

Drake Passage Effect

In the previous subsections, we refer to the dynamics of the density-driven or thermohaline overturning. However, a wind-driven mechanism may also be important – perhaps even more important – in driving the overturning.

The mechanism was discovered in a coarse resolution GCM of the near-global ocean with topography and forcing that was realistic except as follows. Three experiments are run to equilibrium with wind stress south of 30°S multiplied by a factor of 0.5, 1, and 1.5, respectively (Figure 9.13a). These experiments show large differences in the Atlantic overturning (Toggweiler and Samuels, 1995). The peak North Atlantic overturning rises from 15 to 24 Sv, while the volume transport of NADW flowing out of the Atlantic rises even more dramatically from 6.5 to 14 Sv (Figure 9.13b,c).

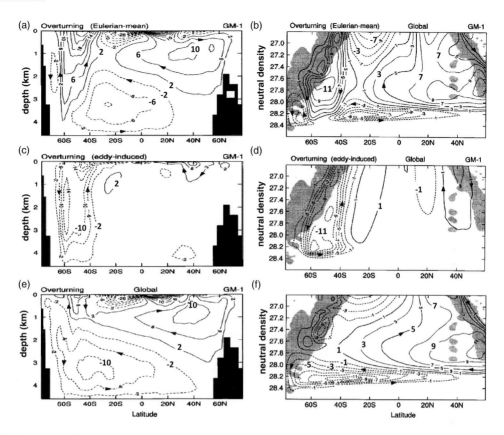

Meridional overturning streamfunction (Sv) in depth (a, c, e) and density (b, d, f) coordinates for a global eddy-suppressing model. The streamfunction is based on advective (a, b) or eddy (c, d) components of velocity, or their sum (e, f). In all panels, solid contours show clockwise circulation. Contour interval is 4 Sv (a, c, e) and 2 Sv (b, d, f). Shading indicates the range of surface density. The model represents eddies using the GM parameterization (compare to Figure 9.10). Based on Hirst and McDougall (1998) Figures 1, 2, 3, 5, 6. ©American Meteorological Society. Used with permission.

Toggweiler and Samuels (1995) relate the wind-driven component of the deep meridional overturning circulation to the Ekman pumping in the Southern Ocean. The westerlies, with a maximum magnitude around 50°S, induce Ekman upwelling south of the maximum, northward Ekman transport in the vicinity of the maximum and Ekman downwelling north of the maximum (Figure 9.14). In Chapter 6 we discussed the idea of meridional overturning cells driven by, and directly under, the local zonal wind stress. We refer to overturning that extends far beyond the region of wind stress forcing as **remote wind-driven overturning**. Toggweiler and Samuels (1995) point to the relevance of the zonal periodicity of the Southern Ocean to the Ekman transport. The southward geostrophic flow needed to balance the northward Ekman transport cannot occur above the minimum depth of the periodic region in those latitudes (2.5 km in the Toggweiler and Samuels, 1995 numerical

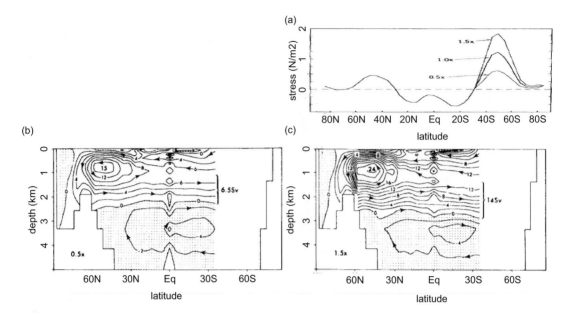

(a) Zonal average zonal wind stress profiles for three values of Southern Ocean wind stress, and meridional overturning streamfunction for (b) 0.5 wind case and (c) 1.5 wind case. Contour interval is 2 Sv. Reprinted from Toggweiler and Samuels (1995), Figures 2 and 3, Copyright 1995, with permission from Elsevier.

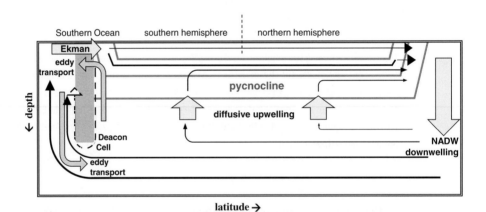

Elements of the Toggweiler and Samuels (1995) concept of wind-driven remote overturning and Gnanadesikan (1999) scaling, showing meridional ocean section with key transports (block arrows), representative streamlines (solid arrows), isopycnals (gray contours), and zonally periodic region (gray patch). The Deacon Cell is shown by the dashed contour that connects to the Ekman transport.

model and arguably 1–1.5 km in the real ocean). Therefore, the southern subpolar cell must extend to deep water (Figure 9.14). Surface water north of the Drake Passage is significantly lighter than water upwelling south of the wind stress maximum, and there is no local process that can make it dense enough. If some of the northward Ekman transport continues traveling far enough to the north, however, it eventually reaches the northern subpolar region where surface cooling and deep mixed layers allow the water to become dense enough (Figure 9.14). In this view, NADW flows from the northern hemisphere to the Southern Ocean to feed the Ekman suction south of the ACC. The Drake Passage effect of zonal periodicity thereby links Southern Ocean wind and the deep meridional overturning circulation.

Remote Wind-Driving Without Periodic Channel

Although the periodicity of the Southern Ocean plays an important role in the sensitivity of the deep meridional overturning circulation to wind, in fact, this periodicity is not required to link the two. In numerical experiments with a closed two-hemisphere basin, sinking in one hemisphere, and zonal wind stress in the other hemisphere, stronger wind stress gives greater basin-wide overturning (Tsujino and Suginohara, 1999). In other words, the Drake Passage effect works without the presence of a Drake Passage. Some of the reasoning of Toggweiler and Samuels (1995) for the Drake Passage can be applied to a closed basin: because meridional Ekman transport must return as a deep flow, some of the overturning takes a long route via a cooling region in the opposite hemisphere. For a closed basin, the depth of the return flow is set by how the pycnocline adjusts to allow the flow rather than by the sill depth of the Drake Passage (Klinger et al., 2004). Nevertheless, the zonal periodicity makes the overturning anomaly induced by a wind anomaly deeper and about twice as strong as in a closed basin.

Meridional Density Gradients

Like the mixing-driven mechanism, remote wind-driven overturning depends on the existence of meridional density variations. In a basin with no horizontal density gradient, imposing wind stress on one hemisphere produces a remote overturning that fades with time (McDermott, 1996). Thus one could argue that both mechanisms represent thermohaline or density-driven circulations, but it is traditional to apply these terms only to the case in which the circulation depends on subsurface mixing. Though Ekman pumping and mixing are very different processes, we can think of them as working on the circulation in an analogous way: by thickening the pycnocline. Subsection 8.2.1 describes how mixing increases pycnocline thickness. The initial effect of strengthening the southern hemisphere westerlies is to compress the pycnocline in the upwelling region on the southern half of the westerlies and thicken the pycnocline in the downwelling region on the northern half. Some of this pycnocline thickening propagates into the rest of the ocean, so that the new steady state has a thicker pycnocline (Klinger and Cruz, 2009; see their Figure 6).

Overturning Scaling Including Wind and Eddies

The contribution of Southern Ocean wind stress to overturning can be added to the advective-diffusive scaling (Subsection 8.2.3) to show how the overturning strength depends both on wind and vertical mixing (Gnanadesikan, 1999; see Subsection 9.3.1). Like the advective-diffusive scaling, it ignores details of the geometry of the basins, the relative densities of different sources of deep water, and other factors.

Gnanadesikan (1999) adds an additional conceptual element to wind and mixing. The Ekman transport in the Drake Passage latitudes is balanced by geostrophic flow but also by the transport velocity associated with mesoscale eddies (Figure 9.14). Here, it is important that the wind-driven overturning is thermodynamically indirect in the southern hemisphere, where it transports warm, light water equatorward and dense, cold water poleward. As mentioned above, this tends to *increase* the density and temperature differences between the equator and Antarctica. Meridional transport associated with eddies in the ACC is thermodynamically direct: by inducing a tendency to relax the sloping isopycnals, it opposes the wind-driven overturning (Subsection 7.2.2) and *reduces* the equator-to-Antarctic gradients. In eddy-suppressing numerical models that use the Gent and McWilliams (1990) parameterization, the eddy amplitude is controlled by the isopycnic eddy diffusivity parameter κ_I (see Subsection 7.3.3).

In the Gnanadesikan (1999) scaling, peak volume transport is equal to the sum of mixing-driven and wind-driven contributions minus an eddy contribution. This balance results in an equation that is cubic in pycnocline depth D, (9.9), and another that shows Ψ (volume transport sinking in North Atlantic) to be proportional to D^2, (9.7). Numerical solution of the cubic for D shows similar sensitivity of the global circulation to wind stress, vertical diffusivity κ_v, and κ_I as GCM experiments (Figure 9.15). Pycnocline depth and NADW sinking increase if Southern Ocean wind stress or vertical diffusivity increases (Figure 9.15a,b) and decrease if eddy diffusivity increases (Figure 9.15c,d). The numerical model of Gnanadesikan (1999) is very similar to that of Toggweiler and Samuels (1995), although he used the Gent and McWilliams (1990) parameterization whereas they used horizontal diffusion.

9.2.3 Elements of Antarctic Circumpolar Current Theory

The Challenge of the Antarctic Circumpolar Current

It is useful to start a discussion of the Antarctic Circumpolar Current (ACC) by comparing it to the subtropical gyres. If a wind-driven subtropical gyre is governed by linear dynamics and does not interact with any sea floor slopes, then the Sverdrup balance gives vertically integrated meridional speed V over most of the subtropical gyre. The gyre maintains a balance in which the ocean absorbs anticyclonic vorticity from the wind throughout the subtropical gyre and expels it through the solid boundaries adjacent to the western boundary current. We can neglect friction over most of the gyre because the Sverdrup balance is independent of friction. Given an impermeable eastern boundary, we can use continuity and V to calculate the corresponding zonal flow U.

In the ACC, the assumptions of the preceding paragraph break down. The latitudes of the Drake Passage (Figure 9.1) form a channel that is periodic in the zonal

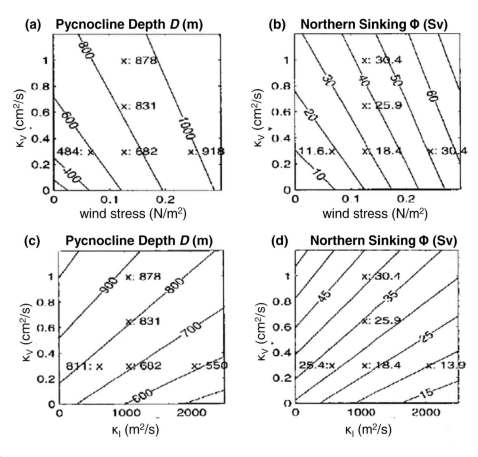

Figure 9.15 Solution to cubic scaling relation (contours) and results from an eddy-suppressing GCM (crosses with printed values) for integral depth scale of Atlantic pycnocline (left panels) and NADW sinking (right panels) as a function of κ_v and Southern Ocean zonal wind stress τ (top panels) or κ_v and κ_I (bottom panels). In top panels, $\kappa_I = 1000\,\mathrm{m^2/s}$, and in bottom panels, $\tau = 0.12\mathrm{N/m^2}$. Republished with permission of the American Association for the Advancement of Science, from Gnanadesikan (1999) Figures 2 and 3; permission conveyed through Copyright Clearance Center, Inc.

direction. Western boundary currents only exist north or south of the channel (which has no western boundary) and may not be able to remove the vorticity input by the wind.

If strong friction is not confined to the western boundary current, then the details of dissipative processes may be important in setting the ACC strength. Friction is intimately related to mesoscale eddies and other small-scale structures (Chapter 7), so we need to consider their dynamics as well. Similarly, in the subtropical gyres, nonlinearity associated with relative vorticity is confined to the vicinity of the strong western boundary currents (Subsection 3.3.1). Put another way, the Sverdrup balance is a linear theory. Observations show that the Southern Ocean has relatively fast zonal flow, suggesting the importance of nonlinearity for the ACC. The sharp fronts seen in Figure 9.6 may be one consequence of such nonlinearity.

Moreover, the absence of any zonal boundaries means that even if we knew V in the Southern Ocean we could not calculate U (unlike in the gyres). In detail, if we have a zonal current $U(\phi, \theta)$ in the channel that satisfies the continuity equation, we can add any function of latitude θ to U and it will still obey continuity. As zonal speed is the key feature of the ACC that we want to understand, this is a problem.

The Zonal Momentum Balance

In order to gain insight into how U is determined by the westerly wind in the Southern Ocean channel, we turn to the zonal momentum equation. We expect this balance to be more intuitive than the subtropical gyre vorticity balance discussed in Chapter 3. Zonal wind pushes on the water, imparting eastward momentum, and the flow has to take on a speed and spatial distribution that allows the momentum to be transferred in turn into the solid Earth.

Using $\langle \cdot \rangle$ to represent a zonal and temporal average, we have (see Exercise 9.4)

$$ u^2|_W^E + \langle (vu)_y + (wu)_z \rangle - f\langle v \rangle = -(1/\rho_0)\, p|_W^E + \nu_V \langle u_{zz} \rangle + \nu_H \langle \nabla^2 u \rangle. \tag{9.3} $$

In a periodic channel, the western (W) and eastern (E) boundaries are at the same longitude, and so the pressure terms are equal and cancel each other, as pointed out in Subsection 9.2.1. However, the Drake Passage is *not* periodic below about 1000–1500 m (depending on the latitude) because of deep barriers such as the South Sandwich Island ridge (Figure 9.1). This partial blocking of the flow allows differences in pressure to occur between the eastern and western flanks of a ridge (Munk and Palmen, 1951). At a solid boundary, pressure exerts a force perpendicular to the boundary. If the pressure on the upstream side of a ridge is greater than the pressure on the downstream side, there is a net horizontal force on the boundary in the downstream direction (Figure 9.16). The boundary, in turn, exerts an equal and opposite force on the water, opposing the current. A horizontal force exerted from below on a slab of fluid is equivalent to a vertical transport of momentum; here it is a downward transport of the current's momentum. This phenomenon is known as **form drag**. It also occurs in other contexts such as the generation of internal waves by flow over topography (Gill, 1982).

Like form drag, the friction terms in (9.3) can transmit momentum to the solid Earth as well. It has long been known that horizontal friction associated with mesoscale eddies is not strong enough to dissipate the momentum input from the wind. For horizontal friction to be strong enough to be a leading term in the force balance, the current would have to become much faster than in the real world (Exercise 9.4; Munk and Palmen, 1951). Similarly, relying on bottom friction necessitates a larger current than observations show. For this reason, in numerical studies, ACC current speed greatly increases when form drag is eliminated by removing topography (see, for instance, McWilliams et al., 1978). Diagnoses of numerical models show that the form drag is the dominant term balancing wind stress in the depth-integrated zonal momentum equation (McWilliams et al., 1978; Gille, 1997; Gnanadesikan and Hallberg, 2000; Tansley and Marshall, 2001; Olbers et al., 2004).

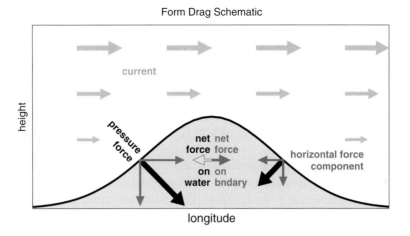

Form Drag Schematic

Figure 9.16 Schematic showing unequal pressure on two sides of ridge creating form drag on current.

Vertical Structure of the Momentum Balance

A closer look at the momentum equation reveals complications associated with the Earth's rotation. The $f\langle v\rangle$ term in (9.3) implies that the *zonal* wind stress may be balanced by the *meridional* current rather than the zonal one. Near the surface, Ekman velocity balances the vertical friction term; integrating the momentum equation vertically over the Ekman layer, the wind stress at the surface is balanced by the Ekman transport. That is precisely why we relate the meridional overturning to the zonal wind in (Subsection 9.2.2).

Does the balance between Ekman transport and surface stress obviate the need for the surface input of momentum to be balanced by an output of momentum at the ocean solid boundaries? No, because in steady state the total meridional volume transport must be zero (neglecting a small amount associated with evaporation and precipitation). As we saw in Subsection 9.2.1 and Subsection 9.2.2, the Ekman transport is associated with a return flow. When integrated over the entire water column, the $f\langle v\rangle$ term cannot contribute to the momentum balance. A few special cases are illustrative of some possible flow states.

If the Southern Ocean was unstratified with a flat bottom, there could be no net geostrophic meridional flow in the unblocked latitudes. The only source of a return flow to compensate the surface Ekman transport would be the bottom Ekman transport (Figure 9.17a). The zonal flow would have to become quite strong in order to generate a bottom transport that is as strong as the surface transport (Exercise 9.4). This is an example of the high-transport flat-bottom case that does not apply to the real ocean.

Now add stratification to this picture. The surface flow is not automatically linked to a bottom flow, raising the question of how the wind input of momentum at the sea surface escapes at the bottom. One solution involves mesoscale eddies. In a stratified channel, eddies drive a net transport of lighter water in one direction and denser water in the other, as in Figure 7.12. In the case of the Southern Ocean, the lighter water flows southward, countering the Ekman transport. In the limit of perfect compensation, this allows all the

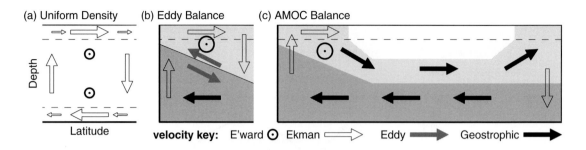

Figure 9.17 Schematic of latitude–depth cross section through ACC for (a) uniform density, (b) impermeable layers interacting via
eddy form stress and (c) layers interacting through global overturning circulation. The sea floor is flat in (a) and has
deep obstructions in (b) and (c). Shading shows density layers, dashed lines mark Ekman layers, arrows represent
meridional and vertical circulation, and circles represent eastward flow. North is to the right. Panels (b) and (c) are
based on Gnanadesikan and Hallberg (2000).

water carried northward as Ekman transport to return within the same density layer due to
the eddy transport (Figure 9.17b). The net meridional volume transport within the layer is
therefore zero.

If eddies cancel the Ekman transport, how is the momentum balance satisfied? The balance is maintained because the eddies also generate a stress on the bottom interface of the
layer. In fact this stress is also a form stress, associated with pressure differences between
upstream and downstream sides of eddies. The denser water in the lowest layer has an eddy-
induced northward flow that needs to be balanced by a southward flow (Figure 9.17b). If a
deep layer has eastern/western boundaries due to ridges, the southward transport can be a
geostrophic flow associated with a zonal pressure differences and form drag (Figure 9.16).
Described in terms of fluxes, the wind inputs momentum at the top, the eddies transmit it
downward, and form stress outputs it into the solid Earth. To put it another way, eddies
increase the effective vertical friction enough that the zonal current at the surface extends
all the way to the bottom (see discussion after (7.53)). In an adiabatic layer model (water
cannot move between layers), this balance prevails (Ward and Hogg, 2011).

The presence of equator-crossing basins adjacent to the Southern Ocean allows another
possibility: the northward Ekman transport sinks to great depth at high northern latitudes
in the adjacent basin(s) and returns below the ridge tops (Figure 9.17c). The geostrophic
return flow is associated with zonal pressure differences between eastern and western
flanks of the ridges, which cause a form drag. This is the NADW cell that Toggweiler
and Samuels (1995) linked to Southern Ocean westerlies. There is some conceptual symmetry in that the Southern Ocean zonal wind can drive a cross-hemispheric overturning
circulation (Subsection 9.2.2), and the overturning circulation plays a role in balancing the
zonal momentum input by the wind.

A related way of thinking about the relationship between overturning and the ACC is
to consider the pycnocline schematic in Figure 9.14. The deep water is all relatively close
to freezing and the surface water has a strong meridional density gradient due primarily to
the surface radiation balance. By the thermal wind relation (Subsection 2.1.2), the strength
of the zonal speed shear increases as the surface density gradient reaches more deeply

into the ocean. Processes that thicken or deepen the pycnocline, such as Southern Ocean wind stress and global diapycnal mixing, contribute to the ACC strength. Thus, wind stress drives the "density-driven" circulation and density forcing drives the "wind-driven" ACC.

We seek a theory linking ACC characteristics, such as volume transport and speed, to extrinsic factors, such as wind stress and topographic features. We also want to understand the role of processes, such as form drag and friction. Because both bottom friction and form drag depend on the vertical structure of the current, a model based on the velocity integrated over the entire depth of the ocean leaves out a key feature. In order to understand ACC strength, we need to understand the three-dimensional structure of the Southern Ocean circulation. This forces us to consider density (Chapter 5), eddy effects on mixing and friction (Chapter 7), and the meridional overturning (Chapter 6 and Chapter 8). Just as the Southern Ocean connects most of the oceans of the world, so too does its dynamics involve most of the topics in this book! In Subsection 9.2.4 we discuss further recent progress in understanding the parameter dependence of the ACC strength.

9.2.4 Sensitivity to Forcing

Relative Strength of Contributions to Overturning

For the range of parameters explored in Figure 9.15, pycnocline depth and overturning strength vary by a factor of about two. What values of overturning and pycnocline depth do the scaling and numerical models predict for the real-world values of the parameters? Both κ_I and κ_v have great geographical variation, however, and there is great uncertainty in their average values. Measurements of κ_v in the pycnocline are around 10^{-5} m^2/s but could be a few times higher or lower. For example, St. Laurent et al. (2012) find values of 2–13×10^{-5} m^2/s in the deep ACC at Drake Passage. Similarly, κ_I is probably in the range from 100 to 3000 m^2/s. For example, LaCasce et al. (2014) find lateral diffusivities of 800 ± 200 m^2/s in Drake Passage. Wind stress is known more precisely; with satellite observations, climatological values are known to probably better than 20% accuracy even in remote areas such as the Southern Ocean (Subsection 3.1.1).

In the real world, the NADW cell carries a volume transport of about 15 Sv (Subsection 8.1.3) and the pycnocline depth is about 570 m (as estimated for the western North Atlantic by Gnanadesikan, 1999). Those values can be reproduced by a model with $(\kappa_v, \kappa_I) \approx (4 \times 10^{-5}, 2000)$ m^2/s or with $(\kappa_v, \kappa_I) \approx (10^{-5}, 1000)$ m^2/s (Figure 9.15c,d). In the former case, the strong vertical mixing and strong compensation of Southern Ocean Ekman transport by eddies means that the majority of upwelling is associated with mixing over much of the ocean. In the latter case, the smaller mixing and less eddy compensation means the majority of upwelling is associated with Southern Ocean wind stress.

Uncertainties in the two mixing parameters make it difficult to state the relative importance of the two processes. If the wind dominates, most of the upwelling should occur in the Southern Ocean, but if mixing dominates, most of the upwelling should occur where mixing is greatest. Simply on the basis of the Pacific's great area, one might expect much of the mixing-driven upwelling to occur there. As we see in Subsection 8.1.3, estimates of deep flow into the Indo-Pacific cover a wide range of values and therefore do not exclude the possibility of either wind or mixing dominating the upwelling. Toggweiler and Samuels (1995) argue that radiometric tracers are not consistent with large upwelling at low latitudes. Nevertheless, Talley et al. (2003) find nearly 8 Sv in the Pacific and 18 Sv in the Indian upwelling to 2 km depth between about 30°S and 30°N, of which at least 6 Sv upwells to less than 1 km (Figure 8.5). For these reasons, at the time of writing, the question of what process dominates the upwelling is still open.

We are treating κ_v and κ_I as external parameters that can be set by decree to an arbitrary value. Such a view is useful for testing dynamical ideas in numerical experiments and is appropriate given uncertainties in the parameters' real-world values. It is unlikely, however, that the diffusivities are truly independent of the ocean circulation. As discussed below, mesoscale eddy strength is probably sensitive to the ACC strength, and so κ_I is probably a function of the large-scale circulation. Similarly, κ_v strength is determined by conversion of wind and tidal energy input to turbulence (Wunsch, 2002). The relationship between the energy input and κ_I may depend on the large-scale ocean state as well.

ACC Strength

Having seen in Subsection 9.2.3 how the zonal momentum balance involves the meridional flow, can we bring this back to the zonal current to constrain the ACC magnitude?

One family of theories is based on assuming a relationship between eddy strength and the zonal speed. Johnson and Bryden (1989) argue that for wind stress at the top to be transmitted down to topography below a kilometer or two, in steady state the vertical transport of momentum has to equal the wind stress X at the surface. If the momentum transport is due to eddies, and if we accept earlier work (Green, 1970; Stone, 1974) relating eddy strength and vertical shear of the zonal speed, then the zonal speed is proportional to $\sqrt{X}$. However, this theory of the ACC current strength is incomplete because it assumes that the stratification $\partial\rho/\partial z$, or, equivalently, the pycnocline depth h, is known. Variants of the Johnson and Bryden (1989) assumptions produce different dependence of zonal speed on X (Karsten et al., 2002; Marshall and Radko, 2003). While it makes sense to impose an externally defined surface buoyancy difference Δb across the width of the ACC, we still need a depth scale h to estimate $\partial\rho/\partial z$. This links the horizontal current strength to the meridional overturning. The Johnson and Bryden (1989) scaling is derived in Subsection 9.3.2.

Another family of theories (pioneered by Straub, 1993) predicts virtually *no* dependence of ACC strength on stress. In this conception, as long as there is enough shear in the ACC to cause baroclinic instability and generate eddies, the eddies carry momentum down fast enough to balance the wind input. A larger wind stress is accompanied by a larger eddy kinetic energy K but not a significantly faster current (Meredith et al., 2012; Munday et al., 2013). For instance, Marshall et al. (2016) claim that "the momentum budget sets its eddy energy...and the eddy energy budget sets its momentum." In their theory, the form stress is proportional to K, so equating the form stress and wind stress implies that K is proportional to X. They argue that K is created at a rate proportional to UK, and dissipated at a rate proportional to K, so if the two rates are equal, the factors of K drop out of the balance and hence U is independent of K and X.

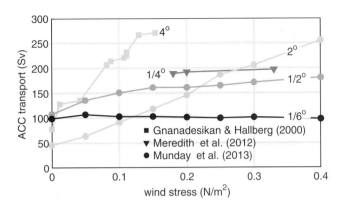

Figure 9.18 Sensitivity of ACC volume transport (Sv) to wind stress (N m^{-2}) for numerical experiments discussed by authors cited in the figure. Each curve represents a series of experiments conducted at a model resolution approximated by the number on the curve.

Testing the sensitivity of Southern Ocean processes to wind stress and other parameters is challenging due to the computational expense of running eddy-resolving models of global processes with long timescales. In eddy-resolving models with domains limited to the Southern Ocean, the ACC strength tends to be insensitive to wind stress (Tansley and Marshall, 2001; Hallberg and Gnanadesikan, 2006; Morrison and Hogg, 2013). Global numerical models give different dependence of the ACC on wind stress for different model resolution, as shown by a compilation of experimental results from various authors (Figure 9.18). Gnanadesikan and Hallberg (2000) look at an eddy-suppressing model that parameterizes eddies by horizontal diffusion, Meredith et al. (2012) examine results from an eddy-permitting climate model, and Munday et al. (2013) conduct eddy-resolving experiments with an idealized basin similar to that shown in Figure 9.8. The experiments show the system becoming insensitive to wind stress as the models resolve eddies better (Figure 9.18). It is noteworthy that even the coarse resolution models show significant ACC flow as the zonal wind stress vanishes, indicating that buoyancy forcing makes a contribution to the flow.

There is some inconsistency in our treatment of eddies for the overturning strength and the ACC strength. In the Gnanadesikan (1999) scaling for overturning, eddies in the Southern Ocean that reduce the large-scale overturning are calculated with a constant diffusivity κ_I (Subsection 9.2.2, Subsection 9.3.1). For the ACC scalings of Straub (1993); Munday et al. (2013); Johnson and Bryden (1989); Karsten et al. (2002) and others, the eddy diffusivity depends on the wind strength in such a way to reduce the sensitivity of the circulation to wind. A lack of sensitivity to wind stress is often referred to as **eddy compensation** for the meridional overturning and **eddy saturation** for the ACC. For instance, Figure 9.18 shows essentially complete eddy saturation for the highest resolution models.

Since the eddy transport counteracts the Ekman transport in driving the remote overturning, strengthening the eddies reduces the sensitivity to the wind (Hallberg and Gnanadesikan, 2006; Farneti et al., 2010). The quantitative sensitivity of overturning to the wind is still not clear. For instance, Morrison and Hogg (2013) find a larger degree of eddy compensation than the previously mentioned studies, and Meredith et al. (2012) argue that saturation should reduce the sensitivity of the ACC more than compensation reduces the sensitivity of the overturning.

In closing this subsection, recall that definitive eddy theories are not yet available (Chapter 7) and that numerical models currently are not finely-resolved enough to simulate all the relevant scales (Subsection 1.2.1). Therefore, the present results on ACC strength are provisional and a definitive conceptual explanation is still pending.

9.3 Theory

9.3.1 Wind-Eddies-Diffusion Overturning Scaling

Gnanadesikan (1999) extended the advective-diffusive scaling (Subsection 8.3.2) to include Southern Ocean wind driving the overturning and mesoscale eddies braking it. The scaling starts with the assumption that the volume transport of sinking in the North Atlantic, Ψ, is the sum of the upwelling transport Ψ_U and the meridional transport in the Southern Ocean Ψ_S:

$$\Psi = \Psi_U + \Psi_S. \tag{9.4}$$

The Ψ_S component consists of Ekman and eddy terms. If the transport velocity associated with eddies is v_E, and the zonal width of the Southern Ocean at a typical latitude of the Drake Passage is L, the transport is given by

$$\Psi_S = \left(\frac{X}{f} - v_E D \right) L, \tag{9.5}$$

where D is pycnocline depth as before and f is evaluated in the Drake Passage. For the Gent and McWilliams (1990) parameterization of eddy transport with diffusivity κ_I over a meridional width M, the transport velocity is given by κ_I/M (Subsection 7.3.3). We can take M to be approximately the meridional width of the ACC.

As before (Subsection 8.3.2), we assume that the upwelling transport is given by the advective-diffusive balance; the only difference is that we have added the Southern Ocean terms. For upwelling area A, we can rewrite the advective-diffusive scaling (8.24) as

$$\Psi_U = \kappa_v A/D. \tag{9.6}$$

As before, we assume Ψ is in thermal wind balance and write the scaling (8.26), (8.30) as

$$\Psi = c\Delta b D^2/f, \tag{9.7}$$

where c is a constant. Gnanadesikan (1999) uses a slightly different argument for this scaling based on the meridional volume transport being restricted to a frictional western boundary current, but the dependence on D is similar.

Substituting the above expressions for Ψ, Ψ_U and Ψ_S in (9.4), we get a cubic expression for D:

$$\frac{c\Delta bD^2}{f} = \frac{\kappa_v A}{D} + \left(\frac{X}{f} - \frac{\kappa_I D}{M}\right)L, \tag{9.8}$$

which implies

$$\frac{c\Delta bD^3}{f} + \frac{\kappa_I L}{M}D^2 - \frac{XL}{f}D - \kappa_v A = 0. \tag{9.9}$$

Once pycnocline depth D is calculated from (9.9), Ψ, Ψ_U, and Ψ_S can be calculated from their respective equations above. The analytic solution for a cubic is complicated and unenlightening. Gnanadesikan (1999) finds a numerical solution to (9.9) and finds that the dependence of D on κ_I, κ_v, and X is broadly in agreement with GCM solutions (Figure 9.15).

If κ_v and κ_I are small enough, $\Psi \approx \Psi_S$ which is simply the northward Ekman volume transport of the southern ocean. For the real world, northern Drake Passage at about $55°$S, we have $\tau \approx 0.12$ N/m^2 (Figure 3.1), $L \approx 2.3 \times 10^7$ m, $|f| \approx 10^{-4}$, so $\Psi \approx 28$ Sv.

9.3.2 Wind-Driven Models of Eddying Antarctic Circumpolar Current

Our goal here is to derive a scale relation for the ACC speed as a function of wind stress and other relevant parameters. The approach is to relate meridional property fluxes to eddies, and to relate the eddies in turn to the mean flow. Be warned that there is mounting evidence from numerical experiments that the scaling derived here is not a good model of the real ocean, but the derivation contains many elements that should be relevant to any theory of ACC strength.

One convenient property is buoyancy $b = -g\rho'/\rho_0$, (8.1). As Figure 9.12 shows, in numerical models the meridional eddy transport in the Southern Ocean is approximately equal and opposite to the overturning due to surface Ekman transport. Therefore, the associated meridional buoyancy fluxes are similar. Using angle brackets to denote averages over all longitudes and depths, and primes to denote departures from these averages (due to eddies),

$$\langle v'b'\rangle \approx \langle|v'|\rangle\langle|b'|\rangle. \tag{9.10}$$

Consider that the average (non-eddying) flow is due to Ekman overturning, with surface Ekman transport returned at a depth H, and the buoyancy difference between surface and return flow is Δb. Then

$$\langle v'b'\rangle \approx \frac{\Delta b}{H}\frac{X}{f}, \tag{9.11}$$

where X is (uniform) zonal wind stress divided by ρ_0. Karsten et al. (2002) make this assumption for an idealized domain with a flat bottom at depth H, in which case the return flow occurs in a bottom Ekman layer.

Johnson and Bryden (1989) considered the more general case in which undersea ridges may help balance the surface stress via form drag. They arrive at the same expression through a different route. They assume that the *vertical* transport of momentum occurs through **interfacial form stress**, which is the pressure force of one isopycnal layer of fluid on the layer below it. Just as there is a form stress at the bottom of a shallow water layer flowing over bottom topography (see Figure 9.16), interfacial form stress represents a similar force between layers of water. For a given interface at height $\eta(x, y)$, above all the bottom topography, the interfacial form stress due to eddies is

$$F = \int p' \frac{\partial \eta'}{\partial x} \, dx, \tag{9.12}$$

again integrating over all longitudes. Integrating by parts,

$$F = [p'\eta']_W^E - \int \eta' \frac{\partial p'}{\partial x} \, dx,$$
$$= - \int \eta' \frac{\partial p'}{\partial x} \, dx, \tag{9.13}$$

because of periodicity. Geometrical considerations show that in the limit of many layers (that is, continuous stratification), $\eta' = -b'/\bar{b}_z$, where the subscript represents a partial derivative. Similarly, geostrophy says that $p'_x = \rho_0 f v'$, (2.3). With these substitutions, (9.13) becomes

$$f \int \frac{b'}{\bar{b}_z} v' \, dx = X, \tag{9.14}$$

which is equivalent to (9.11). Downward transport of momentum by eddies is proportional to lateral transport of buoyancy.

Now relate the eddy transport to the average flow. Johnson and Bryden (1989) use the work of Green (1970) and Stone (1974) to argue that the transport can be written as

$$\langle v'b' \rangle \approx -chR_D \bar{u}_z \bar{b}_y \tag{9.15}$$

(see also Subsection 7.3.3 and Exercise 9.9). Here R_D is the internal deformation radius, which we can write in terms of Δb, the meridional difference in buoyancy across the ACC from (8.2). More accurately, R_D is related to the *vertical* buoyancy range, which we take to be proportional to the *horizontal* surface buoyancy range. The (dimensionless) $c \approx 0.144$ parameter is the correlation coefficient between the v' and b' anomalies.

Using the thermal wind relation (2.10) again, this time as $f\bar{u}_z = \bar{b}_y$, the Johnson and Bryden (1989) eddy parameterization (9.15) becomes

$$\langle v'b' \rangle \approx -c\bar{u}_z^2 h \sqrt{\Delta bh}. \tag{9.16}$$

Combining this equation and (9.11) gives

$$\frac{\Delta b}{H} \frac{X}{f} \approx \bar{u}_z^2 ch \sqrt{\Delta bh} \tag{9.17}$$

(disregarding signs).

If the bottom velocity is small compared to the surface value, we can approximate the surface zonal speed as $\bar{u}_S = h\bar{u}_z$, and

$$\bar{u}_S \approx \left(\frac{X}{cfH}\right)^{1/2} (\Delta bh)^{1/4}. \tag{9.18}$$

This result (9.18) suggests that the surface ACC speed depends on the square root of the wind stress. It therefore shows eddy saturation, as seen in Figure 9.18, at least to some extent (see also Exercise 9.9). It does not show the almost complete indifference to the wind stress as seen in the highest resolution models in Figure 9.18, however. Inserting representative numbers, $X\rho_0 = 0.2 \text{ N m}^{-2}$, $|f| = 1.25 \times 10^{-4} \text{ s}^{-1}$, $H = 3500$ m, $h = 1500$ m, $\Delta b = 1.5 \times 10^{-2} \text{ ms}^{-2}$, yields $\bar{u}_S \approx 0.12 \text{ ms}^{-1}$. This is an average speed, not the peak speed in frontal jets, and is the right order of magnitude compared to the current shown in Figure 9.4. If we take the ACC width to be $M = 10^6$ m, about $10°$ of latitude, the total transport is $\bar{u}_S MH/2 \approx 200$ Sv, a moderately realistic number compared to the observations reported in Subsection 9.1.2. Given the various uncertainties, likely greatest in the eddy parameterization (9.15), this level of agreement suggests that the theory is plausible.

Notice that (9.15) is, essentially, the same as the Green (1970) and Stone (1972) diffusive eddy parameterization from (7.43) (the GS case of Visbeck et al. 1997 seen in Figure 7.14), which reads,

$$\langle v'b' \rangle \approx -\kappa \bar{b}_y, \tag{9.19}$$

where

$$\kappa = \alpha \frac{M^2}{N} \ell^2. \tag{9.20}$$

Inserting $M^2 = \bar{b}_y$ and use of the thermal wind relation $\bar{b}_y = f\bar{u}_z$ gives:

$$\kappa = \alpha \frac{f\bar{u}_z}{N} \ell^2, \tag{9.21}$$

which, recalling that $R_D = Nh/f$, yields

$$\kappa = \alpha \frac{h\bar{u}_z}{R_D} \ell^2. \tag{9.22}$$

This result matches (9.15) if we choose

$$\alpha \ell^2 = cR_D^2, \tag{9.23}$$

and hence, for $\alpha = 0.015$ (from Subsection 7.3.3) and $c = 0.144$ (from above), $\ell \approx 3R_D$; that is, the width of the baroclinic zone ℓ is about three times the deformation radius R_D. This value is reasonable, albeit uncertain: see the discussion about ℓ on page 268.

Thus, Johnson and Bryden (1989), (9.15), essentially apply the Green (1970); Stone (1972) (GS) *diffusive* eddy parameterization. Their work pre-dated the Gent and McWilliams (1990) (GM) *advective* eddy parameterization ideas, and the Visbeck et al. (1997) proposal to unite GS with GM. More recent studies on ACC scaling apply the Visbeck et al. (1997) scheme (see Karsten et al., 2002 and Marshall and Radko, 2003). In these studies the ACC transport varies linearly with wind stress, so they do not exhibit eddy saturation.

Finally, notice two more things: these theories neglect the β effect and flow near the bottom. A more comprehensive theory would incorporate these effects. For a detailed presentation of a related theory, see Subsection 10.3.3, which applies to the Beaufort gyre.

9.4 Excursions: What Wrong Looks Like from the Inside, Part I

Looking back at the history of science, it is tempting to view it as the story of the pure at heart overcoming the prejudices of the less enlightened. There is some validity to this interpretation, but it is much harder to distinguish truth from error in the thick of the debate. When Galileo defended the heliocentric model, he was not just facing the Church, but also Tycho Brahe, the greatest astronomer of his age, who objected to the notion that the Earth moved because he could not detect the motion relative to the stars (Wootton, 2015).

Graduate school in the 1980s taught that the wind drove the Antarctic Circumpolar Current (ACC) and that water sank to the abyss in the North Atlantic because of a global-scale thermohaline circulation. Both numerical and laboratory experiments indicated that surface density differences and subsurface mixing could cause property distributions and deep overturning qualitatively similar to observations. When Robbie Toggweiler found that the overturning responded to the wind in a numerical model, it seemed absurd to say that a buoyancy-driven circulation was responding to the wind. The model he was using had very coarse resolution, which was known to generate spurious mixing, so probably stronger wind increased artificial mixing in the model, spuriously making it look like the wind was influential. Toggweiler is not a physical oceanographer; as a chemist maybe he was putting too much faith in his models. Then models with less spurious mixing reproduced the same effect. At the same time, measurements in the real ocean showed that the turbulent mixing that was needed to help drive the thermohaline circulation was weaker than expected. Now Toggweiler's idea is accepted as at least part of the puzzle.

The idea that the ACC is wind-driven also seemed hard to argue against. Wind-driven theory has had many successes explaining the depth-average flow in gyres, so one might assume the same for the ACC. Numerical models, theory, and analogies with non-rotating flows all argued for wind-forced motion. On the other hand, recent papers are now promoting the view that the ACC is insensitive to the strength of the wind (Subsection 9.2.4), and more sensitive to surface density differences across the current. It is still risky to lay bets on how future oceanographers may explain the ACC.

Exercises

9.1   The Southern Ocean is not the only place where the ocean is zonally unblocked: it also occurs in the Arctic Ocean (Figure 1.9). Following the discussion in Subsection 9.1.1, compare and contrast the zonally unblocked parts of the Southern Ocean and the Arctic Ocean. Discuss the geographical sizes of the two regions, their topography (as in Figure 9.1), and their zonal-average hydrography (as in Figure 8.1) and wind stress (as in Figure 3.2).

9.2 Use Argo, surface drifter, and sea-surface height data for this question (see Appendix A). For each of the subantarctic, polar, and southern ACC fronts (see Subsection 9.1.2):

(a) Show maps of instantaneous SSH and drifter tracks that identify the front.

(b) Estimate and map the flow velocity in the front and include the velocity estimate from your drifters.

(c) Plot hydrographic profiles from Argo floats near the front and hence identify the hydrographic transition at the front. Mark the locations of your Argo profiles on your maps from (a) and (b).

(d) By following surface drifters, estimate the fraction of time spent in the front and the fraction of time spent in slower ambient water. Hence, estimate the time taken for a surface drifter to circle Antarctica once. What is this time for a hypothetical float that always remains in the front?

9.3 Using appropriate bathymetry and climatological hydrography data (see Appendix A, see also Exercise 5.2):

(a) Make a Southern Ocean map of f/H, for Coriolis parameter f and water depth H. Comment on the dynamical significance of the f/H contours for the ACC.

(b) Make maps of planetary potential vorticity, Q, on isopycnals, σ, in the Southern Ocean. Identify the range of (Q, σ) space that connects around Antarctica (for example, on a scatter plot) and estimate the fraction of the ACC that connects this way. Explain the dynamical significance of the reconnection.

(c) Make maps of dynamic height on isopycnals. Hence estimate the transport of the part of the ACC that shows (Q, σ) reconnection around Antarctica (for example, consider the flux across a convenient meridional section). Compare your result with the estimates for the ACC transport and comment.

(d) **Optional extra**: Repeat parts (a)–(c) for the Arctic Ocean, following Exercise 9.1.

9.4 (a) Derive (9.3) stating your assumptions and defining all terms.

(b) Write a scaling relation for the ACC speed if *horizontal* friction is the primary force balancing the zonal wind stress. Estimate the zonal current and volume flux in such a flow and comment.

(c) Write a scaling relation for the ACC speed if *vertical* friction is the primary force balancing the zonal wind stress. Estimate the zonal current and volume flux in such a flow and comment.

(d) Write a scaling relation for the ACC speed if the meridional Ekman transport balances the zonal wind stress. Assume the northward Ekman transport is balanced by a bottom southward Ekman transport, as in Figure 9.17a. Estimate the zonal current and volume flux in such a flow and comment.

9.5   Explain and discuss the following terms giving evidence for (and, where appropriate, against) them: *Deacon Cell, eddy compensation, eddy saturation, form drag, the Drake Passage effect.*

9.6   (a) Using reasonable parameter values, reproduce Figure 9.15 for the Gnanadesikan (1999) model.
　　(b) Discuss the relevance of the other two roots to the cubic equation (9.9) for pycnocline depth.
　　(c) Explain how the scenarios depicted in Figure 9.17b,c are reflected in the Gnanadesikan (1999) model.
　　(d) Discuss to what extent the Gnanadesikan (1999) model shows eddy compensation.

9.7   The Gnanadesikan (1999) scaling for advective-diffusive-Ekman-eddy thermocline balance can be written

$$AD^3 + BD^2 - CD - \kappa_v A_o = 0, \tag{9.24}$$

where $A = c\Delta b/f$, $B = \kappa_I L/M$, $C = XL/f$, and Δb is the surface buoyancy difference ($\Delta b = -g\Delta\rho/\rho$), κ_I is the coefficient controlling thickness advection, (L, M) are the zonal and meridional extents of the Southern Ocean, X is Southern Ocean zonal wind stress, κ_v is vertical density diffusivity and A_o is the area of upwelling.

　　(a) Assuming the flow is geostrophic and hydrostatic, show that a reasonable scaling relationship for T_n (volume transport sinking in northern deep water formation region) is

$$T_n = c_1 \Delta b D^2/f, \tag{9.25}$$

　　　where c_1 is a constant.
　　(b) Using an eddy parameterization, show that the scaling for the volume transport associated with mesoscale eddy bolus velocity in the Southern Ocean is

$$T_{ME} = \kappa_I DL/M. \tag{9.26}$$

　　(c) Now assume that

$$T_n = T_u + T_E - T_{ME}, \tag{9.27}$$

　　　where T_u is volume transport associated with diffusive upwelling over most of the ocean, T_E is volume transport due to Ekman transport in the Southern Ocean. Write down expressions for T_u and T_E in terms of external parameters (such as X, κ_I, etc.) and derive the cubic equation for D above.

9.8   In this problem, assume all the equations in Exercise 9.7 are true. It is clear that we can write $C = C(X)$, with C increasing with X. However, if κ_I depends on the isopycnal slope in the ACC region, then B also increases with X. Also, if Δb decreases as X increases, then so does A. What is the sensitivity of T_n to X (in other words, $\partial T_n/\partial X$)? Does this answer show that T_n is less sensitive to X if κ_I is more sensitive to X? What if Δb is more sensitive to X?

Hint: It is convenient to calculate $\partial D/\partial X$ and use this to draw conclusions about $\partial T_n/\partial X$.

9.9 (a) Explain how the ACC speed scaling (9.18) exhibits eddy saturation. Write a scaling for the ACC transport from the scaling for speed. Compare and contrast this result to the numerical results in Figure 9.18.

(b) Explain how the eddy parameterization (9.15) is related to the eddy parameterization theories in Subsection 7.3.3.

(c) Replace the eddy parameterization (9.15) with the more general form (7.68). Comment on the eddy saturation and the fit to numerical results from Figure 9.18 in this case.

(d) Propose eddy parameterization formulae that reproduce the eddy saturation seen in Figure 9.18 for the highest resolution models.

(e) What salient features of the observed ACC does the wind-driven eddying theory of Subsection 9.3.2 not explain?

Hints: See Subsection 10.3.3 for a detailed solution to a similar problem. Exercise 10.8 is also helpful.

10

Arctic Circulation

Arctic (and Antarctic) oceanography differs from lower latitudes in several respects. Interaction with the atmosphere as well as temperature and salinity distributions are strongly affected by the formation and destruction of ice. This is especially true in the Arctic, a true polar ocean, which is ice-covered for at least part of the year. In high latitudes vertical property gradients are relatively strong compared to the rest of the ocean in some places, and relatively weak in other places. This fact is probably responsible for it being harder to separately analyze "gyre" and "overturning" circulations in the Arctic (and Antarctic). Circulation is influenced by the lack of zonal boundaries – in some latitude ranges the ocean occupies all longitudes – and the relatively small beta effect. For background on Arctic circulation, and access to the primary literature, consult Bluhm et al. (2015) and Rudels (2015).

10.1 Observations

The Arctic Ocean is distinctive for its high-latitude location, broad continental shelves (Figure 10.1a), and extensive ice coverage (Subsection 4.1.1, Figure 4.7). The greater Arctic is often taken to include water bounded by (traveling west around the North Pole) the Bering Strait, Siberia, Scandinavia, the Greenland–Iceland–Scotland Ridge, Greenland, the northern edge of the Labrador Sea, Canada and Alaska. Here we separate the Nordic Seas (Greenland Sea, Norwegian Sea, and Iceland Sea) so that the boundaries of the Arctic proper include the Fram Strait (between the northeast corner of Greenland and the Svalbard archipelago) and the continental slope marking the western edge of the Barents Sea.

Polar observations are difficult, principally due to the weather and the ice, and the high cost of mitigating their threat. Unlike other ocean basins, the initial geographic exploration continued into the twentieth century. Since the late 1970s, satellite observations (especially of sea ice) have rapidly advanced knowledge. In situ hydrographic observations also greatly expanded during the International Polar Year (2007–2008).

10.1.1 Arctic Circulation

Broadly speaking, the wind pattern over the Arctic is characterized by two structures. Over the Canadian Basin, wind blows anticyclonically around the Beaufort High (Figure 10.1b).

(a) Arctic Bathymetry

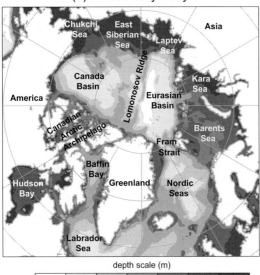

depth scale (m)

-4000   -3000   -2000   -500   -250   -50   0

(b) Ice Trajectories and Sea Level Pressure

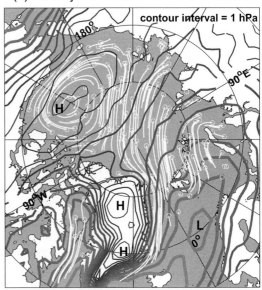

Figure 10.1   Arctic region (a) bathymetry and place-names, and (b) atmospheric sea level pressure (hPa; black contours for positive and gray for negative) and surface circulation (white curves starting at white circles). Data are from ETOPO5 (bathymetry), 1980–2012 average from Polar Pathfinder Sea Ice Motion dataset (Fowler et al., 2013), and NCEP/NCAR reanalysis (pressure, Kalnay et al., 1996). See Appendix A for details on these data.

Over the Eurasian Basin and Siberian continental shelves (Laptev, Kara, and Barents Seas), wind largely blows from Siberia toward the Fram Strait, part of a cyclonic circulation around a low over the Nordic Seas.

Ice cover deters conventional circulation measurements, but also provides an alternative: satellite-tracked trajectories of drifting sea ice roughly indicate the surface ocean flow. Over periods of months to years, sea ice follows surface ocean currents, though on short timescales (days or weeks) it responds more to the wind (as Fridtjof Nansen observed on the *Fram* expedition, inspiring Vagn Walfrid Ekman to develop the boundary layer theory that bears his name; see Subsection 2.1.3 and Exercise 2.5). Internal sea ice stress can also significantly modify the ice motion, especially in places with high concentrations of thick ice like north of the Canadian Arctic Archipelago and Greenland. Nevertheless, despite these difficulties we can use ice drift to track the circulation.

The 1980–2012 average (Figure 10.1b) shows an anticyclonic flow, the **Beaufort gyre**, in the western basin and flow across the pole, called the **trans-polar drift**, towards the Nordic Seas. Typical time-average speeds are 3–5 cm/s in the Beaufort gyre, and slightly smaller in the transpolar drift. There is also flow through the Canadian Arctic Archipelago, west of Greenland, from the Canadian Basin to Baffin Bay and the Labrador Sea. The ice-drift trajectories in Figure 10.1b show flow in Baffin Bay, but miss flows in the Canadian Arctic Archipelago because sea ice is often land fast there (meaning part of it is grounded on the bottom), and the seawater flows beneath it.

Many details of this basic picture are unclear because of logistic challenges in observing the Arctic. The flow also varies significantly from year to year, in response to the Arctic atmosphere. In particular, the strength of the Beaufort gyre and the transpolar drift increase and decrease: When the Beaufort gyre is large and strong, the transpolar drift flows directly from the Eurasian shelves across the pole to Fram Strait in a strong, wide current. At these times, the average current speed is around 10 cm/s approaching, and in, the Fram Strait. When the Beaufort gyre is weak and small, the transpolar drift penetrates further into the Canadian Basin and is also weaker; 1–2 cm/s (Haine et al., 2015).

Flow into Fram Strait in the transpolar drift indicates the importance of flow into and out of the Arctic Ocean at a few ocean "gateways." Access to the Arctic is dominated by the 2600 m deep Fram Strait and the Barents Sea Opening allowing exchange with the Nordic Seas. In comparison, Davis Strait is relatively narrow and shallow and Bering Strait is narrower and barely 50 m deep (Figure 10.2). As Figure 10.2 shows, water flows into the Arctic as the **Norwegian Coastal Current** and the **Norwegian Atlantic Current** in the Barents Sea Opening, the **West Spitsbergen Current** in the eastern parts of Fram Strait, the **West Greenland Current** in the eastern Davis Strait, and the **Alaskan Coastal Current** in the Bering Strait. Flow out of the Arctic occurs in the western parts of Fram and Davis Straits as the **East Greenland Current** and **Baffin Island Current**, respectively.

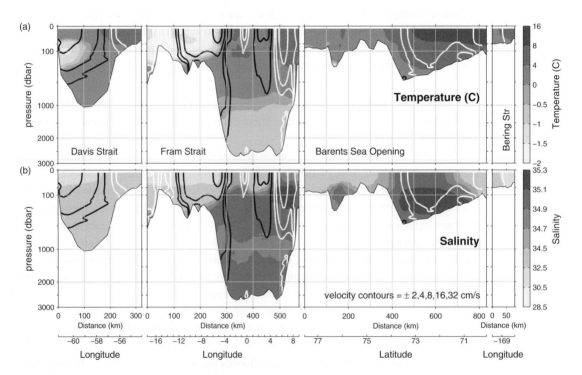

Figure 10.2 Arctic gateway sections with shading showing (a) temperature, and (b) salinity, and with contours showing geostrophic velocity into (white) and out of (black) the Arctic. Depths are proportional to square root of vertical distances. Adapted from Figures 5 and 9c of Tsubouchi et al. (2012). A black-and-white version of this figure appears in some formats. For the color version, please refer to the plate section.

Notice how the flow in Figure 10.2 relates to the bottom slopes: water tends to flow over bathymetry with the shallow water on its right-hand side. Tsubouchi et al. (2012) estimate these gateway currents, using hydrographic data and an inverse method to construct the level-of-no-motion, and hence determine the absolute geostrophic flow. They find that the net flow through Fram Strait is about 1.6 Sv out of the Arctic, with another 3.1 Sv out through Davis Strait. These outflows are balanced by inflows of about 3.6 Sv through the Barents Sea Opening and about 1 Sv through the Bering Strait.

The **Atlantic Water** inflows at Fram Strait and Barents Sea originate in the Norwegian Atlantic Current off the west coast of Norway. The branch entering the shallow Barents Sea flows northeast towards Franz Josef Land and Novaya Zemlya and carries 1–2 Sv. The Fram Strait branch flows northwest towards Svalbard and forms the West Spitsbergen Current which carries about 3 Sv through the eastern side of Fram Strait. On reaching the continental shelf break, the Barents Sea inflow sinks below the cold fresh surface layer, for example in the Saint Anna Trough, and enters the interior Arctic Ocean (see Figures 10.2 and 10.4). The Fram Strait branch enters the interior Arctic Ocean north of Svalbard. Its upper parts are transformed into fresher surface water due to ice melt (Subsection 10.2.1 and Subsection 10.3.2), and the deeper core flows east along the continental slope.

The Fram Strait and Barents Sea inflows feed the principal Arctic subsurface circulation, sometimes called the **Arctic circumpolar boundary current**, which flows cyclonically around the basin along the continental slope (Rudels et al., 1999a). The boundary current is a warm, salty subsurface layer centered at about 400 m depth. It splits at the bathymetric ridges separating the deep basins, with some water recirculating back toward Fram Strait and the Greenland Sea. The complex topography of the deep basins support several bifurcations. Branches flow towards Greenland at the Nansen–Gakkel Ridge and the Lomonosov Ridge (about 3 Sv). The rest of the Atlantic Water circulates around the Makarov and Canada Basins before returning to Fram Strait. Atlantic Water outflow through the western side of Fram Strait occurs in the East Greenland Current (see above) which joins the surface outflow of upper layer water and sea ice into the Nordic Seas. As at the surface, there is significant variability due to eddies, or meanders in the current.

The deep and abyssal waters in the Arctic Ocean (deeper than about 1500 m) circulate slowly and on pathways that remain obscure. Dense water close to the freezing temperature is produced on the Arctic shelves via ice formation and attendant brine release and then drains into the deep basins to replenish the abyss. Exchange through the deep parts of Fram Strait also occurs, which gives the abyssal Arctic waters access to the deep convection site in the Greenland Sea. These deepest waters are obstructed from leaving the Nordic Seas by the Greenland–Iceland–Scotland Ridge, however (see also Subsection 8.1.4).

10.1.2 Arctic Temperature, Salinity, and Freshwater

As mentioned in Subsection 1.1.5, ocean stratification is different in the Arctic than elsewhere. The Arctic surface mixed layer is very fresh and cold. It is close to the freezing point in winter and in summer when sea ice is present (see the hydrographic profiles in Figure 10.3, which show a 20 m thick surface mixed layer). Below the surface layer, first the salinity (Figure 10.3b) and then, at 100 m, the temperature (Figure 10.3a) increase with

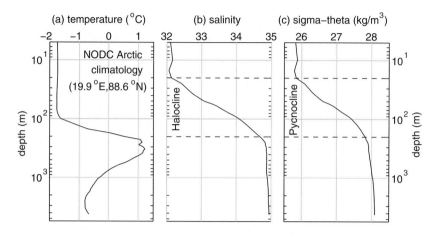

Figure 10.3 Arctic Ocean climatological profiles of temperature, salinity, and potential density, σ_θ, from near the North Pole. Note the logarithmic depth scale. The dataset is the NODC Arctic regional climatology (see Appendix A).

depth. They thus have opposite influences on density. The roughly 3 g/kg salinity increase raises density by over 2 kgm^{-3}, however, overwhelming the decrease of about 0.1 kgm^{-3} due to the temperature increase from $-1.8°C$ to $1°C$. Therefore this is a beta ocean in which stratification is controlled by the (permanent) **halocline** (Subsection 1.1.5). Below the halocline, the water in a roughly 500 m thick layer is much warmer and saltier. Beneath this layer, deeper than about 1500 m, the stratification (here controlled by temperature) is much weaker. This part of the water column is an alpha ocean. Regardless of its origin in temperature or salinity changes, the stably stratified ocean exhibits a pycnocline, separating the relatively light mixed layer water from relatively dense abyssal water (Figure 1.6c and 10.3c).

Why does the Arctic have the unusual feature of warm water *beneath* cooler surface water? The answer lies in the circulation of Atlantic Water under the cold surface layer (Subsection 10.1.1). Anomalous warmth and salinity trace the path of the Atlantic Water across the Arctic that was described in the previous section. Plumes of high salinity at 100 m depth show the flow of Atlantic water in the Norwegian Atlantic Current and its downstream branches in Fram Strait and the Barents Sea, seen in Figure 10.4a. The same figure shows a plume of low-salinity water leaving the Arctic in the East Greenland Current. Mapping the maximum temperature (at each latitude and longitude; Figure 10.4b) shows that the Barents Sea and Fram Strait have the warmest maxima, followed by the region just adjacent to the shelf, with temperature decreasing traveling east around the basin. A section from Norway to Svalbard and around the rim of the basin (Figure 10.4c) shows that this maximum marks the warm layer centered on a depth of a few hundred meters.

Along this pathway, the Atlantic Water loses heat upwards to the cold, fresh surface layer and sideways into the deep basin interiors. Understanding the rates and processes of this heat flux is a major objective of Arctic oceanography (see Subsection 10.2.1 and Subsection 10.3.1) because there is a large amount of heat stored in the Atlantic Water, sufficient to melt the sea ice floating at the surface (see Exercise 10.4). Vertical turbulent

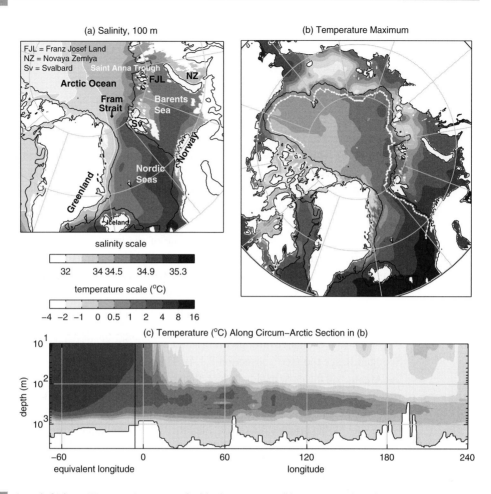

(a) Salinity, 100 m

(b) Temperature Maximum

FJL = Franz Josef Land
NZ = Novaya Zemlya
Sv = Svalbard

salinity scale

32   34 34.5   34.9   35.3

temperature scale (°C)

−4 −2 −1   0 0.5 1   2   4   8 16

(c) Temperature (°C) Along Circum−Arctic Section in (b)

Figure 10.4 Spread of Atlantic Water into Arctic as seen by (a) salinity at 100 m, (b) temperature above freezing at the local subsurface maximum temperature, and (c) temperature section along the white dotted line in (b). In (a) and (b), the black contour marks the 500 m isobath. Data are from the NODC Arctic regional climatology (see Appendix A).

mixing rates in the Arctic Ocean interior are exceptionally weak, however, and a large fraction of the Atlantic Water returns to the Nordic Seas through Fram Strait, without being mixed to the surface and losing its heat to the atmosphere, space, and ice melt. Evidence of the weak vertical mixing comes from CTD data typically showing a "staircase" profile. This phenomenon indicates interleaving of distinct layers with different temperature and salinity, and weak small-scale mixing due to **double diffusion**.

Atlantic Gateway Property Exchanges

At the Atlantic gateways to the Arctic, hydrographic data (Figure 10.2) shows that warm, salty water occurs in the eastern part of Fram Strait and the Barents Sea Opening while cold, fresh water occurs in the western part of Fram Strait (and Davis Strait is similar).

Notice how the circulation is (broadly) correlated with the hydrographic properties: Flow into the Arctic is warm and salty and flow out of the Arctic is cold and fresh. Cold and moderately salty water below 1500 m in the Fram Strait goes both into and out of the Arctic on the eastern and western sides, respectively, of the Strait.

Exchanges at the gateways play an important role in the heat and freshwater balance of the Arctic. For example, the East Greenland Current consists of fresh seawater at the freezing point on the Greenland shelf and above the continental slope (clearly visible in Figure 10.2). This water is less dense than the inflow. Beneath the fresh East Greenland Current is a deep outflow, called Intermediate Water, that is warmer and saltier than the upper part of the East Greenland Current. It is a cooled version of the inflowing water, somewhat fresher, and denser. In terms of the volume transport out of Fram Strait, the East Greenland Current carries about the same flux as the Intermediate Water (Tsubouchi et al., 2012). Finally, at the surface, a large volume of sea ice flows out of the Arctic over the shelf and the continental slope (see Figure 10.1b). Although its volume flux is negligible compared to the currents, sea ice carries a major flux of heat and freshwater (Tsubouchi et al., 2012). Subsections 10.2.1 and 10.3.1 discuss how the Atlantic inflow to the Arctic Ocean bifurcates into two outflows (three if we count the ice), one lighter than the inflow and one denser.

This curious exchange through Fram Strait essentially applies at the Greenland–Iceland–Scotland Ridge too (see Subsection 8.1.4). There is warm salty inflow at the surface from the Atlantic on the eastern side (which then forms the Norwegian Atlantic Current flowing north towards the Fram Strait and Barents Sea). On the western side the outflow is still split into three modes, although they have different strengths than at Fram Strait. There is a fresh, cold, light surface flow (also called the East Greenland Current); a salty, cool, dense overflow, at similar density to the Fram Strait Intermediate Water; and a stream of sea ice, mainly on the east Greenland shelf. Observations show that about 75% of the Atlantic inflow across the Greenland–Iceland–Scotland Ridge returns as overflow water (Rudels, 2010; Hansen and Østerhus, 2000). The water at Fram Strait that is deeper and denser than the overflow water (and Intermediate Water) is blocked by the Greenland–Iceland–Scotland Ridge.

Importance of Freshwater

A key theme in polar oceanography is freshwater, specifically, its distribution, mechanisms, and disposition (Carmack et al., 2016). The reason is that the surface stratification is determined by salinity. The freshwater cycle is intimately connected with sea ice too, which is important for polar and global climate (see Subsection 11.1.4), and mediates the wind driving the surface ocean. For example, the Arctic freshwater cycle is shifting at present so as to store more liquid freshwater in the Arctic Ocean (Haine et al., 2015). The summertime Arctic sea ice has also declined over the last few decades as a result of anthropogenic climate change (Subsection 4.1.1). The Arctic freshwater cycle is therefore an issue of great current interest.

There are large sources of freshwater in the Arctic Ocean (Haine et al., 2015): It receives meteoric water from runoff (mainly on the Eurasian side) at a rate equivalent to a volume flux of 0.4 m/yr over the surface of the entire Arctic. The net precipitation minus evaporation adds about 0.2 m/yr. Flow of relatively fresh seawater from the Pacific through Bering Strait further dilutes the salty Atlantic inflow through Fram Strait and the Barents Sea Opening, at about 0.26 m/yr. The seawater flowing through Bering Strait is fresher than the average Arctic Ocean salinity, which is (near) 34.80 g/kg and is often called the **reference salinity** (Aagaard and Carmack, 1989; see Exercise 10.2). When budgeting the Arctic freshwater cycle it thus traditionally counts as a freshwater source. (See Bacon et al., 2015 for a discussion and a robust proposal to replace the reference salinity. For our purposes, the improvement makes little difference.) Together, these sources (plus about 0.03 m/yr from Greenland) amount to about 0.9 m of freshwater on average being added to the surface of the Arctic Ocean each year. This flux is similar to the highest values of precipitation minus evaporation anywhere in the global ocean (see Figures 11.6 and 11.7).

These sources of Arctic freshwater are (nearly) balanced by marine export through the Canadian Arctic Archipelago and Fram Strait. The net flow through Davis Strait and Fram Strait removes about 0.33 m/yr and 0.28 m/yr, respectively. These flows are of liquid fresh water. Sea ice flux is also important; equivalent to about 0.25 m/yr: 94% of it flows through Fram Strait. In addition, the Barents Sea Opening, and Fury and Hecla Straits (in the Canadian Arctic Archipelago) support small net fluxes of about 0.03 m/yr. The total outflow is therefore about 0.9 m/yr, very close to the total inflows.

Overall, the Arctic freshwater cycle constitutes a sea ice factory: liquid freshwater flows in (from the atmosphere) and liquid freshwater plus sea ice flow out (in the ocean). In addition a large seasonal cycle in sea ice exists (Figure 4.7). For example, about 1.2 m of sea ice forms and melts, on average, over the surface of the Arctic Ocean each year. Moreover, the Arctic freshwater (liquid and solid) drains solely into the subpolar North Atlantic Ocean, where it plays an important and interesting role in the Atlantic Meridional Overturning Circulation (see Subsection 11.2.3 and Subsection 11.2.4).

10.2 Concepts

10.2.1 Arctic Hydrographic Properties

The Nordic Seas and Arctic Ocean are sometimes called the **Arctic Mediterranean Sea**. The reason is, like the Mediterranean Sea itself, the Nordic Seas are small, nearly landlocked basins with currents influenced mainly by temperature and salinity, and exchange with neighbouring basins, rather than by wind forcing (Tomczak and Godfrey, 2003; "Mediterranean" derives from Latin for "in the middle of land"). The exchange of water (both liquid and ice), heat, and salt between it and the subpolar Atlantic and Pacific Oceans is important for climate and climate change. The Arctic and Subarctic circulation is also interesting because the seawater density is controlled by freshwater, including the freezing and melting of sea ice.

Ultimately we would like to understand from first principles the velocity, distribution, and volume transport of the exchange of water between the Arctic and lower latitudes, as well as the temperature, salinity, and other properties of the water. Oceanographers have not yet achieved these objectives. Here we reach for a less ambitious goal: given the volume transport and hydrographic properties of the inflow, what controls the temperature and salinity properties of the outflow, and the sea ice export?

On a global scale the atmosphere carries water from low latitudes to the polar regions, and the Arctic Ocean is surrounded by land so that the effective catchment area is several times larger than the area of the Arctic Ocean itself. Thus it is unsurprising that the warm salty inflow should freshen (see for example Carmack et al., 2016). It is also unsurprising that the warm salty inflow should lose heat to the atmosphere, which in turn loses heat via long wave radiation to space. What is surprising, however, is that the colder and fresher Arctic outflow is split into three distinct modes that are separated in temperature, salinity, density, and depth.

Our discussion of the thermohaline circulation in Chapter 8 shows that the exchange between two water masses depends on density, with lighter water sliding over the heavier water mass. The lightest water has the temperature and salinity of the inflow, the densest water has the temperature and salinity of the densest surface water, and the entire circulation should have temperature and salinity in between these extrema. Why does the Arctic outflow have *three* extrema? The answer to this question rests on an additional element not discussed in Chapter 8: sea ice.

Role of Sea Ice

Sea ice is much less saline than seawater, with salinity of about 4 g/kg. The dissolved salts are therefore expelled from newly forming ice in dense brine at freezing temperature, which drains away from the surface with relatively little mixing. When sea ice melts, the buoyant liquid freshwater from the ice remains at the surface. This annual cycle of freezing and melting thereby stratifies, or un-mixes, the upper Arctic Ocean: about 1.2 m per year of freshwater is frozen and melted on average (Subsection 10.1.2). The process is sometimes called **ice distillation** (Aagaard and Carmack, 1989).

Large surface input of freshwater and sea ice distillation both contribute to a low surface salinity in the Arctic Ocean. Indeed, the Arctic must be (principally) a beta ocean (Subsection 1.1.5) for sea ice to exist: the surface temperature must be near freezing and therefore density stratification must be provided by salinity. Compared to salinity, temperature has little effect on density in the Arctic, which means it is dynamically passive. The reasons are the strong salinity gradients and the fact that the magnitude of the thermal expansion coefficient α becomes small near freezing (see Subsection 1.2.3 and (1.4); for salinities less than about 24, α even changes sign near the freezing point).

The inexorable loss of heat, especially during the polar night, depends mainly on sea ice. Cooling of the warm inflow occurs both by air/sea exchange (the ocean warms the frigid atmosphere and radiates to space) and by melting of sea ice that is blown over the warm water. The Eurasian Basin north of Svalbard and Franz Josef Land is a place where both these processes are important. Interestingly, not all the heat in the warm inflow is extracted. Instead, some of the warm saline water flows beneath the fresh freezing surface layer. The resulting subsurface temperature and salinity maxima comprise the core of the ubiquitous

Atlantic Water that circulates through the Arctic Ocean (Figure 10.4). Understanding how the inflow leaves the surface, and what controls the fraction of heat lost versus the fraction that recirculates, are important open questions.

Sea ice mobility plays an important role in setting the overall heat and freshwater balance. Sea ice moves with the wind and surface currents, and responds to internal stresses in the ice field (Subsection 10.1.1). Little ice is imported to the Arctic, so the ice is produced within the region and exported to the Atlantic (and a small amount to the Pacific). When newly formed ice covers a region, it greatly reduces the heat loss driving the ice formation. Therefore ice production is especially prolific in regions, such as the Siberian shelves, where sea ice drifts away after forming. The accompanying brine production accumulates in subsurface layers or at the bottom. Thus, ice formation on the Arctic shelves makes the Atlantic inflow saltier and denser despite the overall freshening of the inflow by less salty water from the Pacific and fresh water from the atmosphere (near the coast, runoff dominates in 10–20 km wide, 10–20 m thick buoyant fresh plumes called the **riverine coastal domain**: Carmack et al., 2015, 2016).

This ice formation and export also occurs on the continental shelves around Antarctica. In both places water cannot cool below the freezing point, so continued heat extraction produces more ice at the surface and saltier and denser water below. The dense water has the most extreme (freezing, salty, dense) hydrographic properties of any global ocean water mass. It spills off the shallow shelves, entraining the ambient water and hence decreasing density, to fill the deep and abyssal ocean. For these reasons we can say that the polar continental shelves are collectively one of the conceptual end points for the global overturning circulation.

Therefore, the Arctic Ocean exhibits characteristics of a dilution basin (like a fjord) because of the net input of meteoric freshwater. Simultaneously, it exhibits characteristics of a concentration basin, like the Mediterranean Sea, because of the local dominance of ice formation and brine rejection. Unlike the Mediterranean Sea, however, the Arctic transfers a large fraction of fresh water to the floating sea-ice reservoir rather than losing it to the atmosphere. The Arctic Ocean is sometimes called a **double estuary** because of these two characteristics (Rudels, 2010).

This idea is reflected in the two liquid modes being exported through western Fram Strait: the cold fresh light surface East Greenland Current (influenced by dilution), and the warm salty dense deep outflow (influenced by concentration). The balance and interplay of the processes described above control the properties and importance of these modes. A complete quantitative understanding of these processes is not yet available. Nevertheless, the following mixing arguments, based on pioneering work by Bert Rudels (for example, Rudels, 2010), are useful to frame the problem.

Fram Strait Water Properties

Consider the temperature and salinity distribution of Fram Strait hydrographic stations (Figure 10.5). We seek to understand, at least in broad terms, why the hydrographic data are distributed the way they are. In particular, what sets the properties of the East Greenland Current and Intermediate Water? Begin with the properties of the inflowing Atlantic Water.

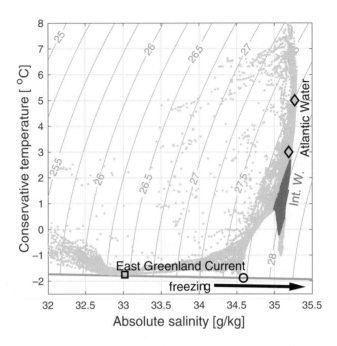

Figure 10.5 Fram Strait temperature and salinity from *RV Polarstern* cruise XVIII (gray dots; Ronski and Budéus, 2010; from PANGAEA, see Appendix A) in August 2002 at 78°N, with gray contours for σ_0 (kgm^{-3}) and freezing temperature at surface pressure (slanting gray line). Atlantic Water, East Greenland Current, and Intermediate Water (dark gray dots) are marked. The symbols come from the mixing model described in the text (see also Rudels, 2010, Figure 6).

The salinity, S_A, is around 35.15–35.25 g/kg. The temperature, T_A, ranges between about 3°C and 7°C. The observations are biased warm compared to the annual average, however, because they were made in summer. Therefore, we take the Atlantic Water temperature to lie between 3 and 5°C (diamonds in Figure 10.5).

First, think about what sets the East Greenland Current salinity. We imagine that the East Greenland Current is formed north of Fram Strait by the interaction of Atlantic Water and sea ice. The Atlantic Water is cooled by (i) heat loss to the atmosphere, and (ii) heat loss to melt sea ice that is blown into contact with the warm water. Let ϕ be the fraction of Atlantic Water heat that melts ice (so a fraction $1 - \phi$ warms the atmosphere). We find that the salinity of the product water, S_p, is

$$S_p \approx \frac{LS_A}{L + \phi c_p \Delta T_A} \tag{10.1}$$

(see Subsection 10.3.1 and (10.9)). Here, L is the latent heat capacity of seawater, c_p is its specific heat capacity, and $\Delta T_A = T_A - T_f$ is the excess Atlantic Water temperature above the freezing temperature (T_f).

This formula shows that as the fraction ϕ of Atlantic Water heat melting ice goes up, the salinity S_p of the product water goes down, which makes sense. Similarly, if the Atlantic Water is warmer (larger T_A), then more ice is melted and the salinity S_p is lower. The lowest salinity one can achieve is therefore for $\phi = 1$ and $T_A = 5°C$, the highest Atlantic

Water temperature. Then, $S_p = 33.0$ g/kg, which is marked with the square in Figure 10.5. This estimate is close to the freshest freezing temperature water in the East Greenland Current. There are some observations of water warmer and fresher, but they are affected by summertime solar warming, an unimportant process in other seasons, and neglected here.

What sets the salinity of the saltiest freezing temperature water in the East Greenland Current? Clearly, we need to use the smallest T_A in (10.1), say 3°C, and the smallest possible $\phi = \phi_{min}$. What might ϕ_{min} be, and why is it greater than zero, which is clear from the hydrographic data in Figure 10.5? It turns out that the lower limit on ϕ is given by:

$$\phi_{min} \approx \frac{\alpha L}{\beta c_p S_A}, \tag{10.2}$$

from (10.34) with α and β the thermal expansion and haline contraction coefficients, respectively (see (1.4) and (1.5)). This result derives from the constraint that during the winter, cooling must be accompanied by melting in order to maintain a stably stratified surface layer (Subsection 10.3.2). Hence, it makes sense that a nonzero fraction of the heat in the Atlantic Water goes to melt ice. Nevertheless, to compute this result we must consider the processes controlling the fresh mixed layer formed when Atlantic Water interacts with sea ice. It turns out that the fraction ϕ_{min} is the value that *minimizes the amount of heat melting ice*.

For now, we insert typical numbers into (10.2). Take α and β as the values at the temperature of the Atlantic Water. We find that $\phi_{min} = 0.35$ for $(T_A, S_A) = (3°C, 35.2$ g/kg). That gives $S_p = 34.6$ g/kg, indicated in Figure 10.5 with the circle and is a good approximation to the observed maximum East Greenland Current salinity. Similar considerations apply for the Barents Sea branch of Atlantic Water and in the Weddell Sea, where warm circumpolar deep water interacts with sea ice (Rudels, 2010; Nicholls et al., 2009).

10.2.2 Beaufort Gyre Circulation

Here we present a conceptual explanation for the surface intensified anti-cyclonic circulation in the Beaufort gyre (Figure 10.1b). The model is due to Manucharyan et al. (2016) (see also Manucharyan and Spall, 2015) and is relevant to the Antarctic circumpolar current (ACC) too (Subsection 9.3.2).

The Beaufort gyre is driven to circulate anticyclonically by the wind, specifically, the Beaufort High in sea level pressure (Figure 10.1b). The anticyclonic wind drives an anticyclonic stress on the ocean surface both by direct contact and by driving around the sea ice which then pushes the water. The anticyclonic stress drives convergence in the Ekman layer and downward Ekman pumping, $w_{Ek} < 0$ (Subsection 2.1.3, (2.13)). As in a subtropical gyre (Chapter 3), this wind forcing raises sea level and depresses the pycnocline (halocline for the Beaufort gyre). Associated with the sloping sea surface is an anticyclonic surface current in geostrophic balance. Associated with the sloping isopycnals (isohaline surfaces) is a thermal wind that causes the current to weaken with depth (see the main panel of Figure 10.9, which is discussed in Subsection 10.3.3). Unlike a subtropical gyre, the beta effect due to the meridional variation in the Coriolis parameter is unimportant. A quick calculation shows that the change in f across the Arctic Ocean is about 6% (between 70°N and the Pole), compared to 63% in the subtropical gyre (between 15°N and 45°N). Polewards of Cape Morris Jessup, the northernmost point on Greenland at 83.5°N, there are no meridional boundaries to the Arctic Ocean. Sverdrup balance and western intensification (Chapter 3) therefore cannot dominate the dynamics (confirmed by the detailed calculations of Yang et al., 2015). These features of recirculating flow with constant Coriolis parameter are common with the ACC.

Instead, a different basic balance must apply to counteract the wind forcing. An influential proposal is that mesoscale eddies provide, on average, a flow that balances that Ekman pumping (see also Subsection 9.2.3). The idea is that the surface-intensified current in the Beaufort gyre is susceptible to baroclinic instability (Subsection 7.2.1), which generates mesoscale eddies (see Figure 7.6). The kinetic energy of the eddies is extracted from the potential energy of the depressed pycnocline and thereby tends to lift it back up, in opposition to the Ekman pumping. If we average over many lifetimes of individual eddies, and around the whole Beaufort gyre, which includes many eddies, then we can solve for the average gyre characteristics, like the depth of the halocline and the current speeds. To do so, we need to exploit a parameterization law for the effects of the eddies themselves, as in Subsection 7.2.3 and Subsection 7.3.3.

In particular, consider axi-symmetric circular flow (as in Figure 10.9). The wind blows round with a surface ocean stress, $X(r)$, that depends on radius r only. The Ekman pumping speed, w_{Ek}, depends on the curl of the stress from (2.13). Because $X(r)$ decreases to negative values from zero in the gyre center (anticyclonic wind forcing), w_{Ek} is also negative (that is, downward pumping). The average equation for buoyancy (and mass) in steady state balances the pumping with upward flow due to the eddies, which are represented with the eddy-induced overturning streamfunction, $\Psi(r)$ (see (7.66)). Hence,

$$\frac{X}{f} + \Psi = 0. \tag{10.3}$$

Use of the Visbeck et al. (1997) parameterization for Ψ (7.68) leads to an expression for the depression of the halocline at the gyre center, Δh. For a stress that increases in magnitude linearly with radius, $X(r) = -\hat{X}r/R$, the Ekman pumping is uniform, and

$$\Delta h = \frac{2R}{3}\sqrt{\frac{\hat{X}}{fk}}, \tag{10.4}$$

where k is the eddy efficiency and the gyre has radius R. Plugging in typical values of $\hat{X}\rho_0 = 0.015 \text{ N m}^{-2}$, $R = 500$ km, $k = \alpha\hat{N}\ell^2 = \alpha\hat{N}(R/5)^2 = 3.7 \times 10^6 \text{ m}^2\text{s}^{-1}$, we find that $\Delta h = 68$ m (see Figure 10.9 and Subsection 10.3.3 for details).

Other useful quantities can be computed using this model, such as the geostrophic velocity field, the amount of freshwater stored in the gyre, and the transient dynamics of the Beaufort gyre. Subsection 10.3.3 explains how to do so for some of these quantities with mathematical details (see also Exercise 10.7 and Manucharyan et al., 2016). For present purposes, consider the total gyre transport streamfunction (for the horizontal flow). It is

$$\Upsilon_{max} \approx \frac{\delta^2 \hat{N}^2 \Delta h}{f}, \tag{10.5}$$

where the stratification has an exponential length scale δ (see (10.42)). For the example in Figure 10.9, $\delta = 200$ m, $\hat{N}/f \approx 120$, and the gyre transport is about 6.5 Sv.

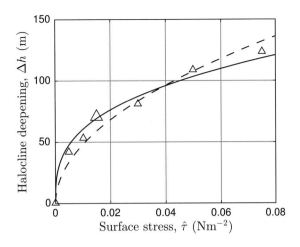

Figure 10.6 Halocline (pycnocline) deepening Δh, in eddy-equilibrated Beaufort gyre, as a function of peak surface stress $\hat{\tau} = \hat{X}\rho_0$ from theoretical predictions (curves) and numerical experiments (triangles; see Manucharyan et al., 2016). Theory includes parametrization in which eddy overturning is proportional to $|\gamma^n|$ (for isopycnal slope γ) with $n = 2$ (dashed curve) and $n = 3$ (solid curve), as discussed in Subsection 10.3.3. See also Figure 10.9 and Exercise 10.7.

Consider the sensitivity of the halocline depression, and via (10.5) the gyre transport, to the wind. Equation 10.4 shows that Δh grows as the square root of the stress. This dependence is seen in Figure 10.6 which shows Δh from the theory (the dashed line) and results from eddy-resolving GCM simulations (triangles). Notice that $\Delta h(\hat{X})$ and Υ_{max} are more sensitive to the stress $\hat{X}$ for small values of stress than for large values. This decreasing sensitivity of the gyre to the stress is called **eddy saturation**, as seen in the ACC. It stems from the eddy streamfunction depending on the isopycnal slopes as a power greater than one in (7.68). Here we use the Visbeck et al. (1997) parameterization which has $n = 2$, leading to the square root dependence of Δh and Υ_{max} on $\hat{X}$. Other choices appear possible, however, such as $n = 3$ (solid line in Figure 10.6). That choice leads to a cube root dependence of halocline depression in stress (see (10.46)) and thus to a stronger eddy saturation effect. This square root dependence of transport on wind is also seen in the ACC theory of Subsection 9.3.2, although that case uses the diffusive GS eddy closure, which is different in detail to the one used here.

10.3 Theory

10.3.1 Eurasian Basin Surface Salinity

Consider the heat and salt budgets for Atlantic Water to melt remotely-formed sea ice in the Eurasian Basin. The aim is to compute the salinity of the surface water thus formed, which is discussed in Subsection 10.2.1 and used in Subsection 10.3.2.

The amount of heat, E, available to melt ice in a mass M_w of Atlantic Water is

$$E = \phi M_w c_p \Delta T_A, \tag{10.6}$$

where ϕ is the fraction of Atlantic Water heat that melts ice, c_p is the specific heat capacity of seawater, $\Delta T_A = T_A - T_f$, and T_f is the freezing temperature of seawater. The ice starts at a temperature T_i, which is typically about $-10°C$ (Perovich et al., 1997). The ice is warmed to the freezing point by the Atlantic Water heat, and then melted. The mass of ice melted, m_i, by this heat is therefore

$$m_i = \frac{E}{L + c_i \Delta T_i}, \tag{10.7}$$

where $\Delta T_i = T_f - T_i$, L is the latent heat of fusion of ice and c_i is its specific heat capacity. Conservation of the mass of dissolved salts tells us that,

$$M_w S_A + m_i S_i = S_p (M_w + m_i), \tag{10.8}$$

where $S_i \approx 4$ g/kg is the sea ice salinity. In this equation the left-hand side is the total mass of dissolved salts from the individual components, and the right-hand side is the total mass of dissolved salts in the product seawater, with unknown salinity S_p. Hence,

$$S_p = \frac{(L + c_i \Delta T_i) S_A + \phi c_p \Delta T_A S_i}{L + c_i \Delta T_i + \phi c_p \Delta T_A} \approx \frac{L S_A}{L + \phi c_p \Delta T_A}, \tag{10.9}$$

where the approximation shows the dominant terms: heat capacity and salinity of the ice are not very important. This formula is (10.1), which we also use at the end of the following section.

10.3.2 Atlantic Water Conversion to the East Greenland Current

Next, think of a simple one-dimensional model for the interaction between warm salty Atlantic Water and Arctic sea ice which is formed remotely and blown into the Atlantic Water by the wind. The model crudely represents the conversion of Atlantic Water into the East Greenland Current. It considers ocean cooling during the winter season (polar night). It builds on the mixed-layer model described in Chapter 4, specifically Subsection 4.3.3, by including sea ice. For various melt rate and heat loss assumptions, we work out how the surface properties evolve over the winter. We focus on the coldest, saltiest properties possible. As shown in Subsection 10.2.1, this gives a reasonably accurate account of the East Greenland current properties flowing south through Fram Strait. In particular, the model predicts the minimum fraction of Atlantic Water heat melting ice, ϕ_{min} in (10.2). It is inspired by the paper by Rudels et al. (1999b) (for a recent summary, see Rudels, 2016).

Consider a one-dimensional water column with ice at the surface, warm Atlantic Water at depth, and a net heat flux, Q_a to the atmosphere (see Figure 10.7). There is melting of ice at a rate M_i (units of meters of ice melted per second), and we neglect evaporation and precipitation. The heat flux *leaving* the water, Q, is

$$Q = Q_a + \rho_i M_i \left[L + c_p(T - T_f) + c_i \Delta T_i \right], \tag{10.10}$$

where T, T_f, and T_i are the temperatures of the surface layer, the freezing point, and the ice, respectively; $\Delta T_i = T_f - T_i$; c_p (c_i) is the specific heat capacity of seawater (ice); ρ_i is

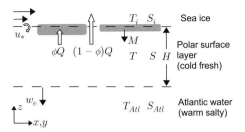

Figure 10.7 Schematic diagram of the one-dimensional winter polar mixing model. The cold fresh surface layer has thickness H, temperature T, and salinity S. Beneath it lies a warm salty Atlantic Water layer of temperature and salinity (T_A, S_A). Heat is lost to the atmosphere at a rate $(1 - \phi)Q$ and to sea ice at a rate ϕQ, where the ice temperature and salinity are (T_i, S_i). Mechanical mixing is parametrized with a friction velocity u_*. The upper layer thickens by entrainment at a speed w_e and sea ice melting at a speed M. The equations for the budgets of volume, heat, and salt are (10.12)–(10.14); the full model equations are (10.21)–(10.27).

the ice density; and L is the latent heat of fusion of ice. Notice that the meaning of the sign of Q here is opposite to that in Chapter 4, where Q is heat *gained* by the mixed layer.

The corresponding formula for the flux of salt, $\mathcal{F}$, leaving the water,

$$\mathcal{F} = -\frac{\rho_i}{\rho} M_i (S - S_i) = -M(S - S_i). \tag{10.11}$$

where the ice salinity is S_i (see Subsection 4.3.2). For future use, we define $M = (\rho_i/\rho)M_i$ in the second equality as the volumetric production rate of seawater (with density ρ) due to melting ice.

We imagine that a homogeneous (mixed), fresh, cold layer evolves at the surface as heat from the Atlantic Water is extracted. The layer is characterized by its depth H, temperature T, and salinity S (Figure 10.7). To begin with, it has almost identical properties as the Atlantic Water itself, but is slightly more buoyant. The layer deepens by entrainment of Atlantic Water at a speed $w_e \geq 0$. That process brings water into the surface layer with a temperature difference and salinity difference of $\Delta T = T_A - T$ and $\Delta S = S_A - S$, respectively. The equations governing the evolution of these properties are therefore:

$$\frac{dH}{dt} = w_e + M, \tag{10.12}$$

$$H\frac{dT}{dt} = w_e \Delta T - \frac{Q}{\rho c}, \tag{10.13}$$

$$H\frac{dS}{dt} = w_e \Delta S - M(S - S_i). \tag{10.14}$$

In this system, we must specify the heat flux Q, the entrainment speed w_e, and the melt rate M.

Entrainment Parameterization

To parameterize the entrainment speed we consider the joint effects of wind mixing and buoyancy gain for the surface layer. The wind adds turbulent kinetic energy to the upper ocean at a rate that is characterized with a friction velocity u_*. Surface buoyancy flux, B, is affected by melting of ice and heat loss to the atmosphere. A model based on the mechanical energy budget expresses the entrainment speed as a competition between wind mixing and surface buoyancy gain (see also Subsection 4.3.3),

$$w_e = \frac{\rho}{g\Delta\rho}\left(\frac{2m_0 u_*^3}{H} - \epsilon B\right), \tag{10.15}$$

$$\epsilon = \begin{cases} 1 \text{ for } B \geq 0, \\ 0.05 \text{ otherwise.} \end{cases} \tag{10.16}$$

The case considered here is for $w_e \geq 0$, so the layer cannot detrain and shrink in thickness (following Rudels et al., 1999b). The detraining case is more complicated: see Exercise 10.5. Physically, that means that the wind mixing is strong enough to overcome the buoyancy gain due to ice melt, and still entrain Atlantic Water. In (10.15), $m_0 = 1.25$ is a nondimensional dissipation factor, derived from laboratory experiments on deepening of a mechanically stirred layer into a stratified fluid (Kato and Phillips, 1969). The ϵ multiplier accounts for the fact that when the surface buoyancy flux is negative (densifying the surface layer), the buoyant convection is relatively inefficient at entrainment. The density difference $\Delta\rho$ between the Atlantic Water and the surface layer above is:

$$\Delta\rho = \rho(T_A, S_A) - \rho(T, S), \tag{10.17}$$

where the full equation of state is used. This density difference is dominated by the salinity difference (fresh surface layer overlying salty Atlantic Water), and so $\Delta\rho \geq 0$ and the water column is stably stratified. The entrainment speed is inversely proportional to the density difference in (10.15). Notice also how the entrainment due to wind mixing becomes less effective as the layer deepens (H increases).

Next, we need an expression for B, the rate of surface buoyancy gain. It is

$$B = g\left[\frac{-\alpha Q}{\rho c_p} + \beta M (S - S_i)\right], \tag{10.18}$$

with α and β the thermal expansion and haline contraction coefficients, respectively (see (1.4) and (1.5)). Remember that α (especially) and β vary with temperature and salinity. The first term in brackets in (10.18) is due to surface heat loss to both ice melt and to the atmosphere. The second term is due to the addition of buoyant melt water. The buoyancy flux is positive when the second term dominates the first,

$$M \geq \frac{\alpha Q}{\rho c_p \beta (S - S_i)}. \tag{10.19}$$

Finally, the melt rate M is:

$$M = \phi\frac{Q}{\rho\left[L + c_p(T - T_f) + c_i\Delta T_i\right]}, \tag{10.20}$$

meaning that a (constant) fraction ϕ of the total heat flux Q leaving the sea surface goes into melting the ice (the $1 - \phi$ remainder goes to the atmosphere, Q_a in (10.10)).

Full Model

Now assemble the system of equations for the surface layer properties

$$\frac{dH}{dt} = w_e + \phi \frac{Q}{\rho \left[L + c_p(T - T_f) + c_i \Delta T_i \right]}, \tag{10.21}$$

$$H\frac{dT}{dt} = w_e \left(T_A - T \right) - \frac{Q}{\rho c_p}, \tag{10.22}$$

$$H\frac{dS}{dt} = w_e \left(S_A - S \right) - \phi \frac{Q\left(S - S_i \right)}{\rho \left[L + c_p(T - T_f) + c_i \Delta T_i \right]}, \tag{10.23}$$

where

$$\Delta \rho = \rho \left(T_A, S_A \right) - \rho \left(T, S \right), \tag{10.24}$$

$$w_e = \max \left[\frac{\rho}{g \Delta \rho} \left(\frac{2 m_0 u_*^3}{H} - \epsilon B \right), 0 \right], \tag{10.25}$$

$$B = \frac{Qg}{\rho} \left\{ \frac{-\alpha}{c_p} + \frac{\phi \beta \left(S - S_i \right)}{\left[L + c_p(T - T_f) + c_i \Delta T_i \right]} \right\}, \tag{10.26}$$

$$\epsilon = \begin{cases} 1 \text{ for } B \geq 0, \\ 0.05 \text{ otherwise.} \end{cases} \tag{10.27}$$

We specify $T_A, S_A, Q, \phi, T_i, S_i$, and u_* (so that $w_e > 0$), plus initial conditions for (H, T, S). The physical parameters are $m_0, L, c_p, c_i, \rho, \rho_i, g, \alpha$ and β. They are chosen to be representative constants or vary with T and S. The solution $(H(t), T(t), S(t))$ depends on the control parameters (Q, ϕ), which are specified constants. We imagine integrating through the cooling season until the surface layer reaches freezing, $T = T_f$, at which point the calculation stops. Mathematically, the problem is a coupled set of nonlinear ordinary differential equations, which are solved with numerical integration software.

Model Solutions

Figure 10.8 shows representative solutions with different (Q, ϕ) values. For a range of sufficiently large (Q, ϕ) the upper layer reaches the **freezing temperature** T_f, melts 1–3 m of sea ice, and freshens by O(0.1) g/kg. Its depth is only slightly changed from the starting value ($H(0) = 50$ m here). This is not the only possible outcome, however, and in this case there are three other possibilities (see the left panel in Figure 10.8):

1 For sufficiently small ϕ there is almost no change in the upper layer temperature or salinity and no melting. These runs end with the upper layer becoming statically **unstable**, meaning that $\Delta \rho \leq 0$ and the upper layer begins to convect into the Atlantic Water. This process may be important in the real polar oceans, for instance in the Greenland Sea, but it is not relevant for the formation of the cold, fresh, light East Greenland Current, so we do not discuss it further.
2 For large ϕ and relatively small Q the heat loss is too weak for the upper layer to reach T_f in a realistic time period (180 days here). These runs take **too long** to reach freezing.
3 For intermediate ϕ the upper layer deepens substantially and becomes unrealistically **too deep** ($H \geq 200$ m here).

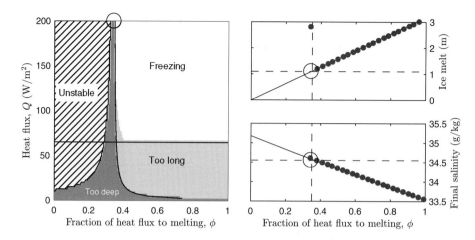

Figure 10.8 Results from the polar mixing model. The integration outcome is shown in the left panel: it stops when either the upper layer becomes too deep ($H \geq 200$ m in this case), becomes unstable and convects into the Atlantic Water ($\Delta\rho < 0$), reaches the freezing temperature ($T = T_f$), or runs out of time ($t = 180$ days in this case). The right panels show the final salinity and the ice melted for those runs that reach the freezing temperature (dots) plotted against the fraction of the heat flux to melting, ϕ. The circles show the properties of the run with maximum salinity. The other lines and curves are explained in the text. The initial thickness, temperature, and salinity of the upper layer are 50 m, 3.01°C, and 35.182 g/kg, respectively (see the lower diamond in Figure 10.5). The Atlantic Water temperature and salinity are 3.0°C and 35.182 g/kg. The sea ice temperature and salinity are −10°C and 4 g/kg. The friction velocity is $u_* = 3.7 \times 10^{-3}$ ms^{-1}, corresponding to a windspeed of 3 ms^{-1}. Results are shown for many combinations of the total heat loss Q and the fractional heat loss to sea ice, ϕ. To integrate the model the thermodynamic properties of seawater and sea ice are computed using the TEOS-10 functions (see Subsection 1.2.3).

Although their values are reasonable, the parameters we choose for Figure 10.8 control the detailed shapes of the different regimes (see Exercise 10.5). We tacitly assume that the system cannot run out of ice to melt. This is also a reasonable choice, but it may not be always true in the Eurasian Basin. We also assume that once the upper layer reaches the freezing temperature the system is more or less preserved in that state. In practice, further winter heat loss will cause ice formation by freezing, and therefore salinification of the upper layer by brine rejection. The assumption is that the heat loss Q is then greatly reduced by the ice cover and the upper layer properties have little opportunity to change before winter ends.

Now look at the right panels in Figure 10.8, which show with dots only the runs that reach the freezing temperature. We see that a maximum salinity exists for intermediate values of ϕ: it is marked with a circle. Correspondingly, there is a minimum thickness of sea ice melted. If less sea ice melts then the surface layer becomes denser than the Atlantic Water below and it overturns (or becomes too deep). The extreme properties of this special run are marked on the Fram Strait temperature–salinity diagram (Figure 10.5) with the circle. They are close to the observed properties of the saltiest waters in the East Greenland Current, encouraging us to think that the basic physics in the one-dimensional polar mixing model may be right.

Extreme Surface Properties

Understanding what sets the properties of this extreme case is the focus of the rest of this section. The argument is technical, but ultimately straightforward; it is also revealing. The first step is to select only the runs that reach the freezing temperature. There are two criteria to do so: the run must not take too long and the layer must not be too deep. The threshold, $Q = Q_{t_{max}}$, for runs that take too long derives from a simple heat budget. It is:

$$Q_{t_{max}} = \frac{\rho c_p H(0) \left[T(0) - T_f \right]}{t_{max}}, \tag{10.28}$$

where t_{max} is the maximum allowed time. The $Q_{t_{max}}$ value is plotted as the horizontal black line in the left panel of Figure 10.8. It agrees well with the outcome of the numerical integrations. Notice that (10.28) assumes that the upper layer does not change its thickness from the initial value $H(0)$ as it cools from temperature $T(0)$ to freezing (it neglects the relatively small increase due to ice melt). This assumption is consistent with the integrations in Figure 10.8.

The threshold for runs that avoid becoming too deep is that $w_e = 0$. Hence, from (10.25),

$$\epsilon B^* = \frac{2 m_0 u_*^3}{H}, \tag{10.29}$$

where we assume that the upper layer remains stratified $\Delta \rho > 0$ (the star superscript indicates the threshold value). Because, from (10.26),

$$B^* = \frac{Q^* g}{\rho} \left\{ \frac{-\alpha}{c_p} + \frac{\phi \beta (S - S_i)}{\left[L + c_p (T - T_f) + c_i \Delta T_i \right]} \right\}, \tag{10.30}$$

we see that (10.29) is a parametric equation for $Q^*(\phi)$. This function defines the edge of the too-deep regime in Figure 10.8. The $Q^*(\phi)$ relationship is of the form:

$$Q^* = \frac{Q_1}{1 + p_2 \phi}, \tag{10.31}$$

with parameters

$$Q_1 = \frac{-2 \rho c_p m_0 u_*^3}{\alpha g H}, \tag{10.32}$$

$$p_2 = -\frac{\beta c_p (S - S_i)}{\alpha \left[L + c_p (T - T_f) + c_i \Delta T_i \right]}. \tag{10.33}$$

Parameter Q_1 is the heat flux to the atmosphere required to balance the mechanically driven entrainment. It is a rearrangement of the formula for the Monin–Obukhov depth from (4.40). Flux Q_1 is negative, meaning that Q_1 corresponds to the rate of heat *gain* required to arrest entrainment in the absence of any melting. We set $\epsilon = 1$ in Q_1 because the upper layer must be gaining buoyancy $B > 0$ in order to counteract the effect of the wind mixing if the entrainment speed vanishes.

Parameters Q_1 and p_2 depend on the state of the upper layer, in particular via T (hence α) and H. A priori, it is hard to guess which are the appropriate choices to define the boundary between the freezing and too-deep regimes in Figure 10.8. Nevertheless, among the family of $Q^*(\phi)$ curves defined by the (Q_1, p_2) parameters, we want the one with smallest Q^* for a given ϕ. That one is the least stringent restriction on the freezing domain in the left panel of Figure 10.8. Therefore, we want the smallest $|Q_1|$ and $|p_2|$ (the modulus signs appear because Q_1 and p_2 are both negative). The black curve in the left panel of Figure 10.8 is this $Q^*(\phi)$ curve for the case of $H = 200$ m, the threshold for layers that are too deep, and $\alpha(T) = \alpha(T_A)$, which is the relevant value here. This $Q^*(\phi)$ curve is a good fit to the boundary of the freezing and too-deep regimes.

Notice that there is a smallest value of $\phi = \phi_{min}$ on this curve for very large Q^*/Q_1. That occurs when the total heat loss is much greater than the heat gained by entrainment; namely, for large heat loss rates and small friction velocities. The smallest value of ϕ is

$$\phi_{min} = -\frac{1}{p_2} = \frac{\alpha \left[L + c_p(T - T_f) + c_i \Delta T_i\right]}{\beta c_p (S - S_i)}, \tag{10.34}$$

which gives (10.2) when approximated. It applies for the largest value of α, which dominates the variation in p_2, and hence the largest temperature, $T = T(0)$. For moderate Q^*/Q_1 values we must have $\phi > \phi_{min}$ and hence a smaller salinity. The value ϕ_{min} is plotted as the dashed vertical line in Figure 10.8. It is a tight lower bound on the smallest ϕ for the model runs that freeze.

Now that we know the smallest value of ϕ for an upper layer that reaches the freezing temperature, we can compute the largest final salinity. The formula to use is (10.9), which follows from the budget of salinity for a warm layer that cools to freezing without any entrainment and melts ice with fractional efficiency ϕ. The salinity $S_p(\phi)$ predicted by (10.9) appears in Figure 10.8 as the diagonal black line and is an excellent fit to the numerical results shown with dots. The circle shows the maximum achievable salinity for water at temperature T_f and $\phi = \phi_{min}$,

$$S_p(\phi_{min}) = \frac{\beta (S - S_i) \left[L + c_i \Delta T_i\right] S + \alpha(T - T_f) \left[L + c_i \Delta T_i + c_p (T - T_f)\right] S_i}{\beta (S - S_i) \left[L + c_i (T_f - T_i)\right] + \alpha(T - T_f) \left[L + c_i \Delta T_i + c_p (T - T_f)\right]},$$

$$\approx \frac{\beta S^2}{\beta S + \alpha(T - T_f)} \tag{10.35}$$

(notice that the upper layer temperature and salinity, T and S, replace the Atlantic Water values used in (10.9)). The result of this formula is also shown on Figure 10.5. The prediction agrees well with the maximum observed salinity in the East Greenland Current. The total ice melted in the model (Figure 10.8) is simply given by (10.7) using the initial upper layer thickness to compute the energy E.

Physically, the maximum salinity is achieved by the upper layer that simultaneously (a) minimizes entrainment of warm Atlantic Water, and (b) minimizes the fraction of heat lost to melt ice. Thus the total mass of ice melted is minimized and so the freshening of the upper layer is minimized. Although entrainment of Atlantic Water increases both the upper layer's temperature and its salinity, the extra heat melts extra ice and (for reasonable parameter choices) the net effect is to freshen, not salinify, the upper layer.

The main point is not that we can match the maximum salinity of the East Greenland Current data in Figure 10.5. Judicious choice of the various parameters allows that. Instead, the main point is that there exists a physical mechanism that limits the maximum salinity. One can imagine an ensemble of model runs with various different initial conditions $(H_{init}, T_{init}, S_{init})$ and forcing conditions (Q, ϕ, u_*). (Rudels has argued that the natural system preferentially selects the case of minimum ice melt, which corresponds to a minimum input of turbulent mechanical energy. The reasoning is different in detail from that presented here; Rudels, 2016, 2010; Rudels et al., 1999b.) The process described here using the simple one-dimensional polar mixed-layer model selects a range of final salinities for water that reaches the freezing temperature. Among this subset of runs there is a limit to the maximum salinity possible, given by (10.35). We observe the same phenomenon in the field measurements.

10.3.3 Beaufort Gyre Circulation: Eddy–Wind Balance

Here we expand the conceptual model of the Beaufort gyre from Subsection 10.2.2. Apart from justifying the results in that section, the aim is to outline a general purpose model of a wind-driven baroclinic flow equilibrated by mesoscale eddies. The argument applies the geostrophic and thermal wind relations as previewed in Subsection 2.1.2. It also showcases a variation of the wind-driven eddying ACC model from Subsection 9.3.2 (see also Exercises 9.9 and 10.8). Specifically, it applies the GM *advective* eddy parameterization from Subsection 7.3.3 in its general form, (7.68). In the eddy-equilibrated ACC model of Subsection 9.3.2, we used the GS *diffusive* eddy closure. That ACC model shows eddy saturation; the present Beaufort gyre model also shows eddy saturation, with the details dependent on the precise form of (7.68). A sample Beaufort gyre model solution appears in Figure 10.9.

Begin by assuming that the wind and sea ice impart an azimuthal stress $X(r)$ that depends on radius r only, up to some maximum radius R. Beyond the gyre radius R the stress vanishes and the density field is simply a function of depth, which we specify below. The characteristic stress is $\hat{X}$.

We assume the fluid has a flat bottom. Alternatively, we can think of the stratification as strong enough to shield any deep bathymetry from the flow in the upper layers. This assumption may be realistic for the Beaufort gyre, but be aware that it is unrealistic in other contexts, such as the Antarctic Circumpolar Current.

The task is to exploit the essential balance in the buoyancy equation,

$$\frac{X}{f} + \Psi = 0, \tag{10.36}$$

to infer the density field $\rho(r, z)$ in the gyre and hence the horizontal circulation from geostrophy in the region of forcing. Notice that the eddy-induced streamfunction Ψ from (10.36) is a function of r only, because X only depends on radius. The eddy-induced speed is purely vertical (upwards) to oppose the Ekman pumping. We must therefore acknowledge that there are boundary layers somewhere near the surface and in deep water that are missing from the model. These boundary layers ensure that the eddy-induced overturning circulation respects the impermeability of the sea surface and sea floor. We assume that the details of the boundary layers do not affect the principal balance in (10.36), but just

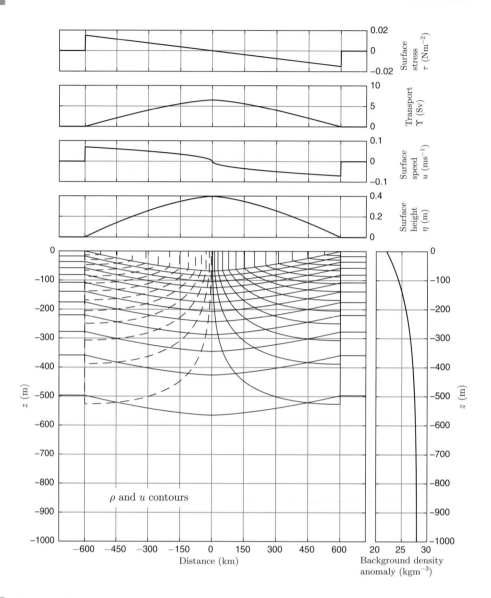

Figure 10.9 Solution for the eddy-equilibrated Beaufort gyre. The model parameters are: $\delta = 200\,\text{m}$, $(\rho_s, \rho_d) = (1022, 1028)\,\text{kgm}^{-3}$, $R = 600\,\text{km}$, $\hat{\tau} = \hat{X}\rho_0 = 0.015\,\text{N m}^{-2}$, $k = 3.7 \times 10^6\,\text{m}^2\text{s}^{-1}$, and $f = 1.4 \times 10^{-4}\,\text{s}^{-1}$. The case of an unstratified upper layer is shown (from Equation 10.61). The horizontal speed contour spacing is $0.005\,\text{ms}^{-1}$ (dashes indicate flow into the page) and the density contour spacing is $0.5\,\text{kgm}^{-3}$.

act to recirculate the eddy-induced steamfunction in a physically reasonable way. Notice also that this essential balance assumes the adiabatic limit (see Subsection 1.3.3 and Subsection 7.3.3), whereby diabatic effects are neglected. In other words, the cross-isopycnic mixing is assumed to be negligible; a convenient idealization.

Proceeding, we specify the eddy-induced streamfunction using (7.68), and so

$$\frac{X}{f} + k|\gamma^{n-1}|\gamma = 0 \tag{10.37}$$

for isopycnal slope $\gamma = \rho_r/\rho_z$ and eddy efficiency k. This equation is solved for $\rho(r, z)$ using the method of characteristics, as follows. Differentiate $\rho(r, z)$ with respect to a new coordinate s (whose meaning will become clear),

$$\frac{d\rho}{ds} = \left(\frac{\partial z}{\partial s}\right)\frac{\partial \rho}{\partial z} + \left(\frac{\partial r}{\partial s}\right)\frac{\partial \rho}{\partial r}. \tag{10.38}$$

From (10.37),

$$\text{sign}(X)\left(\frac{|X|}{fk}\right)^{\frac{1}{n}}\frac{\partial \rho}{\partial z} + \frac{\partial \rho}{\partial r} = 0. \tag{10.39}$$

Now define

$$\frac{\partial z}{\partial s} = \text{sign}(X)\left(\frac{|X|}{fk}\right)^{\frac{1}{n}}$$

$$\frac{\partial r}{\partial s} = 1, \tag{10.40}$$

hence

$$\frac{d\rho}{ds} = 0, \tag{10.41}$$

from (10.38), meaning that as s changes ρ is constant. Therefore, the s variable measures the radial distance along a ρ contour from the edge of the gyre inwards. At the edge, $r = R$, $s = 0$, and the density equals the background density of the fluid, $\rho(R, z)$, which applies around the gyre. We assume that the background density increases exponentially with depth as,

$$\rho(r \geq R, z) = \rho_d + (\rho_s - \rho_d)\exp\left(\frac{z}{\delta}\right), \tag{10.42}$$

where ρ_s and ρ_d are the surface ($z = 0$) and deep densities ($\rho_d - \rho_s = \Delta\rho > 0$) and δ is the characteristic pycnocline thickness (which we must specify in this theory). Therefore, the background buoyancy frequency N is given by

$$N^2 = \hat{N}^2 \exp\left(\frac{z}{\delta}\right), \tag{10.43}$$

with the characteristic value

$$\hat{N} = \sqrt{\frac{g\Delta\rho}{\delta\rho_0}}. \tag{10.44}$$

The solution $\rho(r < R, z)$ is found by integrating (10.40) inwards with a particular starting value of z and $\rho(R, z)$. For example, start at the surface at the gyre edge where $\rho = \rho_s$. We find that the isopycnal ρ_s is displaced by,

$$\Delta z(r) = \int_r^R \text{sign}(X)\left(\frac{|X|}{fk}\right)^{\frac{1}{n}} dr, \tag{10.45}$$

which is negative for anticyclonic forcing $X < 0$, indicating that the isopycnal is depressed as in the Beaufort gyre. Because the integrand in this expression is only a function of radius, not density, this formula actually applies for *all* density contours. Therefore, the stratification in the gyre identically matches the background stratification, just displaced in the vertical direction by Δz.

For the case where the Ekman pumping speed is uniform, $X(r) = -r\hat{X}/R$ for $r \leq R$ (and zero otherwise) the density contours at the gyre center are depressed by a (positive) distance

$$\Delta h = \int_0^R \left(\frac{\hat{X}r}{fkR}\right)^{\frac{1}{n}} dr. \tag{10.46}$$

Assuming that the Visbeck et al. (1997) eddy closure applies (for which $n = 2$, see (7.68)), we find

$$\Delta h = \frac{2R}{3}\sqrt{\frac{\hat{X}}{fk}}, \tag{10.47}$$

as in (10.4). The solution shown in Figures 10.9 and 10.6 is a particular example with these assumptions and parameters relevant to the Beaufort gyre and is discussed below.

Now let us find the horizontal circulation in the gyre. The geostrophic current $u(r, z)$ satisfies thermal wind balance with the sloping isopycnals,

$$f\frac{\partial u}{\partial z} = -\frac{g}{\rho_0}\frac{\partial \rho}{\partial r}, \tag{10.48}$$

(from Subsection 2.1.2). Note that $\hat{z} \times \nabla_p \rho$ in (2.9) in cylindrical coordinates equals $\left(-\frac{1}{r}\frac{\partial \rho}{\partial \phi}, \frac{\partial \rho}{\partial r}, 0\right)$. Hence, (10.48) applies for density with no azimuthal dependence. We assume that the deep circulation vanishes and integrate the thermal wind equation upwards to find

$$u(r, z) = -\int_{-\infty}^z \frac{g}{f\rho_0}\frac{\partial \rho}{\partial r} dz',$$

$$= \int_{-\infty}^z \frac{\gamma N^2}{f} dz',$$

$$= \mathrm{sign}(X)\left(\frac{|X|}{fk}\right)^{\frac{1}{n}}\frac{\hat{N}^2}{f}\int_{-\infty}^z \exp\left(\frac{z' - \Delta z(r)}{\delta}\right) dz',$$

$$= \mathrm{sign}(X)\left(\frac{|X|}{fk}\right)^{\frac{1}{n}}\frac{\hat{N}^2\delta}{f}\exp\left(\frac{z - \Delta z(r)}{\delta}\right), \tag{10.49}$$

where $\Delta z(r)$ is the isopycnal displacement from (10.45). The displacement of the sea surface η can be computed from this formula for $z = 0$ after applying geostrophic and hydrostatic balance, namely.

$$\eta(r) = -\frac{f}{g}\int_r^R u(r', 0)\, dr', \tag{10.50}$$

where we assume for convenience that the displacement vanishes at the edge of the gyre. Also, the streamfunction for the depth-integrated horizontal flow, $\Upsilon(r)$, is given by

$$\frac{\partial \Upsilon(r)}{\partial r} = \int_{-\infty}^{0} u(r, z') \, dz'. \tag{10.51}$$

For the case of uniform Ekman pumping and the Visbeck et al. (1997) closure, we compute

$$\Delta z(r) = -\frac{2}{3} \sqrt{\frac{\hat{X}}{fkR}} \left(R^{\frac{3}{2}} - r^{\frac{3}{2}} \right). \tag{10.52}$$

Therefore,

$$\frac{\partial \Upsilon(r)}{\partial r} = -\sqrt{\frac{\hat{X}r}{fkR}} \frac{\hat{N}^2 \delta^2}{f} \exp\left[\frac{-\Delta z(r)}{\delta} \right], \tag{10.53}$$

from (10.49) and so

$$\Upsilon(r) = \int_{r}^{R} \sqrt{\frac{\hat{X}r'}{fkR}} \frac{\hat{N}^2 \delta^2}{f} \exp\left[\frac{-\Delta z(r')}{\delta} \right] dr', \tag{10.54}$$

assuming that $\Upsilon(R) = 0$ at the gyre edge. Integrating, we find

$$\Upsilon(r) = \frac{\delta^3 \hat{N}^2}{f} \left\{ \exp\left[\frac{(R^{3/2} - r^{3/2})}{b^{3/2}} \right] - 1 \right\}, \tag{10.55}$$

with length-scale parameter

$$b = \left(\frac{9fkR\delta^2}{4\hat{X}} \right)^{\frac{1}{3}} = R \left(\frac{\delta}{\Delta h} \right)^{\frac{2}{3}}. \tag{10.56}$$

The overall gyre strength is quantified by $\Upsilon(r = 0)$, the value of the horizontal streamfunction at the center. In practice, we find that $b > R$ (for example, $b/R \approx 2$ in Figure 10.9). Therefore the argument to the exponential function in (10.55) is small, and to reasonable accuracy the total gyre transport is

$$\Upsilon(0) \approx \frac{2\delta^3 \hat{N}^2}{3f} \sqrt{\frac{\hat{X}}{fk}} = \frac{\delta^2 \hat{N}^2 \Delta h}{f}. \tag{10.57}$$

We assume above that the stratification is given by the exponential relation (10.43) even for positive values of $z - \Delta z$ in (10.49). That means that the upper layers in the model Beaufort gyre are more strongly stratified than the background water surrounding the gyre. The switchover to exponential behavior in (10.55) when $R \gtrsim b$, namely when $\Delta h > \delta$, is controlled by this stratified upper layer.

An alternate assumption, perhaps more realistic, is to say that the upper layers in the gyre have come from Ekman convergence of the surface water and therefore have constant density equal to ρ_s. In this lens of unstratified (fresh) water the thermal wind vanishes and therefore the surface speed and the total gyre transport $\Upsilon(0)$ are diminished compared to the stratified case (10.55). The lens is defined by the water shallower than $\Delta z(r)$. To compute the transport $\Upsilon(r)$ we proceed as before except that the integrand in the third line of (10.49) is set to zero whenever $z' - \Delta z(r) > 0$. Therefore, (10.51) becomes,

$$\frac{\partial \Upsilon(r)}{\partial r} = \int_{-\infty}^{\Delta z(r)} u\left(r, z'\right) \, dz' + \int_{\Delta z(r)}^{0} u\left(r, \Delta z(r)\right) \, dz'. \tag{10.58}$$

We compute that

$$\frac{\partial \Upsilon(r)}{\partial r} = -\sqrt{\frac{\hat{X}r}{fkR}} \frac{\hat{N}^2 \delta^2}{f} + u\left(r, \Delta z(r)\right) \Delta z(r),$$

$$= -\sqrt{\frac{\hat{X}r}{fkR}} \frac{\hat{N}^2 \delta \left[\delta - \Delta z(r)\right]}{f}. \tag{10.59}$$

Hence, from (10.52),

$$\Upsilon(r) = \frac{2\delta^2 \hat{N}^2}{3f} \sqrt{\frac{\hat{X}}{fkR}} \left[\left(R^{3/2} - r^{3/2}\right) + \frac{1}{3\delta}\sqrt{\frac{\hat{X}}{fkR}} \left(R^3 - 2R^{3/2}r^{3/2} + r^3\right)\right],$$

$$= \frac{\delta^2 \hat{N}^2 \Delta h}{fR^{3/2}} \left[\left(R^{3/2} - r^{3/2}\right) + \frac{\Delta h}{2\delta R^{3/2}} \left(R^3 - 2R^{3/2}r^{3/2} + r^3\right)\right], \tag{10.60}$$

and the total gyre transport is

$$\Upsilon(0) = \frac{\delta^2 \hat{N}^2 \Delta h}{f} \left(1 + \frac{\Delta h}{2\delta}\right). \tag{10.61}$$

This result is very similar to the case of a stratified upper layer (10.57) for weak halocline depression compared to the halocline thickness, $\Delta h/\delta \ll 1$. It is the basis for (10.5) in Subsection 10.2.2. The transport $\Upsilon(0)$ is essentially proportional to halocline depression Δh. The model can therefore exhibit a degree of eddy saturation, as discussed in Subsection 10.2.2, because Δh goes to the $1/n$-th power of $\hat{X}$ from (10.46). Figure 10.6 shows the $n = 2$ and $n = 3$ cases, for example.

Figure 10.9 shows the solution for this eddy-equilibrated model of the Beaufort gyre with an unstratified upper layer. We choose representative parameters, in particular the peak stress on the ocean, at the gyre edge, is $\hat{X}\rho_0 = 0.015 \, \mathrm{N\,m^{-2}}$, which is relatively weak compared to typical wind stresses elsewhere (see Figures 3.1 and 3.2). The anticyclonic current peaks at about $7 \, \mathrm{cms^{-1}}$ and the total transport is $\Upsilon(0) \approx 6.5$ Sv. The halocline depression is $\Delta h \approx 68$ m, which is reasonably realistic. Notice that the model (either (10.55) or (10.60)) predicts the transport $\Upsilon(r)$ given $\Delta h(r)$ and information on the stratification, which can be estimated from hydrographic observations.

Here, we take a different approach. We estimate the eddy efficiency k by comparing the model predictions with results from eddy-resolving calculations using an idealized GCM (see Manucharyan et al., 2016 for details). Figure 10.6 shows the halocline depression for seven eddy-resolving simulations with varying surface stress. The large triangle shows the case of $\hat{X}\rho_0 = 0.015 \, \mathrm{N\,m^{-2}}$, as in Figure 10.9. The lines on Figure 10.6 show the theoretical predictions for $n = 2$ (dashed) and $n = 3$ (solid). The model presented above for $n = 2$ (in particular, (10.47)) is a good fit to the $\Delta h(\hat{X})$ values from the eddy-resolving runs, if we pick $k = 3.7 \times 10^6 \text{ m}^2\text{s}^{-1}$, which gives the dashed line in Figure 10.6. Referring to the $n = 2$ closure for k from (7.68), we see that

$$\ell = \sqrt{\frac{k}{\alpha \hat{N}}} \approx R/5 \tag{10.62}$$

(recall that α is a dimensionless baroclinic instability parameter here).

It is unclear why the baroclinic zone width is only about 20% of the gyre radius in this system. The surface current falls to zero in the gyre center, so it makes sense that the baroclinic zone width should be less than the radius. The question is, how much less? A complete theory would include a priori arguments about the value of ℓ and its dependence on the parameters in the problem (a complete theory would also account for the background stratification $\rho(R, z)$). Moreover, the results from the $n = 3$ eddy closure are about as skillful as those from the $n = 2$ closure. These results are encouraging, although the agreement is necessary, but not sufficient, for the simple model to be relevant to the real Beaufort gyre. This statement means that other effects in the real Beaufort gyre may be important and dominate. Still, the simple model has value because of its transparency and applicability to other circumstances. It is a good working hypothesis for the basic dynamics of the real Beaufort gyre.

10.4 Excursions: What Wrong Looks Like from the Inside, Part II

The Nordic Seas provide another example of conventional wisdom being replaced by a new idea. On global scales, warm water becomes colder and denser as it flows to polar regions, where the densest water sinks as part of the process of deep water formation. Therefore, it was easy to accept a conceptual model of deep water formation in the Nordic Seas in which surface water enters from the Atlantic and cools as it flows into central regions of the basin, and part of the resulting water mass returns to the Atlantic in the dense overflows at the Greenland–Iceland–Scotland Ridge. Some of the densest surface water in the basin occurs in that area, which undergoes intense winter heat loss to the atmosphere and dramatic convection. Numerical models of convection (Killworth, 1983) show how an isolated region of cooling creates a subsurface tower of dense water that carries water downward when it slumps due to eddy generation.

This model of Nordic Seas circulation was overturned by Norwegian oceanographer Cecilie Mauritzen in her PhD dissertation (Mauritzen, 1996, and subsequent papers). She showed that the water took a longer route from the inflow to the outflow, with most of the cooling occuring in boundary currents around the periphery of the Nordic Seas and in the Arctic Ocean. Much of the large heat loss in the central basin was merely part of the seasonal cycle of temperature change of a thick layer of water, with relatively small volume transports exchanged with incoming or outgoing water. Later, numerical experiments such as Spall (2004) showed that a fluid exchange between cooling basin and the rest of the ocean mostly occurs around the basin boundaries, with cold but relatively quiescent water in the middle.

Mauritzen is soft-spoken but fearless. Long after her PhD, she sailed across part of the South Pacific in a small wooden raft as chief scientist on the first leg of the Kon-Tiki2 Expedition (www.kontiki2.com). On the second leg, the raft was struck by 8 meter waves and eventually crew and raft had to be rescued due to cumulative damage to the raft.

Exercises

10.1 (a) You place a container holding 3kg of freshwater at 15°C in a freezer that removes heat from the water at 40W ($1W = 1Watt = 1Js^{-1}$). What is the temperature of the water after 12 hours?

(b) You place a container holding 10kg of seawater at 15°C and salinity 34.80 g/kg in a freezer that removes heat from the water at 40W. What is the salinity of the liquid water after 12 hours?

(For simplicity, neglect the change of freezing temperature with salinity.)

10.2 The **liquid freshwater content** m relative to a reference salinity S_{ref} of a salinity profile $S(z)$ is given by:

$$ m = \int_{z_{ref}}^{0} \frac{S_{ref} - S(z)}{S_{ref}} \, dz, \qquad (10.63) $$

where $S(z_{ref}) = S_{ref}$.

(a) Explain why m measures the total thickness of a layer of freshwater that has been hypothetically mixed into a water column of uniform salinity S_{ref}.

(b) Imagine removing 15m of water from the top of the upper Arctic ocean which has a uniform salinity of 34.8 g/kg. Replace it with 15m of freshwater and then completely mix the upper 50m of the water column. Draw the salinity profile as a function of depth. What is the liquid freshwater content?

(c) Estimate the liquid freshwater content from the Arctic salinity profile in Figure 10.3, making sensible simplifying assumptions that you clearly state.

10.3 Use, for example, the NODC Arctic regional climatology for this question (see Figure 10.4 and Appendix A). You may also find the freshwater budget in Haine et al. (2015) or Carmack et al. (2016) useful.

(a) Write software code to compute the liquid freshwater content of an arbitrary hydrographic profile, using (10.63).

(b) Compute m from your hydrographic climatology as a function of latitude and longitude, and make a map. Where is the greatest and least liquid freshwater stored in the Arctic Ocean?

(c) Integrate m over the Arctic ocean to compute the volume of the liquid freshwater reservoir (see Subsection 10.1.1) in cubic kilometers. Compare and contrast your answer to the volume of liquid freshwater stored in sea ice, which you should estimate from the information in Subsection 10.1.1 (assume the area of the Arctic Ocean is 9.7×10^6 km^2).

(d) Using the values for the fluxes in Subsection 10.1.1, estimate the residence time for liquid freshwater in the Arctic Ocean. Make simplifying assumptions that you clearly state.

10.4 Estimate the thickness of ice that would melt if all the heat in the Atlantic Water supplied melting. Clearly state your assumptions and comment on your result compared to the typical thickness of Arctic sea ice.

10.5 Using the polar mixed layer model of Subsection 10.3.2:

(a) Consider the $Q^*(\phi)$ curve separating the unstable regime from the too-deep regime. Estimate this function, compute it for the example shown in Figure 10.8, and comment on the agreement.

(b) Now consider that detrainment can occur, namely w_e is not clipped to nonnegative values (remove the maximum function in (10.25)). Investigate how the results in Figure 10.8 change. Explain your findings using simple theory.

(c) Explore other solutions for a range of parameter settings, such as wind forcing, initial surface layer depth and properties. What happens if there is not enough sea ice to melt? What happens if new sea ice forms as cooling continues once the surface layer reaches freezing?

Hint: Begin by constructing a numerical model to solve the polar mixing model equations (10.21–10.27). Reproduce Figure 10.8, then proceed to answer the questions.

10.6 Explore the behavior of the polar mixing model solutions in temperature-salinity space, for instance in Figure 10.5. Namely, consider how the properties of the upper layer evolve for different (Q, ϕ) parameters. Hence, explain the range of final salinities possible for parameters in the freezing regime of Figure 10.8.

10.7 Using the Beaufort gyre model:

(a) Repeat the analysis of Subsection 10.3.3 for the GM eddy closure with constant κ, rather than the Visbeck et al. (1997) closure. What are the main conceptual differences between these models of the Beaufort gyre?

(b) Compute the total freshwater storage in the gyre as a function of the wind forcing and stratification for both the GM and the Visbeck et al. (1997) closures. Thus, explain how eddy saturation affects the freshwater storage capacity in these models.

10.8 Compare and contrast the Beaufort gyre model (Subsection 10.3.3) with the wind-driven eddying model of the ACC (Subsection 9.3.2). Identify the similarities and differences between the two models. Comment on the statement that the Beaufort Gyre is dynamically similar to the ACC.

Hint: Exercise 9.9 is helpful.

11

Heat Flux, Freshwater Flux, and Climate

Generally speaking, **climate** refers to slowly changing (as in Subsection 1.1.1) distributions of atmospheric temperature, precipitation, wind, and more broadly other elements of the Earth system that interact with these fields. Ocean circulation influences the Earth's climate through physical and chemical interactions with the atmosphere and biosphere, and through its influence on sea ice, which affects the reflectivity of Earth. Here we explore physical interactions between the ocean and the atmosphere.

To a great extent, climate is the consequence of the Earth's energy budget attaining a balance between incoming and outgoing energy. We discuss oceanic transports of heat and freshwater and the role they play in the Earth's energy budget. The ocean directly influences atmospheric temperature via ocean heat transport. Atmospheric freshwater transport may allow the deep meridional overturning overturning to have multiple stable equilibrium states for a given set of boundary conditions. If such multiplicity exists, it raises the possibility that transient changes may lead to abrupt and long-term changes in climate.

11.1 Observations

11.1.1 Equilibrium Energy Balance and Heat Transport

Energy Balance and Equilibrium Temperature

In other chapters, we determine the circulation state of the ocean for a *given* sea surface temperature (SST) distribution imposed from *outside*. Because the atmosphere and ocean both influence each other, we need to step back and recognize that it is actually *solar radiation* that is imposed from outside. A key theme of climate science is how energy transport, including solar radiation, influences atmospheric temperature, circulation, cloud formation, and precipitation. The ocean, in turn, influences energy transport. Here we examine the direct effect of the ocean on the energy budget of the climate system.

We begin with solar radiation. The Earth intercepts solar energy at a rate $E = 175$ PW, where the prefix "P" stands for "peta" or 10^{15}. Averaging over the entire surface area A of the Earth (including the night side), the energy flux from the sun is $E/A = Q/4 = 342$ W m^{-2} at the **top of the atmosphere**. In steady state, energy input E and the energy leaving the Earth must balance. (In reality, the Earth has warmed over the last century by about 1°C but the heating of the ocean reflects an imbalance of less than 0.5% of incoming radiation, Pachauri et al., 2014.) Energy export from Earth occurs through reflection of solar radiation and emission of radiation. The **albedo**, α (not the thermal expansion coefficient), is the fraction of incoming radiation reflected back into space. The albedo depends on cloud and ice distributions and other factors, but currently for the Earth the average value is $\alpha = 0.3$.

Energy flux $\mathcal{H}$ of radiation from an object with $\alpha = 0$ (a black body) at temperature T (in Kelvin) is given by the Stefan–Boltzmann law, $\mathcal{H} = \sigma T^4$ (see (4.8) with $Q_* = 1$; σ here is the Stefan–Boltzmann constant). Radiation emitted from Earth's surface is absorbed and re-emitted by the atmosphere, specifically, by so-called greenhouse gases such as water vapor and carbon dioxide. This interaction depends on the wavelength of the light. Thus the relationship between $\mathcal{H}$ at the top of the atmosphere and T at the surface is more complicated than the pure Stefan–Boltzmann law, necessitating $Q_* < 1$ in (4.8). The global mean surface temperature of Earth is 288 K = 14.8°C for the twentieth century (NOAA, 2015), significantly warmer than for a pure black body (see Exercise 11.1).

Latitude Dependence and Meridional Heat Transport

With these facts in mind, consider that the top of the atmosphere incoming radiation depends on latitude, day of the year, and time of day. If net incoming energy balanced the emitted radiation at each location, the temperature would vary over space and time much more than is observed. The relatively small temperature variation is due to the transfer of thermal energy by the atmosphere and ocean, which is shown schematically in Figure 11.1. In the ocean and the atmosphere, the meridional temperature difference drives general circulations that tend to homogenize the Earth's temperature by carrying energy poleward. As discussed below, the ocean and the atmosphere absorb energy at low latitudes, carry it poleward, and release it at high latitudes.

We can visualize the relationship between horizontal and meridional heat transport by considering energy budgets in an atmosphere and ocean divided into boxes (control

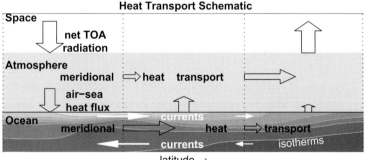

Figure 11.1 Schematic of ocean–atmosphere energy balance in a single hemisphere, showing vertical energy fluxes (vertical arrows), horizontal energy transports (horizontal unfilled arrows), ocean meridional velocity (horizontal filled arrows), and temperature (shading and contours; colder water is darker). The equator (pole) is on the left (right) and TOA means top of the atmosphere.

volumes) corresponding to different latitude ranges (Figure 11.1). In steady state, the time-average energy into and out of each box balances. The ocean heat flux influences the radiation leaving the top of the atmosphere box by modifying the atmospheric temperature. Atmospheric temperature, in turn, influences atmospheric circulation and hence other atmospheric characteristics.

The steady-state balance between meridional energy transport $E(\theta)$ at latitude θ (in W) and zonal-average downward heat flux Q (in W m^{-2}) shown in Figure 11.1 is given by

$$E(\theta_N) - E(\theta_S) = 2\pi R_E^2 \int_{\theta_S}^{\theta_N} Q \cos \theta' d\theta', \tag{11.1}$$

where (θ_N, θ_S) are two latitudes (θ_N north of θ_S) and $2\pi R_E \cos \theta'$ is the length of the latitude circle at latitude θ'. Here Q can represent zonal-average top of the atmosphere net downward heat flux (including all components), in which case E represents the combined meridional heat transport of the atmosphere-ocean system. Or Q can instead represent zonal-average net downward heat flux at the sea surface (including all components), in which case E represents ocean meridional heat transport. If θ_S is the South Pole, then $E(\theta_S) = 0$, and (11.1) relates $Q(\theta)$ to $E(\theta)$.

As discussed in Subsection 6.3.3, the time-average northward ocean heat transport can be estimated from fields of velocity $v(\phi, \theta, z, t)$ and potential temperature $T(\phi, \theta, z, t)$ for longitude ϕ. Taking the average over a time period τ (usually years or decades), the heat transport at a given latitude is given by

$$E(\theta) = \frac{1}{\tau} \int_0^\tau \int_{\phi_W}^{\phi_E} \int_{z_B}^{z_T} c_p \rho T v \, dz \, R_E \cos \theta \, d\phi \, dt, \tag{11.2}$$

where the ocean boundaries are given by (ϕ_W, ϕ_E) in longitude and $(z_B(\phi, \theta), z_T(\phi, \theta))$ in depth. Ocean meridional heat transport occurs when currents carry warm water in one direction and cold water in the opposite direction (Figure 11.1). Although the schematic shows the common occurence of warm currents above cold currents, it is also possible for the warm current to be at the same depth as the cold current but at a different longitude (see Figure 10.2, for instance).

In (11.2), we usually assume that the meridional mass transport

$$M(\theta) = \frac{1}{\tau} \int_0^\tau \int_{\phi_W}^{\phi_E} \int_{z_B}^{z_T} \rho v \, dz \, R_E \cos \theta \, d\phi \, dt \tag{11.3}$$

is zero. M can be quite large in an open basin such as the South Pacific, which has a roughly 15 Sv circulation connecting it to the Indian Ocean via flow around Australia (Figure 3.23). The Atlantic and the Indo-Pacific basins also have $M \neq 0$ because there is a circulation around Eurasia-Africa and/or the Americas and hence $M \neq 0$ at various latitudes in either basin, but this flow is only about 1 Sv as measured in the northward flow through the Bering Strait. As other chapters discuss, typical ocean volume transports of interest are tens of Sv. In any basin, up to about 1 Sv (Subsection 11.1.4) travels meridionally in the atmosphere from regions of net evaporation to regions of net precipitation, and an equal M must complete the circuit in the ocean.

Nonzero meridional mass flux M complicates the interpretation of the meridional heat flux E. The primary reason for thinking about meridional heat transport is to consider its effect on the atmosphere, which occurs via surface heat fluxes, which are related to the divergence of the meridional heat transport. A nonzero M weakens the connection between E and Q. An idealized example that illustrates this is an ocean with uniform temperature T and $Q = 0$ everywhere. Because T is uniform, $E = Mc_p\rho T$, which can be quite large even though it has no impact on the atmosphere. The ocean region has an active thermodynamic role measured by (11.2) when it exchanges warm water for cold (or *vice versa*) with another region. When $M \neq 0$, E measures heat carried both by the exchange and a throughflow that may be thermodynamically irrelevant.

Local and Global View of Surface Heat Flux

The two expressions for heat transport, (11.1) and (11.2), view heat transport in two complementary ways. For instance, a heat flux exists from the ocean to the atmosphere (as in the northern two latitude bands of Figure 11.1) when ocean SST exceeds surface air temperature, which is usually measured 2 m above the surface. (Part of the ocean heat loss is through evaporation, (4.7), which can cool the ocean even when it is colder than the air.) This is the local view of heat flux, relating the flux to local values of SST, surface air temperature, humidity, and so on. In steady state, however, the air–sea heat transport must be the difference between the meridional heat transport just to the north and just to the south. By (11.2), this is determined by temperature and water velocity, which are part of the large-scale patterns (including below the surface) that we have been describing throughout the book. In that sense the heat flux is *not* determined by local features; or to put it another way, the local features themselves are also influenced by the system as a whole; see Subsection 11.2.2.

Similarly, regions of surface heat loss are marked by surface water cooling as it travels, as shown for example in the central latitude band of Figure 11.1. Therefore, in the warm limb of the circulation, the *outward* flow across one boundary (to the north in this example) is cooler than the *inward* flow across another boundary (to the south here), ensuring that the ocean heat transport out of the region is less than the heat transport into the region, as it must be if some heat is lost from the surface.

11.1.2 Observations of Energy Transport

We now use the framework of Subsection 11.1.1 to diagnose the meridional energy transports in the real world. The Earth Radiation Budget Experiment used satellites to measure the top of the atmosphere energy flux (Barkstrom et al., 1990). The net radiation Q (downward minus upward) is positive in the tropics and negative at high latitudes (Figure 11.2a), with typical magnitudes of less than 20% of the global average solar input. Applying (11.1) to Q gives the meridional heat transport (Figure 11.2b), which is poleward with peak values of about 6 PW in each hemisphere (Trenberth and Caron, 2001).

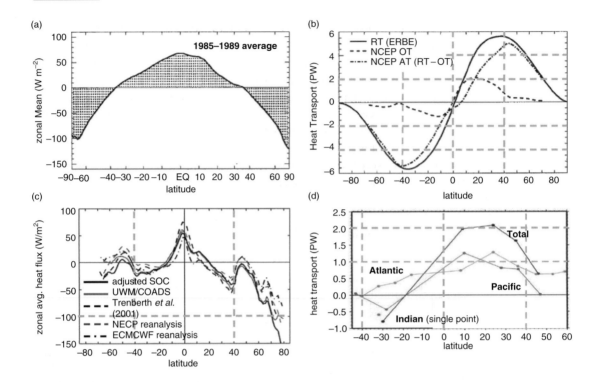

Figure 11.2 (a) Zonal average top of the atmosphere net vertical energy flux from Earth Radiation Budget Experiment observations. (b) Meridional energy transport for (solid) climate system (from energy flux in (a)), (dash-dot) atmosphere (from reanalysis), and (dashed) ocean (residual of climate and atmosphere estimates). (c) Ocean zonal-average surface heat-flux climatologies. (d) Meridional heat transport for individual ocean basins from direct estimates. (a) and (b) from Trenberth and Caron (2001), Figures 1 and 2, (c) adapted from Grist and Josey (2003), Figure 8, and (d) from Talley 2003, Fig. 1. ©American Meteorological Society. Used with permission.

The ocean's share of this meridional heat transport can be estimated from each of three different sets of observations. The **direct method** measures T and v as a function of longitude and depth at some latitude, and calculates heat transport via (11.2). The **heat flux method** calculates the ocean surface heat flux from observed surface characteristics (Chapter 4) and uses (11.1) to infer the ocean meridional heat transport. Finally, the **residual method** subtracts the atmospheric meridional heat transport from the total (atmosphere plus ocean). Because of the importance of atmospheric eddies (weather) to heat transport, observations of atmospheric variables are not dense enough to measure heat transport directly. Instead, the data comes from a reanalysis (Appendix A) in which data is assimilated into a weather model in order to estimate atmospheric variables.

The residual method shows that the atmospheric energy transport accounts for most of the meridional heat transport at most latitudes, with peak values near 45°N and 40°S (Figure 11.2b, Trenberth and Caron, 2001). The ocean heat transport dominates in the deep tropics and has peak values of around 2 PW and 1 PW in the northern and southern hemispheres, respectively (Figure 11.2b). Estimates of the ocean heat transport with the direct method gives values that are similar, though a bit higher in the northern hemisphere (Wunsch, 2005).

The flux at the ocean surface (Figure 11.2c) shows a somewhat more complicated structure than the top of the atmosphere energy flux (Figure 11.2a). The ocean absorbs heat close to the equator and expels heat in subtropical bands at about 20°–40° latitude in each hemisphere and in subpolar bands poleward of about 50°. In the Arctic, the large heat flux corresponds to relatively small meridional heat transport because of the small zonal extent of the oceans at high latitudes.

Individual ocean basins show a much stronger asymmetry between hemispheres than the zonal average of the global ocean, which in turn shows more asymmetry than top of the atmosphere fluxes. The residual method can be adapted to split apart the meridional heat transport carried by individual ocean basins (Trenberth and Caron, 2001). This shows the surprising result (Figure 11.2d) that the Atlantic heat transport is northward at all latitudes—even in the southern hemisphere where heat is being transported from the cold south towards the warm equator!

There is considerable observational uncertainty about the strengths of the southern subtropical Pacific and Indian heat transport. As mentioned in the previous section, there is the problem of nonzero volume transport around Australia. While there is disagreement on the direction of Pacific heat transport at 30°S, the sum of Indian and Pacific transports is roughly 1 PW. Integrating the surface heat flux north of 30°S, about 0.8 PW is absorbed by the Pacific and only 0.1 PW in the Indian (Ganachaud and Wunsch, 2003). The large Indian Ocean export of heat at 30°S implies a large import from the Pacific at Indonesia.

The different behavior of the Atlantic and Pacific suggests the importance of different processes in each, as we discuss in Subsection 11.2.1. Nevertheless, the maximum northern hemisphere heat transport is similar in both the Atlantic and Pacific, about 1 PW near 25°N (Figure 11.2d). As discussed in Subsection 11.2.1 for a simulation, meridional heat transport is largely associated with flow within the top 500 m of the Pacific but involves an exchange between northward flow in the top kilometer and southward flow below.

Mesoscale eddies contribute to meridional heat transport, but the dense sampling necessary to estimate the average eddy heat transport typically does not exist to date. Indirect estimates in the Southern Ocean indicate that eddies account for most of the heat transport there, about 0.3 PW southward, (Volkov et al., 2008; Stammer, 1998). Narrow bands near the poleward edge of the subtropical gyres of both hemispheres appear to have strong poleward heat transport (Stammer, 1998). In contrast, the subtropical North Atlantic (Johns et al., 2011) and North Pacific (Roemmich and Gilson, 2001; Qiu and Chen, 2005) may each have eddy transport only about 10% of the contribution from the time-average circulation.

11.1.3 Distribution of Surface Heat Flux

The distribution of the surface heat flux in longitude and latitude illuminates the mechanisms behind ocean heat transport. Here we use monthly flux estimates from the Objectively analyzed air–sea Fluxes (OAFlux) project (Yu and Weller, 2007, see Appendix A). Other estimates of surface heat flux such as that of the National Oceanography Center (Kent and National Center for Atmospheric Research Staff, 2016) or the International Comprehensive Ocean–Atmosphere Data Set (ICOADS, Freeman et al., 2017) give patterns that are similar to OAFlux, although in general, accurate estimates of air/sea heat fluxes are still challenging, with uncertainties in the range 5–30 W m^{-2} (Valdivieso et al., 2017).

In Subsection 4.3.1 we show that surface heat flux is the sum of shortwave and longwave radiation and turbulent fluxes of latent and sensible heat (see (4.6)). The individual components have distinct patterns, signs, and magnitudes (Figure 11.3). On timescales of years, the downward shortwave radiation from the sun is approximately balanced by upward flux of all the other components. Therefore we use a sign convention in which shortwave radiation is positive if downward and all other components are positive if upward (as in Subsection 4.3.1).

While the annual-average top of the atmosphere solar radiation depends on latitude only, the net radiation has a more complex distribution due to albedo variations from clouds and sea ice (Figure 11.3a). Thus the overall meridional trend in net solar radiation from roughly 100 W/m^2 at high latitudes to about 250 W/m^2 in the tropics is interrupted by lower values in cloudy regions such as the **Intertropical Convergence Zone** just north of the equator in the Pacific and Atlantic, the upwelling zones in the eastern tropical Pacific and Atlantic, and the Indian Ocean near Indonesia. The other components have broadly similar patterns, with latent heat providing the largest upward component (Figure 11.3c), longwave radiation about half as big (Figure 11.3b), and sensible heat smaller still (Figure 11.3d). Significant details of the distribution of these components are different from the incoming shortwave, suggesting the possible influence of ocean currents. For instance, latent and sensible heat flux have a narrow band of low values in the equatorial eastern Pacific (and Atlantic, though to a lesser extent). Latent and sensible heat flux are especially high in the vicinity of the Gulf Stream, Kuroshio, Agulhas, and East Australian Current.

The net heat flux (downward shortwave minus the other three components) looks different from the components (Figure 11.4). As expected, from the poleward ocean heat transport discussed in the previous sections, the heat flux into the ocean is largely positive in the tropics and negative at high latitude. Other features hint at the influence of specific ocean circulation structures in the heat transport. Generally speaking the ocean is gaining heat in the eastern sides of basins and losing it in the west. Tropical heat gain is especially intense within a few degrees of the equator, with values reaching above 160 W/m^2 compared to about 40 W/m^2 over much of the tropics. Regions of heat loss correspond to the subtropical gyre western boundary currents. The northeastern Atlantic and Nordic Seas also lose heat. The Indian Ocean has some zonal features reversed from the other basins, with equatorial heat gain strongest in the *west* and a prominent region of heat loss in the *east* along western Australia.

Some of the features shown in Figure 11.4 correspond to features of SST shown in Figure 4.1. Heat gain occurs in relatively cold regions such as the equatorial cold tongue and the eastern edges of basins. As discussed in Chapter 6, these regions are cooled by intense upwelling, as is the Indian Ocean off the coast of Somalia. The regions of heat loss are largely associated with poleward-flowing western boundary currents. The North Atlantic subpolar heat loss occurs near warm northward flow into regions of deep water formation associated with the Atlantic Meridional Overturning Circulation (AMOC; Chapter 8).

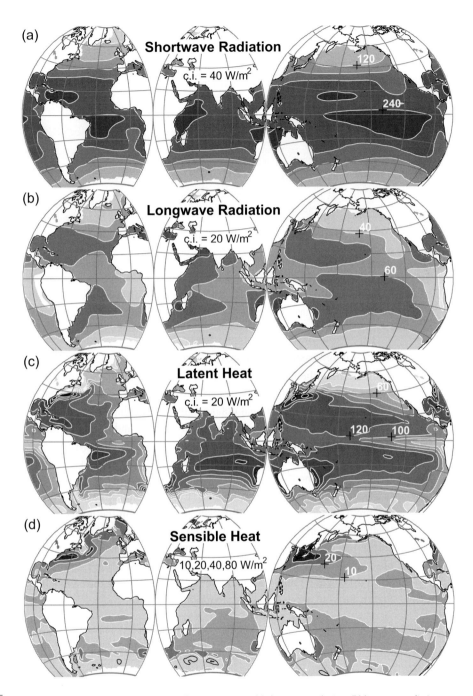

Figure 11.3   Climatological annual average air–sea heat flux components: (a) shortwave radiation, (b) longwave radiation, (c) latent heat flux, and (d) sensible heat flux. Darker shades represent higher magnitudes. Contour intervals are $40 \, W/m^2$ for shortwave component, $20 \, W/m^2$ for longwave and latent components, and nonlinearly spaced at $\pm 10, 20, 40, 80 \, W/m^2$ for sensible heat. Data from OAFlux (see Appendix A), smoothed by $5° \times 5°$ running mean.

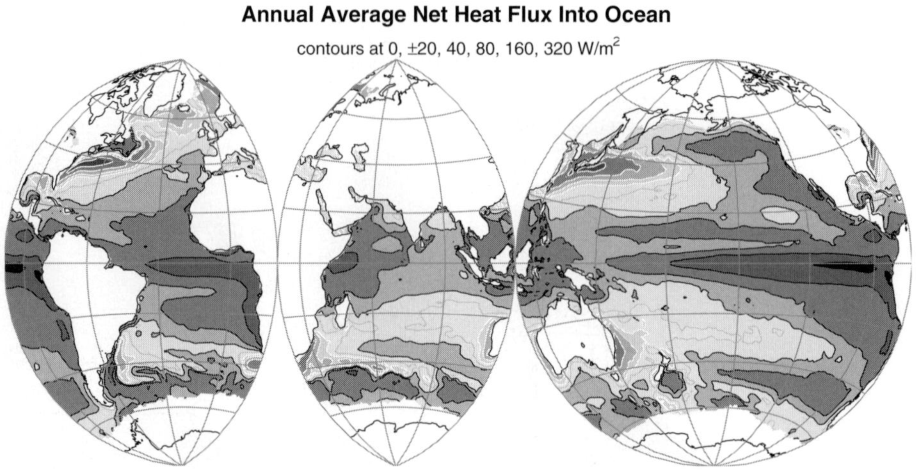

Annual Average Net Heat Flux Into Ocean

contours at 0, ±20, 40, 80, 160, 320 W/m^2

Figure 11.4 Climatological annual average net heat flux into the ocean. Contours at $\pm$ 20, 40, 80, 160, 320 W/m^2; contours black for heat gain, white for heat loss, gray for zero line. Data from OAFlux (see Appendix A). A black-and-white version of this figure appears in some formats. For the color version, please refer to the plate section.

Several asymmetries are apparent in the net heat flux. Heat loss appears to be greater in the northern hemisphere than in the southern, though we must note that in situ data is much sparser in the high latitude southern hemisphere than elsewhere. The equatorial Cold Tongue has much greater heat uptake in the Pacific than in the Atlantic. However, heat loss is about the same magnitude in the North Atlantic and North Pacific.

Finally, remember that the heat flux has a large annual cycle. This is captured by subtracting the January–February–March average from the July–August–September average (Figure 11.5). The difference between summer and winter is larger than the annual mean in most of the ocean, with the weakest cycle in the tropics. As with the annual SST cycle (Figure 4.3), the annual heat flux cycle is stronger in the northern hemisphere than in the southern, primarily in the vicinity of the western boundary currents. Weather-related variability can bring even more intense extremes, with heat loss of more than 1000 W/m^2 observed during outbreaks of cold continental air over the Labrador and Nordic Seas. The April–May–June minus October–November–December difference has a similar pattern to Figure 11.5 but about half the magnitude.

11.1.4 Freshwater Forcing

Interactions at the sea surface modify ocean salinity via evaporation and precipitation, coastal runoff from land, and ice freezing and melting at high latitudes. Dilution from precipitation, runoff, and ice melt decreases surface salinity, and concentration from evaporation and freezing increases it. Salinity does not directly affect atmospheric climate as temperature does, but it influences ocean density and ocean circulation and thus affects climate.

Net Heat Flux Into Ocean: JAS minus JFM

contour interval 80 W/m^2

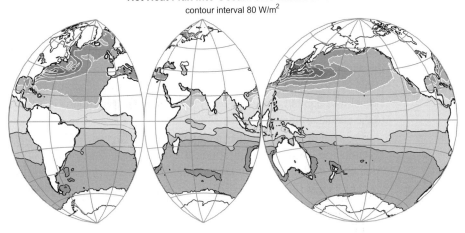

Figure 11.5   Climatological net heat flux into the ocean for July–August–September minus January–February–March averages.
Contour interval of 80 W/m^2 with (white, black, gray) contours representing (positive, negative, zero) values. Data
from OAFlux (see Appendix A).

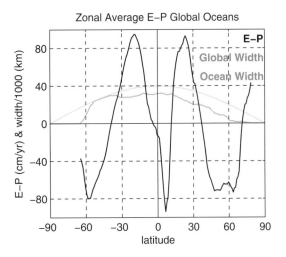

Figure 11.6   Global zonal average of 1987–2006 climatology of $E - P$ from OAFlux and GPCP data (Appendix A). The GPCP data is
interpolated from a 2.5°-resolution grid to the 1°-resolution grid of OAFlux. Annual average $E - P$ (black) in cm/yr,
global width (light gray) and ocean width (dark gray) in 10^3 km.

We examine precipitation P, evaporation E (not energy flux in this subsection), and
net surface freshwater flux $E - P$, using the 1987–2006 time average from the Global
Precipitation Climatology Project (GPCP, Adler et al., 2003, Huffman et al., 2009; see
Appendix A) and OAFlux (see Subsection 11.1.3). The global zonal average $E - P$ is
dominated by net rainfall in the Intertropical Convergence Zone, net evaporation in the
subtropics, and precipitation in subpolar regions (Figure 11.6). For constant ocean mass, E,

P, and runoff R balance. In fact, estimates of global mean sea level (Rhein et al., 2013) show a rise of 17 ± 1 cm/century over the twentieth century and 32 ± 4 cm/century over 1993–2010, much of it related to terrestrial ice melt due to global warming. The volume transport associated with sea level rise (≤ 0.01 Sv) is negligible compared to total precipitation on the sea of 12.2 ± 1.2 Sv, evaporation of 13.0 ± 1.3 Sv, and runoff of 1.25 ± 0.1 Sv (Schanze et al., 2010). Seasonal averages of zonal average $E - P$ are qualitatively similar to the annual average, but at a given latitude the seasonal variation is of comparable magnitude to the annual average.

The zonal average simplifies a complex spatial dependence of rainfall and evaporation. Precipitation has a strong Intertropical Convergence Zone rainfall in the Pacific and the Atlantic as well as a kind of southern hemisphere Intertropical Convergence Zone in the western Pacific (Figure 11.7a). There are strong zonal gradients in rainfall, typically with more rain on the western side (except the Indian, where on the equator it is strongest in the east). Evaporation does not have as strong variation except for some bands of high values near subtropical western boundary currents, where warm poleward-flowing water meets cooler air, and the eastern equatorial Pacific, where the cold tongue has low evaporation (Figure 11.7b). Like precipitation, evaporation is generally stronger on the western sides of basins.

The $E - P$ field (Figure 11.7c) is more zonally uniform than the individual P and E components, though the net precipitation of the "Southern Intertropical Convergence Zone" in the western Pacific is balanced by net evaporation in the eastern Pacific at about the same latitude. A surface integral of $E - P$ over a latitude band of net evaporation (such as between 40°S and 5°S) or net precipitation (such as between 40°N and 70°N) is typically about one Sverdrup. These net freshwater transports are about ten times smaller than the individual total precipitation and evaporation rates (see above). Thus, in any latitude band, much of the water leaving the sea surface as evaporation returns as precipitation in the same band.

As with surface heat flux, zonal average surface freshwater fluxes imply meridional transports, at least if the system (averaged over a few years) is in a steady state. In Figure 11.6, the net freshwater flux into the ocean poleward of about 40° travels equatorward where it leaves the ocean surface as water vapor. Similarly, the rainfall just north of the equator must contribute a net freshwater transport towards one or both of the poles. The O(1) Sv meridional freshwater transports are small compared to the mass transports associated with the gyres, subtropical cells, or deep meridional overturning circulation.

11.1.5 Atlantic Overturning and Paleoclimate Change

We turn to the geological record to find past instances in which a change in ocean circulation – principally, the AMOC – may have induced a climate change. Typically, such geological data involve variations in chemical, isotopic, or species composition of fossils, ice, and sediments. These features that reflect environmental variables may be used as **proxies** for the variables. Proxies whose creation dates can be estimated preserve the

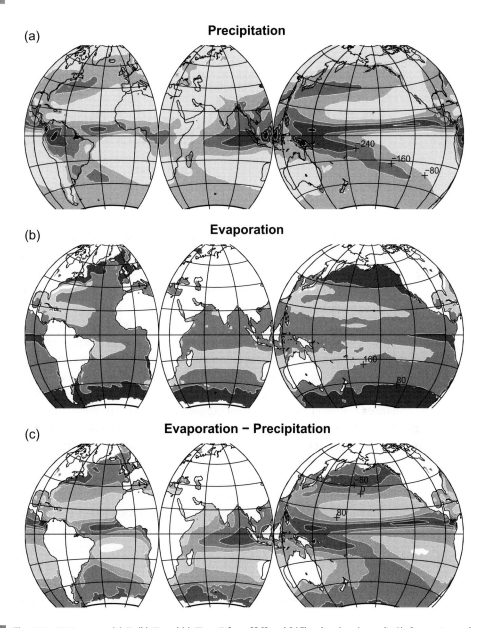

Figure 11.7 The 1987–2006 average (a) P, (b) E, and (c) $E - P$ from GPCP and OAFlux data (see Appendix A). Contour interval is 80 cm/yr. Darker shades represent regions with more precipitation and/or less evaporation, and lighter shades represent less P and/or more E.

history of the variables. Interpreting the physical significance and dating of different proxies is not straightforward, however, and one must be cautious in treating paleodata as an actual time record of a given climate parameter. Here we confine ourselves to a few of the best-known events in paleoceanography.

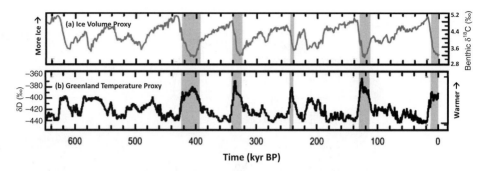

Figure 11.8 Time series of (a) global glacial ice volume fluctuations inferred from benthic $\delta^{18}O$ marine records (Lisiecki and Raymo, 2005) and (b) Greenland temperature fluctuations based on deuterium (δD) found in bubbles from Antarctic ice cores (Petit, 1999). Shaded periods represent interglacials. Figure based on Figure 6.3 from Jansen, E., J. Overpeck, K. R. Briffa, et al. 2007: Palaeoclimate. In: Climate Change 2007: The Physical Science Basis. Contribution of Working Group I to the Fourth Assessment Report of the Intergovernmental Panel on Climate Change [Solomon, S., D. Qin, M. Manning, et al. (eds.)]. Cambridge University Press, Cambridge, United Kingdom and New York, NY, USA.

For hundreds of thousands of years (kyr), the Earth has experienced ice ages. They are distinguished by global mean temperatures about 5°C colder than today and O(1) km thick glaciers covering northern regions of Eurasia and North America (Jansen et al., 2007). These ice ages have been interrupted by 5–20 kyr-long **interglacial** periods of glacial retreat (Figure 11.8a) and warmer climate (Figure 11.8b). We live in one such interglacial, the Holocene, which began about 12 kyr BP (Before Present). Variations in the Earth's orbit and spin appear to set the 100 kyr intervals between interglacials as well as other periodicities of 10s of kyr (Figure 11.8). Superimposed on this externally driven multi-kyr climate variability, there are climate events of about 1 kyr duration (Figure 11.9a) which may be partially caused by ocean physical processes.

The end of the most recent **glacial** period occurred in fits and starts. Ice coverage reached its greatest extent at the **Last Glacial Maximum**, approximately 20 kyr BP. The stack of annual layers observable in the Greenland icecap provides a proxy time series of local temperature (Figure 11.9a) which shows an abrupt warming from glacial conditions at about 15 kyr BP. This **Bølling-Allerød** period persisted until 12.7 kyr BP, when Earth's climate cooled for the O(1) kyr **Younger Dryas** period. The Younger Dryas, in turn, was replaced by the Holocene interglacial (Hemming, 2004; McManus et al., 2004; Alley, 2007).

Before the Bølling-Allerød there was a **Heinrich Event**, H1 (Figure 11.9a), one of many that occurred over the last 100 kyr (McManus et al., 2004). A Heinrich event is marked by a layer of **ice rafted debris** on the seafloor of the midlatitude North Atlantic. This indicates that a pulse of icebergs emerged from the ice cap, scraped rocks off the continents, carried them equatorward, and deposited them on the seafloor upon melting. There is evidence that at least some of the other Heinrich events occurred concurrently with the cold stage following sudden warmings in the ice record known as **Dansgaard–Oeschger** (D-O) events (Figure 11.9a; see Alley, 2007).

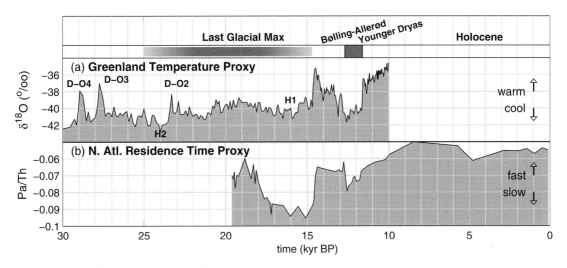

Time series of (a) Greenland Ice Sheet Project 2 $\delta^{18}O$ record as a proxy for local temperature and (b) $^{231}Pa/^{230}Th$ proxy for North Atlantic meridional overturning circulation, with geological time periods marked at top and Heinrich events (H1, H2) and Dansgaard–Oeschger events (D-02, D-03, D-04) marked in (a). The area under the time series are shaded for visibility. Data from (a) Blunier and Brook (2001), and (b) McManus et al. (2004).

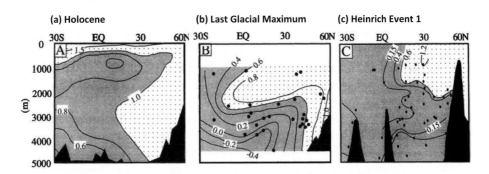

Benthic $\delta^{13}C$ data as proxy for water masses during (a) Holocene, (b) last glacial maximum and (c) Heinrich Event 1, from Figure 3 of Alley et al. (1999) ©1999 American Geophysical Union. All rights reserved. Used with permission.

Water mass distributions give insight into the deep flow of the Atlantic over the millennia. Northern-source water such as North Atlantic Deep Water (NADW) and southern-source water such as Antarctic Bottom Water (AABW) have distinct chemical and isotopic signatures, like cadmium/calcium ratios and $^{13}C/^{12}C$ ratios. These signatures are reflected in the chemical composition of **benthic foraminifera** (forams), millimeter-scale organisms that lived on the sea floor and were fossilized in marine sediments. Benthic forams from a range of latitudes and sea-floor depths provide an estimate of past distributions of NADW and AABW. For the Holocene, this analysis shows the familiar pattern of a tongue of NADW stretching southward at about 2500 m depth and tongues above and below southern-source water extending northward (Figure 11.10a; compare to Figure 8.1).

During the last glacial maximum, the northern-source water was confined to shallower layers, with deep water dominated by AABW (Figure 11.10b). During Heinrich Event 1 (H1), southern-source water dominated even more (Figure 11.10c).

These water mass distributions suggest different overturning circulations. Indeed, proxy measurements of deep water residence time in the North Atlantic suggest that during the Holocene NADW circulated fastest, during the last glacial maximum and the Bølling-Allerød it was slower, and during the Younger Dryas and H1 it was slowest (Figure 11.9b; see Lynch-Stieglitz, 2017). This evidence is consistent with the idea that Heinrich pulses of ice covered the northern North Atlantic with fresh, low-density water reducing or stopping NADW formation.

The Younger Dryas NADW shut-off may have been triggered by a similar freshwater pulse. The melting of the North American ice sheet formed a body of water known as Lake Agassiz. Rapid drainage of Lake Agassiz could have suppressed NADW formation by covering the northern North Atlantic with fresh water (Rooth, 1982; Broecker et al., 1985; Broecker, 2006). It is also possible that meltwater outflows around Antarctica reduced the AABW source and hence strengthened NADW formation (Weaver et al., 2003).

As discussed in Subsection 11.1.2, the NADW overturning cell is associated with a significant northward heat transport that warms the North Atlantic. An NADW shut-down following a meltwater event would cool the northern North Atlantic. Thus, the ocean not only *reacts* to (natural) climate change, but also *causes* climate change, such as the return to ice-age conditions during the Younger Dryas.

A millennial change in overturning could be a response to a millennial change in forcing, or to a briefer forcing change. For instance, the AMOC-off circulation state may have persisted for centuries after Lake Agassiz stopped releasing freshwater. In other words, the AMOC might persist in an off state long after the forcing has returned to normal. This can occur if more than one circulation state exists for a given forcing. An event such as a meltwater pulse could push the AMOC from one state to another, where it would remain even after the forcing returned to the previous configuration. Aside from explaining aspects of past climate change, the possibility of multiple overturning states is an intriguing question about the general circulation: we discuss it in Sections 11.2 and 11.3. Remember that interpreting these paleoceanographic data is not straightforward. Indeed, other explanations exist for the Younger Dryas cold period (Alley, 2007; Lynch-Stieglitz, 2017).

11.2 Concepts

11.2.1 Dynamical Analyses of Heat Transport

The distribution of ocean meridional heat transport described in Subsection 11.1.2 raises the question: What processes are responsible for heat transport? In addition to linking the heat transport to circulation, we want to understand how it is determined by external conditions.

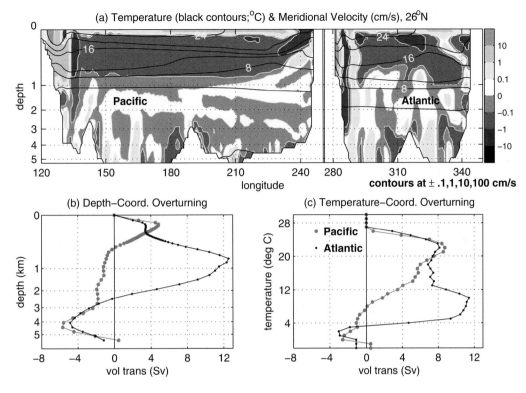

Figure 11.11 GCM results at 26°N for (a) longitude–depth section of v (shading) and T (black contours); overturning streamfunction calculated in (b) depth-coordinates and (c) temperature-coordinates. In (a), v contours are set at 0 and ± 0.1, 1, 10, 100 cm/s, and T contour interval is 4°C. In (a) and (b), the ordinate is the square root of the depth (km). See also Exercise 11.6.

Anatomy of Subtropical Heat Transport

To gain insight into some processes transporting heat, we examine the behavior of a general circulation model (Figure 11.11). The model is from the Gnanadesikan et al. (2006) climate model, with roughly 1° resolution and forcing from observed monthly climatologies of atmospheric conditions and sea surface salinity. The model generates a meridional heat transport that is similar to the observed distribution (Figure 11.2d) although weaker in the northern hemisphere.

At 26°N, the Pacific and Atlantic show similar potential temperature (T) and meridional velocity (v) fields (Figure 11.11a). Both oceans have a western boundary current flowing northward in the top kilometer (Chapter 3 and Chapter 5), recirculating southward flow offshore of the western boundary current (Chapter 3), northward Ekman transport near the surface (Chapter 6), and southward geostrophic Sverdrup flow mostly in the top 500 m (Chapter 5). The difference between the North Atlantic and the Pacific is the presence of the southward-flowing Deep Western Boundary Current in the Atlantic, adjacent to and east of the Gulf Stream (the Pacific also has deep southward flow in the west, but it is an order of magnitude weaker).

The depth-coordinate streamfunction (Figure 11.11b) emphasizes the circulation difference between the two basins. The Atlantic Meridional Overturning Circulation (AMOC) consists of northward flow in the top kilometer, southward flow from about 1–4 km, and weaker northward bottom flow. In the Pacific (Figure 11.11b), upper northward flow is confined to the top 30 m, with return flow restricted to the top 500 m in the subtropical cell. The Pacific also has a bottom cell similar to the Atlantic. Overturning volume transport in the upper cell in the Atlantic is nearly three times greater, over a greater density range, than in the Pacific.

Recall from Subsection 6.3.3, (6.45), and (6.51), that the heat transport $E(\theta)$ can be written in terms of the temperature-coordinate overturning streamfunction, Ψ_T, as:

$$E(\theta) = c_p \rho_0 \int_{T_-}^{T_+} \Psi_T \, dT, \qquad (11.4)$$

where the temperature ranges between T_- and T_+. For the 26°N section, this shows a similar magnitude northward flow of 23–27°C water in both the Atlantic and the Pacific. Figure 11.11a shows that this water is largely confined to within 100 m of the surface. In the Pacific, the southward limb of the cell occurs in a broad temperature range of 6–22°C. In the Atlantic, there is additional northward flow of 10–12°C water (in the top kilometer, penetrating deeper than in the Pacific Ocean) and return flow near 4°C, which is around 2 km deep. The Pacific heat transport is dominated by flow in the thermocline while Atlantic heat transport is dominated by the AMOC.

For a given isotherm in either basin (Figure 11.11a), flow is northward in the western boundary current and southward in the geostrophic mid-ocean region, while the warmest water has a northward Ekman transport. On each isotherm in the Pacific thermocline, the southward Sverdrup flow is a little greater than the northward western boundary current, allowing the thermocline flow to balance the warm northward flow on warmer isotherms. In the Atlantic thermocline, the western boundary current and thermocline flows balance more closely, except around the 12°C isotherm where the volume transport is greater in the western boundary current than in the interior. The model balances are similar to those of the real world for both the Atlantic (Hall and Bryden, 1982) and Pacific (Bryden et al., 1991).

Heat Transport from First Principles?

What determines the magnitude of the heat transport? For the AMOC, the relationship between the circulation structure, temperature structure, and heat transport is dominated by vertical (rather than horizontal) variations in velocity and temperature. Warm thermocline water flows to the north and cold deep water flows to the south. The depth-coordinate volume transport $\Delta\Phi_z$ and ΔT, the difference of zonal-average temperature between the warm and cold limbs of the overturning cell, give a rough estimate of the heat transport, $E = c_p \rho \Delta\Phi_z \Delta T$. Taking the average temperature of the top 700 m (depth range of northward flow according to overturning streamfunction) and of the 2500–1000 m southward flow and using maximum streamfunction value, the estimate is $E \approx (17 - 5°C)(12 \times 10^6 \text{m}^3/\text{s})c_p\rho = 0.6$ PW, which is relatively close to the actual model value of 0.7 PW.

In the Pacific, E depends on the temperature difference between the northward Kuroshio and Ekman flow and the thermocline flow in the Sverdrup interior. The strength of the heat transport identified with the Subtropical Cell can be estimated from the subduction model of the subtropical gyre, as in Chapter 6 (see Figure 6.15), based on the volume transport of water subducted in the subtropical gyre and returning as Ekman transport near the surface (Klinger and Marotzke, 2000).

In contrast, the contribution of the Kuroshio to the heat transport is poorly understood. Simplified models in which ocean temperature and velocity are a function only of latitude and longitude show how air–sea interaction can create a heat transport (see Exercise 11.7). Water in the Sverdrup interior of the gyre (Chapter 3) moves slowly enough to equilibrate with atmospheric forcing, while fast poleward-flowing water in the western boundary current does not have time to equilibrate and retains a higher temperature from heating in the tropics. In a stratified ocean, however, subsurface water flowing adiabatically around the gyre should keep approximately the same temperature until it reaches the mixed layer, which deepens towards the pole.

In Figure 11.11a, only Kuroshio water in the top 100 m or so is warmer than the water to its east. At present, there is no theory for determining the size of this contribution to E, or how it depends on factors such as western boundary current speed or width. In the real world, the fact that the most intense regions of heat loss in all five gyres occur near the western boundary current (Figure 11.4) suggests that the western boundary current heat transports make a large contribution to the total heat transport. In the North Atlantic, the Gulf Stream contribution may be part of the AMOC, because in the northern subtropics, the Gulf Stream is an important part of the shallow limb of the overturning. In the North Pacific the Kuroshio is driven by subtropical wind stress without an overturning component.

The question of western boundary current heat transport is closely related to the difference in meridional extent of the subtropical cell when viewed in z-coordinate or ρ-coordinate streamfunctions (Subsection 6.2.3, especially Figure 6.16). The warm near-surface water flowing northward at 30°N (Figure 6.16b) cannot be carried by the Ekman transport, which is zero at that latitude, so the Kuroshio brings it northwards. If we could understand what determines the ρ-coordinate overturning in the Pacific, we would have a much better understanding of heat transport by the gyre circulation.

Although the way the AMOC carries heat is better understood, that does not mean that Atlantic heat transport is understood. The wind-driven transport operates in the Atlantic as well, though the smaller width of the Atlantic and hence smaller total Sverdrup transport reduces it compared to the Pacific. The subpolar gyre, which connects lower latitudes to regions of large surface heat flux between Europe and Greenland (Figure 11.4), has a large (and poorly understood) temperature difference between warm northward flow in the east and cold southward flow in the west. The deep meridional overturning circulation also includes bottom water flowing from the Southern Ocean into the South Atlantic and South Pacific. This water is probably not directly responsible for much heat transport because its mid-depth southward return flow is only a little warmer than the lower limb. Yet, it indirectly influences heat transport by affecting the AMOC in ways that are not completely understood.

Eddy Heat Transport

The discussion above does not include the effects of mesoscale eddies. Baroclinic insta-bility, which is thought to generate much of the ocean's mesoscale variability, releases potential energy that is associated with heat transport across tilting density interfaces (Subsection 7.2.2). In the mid-latitude atmosphere, eddies dominate the meridional heat transport. In the ocean, gyre and overturning circulations carry heat, probably more than the eddies. In the Southern Ocean, the lack of zonal boundaries at the latitudes of the Drake Passage impedes steady meridional flow and increases the importance of eddy fluxes there (Chapter 9). Eddy-resolving (Wolfe and Cessi, 2009; Bishop and Bryan, 2013) and eddy-permitting (Thompson, 1993; Jayne and Marotzke, 2002) models indicate that the Southern Ocean has the largest eddy contribution to meridional heat transport, although there is also a strong eddy transport due to Tropical Instability Waves (Jayne and Marotzke, 2002; Subsection 7.1.1).

11.2.2 Interactive Surface Conditions

Weak Surface Restoring

Rather than forcing SST to be close to an externally imposed target temperature, as in Subsection 4.3.1, we want a surface boundary condition that allows SST and surface atmo-spheric temperature to influence each other. Equation 4.14 approximates the surface heat flux into the ocean (Q) as a restoring law that pushes sea surface T to a given temperature T_*:

$$Q = \lambda(T_* - T), \tag{11.5}$$

where λ determines the restoration timescale. According to (11.5), places where the ocean is *absorbing* heat (such as the Cold Tongue) have SST *colder* than T_* and regions of ocean heat *loss* (such as the Gulf Stream) have SST *warmer* than T_*. If Q is fixed by other factors in the system, then for stronger λ, SST will stay closer to T_*. If we are given atmospheric conditions including near-surface temperature, humidity, wind, and shortwave radiation, then bulk formulae such as (4.9) and (4.10) allow us to calculate T_* and λ. For typical oceanic values, $\lambda \approx 50$ W m^{-2}K^{-1}(Schopf, 1983; Han, 1984), which restores a 30 m thick layer to T_* with an e-folding time $\tau \approx 45$ days.

If some factor, such as wind stress or a distant surface condition, changes the circulation in a way that changes Q at a given location, (11.5) says that for $\lambda = 50$ W m^{-2}K^{-1}, every 50 W m^{-2} change in Q implies a 1 K change in T. The warmer or cooler ocean, in turn, causes the near-surface *atmospheric* temperature T_A to warm or cool. Therefore we cannot assume that T_A is an external parameter for the ocean. The target temperature T_* depends on T_A in a somewhat complicated way (Subsection 4.3.1), but generally it increases/decreases in the same sense as T_A. If we write (11.5) as $T = T_* - Q/\lambda$, it is clear that T will change more in response to a change in Q if T_* also changes (in the same direction as T is changing) than if T_* is fixed.

The restoring temperature T_* is thus determined by processes independent of the ocean (such as absorbing solar radiation and emitting longwave radiation) as well as by ocean surface temperature. If $Q = 0$, these processes make T_* a slightly different value T_0. Think of the restoring law (11.5) as a spring that pulls T away from T_* under the "force" of ocean heat transport. Similarly, think of T_* as being held by *another* "spring" to T_0 and being pulled away from it by the ocean. This is an example of a **coupled ocean–atmosphere** interaction. Schopf (1983) shows that this coupled process can be modeled with an equation that looks like (11.5), but with T_* replaced by T_0 and with much weaker restoring corresponding to $\lambda \approx 3$ W m^{-2}K^{-1}. We demonstrate this result in Subsection 11.3.1.

Rahmstorf and England (1997) produce an interesting example of circulation effects caused by air–sea interaction. Toggweiler and Samuels (1995) increase the AMOC volume transport by strengthening Southern Ocean wind stress (see Subsection 9.2.2). This in turn increases the heat transport into the northern North Atlantic (Subsection 11.2.1). If SST is set by weak surface restoring, the increased heat transport makes the sea surface in the NADW formation region warmer and lighter. Decreasing the surface density difference between the North Atlantic and the rest of the ocean *weakens* the AMOC. This negative feedback reduces the sensitivity of the AMOC to the wind.

Ocean Heating and Cooling of the Atmosphere

The atmospheric interaction discussed above allows SST to be changed by ocean circulation more than expected from bulk formulae alone. This also influences atmospheric temperature. For instance, the SST difference of about 5 K between the eastern and western ends of the equatorial Pacific (Figure 4.1) is largely due to the strong upwelling of cold Equatorial Undercurrent water in the eastern third of the basin (Figure 11.4). Without upwelling, the entire equator would maintain the SST of the western Warm Pool. This happens during an **El Niño** event, when the equatorial wind weakens, upwelling dies down, and SST becomes almost uniformly warm along the length of the equator in the Pacific. Heat fluxes associated with other processes, such as the regions of intense ocean heat loss in the North Atlantic and North Pacific (Figure 11.4), have not been observed to undergo such large changes as the Pacific cold tongue, but they are likely to affect temperature as well.

When considering the factors that set surface air temperature on Earth, it is helpful to think of the following sequence. To a crude approximation, the radiative balance with space (mediated by albedo and atmospheric chemistry) determines the temperature. The temperature is modified by atmospheric heat transport, and further modified by ocean heat transport.

The direct effect of SST on T_* is not the only air–sea feedback, because a change in T_A alters the atmospheric state in several ways that affect the ocean. A change in T_A changes atmospheric circulation. For example, a warmer sea surface drives greater ocean heat loss, and a deeper atmospheric marine boundary layer. That affects the sea-level wind because the momentum of stronger winds aloft is mixed down to the surface. In turn, a changed wind speed near the sea surface affects latent and sensible heat fluxes. A circulation change can alter cloudiness and the distribution of water vapor (a greenhouse gas) and hence the radiative balance. A changed atmospheric circulation can also alter the atmospheric heat transport to the location in question. Changes in either ocean or atmospheric circulation

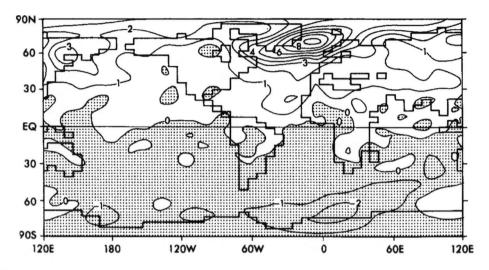

Figure 11.12 Surface air temperature in atmosphere-ocean GCM for AMOC experiment minus no-AMOC experiment. Contour interval 1 K, dotted regions are negative. From Manabe and Stouffer (1988), Figure 22, ©American Meteorological Society. Used with permission.

can change the distribution of ice, which affects the albedo (Subsection 11.1.1) of the Earth and the air–sea heat exchange. All these changes further modify T_* and λ in a way that intrinsically couples the ocean and atmosphere.

Atmosphere-ocean GCMs reveal the effect on the atmosphere of removing or altering heat exchange with the ocean from all the mechanisms mentioned above. The experiments of Manabe and Stouffer (1988) compare two climate states, an AMOC state with 12 Sv of sinking in the North Atlantic and a no-AMOC state. Subtracting the no-AMOC state surface air temperature from the AMOC state isolates the influence of AMOC on surface air temperature (Figure 11.12). The AMOC is associated with a warming centered on the northern Atlantic, with a maximum of over 9 K in the Nordic Seas, and over 1 K over much of the northern hemisphere. There is a weaker cooling in the Southern Hemisphere, with a maximum of over 2 K in the Atlantic sector of the Southern Ocean. More recent experiments with increasingly sophisticated GCMs have shown similar temperature effects from the AMOC.

Freshwater Fluxes

Unlike SST, sea surface salinity (SSS) is *not* directly associated with any physical processes that restores it to any particular value. The salinity is altered by the net surface freshwater flux (evaporation minus precipitation and runoff, plus ice effects). The freshwater flux is a function of the atmospheric state (temperature, humidity, circulation, radiation), which can be influenced by ocean SST. And SST is indirectly affected by SSS because both variables contribute to seawater density and hence the mixed layer properties (Chapter 4). But no simple relationship exists between surface ocean salinity and the components of the air/sea freshwater flux. In the absence of any such relationship it is natural to set the freshwater flux to a constant value in ocean-only models. This contrasts to the surface heat flux, which is very sensitive to SST.

11.2.3 Flux Boundary Conditions and Multiple Circulation States

Surface Properties and Surface Fluxes

Many times in this book we refer to *the* circulation that results from a certain forcing and geometry. The tacit implication is that only one circulation is possible. This is not necessarily the case, however. If the circulation has multiple (stable) equilibria, then a temporary change in forcing can "push" it into another state that persists even if the forcing reverts to its original form. Observations in the real world of the contemporary circulation do not preclude the possibility that past or future circulations may be very different under otherwise similar conditions as today.

The deep meridional overturning circulation can exist in multiple states because of the surface forcing and the fact that the circulation carries heat and salt. If we specify temperature and salinity (T and S) and hence density at the ocean surface, we know that the densest surface water will convect to the bottom, vertical jumps in density will be broadened by turbulent mixing, horizontal density gradients will produce horizontal pressure gradients, and the water will circulate.

Specifying the surface density is problematic, however, because it is influenced by the circulation. It is reasonabke to ask, "*if* a certain surface (T, S) exists, *then* what is the circulation and subsurface (T, S) field?" But if we ask how conditions *external* to the ocean determine the circulation, specifying (T, S) is not quite right.

Specifying SST has some justification because a physically realistic formulation of the surface heat flux has the form of a restoring equation that relaxes SST towards externally specified values. SSS is not specified (Subsection 11.2.2) but is influenced by the surface freshwater flux, which *is* specified. Mathematically, this is a flux boundary condition. As the freshwater flux can be mimicked with a salt flux (Subsection 4.3.2), and because salt flux is easier to work with, it is convenient to model the freshwater flux with a virtual salt flux. Mathematically, specifying the salt flux is equivalent to specifying $\kappa \partial S / \partial z$ (κ is vertical salt diffusivity) at the sea surface.

The relationship between S and freshwater flux F is illustrated in the idealized example shown in Figure 11.13. The circulation moves a parcel of water at the surface anticyclonically, as shown. Freshwater flux is into the ocean on the northern limb of the circulation, out of the ocean on the southern limb, and zero in the eastern and western segments. Thus the water gains salinity in the south and becomes fresher in the north. For this idealized scenario, the change in salinity is inversely proportional to the water speed, so that circulation (as well as F) determines the surface salinity gradient.

Multiple Overturning States, Stability, and Oscillations

An idealized model due to Henry Stommel (Stommel, 1961) shows how **multiple equilibria** may come about in this system. Stommel's paper attracted little notice until Claes Rooth (Rooth, 1982) revived interest just as numerical models began to show similar behavior. These models highlight the role of **mixed boundary conditions**, in which the salinity has a flux boundary condition and temperature has a boundary condition that either specifies SST or restores it to a specified value.

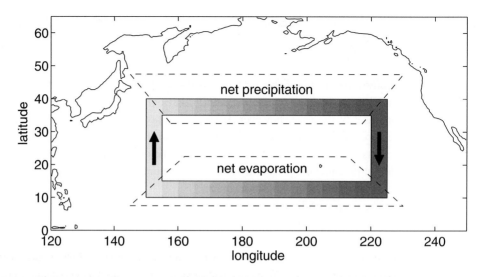

Figure 11.13 Idealized rectangular path with circulation direction shown by arrows, net precipitation and evaporation regions shown by dashed boundaries, and shading showing salinity along path (dark for fresher). See also Exercise 11.7.

The Stommel model simplifies the continuous T and S distribution of the real ocean to an imaginary single-basin, single-hemisphere ocean described by T and S at two locations: low latitude and high latitude (Figure 11.14a). Density is controlled by temperature and salinity. The equations defining the model are defined and solved in Subsection 11.3.2. The property values at each location represent a "box"; that is, a region of homogeneous fluid that exchanges fluid with the neighboring box. This arrangement can be realized in the laboratory by separating the boxes with a wall, mixing the water in each box with a rotor or other mechanical device, and allowing water to flow between the boxes with a pair of small holes cut in the wall. This is a crude model of the ocean, which has strong vertical stratification and continuous meridional gradients, but it does allow us to study the effects of T and S differences between high and low latitudes.

Surface temperature restoring is represented in the model by specifying T in each box: warm in Box 1 and cold in Box 2. For reasons discussed previously, we specify freshwater fluxes rather than S. Box 1 has net evaporation like the real-world subtropics and Box 2 has net precipitation like the subpolar ocean (Figures 11.6 and 11.7c). Thus there is a freshwater transport from Box 1 to Box 2. Presumably, this occurs in the atmosphere, which is not explicitly included in the model. The freshwater transport is modelled with a virtual salt transport going in the other direction (Figure 11.14a). The model does not resolve the real world's freshwater transport between the subtropics and the near-equatorial rain belts, which are contained within Box 1.

The other link between the two boxes is through the ocean circulation, which is specified by a single variable representing the exchange of water between the two boxes. The wind-driven gyre flow is ignored and we assume that the exchange is controlled by a buoyancy-driven overturning circulation. In the real ocean the strength of the deep meridional overturning depends on surface density and other properties (Chapter 8). The model

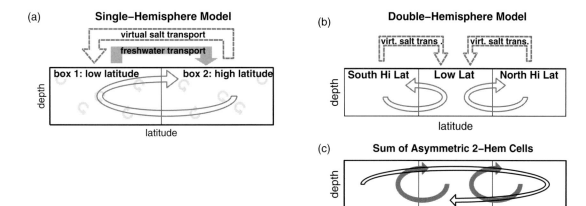

Figure 11.14 Schematic of thermohaline box models for (a) one and (b) two hemispheres. Each box represents a region defined by a single set of parameters (T, S, and ρ). Arrows above boxes show the direction of freshwater transport and the equivalent virtual salt transport. Large curved arrows represents the overturning exchanging fluid between the boxes, and small curved arrows represents small-scale mixing homogenizing properties within each box. Warming (cooling) occurs at low (high) latitude. (c) Schematic overturning in a pair of boxes in each hemisphere and the total basin-scale overturning.

simply sets the volume transport q (not potential vorticity in this chapter) of the exchange to be proportional to the density difference between the two boxes. The behavior is governed by the density-q relation, an equation of state determining density from T and S, and an equation for the evolution of ΔS, the salinity difference between the two boxes (Subsection 11.3.2).

If we start the system with equal S in both boxes, the temperature difference between the two boxes creates a density difference (Box 2 denser than Box 1) that drives the overturning circulation. The atmospheric "salt transport" freshens high-latitude Box 2 and salinifies low-latitude Box 1. The salinity difference reduces the density difference between the boxes and weakens the overturning. The overturning carries relatively salty water from Box 1 to Box 2 and relatively fresh water the other way, thus transporting salt in the opposite direction to the atmospheric salt transport. The system can attain an equilibrium if the salt transport in the ocean is equal to the imposed transport. It turns out that the two-box model can attain this balance in *three* different ways (Subsection 11.3.2).

In one "temperature-dominated" state, temperature drives a strong overturning that virtual salt flux weakens slightly (compared to a no-salt flux case). This state has a relatively big q and small ΔS. The other two states have a small q and a big ΔS. The small q is due to a small density difference $\Delta \rho$ between boxes, which occurs because the effect of ΔS on density nearly cancels that of the temperature difference. One such state is also temperature-dominated, with $\Delta \rho$ the same sign as the small-ΔS state. In the other, "salinity-dominated," case, ΔS reverses the density gradient and the overturning. Deep water formation occurs in the relatively warm, salty, low-latitude box rather than the cool, fresh, high-latitude box. Even though the circulation reverses, the oceanic salt transport stays the same. In Figure 11.14a, the circulation is clockwise in temperature-dominated states and counterclockwise in salinity-dominated states.

The time-evolution of the model from a non-equilibrium state is also interesting. Given a certain initial condition, how do ΔS and q develop over time? If ΔS is less than its value in all three equilibrium states, than one can show that ocean virtual salt transport will be too small, and the surface flux increases ΔS until the system reaches equilibrium. If ΔS is greater than in all three equilibria, then the virtual salt flux due to the circulation will be greater than the imposed flux and ΔS decreases until it falls into the low-latitude-sinking state. The behavior for an intermediate ΔS is more complicated, but Subsection 11.3.2 shows that if the initial condition is not one of the equilibrium states, then the system will *never* evolve towards the weak salt-dominated state, and will always evolve towards one of the other two equilibria. We say that the intermediate equilibrium is **unstable** and the other two states are **stable**.

GCMs compute the evolution of an ocean state from a given initial condition for three-dimensional distributions of T and S. As in the two-box model, starting at different initial conditions may allow the system to reach a new equilibrium. But if the equilibria states are not known, it is unclear if some other initial condition would evolve to a different equilibrium. Unstable equilibria are never found by GCMs, because even a small departure from the equilibrium state will cause the difference to grow over time. Often we are less interested in unstable states anyway, because the real world likewise does not stay in an unstable equilibrium state. Still, the unstable states can give insight into the general behavior of the system.

A major difference between the instability in the two-box model and those that give rise to mesoscale eddies (Subsection 7.1.1, 7.2.1) is that the two-box state evolves until it reaches a new steady state. GCM experiments reveal time-variability of the overturning circulation and the temperature/salinity distribution on times scales of years to centuries. The time variability caused by instability of the overturning is an important topic but beyond the scope of this chapter.

11.2.4 Multiple States in General Circulation Models

Single Basin

The two-box model shows how a physically plausible mechanism can generate multiple ocean states. But the solutions do not give a very accurate description of the solutions for a GCM with a similar idealized configuration (one hemisphere, T specified, simplified virtual salt flux). The GCM does have multiple equilibria, but the difference between the states is more complex and the system can oscillate. For instance, there is a "flush" cycle in which the overturning becomes confined to the top third of the basin and mixing warms the deep ocean until there is a dramatic blast of deep water formation with overturning of over 80 Sv (Weaver et al., 1993).

A three-box model spanning both hemispheres (Figure 11.14b) better resembles GCMs (Rooth, 1982; Thual and Mcwilliams, 1992). This system can be thought of as separate two-box systems. Each hemisphere can have subpolar sinking or tropical sinking. If both hemispheres are in the same circulation state, than the two hemispheres are simply mirror images of each other. However, if the system has sinking at different latitudes – one in the subpolar box, one in the tropical box – then the solution has opposing tropical flows that cancel. A single circulation loop then exists with water sinking in one hemisphere and rising in the other (Figure 11.14c). This solution is interesting because even with geography and forcing symmetric between the hemispheres, the ocean can circulate in an asymmetric way. The asymmetric flow is somewhat reminiscent of the asymmetric AMOC in the real world. Also reminiscent of the real Atlantic, the hemisphere with sinking is saltier than the one with upwelling.

A GCM in a single-basin, two-hemisphere configuration can show both the symmetric and asymmetric overturning states. Bryan (1988) drives the system with a symmetric pattern of zonal wind and a heat flux consisting of restoring to a hemispherically symmetric temperature based on the real-world global zonal-average SST. To determine an appropriate virtual salt flux, he first runs the experiment restoring SSS to a target salinity with high subtropical values and low polar values. The surface virtual salt flux is diagnosed from that experiment, and the model is run again with identical parameters except the salt flux is fixed rather than depending on SSS. The GCM maintains the symmetric circulation state in which each hemisphere has sinking at high latitudes and upwelling at low latitudes, similar to Figure 8.12.

The model is also run starting from an initial state in which the high latitude southern hemisphere surface is made 2 g/kg saltier than before. The denser water at high latitude intensifies the southern hemisphere overturning, which then extends into the northern hemisphere, bringing more low-latitude high-salinity water into the southern hemisphere. The southern overturning grows in strength and extent until the basin is almost completely filled with a single overturning cell with water sinking near the southern boundary, similar to Figure 8.14.

Multiple Basins

"Why is no deep water formed in the North Pacific?" asks the title of a classic paper (Warren, 1983). This refers to the observation that water sinks to the bottom in the northern North Atlantic and near Antarctica, but not in the northern North Pacific (Chapter 8). The proximate reason is that the North Pacific surface water is less dense than in the North Atlantic or the Southern Ocean (Figure 4.6). The low Pacific density is due to salinity, which is lower in the North Pacific than in the subpolar Atlantic and Southern Ocean (Figure 4.4). The North Pacific has bigger changes in both salinity and density from mid-depths to the surface than the North Atlantic (Figure 8.1). It is not obvious what determines the salinity difference between the North Pacific and elsewhere, however (Subsection 11.2.3).

The existence of multiple states raises the possibility that the no-Pacific-sinking state is not a necessary feature of the circulation and other possible states exist. Marotzke and Willebrand (1991) conducted a series of numerical experiments showing that even with "Pacific" and "Atlantic" basins with identical geometry (Figure 11.15a) and surface forcing, an ocean can lose its sinking in the north of one basin. As in the Bryan (1988)

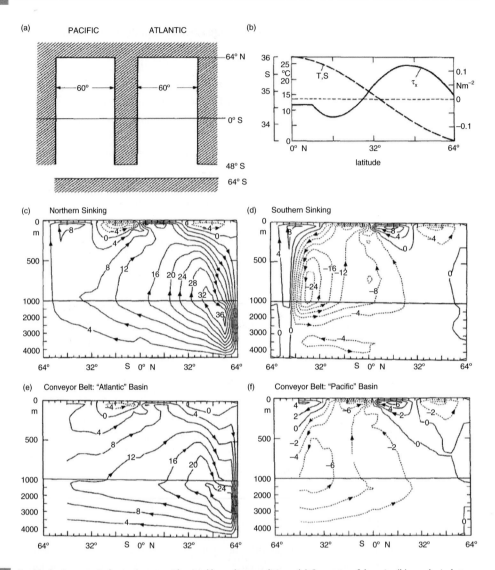

Figure 11.15 Double-basin numerical experiments with mixed boundary conditions. (a) Geometry of domain, (b) zonal wind stress (solid) and restoring profiles for surface temperature and salinity (dashed), and meridional overturning streamfunction for (c) each basin, northern sinking state, (d) each basin, southern sinking state, (e) "Atlantic" basin, conveyor belt state, and (f) "Pacific" basin, conveyor belt state. For streamfunction, contour interval is 4 Sv and solid curves represent clockwise flow. From Marotzke and Willebrand (1991), Figures 1, 2, 6a, 8a, 7a, 7b; ©American Meteorological Society. Used with permission.

experiment, the system is first forced with a restoring boundary condition in T and S that is symmetric about the equator, and in this case is also the same in each basin (Figure 11.15b). After equilibrating, the freshwater flux is diagnosed from the first experiment and used to replace the salinity restoring in subsequent experiments. By temporarily changing the freshwater flux and then returning it to the original value, Marotzke and Willebrand (1991)

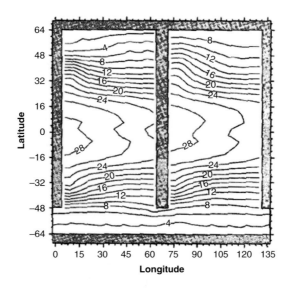

Figure 11.16   SST in GCM experiment with mixed boundary conditions and weak temperature salinity restoring, for "conveyor belt" state with deep water formation in north of basin on right but not on left. Contour interval is 2°C. From Rahmstorf and Willebrand (1995) Figure 8, ©American Meteorological Society. Used with permission.

are able to allow the same forcing to produce deep sinking in the north of both basins (Figure 11.15c), deep sinking in the south of both basins (Figure 11.15d), and a "conveyor-belt" circulation with sinking in one basin (Figure 11.15e) and no deep sinking in the other (Figure 11.15f). Each overturning state has a correspondingly different salinity distribution with salinity greater and vertically homogeneous in the sinking region and more stratified and fresher on the surface of subpolar boundaries with no sinking.

In the real world, another important difference between the Atlantic and the Pacific is that the Atlantic SST is about 3°C warmer than the Pacific at the same latitude (Figure 4.1). The model of Marotzke and Willebrand (1991) does not reproduce this difference because the surface temperature restoring is too strong to allow much change in SST with circulation state. When SST is driven by weak restoring, such as that described in Subsection 11.2.2, a more realistic SST difference is found (Figure 11.16).

11.3 Theory

11.3.1 Surface Heat Flux with Atmospheric Interaction

As discussed in Subsection 11.2.2, formulating the ocean surface boundary condition as restoring to a target temperature based on given atmospheric conditions is potentially misleading because ocean circulation itself influences the atmosphere. A wide range of complex interactions can be simulated with an atmosphere–ocean GCM. Sometimes we want to isolate the ocean while capturing the direct influence of SST on cooling or warming

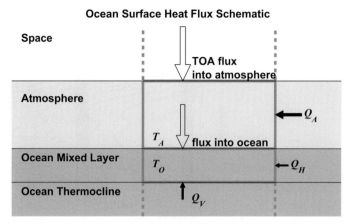

Figure 11.17 Schematic of model of a control volume with two active layers (atmosphere and mixed layer) showing heat transports within each layer (solid arrows) and heat transport between components of the system (open arrows). TOA means top of the atmosphere.

the atmosphere. A two-layer model (Schopf, 1983; see also Dickinson, 1981) consisting of an ocean mixed layer and the lower atmosphere (Figure 11.17) can accomplish this task.

As in Subsection 4.3.1, we start with the bulk and radiative formulae for air–sea heat flux. We also include horizontal and vertical components of ocean advective heat flux divergence, denoted Q_H and Q_V, respectively. We relate the heat content (per surface area) of the mixed layer to ocean temperature with $C_O T$, where $C_O \equiv H c_p \rho$ is the heat capacity over the mixed layer whose thickness is H.

For the atmospheric layer (Figure 11.17), heat that *enters* the ocean must *leave* the atmosphere, so the heat budget for the atmospheric layer has a term that is minus the air–sea term in the ocean budget. The atmosphere also has a horizontal heat transport term Q_A and a heat capacity factor C_A that are analogous to the ocean values. Finally the atmosphere has a radiative interaction with space with long wave, $-\mu T_A$, and short wave, Q_1, terms as in the ocean heat budget.

In general, the ocean mixed layer is nearly homogeneous (Chapter 4), but the atmosphere has great vertical temperature variations. The relationship between the surface temperature T_A and radiative emissions into space is not straightforward as it involves clouds, water vapor, aerosols, and the atmospheric circulation. Nevertheless, we assume that these processes can be approximated with a linear dependence on T_A and proportionality constant μ. This reasoning says that:

$$C_O \frac{dT}{dt} = \lambda(T_A - T) + Q_H + Q_V, \tag{11.6}$$

$$C_A \frac{dT_A}{dt} = \lambda(T - T_A) + Q_1 - \mu T_A + Q_A. \tag{11.7}$$

If there is no heat flux between the atmosphere and ocean, and the atmosphere is in equilibrium (steady state) with T_A equal to an equilibrium temperature T_R, then (11.7) becomes

$$\mu T_R = Q_1 + Q_A, \tag{11.8}$$

which defines T_R and allows us to rewrite (11.6)–(11.7) as

$$C_O \frac{dT'}{dt} = \lambda(T'_A - T') + Q_H + Q_V, \tag{11.9}$$

$$C_A \frac{dT'_A}{dt} = \lambda(T' - T'_A) - \mu T'_A, \tag{11.10}$$

where $T' = T - T_R$ and $T'_A = T_A - T_R$.

Now eliminate T'_A from the expression for air–sea heat flux so that the ocean is forced by atmospheric parameters that are not influenced by the ocean state. For a steady state, the left-hand side of (11.9) is zero and we can find T'_A in terms of T and the parameters. If the system evolves due to changing forcing (for instance changing Q_V due to changing wind), then a scale analysis shows that $C_A dT'_A/dt$ is negligible (see Exercise 11.8) and thus (11.9) becomes a single equation for T:

$$C_O \frac{dT}{dt} = \mu'(T_R - T) + Q_H + Q_V, \tag{11.11}$$

where

$$\mu' = \frac{\mu}{1 + \mu/\lambda}, \tag{11.12}$$

as discussed in Subsection 11.2.2. The parameter λ represents the net effect of all heat flux terms at the sea surface (Subsection 4.3.1, (4.13)), including radiation and turbulent heat fluxes. According to (11.8), μ represents the sensitivity of atmospheric surface temperature to changes in solar radiation or horizontal heat flux. If the top of the atmosphere outgoing radiation flux was given purely by the black body law for T_A, we would know the value of μ from the Stefan–Boltzmann law (4.8). Because of greenhouse gases and all the other factors that affect radiative emissions, we can only estimate μ empirically from data or from atmospheric general circulation models. The sensitivity of the Earth's climate is is still not known to better than about a factor of two (Roe and Baker, 2007; Bindoff et al., 2013). It is easy to show that the sensitivity of energy flux to temperature is much less than the 50 W m^{-2}K^{-1} derived from bulk formulae for fixed atmospheric properties. It is closer to its black-body value, 3 W m^{-2}K^{-1}, for typical values of T_A.

11.3.2 Two-Box Model with Multiple Equilibria

Subsection 11.2.3 describes the solutions to a box model of a single-hemisphere thermohaline circulation. Here, we derive these results mathematically. Quantities refer to Box n ($n = 1, 2$) in Figure 11.14a, and we consider the temperatures T_n as externally specified parameters. Although the restoring boundary condition gives some justification for specifying SST, ignoring the basin-average temperature dependence on circulation is more dubious. When surface restoring mainly determines the temperature difference between the low-latitude and high-latitude boxes, this approximation may still be reasonable.

The system evolves according to a budget for ocean salinity. The variable $V = V_1 = V_2$ is the volume of each box, S_n is salinity, $\tilde{q}$ is the seawater volume transport between the two boxes due to the circulation, $\tilde{H}_S$ is the virtual salt transport (units of g kg^{-1} m^{3}s^{-1}). The salt budget is then

$$V\frac{dS_1}{dt} = \tilde{H}_S + |\tilde{q}|(S_2 - S_1), \tag{11.13}$$

$$V\frac{dS_2}{dt} = -\tilde{H}_S - |\tilde{q}|(S_2 - S_1). \tag{11.14}$$

We are interested in the overturning direction: shallow water poleward and deep water equatorward, or the reverse. The model is so simple that shallow and deep water have the same salinity, and the salinity flux between boxes is independent of flow direction. Thus, we take the absolute value of $\tilde{q}$ in the equations above. We define $\tilde{q} > 0$ to be the state with sinking in Box 2. With salt transport per volume $H_S \equiv \tilde{H}_S/V$ and volume transport per volume $q' \equiv \tilde{q}/V$ (where $1/q'$ is the characteristic box flushing time), the equations become

$$\frac{dS_1}{dt} = H_S + |q'|(S_2 - S_1), \tag{11.15}$$

$$\frac{dS_2}{dt} = -H_S - |q'|(S_2 - S_1). \tag{11.16}$$

We assume a linear equation of state (Subsection 1.2.3, (1.6)). Finally, assume that the overturning is proportional to the density difference between the boxes:

$$q' = -\frac{k\Delta\rho}{\rho_0} = k(\alpha\Delta T - \beta\Delta S), \tag{11.17}$$

where k (units of s^{-1}) is a proportionality constant and $\Delta S = S_1 - S_2$, and so on. This is a crude description of the relation between overturning strength and the density structure, but it captures the fact that q' increases with meridional density gradient (as in the thermohaline scaling in Subsection 8.3.2, (8.30)).

Combining (11.15) and (11.17), the overturning equation for the salinity gradient is

$$\frac{d\Delta S}{dt} = 2H_S - 2k|\alpha\Delta T - \beta\Delta S|\Delta S. \tag{11.18}$$

We have reduced the governing equations for the thermohaline circulation of a single-hemisphere basin to a single equation for the meridional salinity gradient. The equation is nonlinear, because of the quadratic dependence on ΔS and the absolute value sign. The nonlinearity arises from the salinity advection because the flow speed depends on salinity (advective nonlinearity is always quadratic for this reason). The nonlinearity does not derive from nonlinearity in the equation of state (which is not quadratic in general, and is linear here).

The equation is simplified by nondimensionalizing as follows:

$$s = \frac{\beta\Delta S}{\alpha\Delta T}, \tag{11.19}$$

which represents the relative effect of salinity and temperature on the density gradient. Nondimensional time,

$$\tau = (2k\alpha\Delta T)t, \tag{11.20}$$

represents time normalized by (half) the time to fill the box at a rate given by the $\Delta S = 0$ volume transport. And q' is nondimensionalized as

$$q = \frac{q'}{k\alpha \Delta T} = 1 - s. \tag{11.21}$$

The scale to normalize volume transport is the volume transport for zero salinity gradient. The nondimensionalized equation has only one parameter,

$$E = \frac{\beta H_S}{k(\alpha \Delta T)^2}, \tag{11.22}$$

which is the buoyancy flux due to the virtual salt transport normalized by the $\Delta S = 0$ buoyancy flux due to the zero-salinity-gradient circulation. Inserting the above definitions into (11.18) yields

$$\frac{ds}{d\tau} = E - |1 - s|s. \tag{11.23}$$

The system is in an equilibrium state (meaning a steady state) when $ds/d\tau = 0$. Plotting $ds/d\tau$ (horizontal axis) as a function of s (vertical axis) for the example of $E = 0.15$ (Figure 11.18a), we see that there are three values of s that are equilibria. For this system, changing the value of E has a simple effect on the graph: it shifts the $ds/d\tau$ curve to the right (larger E) or to the left (smaller E).

Now solve for the steady states, denoted by s_E and q_E, by finding the roots of

$$E - |1 - s_E|s_E = 0. \tag{11.24}$$

If $s_E < 1$, then temperature dominates density and (11.21) says that $q_E > 0$, so we have high-latitude sinking. In that case, $|1 - s_E| = 1 - s_E$ and the solution to the quadratic equation (11.24) is

$$s_E = \frac{1}{2} \pm \sqrt{\frac{1}{4} - E}. \tag{11.25}$$

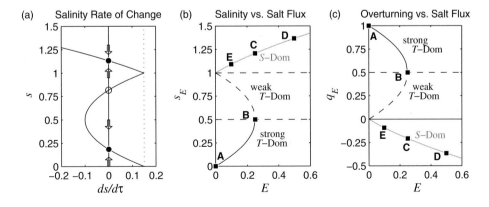

Figure 11.18 Two box model solution (nondimensional). (a) salinity tendency, $ds/d\tau$, versus salinity, s, from (11.23) for $E = 0.15$. Circles show stable (filled) and unstable (unfilled) equilibria; arrows indicate possible $s(\tau)$ paths. Steady state solutions (b) s_E and (c) q_E as functions of E. See text for letters.

To be physical, we require real values of s_E, so we require $E \leq 1/4$ for these solutions to exist.

If $s_E > 1$, then salinity dominates density, $q_E < 0$, and $|1 - s_E| = s_E - 1$, so we have low-latitude sinking, and

$$s_E = \frac{1}{2} + \sqrt{\frac{1}{4} + E} \qquad (11.26)$$

(discarding the negative root because it violates $q_E < 0$). Thus, we are left with either one or three possible solutions for a given value of E.

It is instructive to see how the equilibrium solutions depend on E, so we plot $s_E(E)$ and $q_E(E)$ (Figure 11.18b,c). The form of (11.23) is so simple that $E(s)$ has a very similar form as $ds/d\tau$. For $E \leq 1/4$, there are three physically valid solutions: strong T-dominant (small s_E, large q_E), weak T-dominant (large s_E, small q_E), and S-dominant ($s_E > 1$, $q_E < 0$). For $E > 1/4$, there is only the S-dominant solution.

Starting with $E = 0$, consider how the equilibrium state changes when E increases. In a physical system (or a numerical model) this could be accomplished by changing E over time. The change keeps pushing the system out of equilibrium, but if the change is slow enough, the system continually approaches the equilibrium state before E changes much. Start at the solution marked **A** on the $s_E(E)$ plot (Figure 11.18b). Increasing the salt flux increases s_E. Figure 11.18c shows that $q_E = 1$ when $E = 0$, and q_E decreases with rising E as the density gradient due to salinity increasingly opposes the density gradient due to temperature. As E approaches $1/4$, both s_E and q_E change more rapidly, but at $E = 1/4$ (point **B**) a dramatic change occurs. The system "jumps" to point **C** on the S-dominant solution branch with very different values of s_E and q_E. The real-world equivalent of this would be if a small acceleration of the hydrological cycle made the AMOC reverse direction! Continuing to increase E makes both overturning and salinity gradient strengthen more (point **D**). Decreasing E, the system remains on the S-dominant solution path even when E goes below $1/4$, as at point **E**. This is known as **hysteresis**: the state of the system depends on the forcing parameters (in this case, parameter E), and also depends on the history of the system.

Given a particular E, what happens if the system is not in an equilibrium state? In that case, $s \neq s_E$, and changes according to (11.23). Looking again at Figure 11.18a, several possibilities exist, depending on the initial value of s. If s is less than the lowest equilibrium s_E (strong T-dominated state), then $ds/d\tau > 0$ and hence s increases until it reaches the equilibrium state. This is indicated by an arrow on the $ds/d\tau$ axis. If s is between the strong T-dominated state and the weak T-dominated state, $ds/d\tau < 0$ and s will decrease until the system reaches the T-dominated state. Similarly, if the initial s exceeds the weak S-dominated state, the system evolves to the S-dominated equilibrium.

If s makes any departure from the strong T-dominated or S-dominated state, it will return to the original equilibrium (provided the perturbation does not push s past the weak T-dominated state). If s makes any departure, no matter how small, from the weak T-dominated state, s will evolve to one of the other equilibria. In other words, the S-dominated and strong T-dominated states are stable equilibria, whereas the weak T-dominated state is unstable. Exercise 11.9 gives details; in particular, it is possible to solve the time dependent problem analytically.

The idealized system here is one of many configurations in which nonlinear equations allow for multiple equilibria. Both thermohaline and wind-driven circulations may exhibit such behavior. The textbook by Dijkstra (2005) is a comprehensive examination of these phenomena.

Exercises

11.1 (a) Using a global energy balance, estimate the average temperature for the Earth if it had no atmosphere. Make and state sensible simplifying assumptions.

(b) Compare and contrast your estimate to the average surface temperature of the Moon, explaining your argument.

(c) Hence, compute the excess surface air temperature on Earth due to the greenhouse effect, explaining your argument.

(d) Estimate the average surface temperature at Earth's equator and poles in the absence of ocean/atmosphere heat transport (use Figure 11.2a to estimate the average solar insolation at equator and poles).

(e) **Optional extra**: Estimate the average temperature for the Earth if it was covered in ice, making reasonable assumptions. Comment on the realism of this state. What would you investigate next to explore this possibility, and why?

11.2 Estimate meridional transports with the **direct method**: Using the zonal hydrographic section you picked in Exercise 2.11:

(a) Compute and plot the total meridional streamfunction in temperature coordinates. Split your streamfunction into components (and plot them) due to: the Ekman transport, the western boundary current, and the mid-ocean geostrophic flow.

Hints: See Figure 11.11 for an example of what to compute and plot. See Exercise 3.1 on wind stress data to compute the Ekman transport.

(b) Hence estimate the meridional heat flux across your section carried by the ocean. Compare and contrast your estimate to the total oceanic heat flux at the latitude of your section, and the total atmospheric heat flux at that latitude.

(c) Repeat (a) and (b) for freshwater flux.

State clearly, with brief discussion and justification, any assumptions you use.

11.3 Estimate meridional transports with the **heat flux method**: Using the OAFlux air/sea flux climatology (see Appendix A):

(a) Make a plot like Figure 11.4 of the annual mean climatological net heat flux into the ocean.

(b) Hence compute and plot the total net heat flux into the ocean as a function of latitude for the (i) Atlantic, (ii) Pacific, (iii) Indian, and (iv) global oceans.

(c) Integrate your total net heat fluxes into the ocean from (b) to estimate and plot the total meridional heat flux as a function of latitude carried by the individual ocean basins and by the global ocean. Compare and contrast your results with those in Figure 11.2 and your own results from Exercise 11.2.

(d) Repeat (a), (b), and (c) for freshwater flux. Use the Global Precipitation and Climatology Project data for precipitation, for example, (see Appendix A).

State clearly, with brief discussion and justification, any assumptions you use.

11.4 Consider Figures 11.6 and 11.7 and the associated latent heat exchange between the ocean and atmosphere. Using appropriate data and suitable simplifying assumptions (that you state):

(a) Compute and plot the zonal average evaporation and precipitation rates as functions of latitude, similar to Figure 11.6. State the global totals of these quantities.

(b) Compute and plot the zonal average evaporative heat flux from the ocean and the global total.

(c) Compute and plot the zonal average latent heating of the atmosphere and the global total.

(d) Hence, estimate and plot the meridional latent heat flux in the atmosphere.

(e) Discuss your results compared to other parts of the ocean/atmosphere heat budget (such, as in Figure 11.2).

11.5 Briefly,

(a) Summarize the evidence that the Atlantic Meridional Overturning Circulation (AMOC) has existed in different states in Earth's past.

(b) Explain how AMOC changes may interact with Earth's ice ages.

11.6 Considering the GCM sections of temperature and velocity in Figure 11.11,

(a) Explain how the temperature-coordinate overturning streamfunction relates to the meridional heat flux.

(b) Estimate the meridional heat flux in the Atlantic and Pacific oceans at 26°N. Compare and contrast your estimates to those in Figure 11.2.

(c) Briefly explain and discuss the statement that the North Atlantic meridional heat flux is mainly carried by the overturning circulation, but the North Pacific meridional heat flux is mainly carried by the subtropical gyre.

11.7 Imagine that the Stommel model of a subtropical gyre (Figure 3.12) is augmented with a dynamically passive temperature field, $T(x, y)$. Temperature relaxes back to atmospheric temperature $T_*(y)$ according to (11.5), which only depends on latitude. With this model in mind (see also Figure 7.11), and using appropriate assumptions:

(a) Derive an expression for the T field, explaining your argument.

Hint: Consider the evolution of T following Lagrangian paths (streamlines).

(b) Thus, compute formulae for the air–sea heat flux, $Q(x, y)$ and the zonally averaged meridional heat flux $E(y)$ carried by the gyre.

(c) Sketch T and Q for three cases: (i) rapid, (ii) slow, and (iii) intermediate air/sea exchange rates. Define a nondimensional number to specify these cases.

(d) Compute and sketch the maximum value of $E(y)$ as a function of your nondimensional number. Explain your findings in terms of the three cases from part (c).

(e) Taking typical parameter values and making suitable comparisons with figures in this book, discuss the realism of this model.

(f) **Bonus question**: Consider how this model might apply to salinity.

11.8 Consider the model of air/sea interaction in Subsection 11.3.1.

(a) Assume that the ocean temperature T_O is constant. Hence, solve for the atmospheric temperature T_A as a function of time. State the characteristic timescale for T_A to change.

(b) Now assume that the atmosphere temperature T_A is constant. Hence, solve for the oceanic temperature T_O as a function of time. State the characteristic timescale for T_O to change.

(c) Estimate C_O/C_A using reasonable parameter choices that you justify.

(d) Hence justify the assumption that $C_A dT'_A/dt$ can be neglected in (11.10).

11.9 Solve the evolution equation for the two-box model of Subsection 11.3.2,

$$\frac{ds}{d\tau} = E - |1 - s|s, \tag{11.27}$$

assuming that s is always in the temperature-dominated regime. When solving, set $s(\tau) = s_E + x(\tau)$, where s_E is the temperature-dominated equilibrium solution and x is the anomaly, and solve for $x(\tau)$ rather than $s(\tau)$. Note that in this case we can solve the full nonlinear equation exactly without linearizing. What is the long-time behavior of the solution? Give a physical interpretation of the (dimensional) timescale over which the system evolves. Plot the family of solutions for a representative selection of initial conditions and E parameters.

Optional bonus question: Repeat this procedure for the salt-dominated regime, and hence find the full solution to (11.27).

11.10 Consider a single-basin ocean that is symmetric about the equator, with a symmetric meridional overturning circulation with sinking near the poles and upwelling everywhere else. Suppose we increase the magnitude of the overturning by increasing the ocean's diapycnal diffusivity κ. What do you expect to happen to midlatitude heat transport? What, if anything, do you expect to happen to equatorial SST and surface air temperature? Explain your reasoning.

Appendix A Data Sources

Observational datasets are used throughout this book. They typically take the form of atlases in which measurements taken at various times and locations are interpolated and averaged on to a regular grid in latitude and longitude. As discussed in Chapter 1 (especially Subsection 1.1.4), there are serious issues concerning data sampling. Generally there is a tradeoff between two desirable indicators of quality: high grid resolution and the large number of measurements used to estimate the value of a variable at each grid point. Care must be taken in converting the raw data into a gridded climatology. Care must also be taken in using gridded climatologies, because it is easy to forget that they have nonuniform quality and complex error characteristics (see Wunsch, 2015 and Emery and Thompson, 2001, for instance). Here we discuss the datasets used: see Table A.1. For discussion of different oceanographic instruments, see Subsection 1.1.2; for different observing platforms, see Subsection 1.1.3.

Temperature and Salinity

Much ocean data is managed in the US by National Centers for Environmental Information (NCEI), which until recently was known as the National Oceanographic Data Center, NODC. The World Ocean Database (WOD; Seidov et al., 2014; Boyer et al., 2012) compiles temperature, salinity, and chemical in situ profile measurements from many sources around the world. Other repositories specialize in subsets of these hydrographic profile data along sections or at time series stations with many repeat profiles. A series of annual and monthly climatologies from the WOD (Levitus and Oort, 1977) is available. It includes the World Ocean Atlas (WOA) 2013 (Locarnini et al., 2013 and Zweng et al., 2013) with grid resolution of $1/4°$ (about 27 km) in latitude and longitude.

In WOA13, a particular grid point value is estimated with a weighted average of measurements taken within a given horizontal **radius of influence** of the grid point. The weighting factor for a particular observation decreases with its distance from the grid point. The estimate uses an iterative process in which the influence radius is reduced from 321 km to 214 km, producing an effective resolution that is significantly coarser than the grid resolution. Features are distorted if smaller than $1°$ and virtually undetectable if smaller than $1/2°$.

The CSIRO Atlas of Regional Seas (CARS) is an alternative climatology using most of the same measurements as WOA13. CSIRO is the Commonwealth Scientific and Industrial Research Organization, a corporate entity of the Australian Government. A key advantage of CARS is that the radius of influence varies with the local data density. Thus O(100) km

Table A.1 Data sources used in this book.

Dataset Name	Resolution	URL	Cross-references
T, S, tracers:			
WOA13	0.5°, monthly	www.nodc.noaa.gov/OC5/woa13	Figures 4.4, 4.5, 4.6, 4.8, 1.6, 5.9, 5.6, 6.1, 6.17, 8.1, Exercises 1.10, 1.12, 5.1, 5.2, 6.2, 8.1, 8.11, 9.3
CARS	0.25°, seasonal	www.marine.csiro.au/~dunn/cars2009	Figures 5.1, 5.2, 5.3, 5.4, 5.5, 8.3, 8.4, 8.7, 8.8, 8.9, Exercise 8.3
Arctic Regional Climatology	0.25°, monthly	www.nodc.noaa.gov/OC5/ regional_climate/arctic	Figures 10.3, 10.4, Exercises 5.1, 10.3
WOD	(profile data)	www.nodc.noaa.gov/OC5/WOD/pr_wod.html	Figures 1.6, 1.4, 1.5, 8.7, Exercises 1.9, 5.3, 8.3
Hydrographic sections	(profile data)	www.cchdo.ucsd.edu	Exercises 2.11, 11.2
Eulerian hydrographic data	(profile data)	BATS, HOTS	Exercise 4.1
Lagrangian hydrographic data	(profile data)	www.argodatamgt.org	Exercises 4.2, 8.3, 9.2
PANGAEA	(profile data)	www.pangaea.de	Figure 10.5
Chlorofluorocarbon data	(bottle data)	www.nodc.noaa.gov/ocads/oceans/ bottle_discrete.html	Exercise 8.2
Ocean currents:			
GDP (surface drifters)	0.5°, monthly	www.aoml.noaa.gov/phod/dac/index.php	Figures 6.6, 6.7, 6.9, 7.1, Exercises 6.2, 9.2
OceanSITES	(mooring data)	www.pmel.noaa.gov/gtmba/oceansites	Figure 1.1, Exercise 6.1

Table A.1 *(cont.)*

Dataset Name	Resolution	URL	Cross-references
Sea surface temperature (SST), height (SSH), color:			
OISST SST data	0.25°, daily	www.ncdc.noaa.gov/oisst	Figures 4.1, 4.2, 4.3, Exercise 7.1
AVHRR-Pathfinder SST data	4 km, daily	www.podaac.jpl.nasa.gov/AVHRR-Pathfinder	Figure 6.9
CMEMS SSH data	0.25°, daily	www.marine.copernicus.eu/	Figures 3.3, 3.5, 7.2, 9.4, 9.5, Exercises 3.2, 7.1, 7.2, 9.2
NASA OceanColor Web data	9 km, 8 days	www.oceancolor.gsfc.nasa.gov/data/seawifs	Figure 6.8, Exercise 6.2
Air/sea fluxes:			
SCOW (wind stress)	0.25°, monthly	www.cioss.coas.oregonstate.edu/scow/index.html	Figures 3.1, 3.6, 6.4, 6.7, Exercises 3.1, 6.9
OAFlux (heat flux components)	0.25°, daily	www.oaflux.whoi.edu/index.html	Figures 4.11, 11.3, 11.4, 11.5, 11.6, 11.7, Exercises 11.3, 11.4
GPCP (precipitation)	2.5°, monthly	www.esrl.noaa.gov/psd/data/gridded/data.gpcp.html	Figures 4.11, 11.6, 11.7, Exercises 11.3, 11.4
Miscellaneous:			
Index (ice extent)	25 km, daily	www.nsidc.org/data/G02135	Figure 4.7
Polar Pathfinder (ice motion)	25 km, daily	www.nsidc.org/data/NSIDC-0116	Figure 10.1
ETOPO1 (bathymetry)	1/60°	www.ngdc.noaa.gov/mgg/global/global.html	Figures 8.6, 8.8, 9.1, 10.1
ETOPO5 (bathymetry)	1/12°	www.ngdc.noaa.gov/mgg/global/global.html	Figure 9.1, Exercises 2.11, 9.3, among others
Ifremer (mixed layer)	2°, monthly	www.ifremer.fr/cerweb/deboyer/mld/home.php	Figure 4.9
ECCO (ocean state)	1°, monthly	www.ecco-group.org/index.htm	Figure 3.6
NCEP/NCAR Reanalysis	1.9°, 6-hourly	www.esrl.noaa.gov/psd/data/gridded/data.ncep .reanalysis.html	Figure 10.1

wide features in densely measured regions (for instance, the Gulf Stream) are relatively well-resolved in the atlas (Ridgway et al., 2002), which has a grid resolution of 0.5°. However, the time resolution of the seasonal cycle is inferior to WOA13. Generally, we use the 2009 edition of CARS for annual-average quantities and WOA13 for monthly averages.

Velocity

Near-surface velocity can be inferred from satellite-tracked drifters (Lumpkin and Garraffo, 2005; Lumpkin and Pazos, 2006), which float at the surface and are drogued at 15 m depth. The Global Drifter Program (GDP) Data Assembly Center creates annual and monthly climatologies of this data from drifter location data (Lumpkin and Johnson, 2013). Values on the 0.5° resolution grid are based on interpolation and smoothing from measurements within 2° of each grid point. Velocity measurements are also filtered in time to remove tidal and near-inertial components of the velocity.

For subsurface velocity there are direct observations from moorings, but they are very sparse in general. The best large-scale coverage is in the Tropics with data served by the OceanSITES program. The Tropical Atmosphere Ocean/Triangle Trans-Ocean Buoy Network array in the Pacific contains 50–70 moorings spaced a few hundred kilometers apart in the meridional direction and about 1000 km apart in the zonal direction.

Sea Surface Temperature, Height, Chlorophyll

Satellite instruments have made near-continuous measurements of global surface ocean properties in the last few decades. The NCEI Optimum Interpolation Sea Surface Temperature (OISST) is a gridded 1/4°, daily analysis from 1981 to the time of writing (Banzon et al., 2016). It combines satellite observations with in-situ data. The Advanced Very-High Resolution Radiometer (AVHRR) is the name of a family of satellite SST instruments, which produce the AVHRR-Pathfinder dataset. AVHRR-Pathfinder SST data are at 4 km resolution and are available daily since 1981. The AVHRR instruments observe infrared-radiation, and therefore cannot observe SST through clouds.

Global high quality measurements of sea-surface topography began in the early 1990s. We use the gridded Archiving, Validation and Interpretation of Satellite Oceanographic (AVISO) product, which is now distributed by the European Copernicus Marine Environment Monitoring Service (CMEMS). This is a gridded field of sea surface height available at 1/4° resolution daily since 1993. The product merges all available satellite altimeter data.

Surface ocean chlorophyll data are taken from the Sea-Viewing Wide Field-of-View Sensor (SeaWiFS). This instrument operated from September 1997 to December 2010. Data from the NASA OceanColor website are available at 9 km resolution most days of the mission. Eight-day composites give near global coverage.

Wind Stress

Scatterometer Climatology of Ocean Winds (SCOW) is a monthly wind stress climatology based on eight years of satellite data (September 1999 to August 2007). Radar backscatter from the sea surface is influenced by the roughness of the sea surface, which in turn can be related to the direction and magnitude of the wind stress. The accuracy of the measurements is equivalent to about ±1.5 m/s in the wind speed (at 10 m height) and $14°$ in direction (more for wind speeds less than 6 m/s). The SeaWinds scatterometer on the QuikSCAT satellite covered about 90% of the ocean each day for the period July 1999 to November 2009 (Risien and Chelton, 2008). The climatology is calculated on a $1/4°$ resolution grid, with interpolation and smoothing at a given grid point using observations within 70 km.

Air/Sea Fluxes

We use the ocean–atmosphere heat flux (OAFlux), evaporation rate, and surface windspeed climatology (Yu and Weller, 2007, there are several other products; Valdivieso et al., 2017 compares them, for example). OAFlux combines satellite and weather model analyses to estimate air/sea fluxes of latent heat flux, sensible heat flux, and evaporation rate. It blends these data with in-situ measurements where possible. Also available in the same format are downward shortwave and longwave radiation, SST, 10 m windspeed, wind stress, and wind stress curl. The fields have $1/4°$ ($1°$) daily resolution since July 1985 (January 1958).

OAFlux does not contain precipitation data, so we use the Global Precipitation Climatology Project (GPCP) product (Adler et al., 2003; Huffman et al., 2009). This dataset is on a $2.5°$ grid with monthly resolution since January 1979. The principal source of information in the GPCP data is from satellites (microwave and infrared) and rain gauges on land.

Sea Ice

The National Snow and Ice Data Center (NSIDC) sea ice index data gives daily sea ice extent (Fetterer et al., 2016) based on satellite observations. For sea ice motion (velocity) we use the Polar Pathfinder data (Fowler et al., 2013; Tschudi et al., 2016). It also uses satellite sea ice imagery as well as in-situ platforms. Both datasets are provided on a 25 km resolution polar stereographic grid and start in October 1979.

Miscellaneous

Various miscellaneous datasets are used in the book. They include: global ocean bathymetry, which is a synthesis of satellite data and ship soundings (Amante and Eakins, 2009); a climatology of mixed layer depth (de Boyer Montegut et al., 2004); a global ocean state estimate from the Estimating Circulation and Climate of the Ocean (ECCO) project (Forget et al., 2015); and, the NCEP/NCAR atmospheric reanalysis (Kalnay et al., 1996), which is a consistent merger of atmospheric data and simulations from weather models. Alternative datasets exist.

Appendix B Vector Calculus and Spherical Coordinates

When written in component form, the equations of motion are relatively simple in Cartesian coordinates $\mathbf{x} = x\hat{\mathbf{x}} + y\hat{\mathbf{y}} + z\hat{\mathbf{z}}$, with the usual meanings. Cartesian coordinates also most readily expose the physics the equations encode. As the Earth is (nearly) a sphere, however, it is often useful to write the equations in spherical coordinates. The main results are summarized in this Appendix; see Arfken and Weber (1995) or Boas (1983) for more on vector calculus in general and Kundu et al. (2012), Vallis (2006) or Batchelor (1967) for more on the applications to fluid dynamics.

B.1 Vector Operators

We can see the connection between spherical-coordinate components and various vector derivatives by starting with coordinate-free definitions (see also Subsection 2.1.4, Subsection 2.2.3, and Figure B.1):

The **divergence** div of a vector function $\mathbf{u}(\mathbf{x})$ is defined by means of a region which has bounding surface S and volume V. It is the average value of the component of $\mathbf{u}$ normal to the bounding surface as V vanishes:

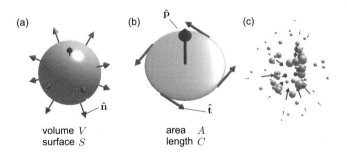

Figure B.1 Schematic of vector calculus operators. (a) The divergence of a vector field is the average value of the vector field on the surface of a vanishing sphere in the direction $\hat{\mathbf{n}}$ pointing outwards; Equation B.1. (b) The curl of a vector field in direction $\hat{\mathbf{p}}$ is the average value of the vector field twisting around the perpendicular vanishing disk in tangent direction $\hat{\mathbf{t}}$; Equation B.2. (c) The gradient of a scalar field is the directional derivative vector pointing to larger values of the field; Equation B.3. Here, the magnitude of the scalar field is shown by the size of the spheres and the gradient vector is shown by the arrows.

$$\text{div } \mathbf{u} \equiv \nabla \cdot \mathbf{u} = \lim_{V \to 0} \frac{1}{V} \oiint_S \mathbf{u} \cdot \hat{\mathbf{n}} \, dS, \tag{B.1}$$

where $\hat{\mathbf{n}}(\mathbf{x})$ is the unit vector normal to the surface S and pointing outward. Note that the **divergence theorem** has a similar form as (B.1), except that V is finite and $V\nabla \cdot \mathbf{u}$ is replaced by the volume integral of $\nabla \cdot \mathbf{u}$ (see (2.56)). Although we present the three-dimensional case here, the divergence operator applies to a vector function $\mathbf{u}$ of any dimension.

The **curl** of a (three-dimensional) vector function $\mathbf{u}(\mathbf{x})$ is defined using a simply connected planar surface of area A bounded by a contour of length C with unit tangent vector $\hat{\mathbf{t}}$. The orientation of the surface is defined by the unit normal vector $\hat{\mathbf{p}}$, which is oriented so that $\hat{\mathbf{t}}$ circulates C clockwise when viewed in direction $\hat{\mathbf{p}}$. The curl is related to the average of the component of $\mathbf{u}$ along the boundary:

$$(\text{curl } \mathbf{u}) \cdot \hat{\mathbf{p}} \equiv (\nabla \times \mathbf{u}) \cdot \hat{\mathbf{p}} = \lim_{A \to 0} \frac{1}{A} \oint_C \mathbf{u} \cdot \hat{\mathbf{t}} \, ds, \tag{B.2}$$

where s measures distance around C. This formula defines the component of the three-dimensional curl vector in the direction $\hat{\mathbf{p}}$. The full curl vector is constructed by taking $\hat{\mathbf{p}}$ in three different mutually orthogonal directions. The curl formula (B.2) calls to mind **Stokes' theorem**, which is obtained for nonvanishing A when $A(\nabla \times \mathbf{u}) \cdot \hat{\mathbf{p}}$ is replaced with the area integral of $(\nabla \times \mathbf{u}) \cdot \hat{\mathbf{p}}$ (see (2.58)).

Finally, the **gradient** grad of a scalar function $F(\mathbf{x})$ is a vector function for which, for any unit vector $\mathbf{x_D}$, $(\text{grad } F) \cdot \mathbf{x_D}$ is the **directional derivative** in the direction given by $\mathbf{x_D}$:

$$(\text{grad } F) \cdot \mathbf{x_D} \equiv \nabla F \cdot \mathbf{x_D} = \lim_{h \to 0} \frac{F(\mathbf{x} + h\mathbf{x_D}) - F(\mathbf{x})}{h}. \tag{B.3}$$

The gradient operator just generalizes the idea of a function derivative in one dimension to a vector of derivatives in more than one dimension. The corollary of the gradient formula (B.3) for nonvanishing h is the **fundamental theorem of calculus** applied to the line integral (along any path) between $\mathbf{x}$ and $\mathbf{x} + h\mathbf{x_D}$.

If we apply these definitions to regions in the shape of a cube (for div), a square (for curl), and a line (for grad), we can recover the form of each of these quantities in terms of partial derivatives in Cartesian coordinates. Here we set $\mathbf{u} = u\hat{\mathbf{x}} + v\hat{\mathbf{y}} + w\hat{\mathbf{z}}$, which gives

$$\nabla \cdot \mathbf{u} = \frac{\partial u}{\partial x} + \frac{\partial v}{\partial y} + \frac{\partial w}{\partial z}, \tag{B.4}$$

$$\nabla \times \mathbf{u} = \begin{vmatrix} \hat{\mathbf{x}} & \hat{\mathbf{y}} & \hat{\mathbf{z}} \\ \partial_x & \partial_y & \partial_z \\ u & v & w \end{vmatrix} = (w_y - v_z)\hat{\mathbf{x}} + (u_z - w_x)\hat{\mathbf{y}} + (v_x - u_y)\hat{\mathbf{z}}, \tag{B.5}$$

$$\nabla F = \hat{\mathbf{x}}\frac{\partial F}{\partial x} + \hat{\mathbf{y}}\frac{\partial F}{\partial y} + \hat{\mathbf{z}}\frac{\partial F}{\partial z}, \tag{B.6}$$

where we use the notation $\partial_x \equiv \partial/\partial x$ etc.

Material Derivative

The **material** (or total or advective) **derivative**, D/Dt, follows from these ideas. For a scalar field $F(\mathbf{x}, t)$,

$$\frac{dF}{dt} = \frac{\partial F}{\partial t} + \frac{d\mathbf{x}}{dt} \cdot \nabla F, \tag{B.7}$$

from the chain rule. As $d\mathbf{x}/dt = \mathbf{u}$, the fluid velocity, we see that

$$\frac{DF}{Dt} = \frac{\partial F}{\partial t} + \mathbf{u} \cdot \nabla F, \tag{B.8}$$

which defines the material derivative operator,

$$\frac{D}{Dt} \equiv \frac{\partial}{\partial t} + \mathbf{u} \cdot \nabla, \tag{B.9}$$

namely, the time rate of change moving with the flow (the Lagrangian time rate of change). In Cartesian coordinates, $\mathbf{u} \cdot \nabla = u\partial_x \hat{\mathbf{x}} + v\partial_y \hat{\mathbf{y}} + w\partial_z \hat{\mathbf{z}}$, so that

$$\mathbf{u} \cdot \nabla F = u\frac{\partial F}{\partial x}\hat{\mathbf{x}} + v\frac{\partial F}{\partial y}\hat{\mathbf{y}} + w\frac{\partial F}{\partial z}\hat{\mathbf{z}}, \tag{B.10}$$

$$\begin{aligned}
\mathbf{u} \cdot \nabla\mathbf{u} = {} & \left(u\frac{\partial u}{\partial x} + v\frac{\partial u}{\partial y} + w\frac{\partial u}{\partial z} \right)\hat{\mathbf{x}} \\
& + \left(u\frac{\partial v}{\partial x} + v\frac{\partial v}{\partial y} + w\frac{\partial v}{\partial z} \right)\hat{\mathbf{y}} \\
& + \left(u\frac{\partial w}{\partial x} + v\frac{\partial w}{\partial y} + w\frac{\partial w}{\partial z} \right)\hat{\mathbf{z}}.
\end{aligned} \tag{B.11}$$

Laplacian Operator

The **Laplacian** operator del-squared, ∇^2, is defined by $\nabla^2 F = \nabla \cdot (\nabla F)$ for scalar field F, and by $\nabla^2\mathbf{u} = \nabla (\nabla \cdot \mathbf{u}) - \nabla \times (\nabla \times \mathbf{u})$ for vector field $\mathbf{u}$. Hence, in Cartesian coordinates,

$$\nabla^2 F = \frac{\partial^2 F}{\partial x^2} + \frac{\partial^2 F}{\partial y^2} + \frac{\partial^2 F}{\partial z^2}, \tag{B.12}$$

$$\nabla^2\mathbf{u} = \left(\nabla^2 u \right)\hat{\mathbf{x}} + \left(\nabla^2 v \right)\hat{\mathbf{y}} + \left(\nabla^2 w \right)\hat{\mathbf{z}}. \tag{B.13}$$

The Laplacian of a field measures its curvature.

B.2 Components in Curvilinear Coordinates

We can write the components of these functions in terms of an arbitrary set of orthogonal coordinates $(\hat{\mathbf{x}}_1, \hat{\mathbf{x}}_2, \hat{\mathbf{x}}_3)$ so that a location in space is given by $\mathbf{x} = x_1\hat{\mathbf{x}}_1 + x_2\hat{\mathbf{x}}_2 + x_3\hat{\mathbf{x}}_3$. In this coordinate system, our vector function $\mathbf{u}$ is given by $\mathbf{u} = u_1\hat{\mathbf{x}}_1 + u_2\hat{\mathbf{x}}_2 + u_3\hat{\mathbf{x}}_3$. Such coordinate systems are characterized by a scale factor s_n for each unit vector $\hat{\mathbf{x}}_n$ ($n = 1, 2,$ or 3). The scale factor converts a given change in variable x_n to the distance as measured in Cartesian coordinates, and is given by

$$s_n = \sqrt{\left(\frac{\partial x}{\partial x_n}\right)^2 + \left(\frac{\partial y}{\partial x_n}\right)^2 + \left(\frac{\partial z}{\partial x_n}\right)^2}. \tag{B.14}$$

Consideration of integration regions appropriate to each operator and to the particular coordinate system gives

$$\nabla \cdot \mathbf{u} = \frac{1}{s_1 s_2 s_3} \left[\partial_{x_1}(u_1 s_2 s_3) + \partial_{x_2}(u_2 s_1 s_3) + \partial_{x_3}(u_3 s_1 s_2)\right], \tag{B.15}$$

$$\nabla \times \mathbf{u} = \frac{1}{s_1 s_2 s_3} \begin{vmatrix} s_1 \hat{\mathbf{x}}_1 & s_2 \hat{\mathbf{x}}_2 & s_3 \hat{\mathbf{x}}_3 \\ \partial_{x_1} & \partial_{x_2} & \partial_{x_3} \\ s_1 u_1 & s_2 u_2 & s_3 u_3 \end{vmatrix}, \tag{B.16}$$

$$\nabla F = \sum_{n=1}^{3} \frac{1}{s_n} \frac{\partial F}{\partial x_n} \hat{\mathbf{x}}_n. \tag{B.17}$$

The material derivative and Laplacian operators are not needed in general curvilinear orthogonal coordinates in this book, but see Batchelor (1967) for details.

B.3 Vector Operators in Spherical Coordinates

Spherical coordinates are one example of such a curvilinear coordinate system. The coordinates are longitude ϕ, latitude θ, and radial distance r (see Figure B.2). We can write Cartesian coordinates in terms of the spherical coordinate variables using

$$x = r \cos\theta \cos\phi, \tag{B.18}$$

$$y = r \cos\theta \sin\phi, \tag{B.19}$$

$$z = r \sin\theta \tag{B.20}$$

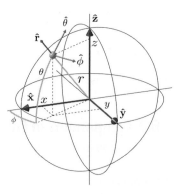

Figure B.2   Spherical and Cartesian coordinates. The position of the point of interest is $\phi\hat{\phi} + \theta\hat{\theta} + r\hat{\mathbf{r}}$ in spherical coordinates for longitude ϕ, latitude θ, and radius r. It is $x\hat{\mathbf{x}} + y\hat{\mathbf{y}} + z\hat{\mathbf{z}}$ in Cartesian coordinates. Notice how the unit vectors in spherical coordinates (east, north, up) change direction with the point of interest. Also, the Cartesian coordinate system here is different to the local Cartesian coordinate system aligned with $(\hat{\phi}, \hat{\theta}, \hat{\mathbf{r}})$.

and then inserting (B.18)–(B.20) into (B.14). We get $s_r = 1$, $s_\theta = r$, and $s_\phi = r\cos\theta$. Here we have substituted variable names for the subscript numbers according to $(1, 2, 3) \Leftrightarrow (\phi, \theta, r)$. We consider a vector $\mathbf{u} = u\hat{\phi} + v\hat{\theta} + w\hat{\mathbf{r}}$, where (u, v, w) now represent components in spherical coordinates. If we simultaneously consider components of $\mathbf{u}$ in both Cartesian and spherical coordinates, we need separate symbols for the set of components in each coordinate system, but it should not cause a problem in the discussion below.

In spherical coordinates, setting $B = \cos\theta$, the three functions are

$$\nabla \cdot \mathbf{u} = \frac{1}{rB}\left[\partial_\phi u + \partial_\theta(vB)\right] + \frac{1}{r^2}\partial_r(r^2 w), \tag{B.21}$$

$$\nabla \times \mathbf{u} = \frac{1}{r^2 B}\begin{vmatrix} rB\hat{\phi} & r\hat{\theta} & \hat{\mathbf{r}} \\ \partial_\phi & \partial_\theta & \partial_r \\ rBu & rv & w \end{vmatrix},$$

$$= \frac{1}{r}[\partial_\theta w - \partial_r(rv)]\hat{\phi} - \frac{1}{rB}\left[\partial_\phi w - \partial_r(rBu)\right]\hat{\theta} +$$

$$\frac{1}{r^2 B}\left[\partial_\phi(rv) - \partial_\theta(rBu)\right]\hat{\mathbf{r}}, \tag{B.22}$$

$$\nabla F = \frac{1}{rB}\frac{\partial F}{\partial\phi}\hat{\phi} + \frac{1}{r}\frac{\partial F}{\partial\theta}\hat{\theta} + \frac{\partial F}{\partial r}\hat{\mathbf{r}}. \tag{B.23}$$

The material derivative for a scalar field is:

$$\frac{DF}{Dt} = \partial_t F + \mathbf{u}\cdot\nabla F = \partial_t F + \frac{u}{rB}\partial_\phi F + \frac{v}{r}\partial_\theta F + w\partial_r F. \tag{B.24}$$

There is an additional complication when using spherical coordinates for the momentum equation (Subsection 1.3.1). The reason is that in the material derivative for the vector velocity

$$\frac{D\mathbf{u}}{Dt} = \frac{D}{Dt}(u\hat{\phi} + v\hat{\theta} + w\hat{\mathbf{r}}) \tag{B.25}$$

the unit coordinate direction vectors $(\hat{\phi}, \hat{\theta}, \hat{\mathbf{r}})$ change with position, which is not true in Cartesian coordinates. Therefore:

$$\frac{D\mathbf{u}}{Dt} = \frac{Du}{Dt}\hat{\phi} + \frac{Dv}{Dt}\hat{\theta} + \frac{Dw}{Dt}\hat{\mathbf{r}} + u\frac{D\hat{\phi}}{Dt} + v\frac{D\hat{\theta}}{Dt} + w\frac{D\hat{\mathbf{r}}}{Dt}. \tag{B.26}$$

These extra **metric terms** are

$$u\frac{D\hat{\phi}}{Dt} + v\frac{D\hat{\theta}}{Dt} + w\frac{D\hat{\mathbf{r}}}{Dt} = \frac{1}{r}\Big[(-uv\tan\theta + uw)\hat{\phi}$$
$$+ (u^2\tan\theta + vw)\hat{\theta}$$
$$- (u^2 + v^2)\hat{\mathbf{r}}\Big] \tag{B.27}$$

(Kundu et al., 2012; Vallis, 2006; Batchelor, 1967).

In spherical coordinates, the scalar and vector Laplacian operators are:

$$\nabla^2 F = \frac{1}{r^2 B^2}\frac{\partial^2 F}{\partial\phi^2} + \frac{1}{r^2 B}\frac{\partial}{\partial\theta}\left(B\frac{\partial F}{\partial\theta}\right) + \frac{1}{r^2}\frac{\partial}{\partial r}\left(r^2\frac{\partial F}{\partial r}\right), \tag{B.28}$$

$$\nabla^2 \mathbf{u} = \left(\nabla^2 u - \frac{u}{r^2 B^2} + \frac{2}{r^2 B} \frac{\partial w}{\partial \phi} + \frac{2C}{r^2 B^2} \frac{\partial v}{\partial \phi} \right) \hat{\phi}$$
$$+ \left(\nabla^2 v - \frac{v}{r^2 B^2} - \frac{2}{r^2} \frac{\partial w}{\partial \theta} - \frac{2C}{r^2 B^2} \frac{\partial u}{\partial \phi} \right) \hat{\theta}$$
$$+ \left(\nabla^2 w - \frac{2w}{r^2} + \frac{2}{r^2 B} \frac{\partial (vB)}{\partial \theta} - \frac{2}{r^2 B} \frac{\partial v}{\partial \phi} \right) \hat{\mathbf{r}}, \tag{B.29}$$

where $C = \sin \theta$ (Kundu et al., 2012; Batchelor, 1967).

B.4 Thin-Shell and Traditional Approximations

While spherical coordinates complicate the expressions for vector operators and the governing equations, some of the new terms are much smaller than others and can be safely neglected. As the Earth's radius $R_E \approx 6371$ km is about 1000 times greater than typical ocean depths, we rewrite radial distance as $r = R_E + z$, where z is the vertical distance from a convenient reference point such as the sea surface. Then we make the **thin-shell** approximation, in which $r \approx R_E$ and $\partial_r = \partial_z$. This allows us to rewrite the vector operators (B.21)–(B.23) as

$$\nabla \cdot \mathbf{u} = \frac{1}{R_E B} \left[\frac{\partial u}{\partial \phi} + \partial_\theta (vB) \right] + \frac{\partial w}{\partial z}, \tag{B.30}$$

$$\nabla \times \mathbf{u} = \left(\frac{1}{R_E} \frac{\partial w}{\partial \theta} - \frac{\partial v}{\partial z} \right) \hat{\phi} - \frac{1}{B} \left[\frac{1}{R_E} \frac{\partial w}{\partial \phi} - \partial_z (Bu) \right] \hat{\theta} +$$
$$\frac{1}{R_E B} \left[\frac{\partial v}{\partial \phi} - \partial_\theta (Bu) \right] \hat{\mathbf{z}}, \tag{B.31}$$

$$\nabla F = \frac{1}{R_E B} \frac{\partial F}{\partial \phi} \hat{\phi} + \frac{1}{R_E} \frac{\partial F}{\partial \theta} \hat{\theta} + \frac{\partial F}{\partial z} \hat{\mathbf{z}}. \tag{B.32}$$

We also can drop terms from the material derivative and the Coriolis force in the momentum equation, $D\mathbf{u}/Dt + 2\mathbf{\Omega} \times \mathbf{u}$. This is the so-called **traditional approximation**. A scale analysis (see Vallis, 2006) shows that terms proportional to w are negligible, and so the horizontal momentum equations are

$$\frac{Du}{Dt} - fv - \frac{uv}{R_E} \tan \theta = -\frac{1}{\rho R_E \cos \theta} \frac{\partial p}{\partial \phi} + D_\phi, \tag{B.33}$$

$$\frac{Dv}{Dt} + fu + \frac{u^2}{R_E} \tan \theta = -\frac{1}{\rho R_E} \frac{\partial p}{\partial \theta} + D_\theta, \tag{B.34}$$

$$\frac{D}{Dt} = \frac{\partial}{\partial t} + \frac{u}{R_E \cos \theta} \frac{\partial}{\partial \phi} + \frac{v}{R_E} \frac{\partial}{\partial \theta} + w \frac{\partial}{\partial z}, \tag{B.35}$$

where $f = 2\Omega \sin \theta$ is the Coriolis parameter, and D_ϕ and D_θ are the momentum mixing parameterizations.

Appendix C Tables of Notation and Useful Values

C

Table C.1 Ocean circulation acronyms. See also Appendix A and Table A.1 in particular.		
Acronym	Name	Cross-reference
AABW	Antarctic bottom water	Subsection 8.1.1
AAIW	Antarctic intermediate water	Subsection 8.1.1
ACC	Antarctic circumpolar current	Subsection 3.1.2
ADCP	Acoustic doppler current profiler	Subsection 1.1.2
AMOC	Atlantic meridional overturning circulation	Subsection 11.1.3
CFC	Chlorofluorocarbon	Subsection 1.1.2
CTD	Conductivity–temperature–depth	Subsection 1.1.3
EUC	Equatorial undercurrent	Subsection 6.1.1
GCM	General circulation model	Subsection 1.2.2
GM	Gent–McWilliams (eddy parameterization)	Subsection 7.2.3
GS	Green–Stone (eddy parameterization)	Figure 7.14
LPS	Luyten–Pedlosky–Stommel (ventilated thermocline model)	Subsection 5.2.1
ML	mixed layer	Chapter 4
NADW	North Atlantic deep water	Subsection 8.1.1
PV	potential vorticity	Subsection 2.1.5
SSH	Sea-surface height	Subsection 1.1.2
SSS	Sea-surface salinity	Subsection 1.1.2
SST	Sea-surface temperature	Subsection 1.1.2
STC	subtropical cell	Chapter 6
TEOS-10	2010 Thermodynamic equation of state	Subsection 1.2.3
WBC	western boundary current	Chapter 3

Table C.2 Frequently used variable names (see also the Index). Be aware that some variables have multiple meanings.			
Symbol	Units	Description	Cross-reference
$\mathbf{x} = (x, y, z)$	m	position vector	Table 1.1, Subsection 1.1.3, Figure B.2
ϕ, θ	°	longitude, latitude	Table 1.1, Subsection 1.1.3, Figure B.2
t	s	time	Subsection 1.1.1, 1.2.1
$(\hat{\mathbf{x}}, \hat{\mathbf{y}}, \hat{\mathbf{z}})$		Cartesian unit vectors	Subsection 2.2.3

Table C.2 *(cont.)*			
Symbol	Units	Description	Cross-reference
$\hat{\mathbf{n}}, \hat{\mathbf{t}}$		normal & tangent unit vectors	Figures 3.21, B.1
$\mathbf{u} = (u, v, w)$	m s^{-1}	velocity vector	Subsection 1.1.2
$\mathbf{u}_h = (u, v)$	m s^{-1}	horizontal velocity vector	Subsection 1.1.2
$\mathbf{u}_G = (u_G, v_G)$	m s^{-1}	geostrophic velocity vector	Equation (2.2)
T, θ	°C, K	temperature, potential temperature	Subsection 1.1.2
p	N m^{-2}	pressure	Subsection 1.1.2
S	g/kg	salinity	Subsection 1.1.2
ρ, ρ_0	kg m^{-3}	density, reference density	Subsection 1.1.2, 1.2.3
σ	kg m^{-3}	density anomaly	Subsection 1.1.2, 1.2.3
σ_n	kg m^{-3}	potential density anomaly referenced to n km depth	Subsection 1.2.3
α	°C^{-1}	thermal expansion coefficient	Equation (1.4)
β		haline contraction coefficient	Equation (1.5)
χ	Mol kg^{-3}	tracer concentration	Table 1.1, Subsection 1.3.1
$\boldsymbol{\tau} = (\tau, \sigma)$	N m^{-2}	stress	Subsection 2.1.3
$\mathbf{X} = (X, Y)$	m^2 s^{-2}	stress/density ($= \boldsymbol{\tau}/\rho$)	Subsection 2.1.3
Q	W m^{-2}	heat flux	Subsection 4.3.1
h	m	layer thickness	Subsection 2.1.5
H	m	water depth, mixed layer depth	Subsection 1.1.6
η	m	sea-surface height	Subsection 1.1.2
$\mathbf{U} = (U, V)$	m^{2}s^{-1}	depth-integrated velocity	Subsection 2.2.3
$\mathbf{U}_{Ek}$	m^{2}s^{-1}	Ekman depth-integrated velocity	Equation (2.11), Figure 2.4
w_{Ek}	ms^{-1}	Ekman pumping/suction speed	Equation (2.13), Figure 2.4
H_{Ek}	m	Ekman layer thickness	Equation (2.40)
w_e	ms^{-1}	entrainment/detrainment speed	Equation (4.26)
w_*	ms^{-1}	interfacial vertical speed	Subsection 2.2.4
b	m s^2	buoyancy ($= -g(\rho - \rho_0)/\rho_0$)	Subsection 8.2.1
N	s^{-1}	Brunt–Väisälä (buoyancy) frequency	Equation (1.1)
$\mathcal{D}_\theta, \mathcal{D}_S, \mathcal{D}_\mathbf{u}$	m s^{-2}	dissipation operator on $\theta, S, \mathbf{u}$	Subsection 1.3.1
ν_h, ν_v	m^{2}s^{-1}	horizontal, vertical viscosity	Equation (2.72)
κ_h, κ_v	m^{2}s^{-1}	horizontal, vertical diffusivity	Subsection 7.3.2, Subsection 5.3.1
f	s^{-1}	Coriolis parameter	Equation (2.1), Figure 2.1
$\beta = df/dy$	m^{-1}s^{-1}	Coriolis parameter gradient	Equation (2.63)
Q	m^{-1}s^{-1}	Ertel potential vorticity	Equation (2.24)
q	m^{-1}s^{-1}	potential vorticity	Equation (2.22)
Ψ, Υ	m^{3}s^{-1} (sometimes m^{2}s^{-1})	streamfunction	Subsection 2.1.4, 10.2.2, Figure 2.8
Ψ_{Sv}	m^{3}s^{-1}	Sverdrup streamfunction	Equation (3.34)

Ψ	m^2s^{-2}	dynamic height	Subsection 2.2.2, Equation (2.48), Figure 5.6
$\boldsymbol{\zeta} = \nabla \times \mathbf{u}$	s^{-1}	absolute vorticity vector	Subsection 2.1.4, 2.1.5, Figure 2.7, Equation (2.15)
$\zeta = \boldsymbol{\zeta} \cdot \hat{\mathbf{z}}$	s^{-1}	vertical component of vorticity	Subsection 2.1.4, 2.1.5, Figure 2.7, Equation (2.15)
$D = \nabla \cdot \mathbf{u}$	s^{-1}	velocity divergence	Subsection 2.1.4, 2.2.3, 2.2.3, Equation (2.14), Figure 2.5

Table C.3 Geophysical constants.

Symbol	Value	Units	Description
T_S	86400	s	length of solar day
T_D	86164	s	length of sidereal day
T_A	3.16×10^7	s	year ($= 365.25 T_S$)
$\Omega = 2\pi/T_D$	7.29×10^{-5}	rad s^{-1}	Earth angular rotation rate
2Ω	1.46×10^{-4}	s^{-1}	maximum Earth Coriolis parameter
	12.42	h	period semidiurnal tide
R_E	6371	km	mean Earth radius
R_{Ee}/R_{Ep}	1.0034		Earth (equatorial radius)/(polar radius)
$2\Omega/R_E$	2.29×10^{-11}	m^{-1}s^{-1}	maximum Earth Coriolis gradient
$4\pi R_E^2$	5.10×10^8	km^2	Earth surface area
	0.71		(ocean surface area)/(Earth surface area)
g	9.81	m s^{-2}	acceleration due to gravity
$\overline{Q_S}$	340	W m^2	average top-of-atmosphere solar energy flux
$\overline{H}$	3800	m	average ocean depth
p_0	1.013×10^5	N m^{-2}	mean sea-level atmospheric pressure
ρ_0	1026	kgm^{-3}	seawater reference density
ρ_a	1.23	kgm^{-3}	mean sea-level atmospheric density

Table C.4 Physical constants. For thermodynamics, see TEOS-10 (Subsection 1.2.3, IOC, SCOR, and IAPSO, 2010). Viscosity and diffusivities are from Gill (1982).

Symbol	Value	Units	Description
T_f	$(0, -1.91)$	°C	surface freezing point, $S = (0, 35)$ g/kg
c_p	3992	J kg^{-1}°C^{-1}	specific heat capacity, seawater
c_i	2097	J kg^{-1}°C^{-1}	specific heat capacity, ice
L	3.33×10^5	J kg^{-1}	latent heat of fusion for water
ρ	1027.0	kg m^{-3}	density, $\theta = 10$°C, $S = 35.25$ g/kg, $p = 0$ db

Table C.4 (*cont.*)

Symbol	Value	Units	Description
ρ_i	917	kg m^{-3}	density of ice at freezing point
ν	10^{-6}	m^2s^{-1}	molecular kinematic viscosity, water
κ_θ	1.4×10^{-7}	m^2s^{-1}	molecular thermal diffusivity, water
κ_S	1.5×10^{-9}	m^2s^{-1}	molecular haline diffusivity, water at 25°C
C_s	1490	m s^{-1}	sound speed (at $\theta = 10$°C, $S = 35$ g/kg)
σ	5.671×10^{-8}	$\text{W m}^{-2}\text{K}^{-4}$	Stefan–Boltzmann constant

Table C.5 Dynamical parameters.

Symbol	Expression	Name	Cross-reference
U		speed scale	Subsection 2.2.1
L		horizontal length scale	Subsection 2.2.1
H		depth scale	Subsection 2.2.1
$\gamma_n, \Delta b$	$g\Delta\rho/\rho_0$	reduced gravity, buoyancy scale	Equation (2.68)
γ	ρ_x/ρ_z	isopycnic slope	Equation (7.25)
R_D	$\sqrt{\Delta b H}/f$	internal deformation radius (two layers)	Equation (8.2)
L_I	$\sqrt{U/\beta}$	inertial length scale, Rhines scale	Equation (3.13)
L_S	r/β	Stommel length scale	Equation (3.17)
L_M	$\left(\nu_h/\beta\right)^{1/3}$	Munk length scale	Equation (3.19)
δ_{Ek}	H_{Ek}/H	nondimensional Ekman layer depth	Subsection 3.3.1
As	H/L	aspect ratio	Subsection 1.1.6
Ro	U/fL	Rossby number	Subsection 2.1.1, Equation (2.32)
Ek	$(1/2)(H_{Ek}/H)^2$	Ekman number	Equation (2.34)
Re	UL/ν	Reynolds number	Subsection 1.3.3
Ma	U/C_s	Mach number	Subsection 1.3.1

Table C.6 Units commonly used in physical oceanography.

Oceanography Unit	SI units	Description
degree latitude (°)	111×10^3 m	Angular unit
nautical mile (nm)	1852 m	1/60th degree latitude
knot (kn)	0.514 ms^{-1}	1 nautical mile/hour
Sverdrup (Sv)	$10^6 \text{ m}^3\text{s}^{-1}$	Volume transport
Pascal (Pa)	N m^{-2}	Pressure (Pa is also SI)
bar	10^5 Pa = 1000 hPa	Pressure
degree Celsius (°C)	K -273.15	Temperature

References

Aagaard, K. and Carmack, E. C. 1989. The role of sea ice and other fresh water in the Arctic circulation. *J. Geophys. Res.*, **94**, 14485–14498.

Adler, R. F., Huffman, G. J., Chang, A., Ferraro, R., Xie, P. et al. 2003. The version 2 global precipitation climatology project (GPCP) monthly precipitation analysis (1979–Present). *J. Hydrometeor.*, **4**, 1147–1167.

Alley, R. B. 2007. Wally was right: predictive ability of the North Atlantic "conveyor belt" hypothesis for abrupt climate change. *Ann. Rev. Earth Planet. Sci.*, **35**, 241–272.

Alley, R. B., Clark, P. U., Keigwin, L. D., and Webb, R. S. 1999. Making sense of millennial-scale climate change. In: Clark, P. U., Webb, R. S., and Keigwin, L. D. (eds), *Mechanisms of Global Climate Change at Millennial Time Scales*. AGU.

Amante, C. and Eakins, B. W. 2009. *ETOPO1 1 Arc-Minute Global Relief Model: Procedures, Data Sources and Analysis*. Tech. rept. NOAA Technical Memorandum NESDIS NGDC-24. National Geophysical Data Center, NOAA.

Andrews, D. G. 2010. *An Introduction to Atmospheric Physics*. 2nd edn. Cambridge and New York, NY: Cambridge University Press.

Andrews, D. G., Holton, J. R., and Leovy, C. B. 1987. *Middle Atmosphere Dynamics*. San Diego, CA: Academic Press.

Arfken, G. B. and Weber, H. J. 1995. *Mathematical Methods for Physicists*. 4th edn. San Diego, CA: Academic Press.

Armi, L. and Zenk, W. 1984. Large lenses of highly saline Mediterranean water. *J. Phys. Oceanogr.*, **14**, 1560–1576.

Arthur, R. S. 1960. A review of the calculation of ocean currents at the equator. *Deep Sea Res.*, **6**, 287–297.

Bacon, S., Fawcett, S., and Madec, G. 2015. Arctic mass, freshwater and heat fluxes: methods and modelled seasonal variability. *Phil. Trans. R. Soc. Lond. A*, **373**.

Banzon, V., Smith, T. M., Chin, T. M., Liu, C., and Hankins, W. 2016. A long-term record of blended satellite and in situ sea-surface temperature for climate monitoring, modeling and environmental studies. *Earth System Science Data*, **8**, 165–176.

Baringer, M. O. and Price, J. F. 1997. Mixing and spreading of the Mediterranean Outflow. *J. Phys. Oceanogr.*, **27**, 1654–1677.

Barkstrom, B. R., Harrison, E. F., Lee., R. B., and ERBE Science Team. 1990. Earth Radiation Budget Experiment. *Eos, Trans., AGU*, **71**(9), 297–304.

Batchelor, G. K. 1967. *An Introduction to Fluid Dynamics*. Cambridge and New York, NY: Cambridge University Press.

Beal, L. M., de Ruijter, W. P. M., Biastoch, A., Zahn, R., and SCOR/WCRP/IAPSO Working Group 136. 2011. On the role of the Agulhas system in ocean circulation and climate. *Nature*, **472**.

Bender, C. M. and Orszag, S. A. 1987. *Advanced Mathematical Methods for Scientists and Engineers*. New York, NY McGraw-Hill.

Bindoff, N. L., Stott, P. A., AchutaRao, K. et al. 2013. Detection and attribution of climate change: from global to regional. In: Stocker, T. F., Qin, D., Plattner, G.-K. et al. (eds), *Climate Change 2013: The Physical Science Basis. Contribution of Working Group I to the Fifth Assessment Report of the Intergovernmental Panel on Climate Change*. Cambridge and New York, NY: Cambridge University Press.

Bishop, S. P. and Bryan, F. O. 2013. A comparison of mesoscale eddy heat fluxes from observations and a high-resolution ocean model simulation of the Kuroshio extension. *J. Phys. Oceanogr.*, **43**, 2563–2570.

Blandford, R. R. 1971. Boundary conditions in homogeneous ocean models. *Deep Sea Res.*, **18**(7), 739–751.

Bluhm, B. A., Kosobokova, K. N., and Carmack, E. C. 2015. A tale of two basins: An integrated physical and biological perspective of the deep Arctic Ocean. *Prog. Oceanogr.*, **139**, 89–121.

Blunier, T. and Brook, K. 2001. *Synchronization of the Byrd and Greenland (GISP2/GRIP) Records*. Tech. rept. Data Contribution Series # 2001-003. IGBP PAGES/World Data Center-A for Palaeoclimatology, NOAA/NGDC Palaeoclimatology Program, Boulder, CO.

Boas, M. L. 1983. *Mathematical Methods in the Physical Sciences*. New York, NY: John Wiley & Sons.

Böning, C. W. 1986. On the influence of frictional parameterization in wind-driven ocean circulation models. *Dyn. Atmos. Oceans*, **10**, 63–92.

Böning, C. W., Holland, W. R., Bryan, F. O., Danabasoglu, G., and McWilliams, J. C. 1995. An overlooked problem in model simulations of the thermohaline circulation and heat transport in the Atlantic Ocean. *J. Climate*, **8**, 515–523.

Boudra, D. B. and Chassignet, E. P. 1988. Dynamics of Agulhas Retroflection and ring formation in a numerical model, part I: the vorticity balance. *J. Phys. Oceanogr.*, **18**, 280–303.

Boyer, T. P., Baranova, O. K., Biddle, M., Johnson, D. R., Mishonov, A. V. et al. 2012. *Arctic Regional Climatology, Regional Climatology Team*. Tech. rept. NOAA/NODC. www.nodc.noaa.gov/OC5/regional_climate/arctic.

Brambilla, E. and Talley, L. D. 2006. Surface drifter exchange between the North Atlantic subtropical and subpolar gyres. *J. Geophys. Res.*, **111**.

Brink, K. H., Halpern, D., and Smith, R. L. 1980. Circulation in the Peruvian upwelling system near 15°S. *J. Geophys. Res.*, **85**, 4036–4048.

Broecker, W. S. 1987. The biggest chill. *Nat. Hist. Mag.*, **97**, 74–82.

Broecker, W. S. 2006. Was the Younger Dryas triggered by a flood? *Science*, **312**, 1146–1148.

Broecker, W. S., Peteet, D. M., and Rind, D. 1985. Does the ocean–atmosphere system have more than one stable mode of operation? *Nature*, **312**, 21–26.

Brown, J. N. and Fedorov, A. V. 2010. Estimating the diapycnal transport contribution to warm water volume variations in the tropical Pacific Ocean. *J. Climate*, **23**, 221–237.

Bryan, F. 1987. Parameter sensitivity of primitive equation ocean general circulation models. *J. Phys. Oceanogr.*, **17**, 970–985.

Bryan, F. 1988. High-latitude salinity effects and interhemispheric thermohaline circulations. *Nature*, **323**, 301–304.

Bryan, F. O., Böning, C. W., and Holland, W. R. 1995. On the midlatitude circulation in a high resolution model of the North Atlantic. *J. Phys. Oceanogr.*, **25**, 289–305.

Bryan, F. O., Hecht, M. W., and Smith, R. D. 2007. Resolution convergence and sensitivity studies with North Atlantic circulation models. Part I: The western boundary current system. *Ocean Modelling*, **16**, 141–159.

Bryan, K. 1963. A numerical investigation of a nonlinear model of a wind-driven ocean. *J. Atmos. Sci.*, **20**, 594–606.

Bryan, K. and Cox, M. 1967. A numerical investigation of the oceanic general circulation. *Tellus*, **19**, 54–80.

Bryan, K. and Lewis, L. J. 1979. A watermass model of the world ocean. *J. Geophys. Res.*, **84**, 2503–2517.

Bryden, H. L. and Brady, E. C. 1985. Diagnostic model of the three-dimensional circulation in the upper equatorial Pacific Ocean. *J. Phys. Oceanogr.*, **15**, 1255–1275.

Bryden, H. L., Roemmich, D. H., and Church, J. A. 1991. Ocean heat transport across 24°N in the Pacific. *Deep Sea Res.*, **38**(3), 297–324.

Bryden, H. L., Johns, W. E., and Saunders, P. M. 2005. Deep western boundary current east of Abaco: Mean structure and transport. *J. Mar. Res.*, **63**, 35–57.

Burkholder, K. C. and Lozier, M. S. 2011. Subtroical to subpolar pathways in the North Atlantic: deductions from Lagrangian trajectories. *J. Geophys. Res.*, **116**.

Canuto, V. M., Howard, A. M., Cheng, Y., Muller, C. J., Leboissetier, A., and Jayne, S. R. 2010. Ocean turbulence, III: New GISS vertical mixing scheme. *Ocean Modelling*, **34**(3-4), 70–91.

Capet, X., McWilliams, J. C., Molemaker, M. J., and Shchepetkin, A. F. 2008. Mesoscale to submesoscale transition in the California Current System, Part I: Flow structure, eddy flux, and observational tests. *J. Phys. Oceanogr.*, **38**, 29–43.

Carmack, E. 2007. The alpha/beta ocean distinction: a perspective on freshwater fluxes, convection, nutrients and productivity in high-latitude seas. *Deep Sea Res., Part II*, **54**, 2578–2598.

Carmack, E., Winsor, P., and Williams, W. 2015. The contiguous panarctic Riverine Coastal Domain: A unifying concept. *Prog. Oceanogr.*, **139**, 13–23.

Carmack, E., Yamamoto-Kawai, M., Haine, T., Bacon, S., Bluhm, B. et al. 2016. Fresh water and its role in the Arctic Marine System: sources, disposition, storage, export, and physical and biogeochemical consequences in the Arctic and global oceans. *J. Geophys. Res.*, **121**, 675–717.

Carr, M. E. and Kearns, E. J. 2003. Production regimes in four eastern boundary current systems. *Deep Sea Res., Part II*, **50**, 3199–3221.

Cember, R. P. 1988. On the sources, formation and circulation of Red Sea deep water. *J. Geophys. Res.*, **93**, 8175–8191.

Cessi, P. 1992. Ventilation of eastern subtropical gyres. *J. Phys. Oceanogr.*, **22**, 683–685.

Chassignet, E., Smith, L., and Bleck, R. 1996. A model comparison: Numerical simulations of the North and Equatorial Atlantic oceanic circulations in depth and isopycnic coordinates. *J. Phys. Oceanogr.*, **26**, 1849–1867.

Chassignet, E. P., Bleck, R., and Rooth, C. G. H. 1995. The influence of layer outcropping on the separation of boundary currents, Part II: The wind- and buoyancy-driven experiments. *J. Phys. Oceanogr.*, **25**, 2404–2422.

Coachman, L. K. and Aagaard, K. 1966. On the water exchange through Bering Strait. *Limnol. Oceanogr.*, **11**, 44–59.

Colin de Verdière, A. and Ollitrault, M. 2016. A direct determination of the world ocean barotropic circulation. *J. Phys. Oceanogr.*, **46**, 255–273.

Cox, M. D. 1989. An idealized model of the world ocean. Part I: The global-scale water masses. *J. Phys. Oceanogr.*, **19**, 1730–1752.

Cresswell, G. R. and Golding, T. J. 1980. Observations of a south-flowing current in the southeastern Indian Ocean. *Deep Sea Res., Part A*, **27A**, 449–466.

Cromwell, T., Montgomery, R. B., and Stroup, E. D. 1954. Equatorial undercurrent in Pacific Ocean revealed by new methods. *Science*, **119**, 648–649.

Cunningham, S. A., Alderson, S. G., King, B. A., and Brandon, M. A. 2003. Transport and variability of the Antarctic Circumpolar Current in Drake Passage. *J. Geophys. Res.*, **108**.

Danabasoglu, G. and McWilliams, J. C. 1995. Sensitivity of the global ocean circulation to parameterizations of mesoscale tracer transports. *J. Climate*, **8**, 2967–2987.

D'Asaro, E. A. 1988. Generation of submescale vortices: a new mechanism. *J. Geophys. Res.*, **93**, 6685–6693.

Davis, R. E. 2006. Contributions to global ocean observations. *Physical Oceanography Developments since 1950*. New York, NY: Springer.

de Boyer Montegut, C., Madec, G., Fischer, A. S., Lazar, A., and Iudicone, D. 2004. Mixed layer depth over the global ocean: an examination of profile data and a profile-based climatology. *J. Geophys. Res.*, **109,** C12003.

de Ruijter, W. 1982. Asymptotic analysis of the Agulhas and Brazil Current System. *J. Phys. Oceanogr.*, **12**, 361–373.

de Ruijter, W. P. M., Biastoch, A., Drijfhout, S. S., Lutjeharms, J. R. E., Matano, R. P. et al. 1999. Indian–Atlantic interocean exchange: dynamics, estimation and impact. *J. Geophys. Res.*, **104**, 20885–20910.

Deacon, G. E. R. 1937. The hydrology of the Southern Ocean. *Discovery Rep.*, **15**, 3–122.

Deacon, M. 1997. *Scientists and the Sea, 1650–1900: A Study of Marine Science*. 2nd edn. Routledge.

Dee, D. P., Uppala, S. M., Simmons, A. J., Berrisford, P., Poli, P. et al. 2011. The ERA-Interim reanalysis: Configuration and performance of the data assimilation system. *Q. J. R. Meteorol. Soc.*, **137**, 553–597.

Dengler, M., Schott, F. A., Eden, C., Brandt, P., Fischer, J., and Zantopp, R. J. 2004. Break-up of the Atlantic deep western boundary current into eddies at 8°S. *Nature*, **432**, 1018–1020.

Dickinson, R. E. 1981. Convergence rate and stability of ocean–atmosphere coupling schemes with a zero-dimensional climate model. *J. Atmos. Sci.*, **38**, 2112–2120.

Dijkstra, H. A. and de Ruijter, W. P. M. 2001. On the physics of the Agulhas Current: steady retroflection regimes. *J. Phys. Oceanogr.*, **31**, 2971–2985.

Dijkstra, H. A., 2005. *Nonlinear Physical Oceanography: A Dynamical Systems Approach to the Large Scale Ocean Circulation and El Nino*, 2nd edition, London and New York, NY: Springer.

Donohue, K. A., Tracey, K. L., Watts, D. R., Chidichimo, M. P., and Chereskin, T. K. 2016. Mean Antarctic Circucmpolar Current transport measured in Drake Passage. *Geophys. Res. Lett.*, **43**, 11760–11767.

Döös, K. 1994. Semianalytical simulation of the meridional cells in the Southern Ocean. *J. Phys. Oceanogr.*, **24**, 1281–1293.

Döös, K. and Webb, D. 1994. The Deacon cell and the other meridional cells of the Southern Ocean. *J. Phys. Oceanogr.*, **24**, 429–442.

Ducet, N., Le Traon, P. Y., and Reverdin, G. 2000. Global high-resolution mapping of ocean circulation from TOPEX/Poseidon and ERS-1 and -2. *J. Geophys. Res.*, **105**, 19477–19498.

Dushaw, B. D. and Menemenlis, D. 2014. Antipodal acoustic thermometry: 1960, 2004. *Deep Sea Res., Part I*, **86**, 1–20.

Edson, J. B., Jampana, V., Weller, R. A., Bigorre, S. P., Plueddemann, A. J. et al. 2013. On the exchange of momentum over the open ocean. *J. Phys. Oceanogr.*, **43**(8), 1589–1610.

Elliot, B. A. and Sanford, T. B. 1985. The subthermocline lens D1, I, Description of water properties and velocity profiles. *J. Phys. Oceanogr.*, **15**.

Emery, W. J., and Thompson, R. E. 2001. *Data Analysis: Methods in Physical Oceanography*. 2nd edn. Elsevier Inc.

England, M. H. 1993. Representing the global-scale water masses in ocean general circulation models. *J. Phys. Oceanogr.*, **23**, 1523–1552.

Fairall, C. W., Bradley, E. F., Rogers, D. P., Edson, J. B., and Young, G. S. 1996. Bulk parameterization of air–sea fluxes in TOGA COARE. *J. Geophys. Res.*, **101**, 3747–3767.

Fairall, C. W., Bradley, E. F., Hare, J. E., and Grachev, A. A. 2003. Bulk parameterization of air–sea fluxes: updates and verification for the COARE algorithm. *J. Climate*, **16**, 571–591.

Farneti, R., Delworth, T. L., Rosati, A. J., Griffies, S. M., and Zeng, F. 2010. The role of mesoscale eddies in the rectification of the Southern Ocean response to climate change. *J. Phys. Oceanogr.*, **40**, 1539–1557.

Fetterer, F., Knowles, K., Meier, W., and Savoie, M. 2016, updated daily, accessed 6 April 2017. *Sea Ice Index, Version 2*. Tech. rept. National Snow and Ice Data Center, Boulder, CO.

Fofonoff, N. P. 1954. Steady flow in a frictionless homogeneous ocean. *J. Mar. Res.*, **13**, 254–262.

Fofonoff, N. P. and Montgomery, R. B. 1955. The equatorial undercurrent in the light of the vorticity equation. *Tellus*, **7**, 518–521.

Forget, G., Campin, J. M., Heimbach, P., Hill, C. N., Ponte, R. M., and Wunsch, C. 2015. ECCO version 4: an integrated framework for non-linear inverse modeling and global ocean state estimation. *Geosci. Mod. Dev.*, **8**, 3071–3104.

Foster, T. D. and Carmack, E. C. 1976. Frontal zone mixing and Antarctic bottom water formation in the southern Weddell Sea. *Deep Sea Res.*, **23**, 301–317.

Fowler, C., Emery, W., and Tschudi, M. 2013. *Polar Pathfinder Daily 25 km EASE-Grid Sea Ice Motion Vectors. Version 2.* Tech. rept. National Snow and Ice Data Center, Boulder, CO.

Fowlis, W. W. and Hide, R. 1965. Thermal convection in a rotating annulus of liquid: effect of viscosity on the transition between axisymmetric and non-axisymmetric flow regimes. *J. Atmos. Sci.*, **22**, 541–558.

Freeman, E., Woodruff, S. D., Worley, S. J., Lubker, S. J., Kent, E. C. et al. 2017. ICOADS Release 3.0: a major update to the historical marine climate record. *International Journal of Climatology*, **37**(5), 2211–2232.

Gagnon, J.-S., Lovejoy, S., and Schertzer, D. 2006. Multifractal earth topography. *Nonlinear Proc. Geophys.*, **13**, 541–570.

Ganachaud, A. and Wunsch, C. 2000. Improved estimates of global ocean circulation, heat transport and mixing from hydrographic data. *Nature*, **408**, 453–457.

Ganachaud, A. and Wunsch, C. 2003. Large-scale ocean heat and freshwater transports during the World Ocean Circulation Experiment. *J. Climate*, **16**, 696–705.

Garrett, C. and Munk, W. 1979. Internal waves in the ocean. *Ann. Rev. Fluid Mech.*, **11**, 339–369.

Gent, P. R. 2011. The Gent–McWilliams parameterization: 20/20 hindsight. *Ocean Modelling*, **39**, 2–9.

Gent, P. R. and McWilliams, J. C. 1990. Isopycnal mixing in ocean circulation models. *J. Phys. Oceanogr.*, **20**, 150–155.

Gent, P. R., Willebrand, J., McDougall, T. J., and McWilliams, J. C. 1995. Parameterizing eddy-induced tracer transports in ocean general circulation models. *J. Phys. Oceanogr.*, **25**, 463–474.

Gill, A. E. 1982. *Atmosphere–Ocean Dynamics*. San Diego, CA: Academic Press.

Gill, A. E. and Bryan, K. 1971. Effects of geometry on the circulation of a three-dimensional southern-hemisphere ocean model. *Deep Sea Res.*, **18**, 685–721.

Gille, S. T. 1997. The Southern Ocean momentum balance: Evidence for topographic effects from numerical model output and altimeter data. *J. Phys. Oceanogr.*, **27**, 2219–2232.

Gnanadesikan, A. 1999. A simple predictive model for the structure of the oceanic pycnocline. *Science*, **283**, 2077–2079.

Gnanadesikan, A. and Hallberg, R. 2000. On the relationship of the circumpolar current to southern hemisphere winds in coarse-resolution ocean models. *J. Phys. Oceanogr.*, **30**, 2013–2034.

Gnanadesikan, A., Dixon, K. W., Griffies, S. M. et al. 2006. GFDL's CM2 global coupled climate models. Part II: The baseline ocean simulation. *J. Climate*, **19**(5).

Godfrey, J. S. 1989. A Sverdrup model of the depth-integrated flow for the world ocean allowing for island circulations. *Geophys. Astrophys. Fluid Dyn.*, **45**, 89–112.

Godfrey, J. S. and Masumoto, Y. 1999. Diagnosing the mean strength of the Indonesian Throughflow in an ocean general circulation model. *J. Geophys. Res.*, **104**, 7889–7895.

Gordon, A. L. 1986. Interocean exchange of thermocline water. *J. Geophys. Res.*, **91**, 5037–5046.

Gordon, A. L. and Haxby, W. F. 1990. Agulhas eddies invade the South Atlantic: evidence from Geosat altimeter and shipboard conductivity-temperature-depth survey. *J. Geophys. Res.*, **95**, 3117–3125.

Gordon, A. L., Sprintall, J., Van Aken, H. M., Susanto, D., Wijffels, S. et al. 2010. The Indonesian throughflow during 2004–2006 as observed by the INSTANT program. *Dyn. Atmos. Oceans*, **50**, 115–128.

Gray, A. R. and Riser, S. C. 2014. A global analysis of Sverdrup balance using absolute geostrophic velocities from Argo. *J. Phys. Oceanogr.*, **44**(4), 1213–1229.

Green, J. S. A. 1970. Transfer properties of the large-scale eddies in the general circulation of the atmosphere. *Q. J. R. Meteorol. Soc.*, **96**, 157–185.

Griffies, S. M. 2004. *Fundamentals of Ocean Climate Models*. Princeton, NJ and Oxford: Princeton University Press.

Grist, J. P. and Josey, S. A. 2003. Inverse analysis adjustment of the SOC air-sea flux climatology using ocean heat transport constraints. *J. Climate*, **16**, 3274–3295.

Haberman, R. 1987. *Elementary Applied Partial Differential Equations*. 2nd edn. Englewood Cliffs, NJ: Prentice-Hall, Inc.

Haine, T., Böning, C., Brandt, P., Fischer, J., Funk, A. et al. 2008. North Atlantic Deep Water formation in the Labrador Sea, recirculation through the subpolar gyre, and discharge to the subtropics. In Dickson, R. R., Meincke, J., and Rhines, P. (eds.), *Arctic-Subarctic Ocean Fluxes: Defining the Role of the Northern Seas in Climate*. Springer-Verlag.

Haine, T. W. N. 2008. What did the Viking discoverers of America know of the North Atlantic environment? *Weather*, **63**, 60–65.

Haine, T. W. N. 2012. Greenland Norse knowledge of the North Atlantic environment. In Hudson, B. (ed.), *Studies in the Medieval Atlantic*. Palgrave-MacMillan.

Haine, T. W. N. and Cherian, D. A. 2013. Analogies of ocean/atmosphere rotating fluid dynamics with gyroscopes—teaching opportunities. *Bull. Amer. Meteor. Soc.*, **94**, 673–684.

Haine, T. W. N. and Fuller, A. 2016. Boundary β-plumes and their vorticity budgets. *Q. J. R. Meteorol. Soc.*, **142**, 2758–2767.

Haine, T. W. N. and Hall, T. M. 2002. A generalized transport theory: Water-mass composition and age. *J. Phys. Oceanogr.*, **32**, 1932–1946.

Haine, T. W. N. and Marshall, J. C. 1998. Gravitational, symmetric and baroclinic instability of the ocean mixed layer. *J. Phys. Oceanogr.*, **28**, 634–658.

Haine, T. W. N., Watson, A. J., Liddicoat, M. I., and Dickson, R. R. 1998. The flow of Antarctic bottom water in the southwest Indian ocean estimated using CFCs. *J. Geophys. Res.*, **103**, 27637–27653.

Haine, T. W. N., Curry, B., Gerdes, R., Hansen, E., Karcher, M. et al. 2015. Arctic freshwater export: Status, mechanisms, and prospects. *Glob. Planet. Change*, **125**, 13–35.

Halkin, D. and Rossby, T. 1985. The structure and transport of the Gulf Stream at 73°W. *J. Phys. Oceanogr.*, **15**, 1439–1452.

Hall, M. M. and Bryden, H. L. 1982. Direct estimates and mechanisms of ocean heat transport. *Deep Sea Res.*, **29**, 339–359.

Hallberg, R. and Gnanadesikan, A. 2006. The role of eddies in determining the structure and response of the wind-driven southern hemisphere overturning: Results from the Modeling Eddies in the Southern Ocean (MESO) project. *J. Phys. Oceanogr.*, **36**, 2232–2252.

Halpern, D. and Weisberg, R. H. 1989. Upper ocean thermal and flow fields at 0°, 28° W (Atlantic) and 0°, 140° W (Pacific) during 1983–1985. *Deep Sea Res.*, **36**, 407–418.

Han, Y.-J. 1984. A numerical world ocean general circulation model Part II. A baroclinic experiment. *Dyn. Atmos. Oceans*, **8**, 141–172.

Haney, R. 1971. Surface thermal boundary condition for ocean circulation models. *J. Phys. Oceanogr.*, **1**, 241–248.

Hansen, B. and Østerhus, S. 2000. North Atlantic-Nordic Seas exchange. *Prog. Oceanogr.*, **45**, 109–208.

Hautala, S. L., Roemmich, D. H., and W. J. Schmitz, Jr. 1990. Is the North Pacific in Sverdrup balance along 24°N? *J. Geophys. Res.*, **99**, 16041–16052.

Hazeleger, W., de Vries, P., and van Oldenborgh, G. J. 2001. Do tropical cells ventilate the Indo-Pacific equatorial thermocline? *Geophys. Res. Lett.*, **28**, 1763–1766.

Hellermann, S., and Rosenstein, M. 1983. Normal monthly wind stress over the world ocean with error estimates. *J. Phys. Oceanogr.*, **13**, 1093–1104.

Hemming, S. R. 2004. Heinrich Events: massive late Pleistocene detritus layers of the N Atl and their Global Climate imprint, Rev Geophys, 42. *Rev. Geophys.*, **42**.

Hickey, B. M. 1979. The California current system–hypotheses and facts. *Prog. Oceanogr.*, **8**, 191–279.

Hill, A. E., Hickey, B. M., Shillington, F. A., Strub, P. T., Brink, K. H. et al. 1998. Eastern ocean boundaries. In Robinson, A. R. and Brink, K. H. (eds.), *The Sea*, vol. 11. New York, NY: John Wiley & Sons.

Hirst, A. C. and McDougall, T. J. 1998. Meridional overturning and dianeutral transport in a z-coordinate ocean model including eddy-induced advection. *J. Phys. Oceanogr.*, **28**, 1205–1223.

Hirst, A. C., Jackett, D. R., and McDougall, T. J. 1996. The meridional overturning cells of a world ocean model in neutral density coordinates. *J. Phys. Oceanogr.*, **26**, 775–791.

Holland, W. R., Keffer, T., and Rhines, P. B. 1984. Dynamics of the oceanic general circulation: the potential vorticity field. *Nature*, **308**, 698–705.

Holton, J. R. and Hakim, G. J. 2013. *An Introduction to Dynamic Meteorology*. 5th edn. San Diego, CA: Academic Press.

Houghton, J. T. 1986. *The Physics of Atmospheres*. 2nd edn. Cambridge: Cambridge University Press.

Huang, R. X. 1990. On the three-dimensional structure of the wind-driven circulation in the North Atlantic. *Dyn. Atmos. Oceans*, **15**, 117–159.

Huang, R.-X. 1998. Mixing and available potential energy in a Boussinesq ocean. *J. Phys. Oceanogr.*, **28**, 669–678.

Huang, R. X. 1999. Mixing and energetics of the oceanic thermohaline circulation. *J. Phys. Oceanogr.*, **29**, 727–746.

Huang, R. X. and Flierl, G. R. 1987. Two-layer models for the thermocline and current structure in subtropical/subpolar gyres. *J. Phys. Oceanogr.*, **17**, 872–884.

Huang, R. X. and Qiu, B. 1994. Three-dimensional structure of the wind-driven circulation in the subtropical North Pacific. *J. Phys. Oceanogr.*, **24**, 1608–1622.

Huang, R. X. and Russell, S. 1998. Ventilation of the subtropical North Pacific. *J. Phys. Oceanogr.*, **24**, 2589–2605.

Huffman, G. J., Adler, R. F., Bolvin, D. T., and Gu, G. 2009. Improving the global precipitation record: GPCP Version 2.1. *Geophys. Res. Lett.*, **36**(L17808).

Hughes, G. O. and Griffiths, R. W. 2008. Horizontal convection. *ARFM*, **40**, 185–208.

Hughes, T. M. C. and Weaver, A. J. 1994. Multiple Equilibria of an Asymmetric Two-Basin Ocean Model. *J. Phys. Oceanogr.*, **24**, 619–637.

IOC, SCOR, and IAPSO. 2010. *The International Thermodynamic Equation of Seawater – 2010: Calculation and Use of Thermodynamic Properties*. Tech. rept. Intergovernmental Oceanographic Commission, Manuals and Guides No. 56, UNESCO (English).

Iorio, D. D. and Yüce, H. 1999. Observations of Mediterranean flow into the Black Sea. *J. Geophys. Res.*, **104**, 3091–3108.

Iselin, C. O. 1939. The influence of vertical and lateral turbulence on the characteristics of the waters at mid-depths. *Eos, Trans., AGU*, **20**, 414–417.

Jansen, E., Overpeck, J., Briffa, K. R., Duplessy, J.-C., Joos, F. et al. (eds.), *Climate Change 2007: The Physical Science Basis. Contribution of Working Group I to the Fourth Assessment Report of the Intergovernmental Panel on Climate Change*. Cambridge and New York, NY: Cambridge University Press.

Jayne, S. R. and Marotzke, J. 2002. The oceanic eddy heat transport. *J. Phys. Oceanogr.*, **32**, 3328–3345.

Jayne, S. R., Hogg, N. G., Waterman, S. N., Rainville, L., Donohue, K. A. et al. 2009. The Kuroshio extension and its recirculation gyres. *Deep Sea Res., Part I*, **56**, 2088–2099.

Jenkins, A. D., and Bye, J. A.T. 2006. Some aspects of the work of V. W. Ekman. *Polar Record*, **42**(01), 15.

Jochem, M. and Murtugudde, R. 2006. *Physical Oceanography Developments since 1950*. New York, NY: Springer.

Johns, W. E., Shay, T. J., Bane, J. M., and Watts, D. R. 1995. Gulf Stream structure, transport, and recirculation near 68 °W. *J. Geophys. Res.*, **100**, 817–838.

Johns, W. E., Baringer, M. O., Beal, L. M., Cunningham, S. A., Kanzow, T. et al. 2011. Continuous, array-based estimates of Atlantic ocean heat transport at 26.5°N. *J. Climate*, **24**, 2429–2449.

Johns, W. E., Brandt, P., Bourles, B., Tantet, A., Papostolou, A., and Houk, A. 2014. Zonal structure and seasonal variability of the Atlantic Equatorial Undercurrent. *Clim. Dyn.*, **43**, 3047–3069.

Johnson, E. S. and Luther, D. S. 1994. Mean zonal momentum balance in the upper and central equatorial Pacific Ocean. *J. Geophys. Res.*, **99**, 7689–7704.

Johnson, G. C. 2008. Quantifying Antarctic bottom water and North Atlantic deep water volumes. *J. Geophys. Res.*, **113**.

Johnson, G. C. and Bryden, H. L. 1989. On the size of the Antarctic circumpolar current. *Deep Sea Res.*, **36**, 39–53.

Johnson, G. C., Sloyan, B. M., Kessler, W. S., and McTaggart, K. E. 2002. Direct measurements of upper ocean currents and water properties across the tropical Pacific during the 1990s. *Prog. Oceanogr.*, **52**, 31–61.

Jones, J. E. and Jones, I. S. F. 2002. *The Western Boundary Current in the Pacific: The Development of Our Oceanographic Knowlege.* Washington, DC: University of Washington Press.

Josey, S. A., Kent, E. C., and Taylor, P. K. 2002. Wind stress forcing of the ocean in the SOC Climatology: Comparisons with the NCEP-NCAR, ECMWF, UWM/COADS, and Hellerman and Rosenstein Datasets. *J. Phys. Oceanogr.*, **32**, 1993–2019.

Joyce, T. M. 1984. Velocity and hydrographic structure of a Gulf Stream warm-core ring. *J. Phys. Oceanogr.*, **14**, 936–947.

Kagimoto, T. and Yamagata, T. 1997. Seasonal transport variations of the Kuroshio: an OGCM simulation. *J. Phys. Oceanogr.*, **27**, 403–418.

Kalnay, E., Kanamitsu, M., Kistler, R., Collins, W., Deaven, D. et al. 1996. The NCEP-NCAR 40-year reanalysis project. *Bull. Amer. Meteor. Soc.*, **77**, 437–471.

Kamenkovich, I. V. and Goodman, P. J. 2000. The dependence of AABW transport in the Atlantic on vertical diffusivity. *Geophys. Res. Lett.*, **27**(22), 3739–3742.

Kanzow, T., Cunningham, S. A., Johns, W. E., Hirschi, J. J.-M., Marotzke, J. et al. 2010. Seasonal variability of the Atlantic Meridional Overturning Circulation at 26.5° N. *J. Climate*, **23**, 5678–5698.

Karsten, R., Jones, H., and Marshall, J. 2002. The role of eddy transfer in setting the stratification and transport of a circumpolar current. *J. Phys. Oceanogr.*, **32**, 39–54.

Kato, H. and Phillips, O. M. 1969. On the penetration of a turbulent layer into a stratified fluid. *J. Fluid Mech.*, **37**, 643–655.

Kawai, Y. and Wada, A. 2007. Diurnal sea surface temperature variation and its impact on the atmosphere and ocean: a review. *J. Ocean.*, **63**, 721–744.

Kelley Jr., J. E. 1991. Juan Ponce de Leon's discovery of Florida: Herrera's narrative revisited. *Revista de Historia de América*, **111**, 31–65.

Kent, E. and National Center for Atmospheric Research Staff. 2016. *The Climate Data Guide: Surface Flux and Meteorological Dataset: National Oceanography Centre (NOC) V2.0.* Tech. rept. National Oceanography Center. Retrieved from `https://climatedataguide.ucar.edu/climate-data/surface-flux-and-meteorological-dataset-national-oceanography-centre-noc-v20`.

Kessler, W. S., Johnson, G. C., and Moore, D. W. 2003. Sverdrup and nonlinear dynamics of the Pacific equatorial currents. *J. Phys. Oceanogr.*, **33**, 994–1008.

Killworth, P. 1983. Deep convection in the world ocean. *Rev. Geophys.*

Klinger, B. A. and Cruz, C. 2009. Decadal response of global circulation to Southern Ocean wind stress perturbation. *J. Phys. Oceanogr.*, **39**, 1888–1904.

Klinger, B. A. and Marotzke, J. 1999. Behavior of double-hemisphere thermohaline flows in a single basin. *J. Phys. Oceanogr.*, **29**, 382–399.

Klinger, B. A. and Marotzke, J. 2000. Meridional heat transport by the subtropical cell. *J. Phys. Oceanogr.*, **30**, 696–705.

Klinger, B. A., Drijfhout, S., Marotzke, J., and Scott, J. 2003. Sensitivity of basinwide meridional overturning to diapycnal diffusion and remote wind forcing in an idealized Atlantic-Southern Ocean geometry. *J. Phys. Oceanogr.*, **33**, 249–266.

Klinger, B. A., Drijfhout, S., Marotzke, J., and Scott, J. 2004. Remote wind-driven overturning in the absence of the Drake passage effect. *J. Phys. Oceanogr.*, **34**, 1036–1049.

Körtzinger, A., Schimanski, J., Send, U., and Wallace, D. 2004. The ocean takes a deep breath. *Science*, **306 (5700)**, 1337.

Koszalka, I. M., Haine, T. W. N., and Magaldi, M. G. 2013. Fates and travel times of Denmark Strait Overflow Water in the Irminger Basin. *J. Phys. Oceanogr.*, **43** (12), 2611–2628.

Kraus, E. B. and Turner, J. S. 1967. A one-dimensional model of the seasonal thermocline. II. The general theory and its consequences. *Tellus*, **19**, 98–105.

Kuhlbrodt, T., Smith, R. S., Wang, Z., and Gregory, J. M. 2012. The influence of eddy parameterizations on the transport of the Antarctic Circumpolar Current in coupled climate models. *Ocean Modelling*, **52–53**(Aug), 1–8.

Kundu, P. K., Cohen, I. M., and Cowling, D. R. 2012. *Fluid Mechanics*. 5th edn. Elsevier Inc.

LaCasce, J. H., Ferrari, R., Marshall, J., Tulloch, R., Balwada, D., and Speer, K. 2014. Float-Derived Isopycnal Diffusivities in the DIMES Experiment. *J. Phys. Oceanogr.*, **44**(2), 764–780.

Landsteiner, M. C., McPhaden, M. J., and Picault, J. 1990. Sensitivity of Sverdrup transport estimates to the specification of wind stress forcing in the tropical Pacific. *J. Geophys. Res.*, **95**, 1681–1691.

Laurindo, L. C., Mariano, A. J., and Lumpkin, R. 2017. An improved near-surface velocity climatology for the global ocean from drifter observations. *Deep Sea Res., Part I*, **124**(jun), 73–92.

Lazar, A., Inui, T., Malanotte-Rizzoli, P., Busalacchi, A. J., Wang, L., and Murtugudde, R. 2002. Seasonality of the ventilation of the tropical Atlantic thermocline in an ocean general circulation model,. *J. Geophys. Res.*, **107**.

Leetmaa, A., Niiler, P., and Stommel, H. 1977. Does the Sverdrup relation account for the mid-Atlantic circulation? *J. Mar. Res.*, **35**, 1–10.

Legeckis, R. 1977. Long waves in the eastern equatorial Pacific Ocean: a view from a geostationary satellite. *Science*, **197**, 1179–1181.

Lenn, Y.-D. and Chereskin, T. K. 2009. Observations of Ekman currents in the Southern Ocean. *J. Phys. Oceanogr.*, **39**, 768–779.

Lentz, S. J. 1992. The surface boundary layer in coastal upwelling regions. *J. Phys. Oceanogr.*, **22**, 1517–1539.

Levitus, S. and Oort, A. H. 1977. Global analysis of oceanographic data. *Bull. Amer. Meteor. Soc.*, **58**, 1270–1274.

Lisiecki, L. E. and Raymo, M. E. 2005. A Pliocene-Pleistocene stack of 57 globally distributed benthic $\delta^{18}O$ records. *Paleoceanography*, **20**(1).

Liu, Z. and Philander, S. G. H. 1995. Patterns affect the tropical-subtropical circulation of the upper ocean. *J. Phys. Oceanogr.*, **25**, 449–462.

Liu, Z., Pedlosky, J., Marshall, D., and Warncke, T. 1993. On the feedback of the Rhines-Young pool on the ventilated thermocline. *J. Phys. Oceanogr.*, **23**, 1592–1596.

Locarnini, R. A., Mishonov, A. V., Antonov, J. I., Boyer, T. P., Garcia, H. E. et al. 2013. *World Ocean Atlas 2013, Volume 1: Temperature*. Tech. rept. 73. NESDIS.

Lorenz, E. N. 1955. Available potential energy and the maintenance of the general circulation. *Tellus*, **7**(2), 157–167.

Lu, P. and McCreary, J. P. 1995. Influence of the ITCZ on the flow of thermocline water from the subtropical to the equatorial Pacific Ocean. *J. Phys. Oceanogr.*, **25**, 3076–3088.

Lu, P., McCreary, J. P., and Klinger, B. A. 1998. Meridional circulation cells and the source waters of the Pacific Equatorial Undercurrent. *J. Phys. Oceanogr.*, **28**, 62–84.

Lu, Y. and Stammer, D. 2004. Vorticity balance in coarse-resolution global ocean simulations. *J. Phys. Oceanogr.*, **34**, 605–622.

Lumpkin, R. and Garraffo, Z. 2005. Evaluating the decomposition of tropical Atlantic drifter observations. *J. Atmos. Oc. Tech.*, **22**, 1403–1415.

Lumpkin, R. and Johnson, G. C. 2013. Global ocean surface velocities from drifters: Mean, variance, El Niño-Southern Oscillation response, and seasonal cycle. *J. Geophys. Res.*, **118**, 2992–3006.

Lumpkin, R. and Pazos, M. 2006. Measuring surface currents with Surface Velocity Program drifters: The instrument, its data, and some recent results. In Griffa, A., Kirwan, A. D., Mariano, A. J., Ozgokmen, T., and Rossby, T. (eds), *Lagrangian Analysis and Prediction of Coastal and Ocean Dynamics (LAPCOD)*. Cambridge and New York, NY: Cambridge University Press.

Lumpkin, R. and Speer, K. 2007. Global ocean meridional overturning. *J. Phys. Oceanogr.*, **37**, 2550–2562.

Luyten, J. R., Pedlosky, J., and Stommel, H. 1983. The ventilated thermocline. *J. Phys. Oceanogr.*, **13**, 292–309.

Lynch-Stieglitz, J. 2017. The Atlantic meridional overturning circulation and abrupt climate change. *Ann. Rev. Mar. Sci.*, **9**, 83–104.

Macdonald, A. M. 1998. The global ocean circulation: a hydrographic estimate and regional analysis. *Prog. Oceanogr.*, **41**, 281–382.

MacKinnon, J. A., Alford, M. H., Ansong, J. K., Arbic, B. K., Barna, A. et al. 2017. Climate Process Team on internal-wave driven ocean mixing. *Bull. Amer. Meteor. Soc.*

Manabe, S. and Stouffer, R. J. 1988. Two stable equilibria of a coupled ocean-atmosphere model. *J. Climate*, **1**, 841–866.

Manley, T. O. and Hunkins, K. 1985. Mesoscale eddies of the Arctic Ocean. *J. Geophys. Res.*, **90, C3**, 4911–4930.

Mann, K. H. and Lazier, J. R. N. 2005. *Dynamics of Marine Ecosystems*. Wiley-Blackwell.

Manucharyan, G. E., and Spall, M. A. 2015. Wind-driven freshwater buildup and release in the Beaufort Gyre constrained by mesoscale eddies. *Geophys. Res. Lett.*, **42**.

Manucharyan, G. E., Spall, M. A., and Thompson, A. F. 2016. A theory of the wind-driven Beaufort Gyre. *J. Phys. Oceanogr.*

Marotzke, J. and Willebrand, J. 1991. Multiple equilibria of the global thermohaline circulation. *J. Phys. Oceanogr.*, **21**, 1372–1385.

Marshall, D. P., Ambaum, M. H. P., Maddison, J. R., Munday, D. R., and Novak, L. 2016. Eddy saturation and frictional control of the Antarctic Circumpolar Current. *Geophys. Res. Lett.*, **44**.

Marshall, J. C. 1986. Wind driven ocean circulation theory–steady free flow. In: Willebrand, J., and Anderson, D. L. T. (eds), *Large-scale transport processes in Oceans and Atmospheres*. Springer., 225–245.

Marshall, J. and Plumb, R. A. 2008. *Atmosphere, Ocean, and Climate Dynamics: An Introductory Text*. Elsevier Inc.

Marshall, J. and Radko, T. 2003. Residual-mean solutions for the Antarctic Circumpolar Current and its associated overturning circulation. *J. Phys. Oceanogr.*, **33**(11), 2341–2354.

Marshall, J. C., Nurser, A. J. G., and Williams, R. G. 1993. Inferring the subduction rate and period over the North Atlantic. *J. Phys. Oceanogr.*, **23**, 1315–1329.

Mata, M. M., Tomczak, M., Wijffels, S., and Church, J. A. 2000. East Australian Current volume transports at 30°S: Estimates from the World Ocean Circulation Experiment hydrographic sections PR11/P6 and the PCM3 current meter array. *J. Geophys. Res.*, **105**, 28509–28526.

Matano, R. P. 1996. A numerical study of the Agulhas Retroflection: the role of bottom topography. *J. Phys. Oceanogr.*, **26**, 2267–2279.

Mauritzen, C. 1996. Production of dense overflow waters feeding the North Atlantic across the Greenland–Scotland Ridge. Part I: Evidence for a revised circulation scheme. *Deep Sea Res.*, **43**, 769–806.

Maximenko, N. A., Bang, B., and Sasaki, H. 2005. Observational evidence of alternating zonal jets in the world ocean. *Geophys. Res. Lett.*, **32**(12).

McCreary, J. P. and Lu, P. 1994. Interaction between the subtropical and equatorial ocean circulations: the subtropical cell. *J. Phys. Oceanogr.*, **24**, 466–497.

McCreary., J. P., Shetye, S. R., and Kundu, P. K. 1986. Thermohaline forcing of eastern boundary currents: with application to the circulation off the west coast of Australia. *J. Mar. Res.*, **44**, 71–92.

McDermott, D. A. 1996. The regulation of northern overturning by southern hemisphere winds. *J. Phys. Oceanogr.*, **26**, 1234–1255.

McDougall, T. J. 1991. Parameterizing mixing in inverse model. In Müller, P. and Henderson, D. (eds.), *Dynamics of Oceanic Internal Gravity Waves*. Proceedings of the Sixth 'Aha Huliko'a Hawaiian Winter Workshop. Hawaii Institute of Geophysics.

McDougall, T. J. and McIntosh, P. C. 2001. The temporal-residual-mean velocity. Part II: Isopycnal interpretation and the tracer and momentum equations. *J. Phys. Oceanogr.*, **31**, 1222–1246.

McDowell, S. E. and Rossby, H. T. 1978. Mediterranean Water: An intense mesoscale eddy off the Bahamas. *Science*, **202**, 1085–1087.

McManus, J. F., Francois, R., Gherardi, J.-M., Keigwin, L. D., and Brown-Leger, S. 2004. Collapse and rapid resumption of Atlantic meridional circulation linked to deglacial climate changes. *Nature*, **428**, 834–837.

McPhaden, M. J. 1986. The equatorial undercurrent: 100 years of discovery. *Eos, Transactions American Geophysical Union*, **67**(40), 762–765.

McWilliams, J. C. 1985. Submesoscale, coherent vortices in the ocean. *Rev. Geophys.*, **23**, 165–182.

McWilliams, J. C. 2006. *Fundamentals of Geophysical Fluid Dynamics*. 1st edn. Cambridge and New York, NY: Cambridge University Press.

McWilliams, J. C., Holland, W. B., and Chow, J. H. S. 1978. A description of numerical Antarctic Circumpolar Currents. *Dyn. Atmos. Oceans*, **2**, 213–291.

MEDOC Group. 1970. Observations of formation of deep water in the Mediterranean Sea, 1969. *Nature*, **227**, 1037–1040.

Meredith, M. P., Naveira Garabato, A. C., Hogg, A. McC., and Farneti, R. 2012. Sensitivity of the overturning circulation in the Southern Ocean to decadal changes in wind forcing. *J. Climate*, **25**, 99–110.

Miller, R. N. 2007. *Numerical Modeling of Ocean Circulation*. 1st edn. Cambridge and New York, NY: Cambridge University Press.

Mills, E. L. 2011. *The Fluid Envelope of our Planet*. University of Toronto Press.

Montgomery, R. B. 1940. The present evidence on the importance of lateral mixing processes in the ocean. *Bull. Amer. Meteor. Soc.*, **21**, 87–94.

Morrison, A. K. and Hogg, A. McC. 2013. On the relationship between Southern Ocean overturning and ACC transport. *J. Phys. Oceanogr.*, **43**, 140–148.

Müller, T. J., Ikeda, Y., Zangenberg, R., and Nonato, L. V. 1998. Direct measurements of western boundary currents off Brazil between 20° S and 28° S. *J. Geophys. Res.*, **103**, 5429–5437.

Munday, D. R., Johnson, H. L., and Marshall, D. P. 2013. Eddy saturation of equilibrated circumpolar currents. *J. Phys. Oceanogr.*, **43**, 507–532.

Munk, W. and Wunsch, C. 1998. Abyssal recipes II: energetics of tidal and wind mixing. *Deep Sea Res., Part I*, **45**, 1977–2010.

Munk, W., Worcester, P., and Wunsch, C. 1995. *Ocean Acoustic Tomography*. Cambridge and New York, NY: Cambridge University Press.

Munk, W. H. 1950. On the wind-driven ocean circulation. *J. Meteorolog.*, **7**, 79–93.

Munk, W. H. 1966. Abyssal recipes. *Deep Sea Res.*, **13**, 707–730.

Munk, W. H. and Palmen, E. 1951. Note on the dynamics of the Antarctic Circumpolar Current. *Tellus*, **3 (1)**, 53–55.

Naveira Garabato, A. C., Williams, A. P., and Bacon, S. 2014. The three-dimensional overturning circulation of the Southern Ocean during the WOCE era. *Prog. Oceanogr.*, **120**, 41–78.

Nelson, G. and Hutchings, L. 1983. The Benguela upwelling area. *Prog. Oceanogr.*, **12**, 333–356.

Nicholls, K. W., Østerhus, S., Makinson, K., Gammelsrød, T., and Fahrbach, E. 2009. Ice-ocean processes over the continental shelf of the southern Weddell Sea, Antarctica: A review. *Rev. Geophys.*, **47**(RG3003).

NOAA. 2015. *State of the Climate: Global Climate Report for May 2015*. Tech. rept. NOAA National Centers for Environmental Information.

Notz, D. and Marotzke, J. 2012. Observations reveal external driver for Arctic sea-ice retreat. *Geophys. Res. Lett.*, **39**.

Olbers, D., Borowski, D., Völker, C., and Wölff, J.-O. 2004. The dynamical balance, transport and circulation of the Antarctic Circumpolar Current. *Antarctic Science*, **16**(4), 439–470.

Orsi, A. H., Whitworth III, T., and Nowlin Jr., Worth D. 1995. On the meridional extent and fronts of the Antarctic Circumpolar Current. *Deep Sea Res., Part I*, **42**(5), 641–673.

Pacanowski, R. C., and Griffies, S. M. 1999. *The MOM3 Manual*. Tech. rept. Princeton, NJ: Geophys. Fluid Dyn. Lab. Ocean Group.

Pachauri, R. K., Allen, M. R., Barros, V. R., Broome, J., Cramer, W. et al. 2014. Climate Change 2014: Synthesis Report. Contribution of Working Groups I, II and III to the Fifth Assessment Report of the Intergovernmental Panel on Climate Change. Geneva: IPCC.

Park, G. Y. and Bryan, K. 2000. Comparison of thermally driven circulations from a depth-coordinate model and an isopycnal-layer model, Part I: Scaling-law sensitivity to vertical diffusivity. *J. Phys. Oceanogr.*, **30**, 590–605.

Pawlowicz, R. 2010. *What Every Oceanographer Needs to Know about TEOS-10 (The TEOS-10 Primer)*.

Pedlosky, J. 1987a. *Geophysical Fluid Dynamics*. 2nd edn. Springer-Verlag.

Pedlosky, J. 1987b. An inertial theory of the equatorial undercurrent. *J. Phys. Oceanogr.*, **17**, 1978–1985.

Pedlosky, J. 1996. *Ocean Circulation Theory*. Springer-Verlag.

Pedlosky, J. 2006. A history of thermocline theory. In *Physical Oceanography Developments since 1950*. New York, NY: Springer.

Pedlosky, J. and Young, W. R. 1983. Ventilation, potential vorticity homogenization and the structure of the ocean circulation. *J. Phys. Oceanogr.*, **13**, 2020–2037.

Pedlosky, J., Iaocono, R., Napolitano, E., and Helfrich, K. 2009. The skirted island: The effect of topography on the flow around planetary scale islands. *J. Mar. Res.*, **67**, 435–478.

Peixoto, J. P. and Oort, A. H. 1992. *Physics of Climate*. Springer-Verlag.

Perovich, D. K., Elder, B. C., and Ricter-Menge, J. A. 1997. Observations of the annual cycle of sea ice temperature and mass balance. *Geophys. Res. Lett.*, **24**(5), 555–558.

Petit, J. R. 1999. Climate and atmospheric history of the past 420,000 years from the Vostok ice core, Antarctica. *Nature*, **399**, 429–436.

Pickart, R. S. and Smethie, W. M. 1993. How does the Deep Western Boundary Current cross the Gulf Stream? *J. Phys. Oceanogr.*, **23**, 2602–2616.

Pickart, R. S., Spall, M. A., Ribergaard, M. H., Moore, G. W. K., and Milliff, R. F. 2003. Deep convection in the Irminger Sea forced by the Greenland tip jet. *Nature*, **424**, 152–156.

Pierce, S. D., Smith, R. L., Kosro, P.M., Barth, J.A., and Wilson, C. D. 2000. Continuity of the poleward undercurrent along the eastern boundary of the mid-latitude north Pacific. *Deep Sea Res., Part II*, **47**, 811–829.

Plumb, R. A. and Mahlman, J. D. 1987. The zonally averaged transport characteristics of the GFDL general circulation-transport model. *J. Atmos. Sci.*, **44**, 298–327.

Pollard, J. and Reid, H. 2007. *The Rise and Fall of Alexandria: Birthplace of the Modern World*. Penguin Books.

Polzin, K. L., Toole, J. M., Ledwell, J. R., and Schmitt, R. W. 1997. Spatial variability of turbulent mixing in the abyssal ocean. *Science*, **276**, 93–96.

Pratt, L. J. and Lundberg, P. A. 1991. Hydraulics of rotating strait and sill flow. *Ann. Rev. Fluid Mech.*, **23**, 81–106.

Pratt, L. J. and Pedlosky, J. 1998. Barotropic circulation around islands with friction. *J. Phys. Oceanogr.*, **28**, 2148–2162.

Pratt, L. J. and Whitehead, J.A. 2008. *Rotating Hydraulics: Nonlinear Topographic Effects in the Ocean and Atmosphere*. Springer-Verlag.

Price, J. F. and Baringer, M. O'Neil. 1994. Outflows and deep water production by marginal seas. *Prog. Oceanogr.*, **33**, 161–200.

Qiu, B. and Chen, S. 2005. Eddy-induced heat transport in the subtropical North Pacific from Argo, TMI, and altimetry measurements. *J. Phys. Oceanogr.*, **35**, 458–473.

Qiu, B. and Huang, R. X. 1995. Ventilation of the North Atlantic and North Pacific: Subduction versus obduction. *J. Phys. Oceanogr.*, **25**, 2374–2390.

Qu, T. and Lindstrom, E. J. 2002. A climatological interpretation of the circulation in the western South Pacific. *J. Phys. Oceanogr.*, **32**, 2492–2508.

Qu, T. and Lukas, R. 2003. The bifurcation of the North Equatorial Current in the Pacific. *J. Phys. Oceanogr.*, **33**, 5–18.

Rahmstorf, S. and England, M. 1997. Influence of southern hemisphere winds on North Atlantic deep water flow. *J. Phys. Oceanogr.*, **27**, 2040–2054.

Rahmstorf, S. and Willebrand, J. 1995. The role of temperature feedback in stabilizing the thermohaline circulation. *J. Phys. Oceanogr.*, **25**, 787–805.

Randall, D. A., Wood, R. A., Bony, S., Colman, R., Fichefet, T. et al. In Solomon, S., Qin, D., Manning, M., Chen, Z., Marquis, M. et al. (eds.), *Climate Change 2007: The Physical Science Basis. Contribution of Working Group I to the Fourth Assessment Report of the Intergovernmental Panel on Climate Change*. Cambridge and New York, NY: Cambridge University Press.

Redi, M. H. 1982. Oceanic isopycnal mixing by coordinate rotation. *J. Phys. Oceanogr.*, **12**, 1154–1158.

Reid, J. L. 1959. Evidence of a south equatorial countercurrent in the Pacific Ocean. *Nature*, **184**, 209–210.

Reynolds, R. W., Smith, T. M., Liu, C., Chelton, D. B., Casey, K. S., and Schlax, M. G. 2007. Daily high-resolution-blended analyses for sea surface temperature. *J. Climate*, **20** (22), 5473–5496.

Rhein, M., Rintoul, S. R., Aoki, S., Campos, E., Chambers, D. et al. 2013. Observations: Ocean. In Stocker, T. F., Qin, D., Plattner, G.-K., Tignor, M., Allen, S. K. et al. (eds.), *Climate Change 2013: The Physical Science Basis. Contribution of Working Group I to the Fifth Assessment Report of the Intergovernmental Panel on Climate Change*. Cambridge and New York, NY: Cambridge University Press.

Rhines, P. B. 1975. Waves and turbulence on the β-plane. *J. Fluid Mech.*, **69**, 417–443.

Rhines, P. B. 1977. The dynamics of unsteady currents. In Goldberg, E. (ed.), *The Sea*, vol. 6. New York, NY: John Wiley & Sons.

Rhines, P. B. 1982. *Basic Dynamics of the Large-Scale Geostrophic Circulation*. Tech. rept. Woods Hole Oceanographic Institution.

Rhines, P. B. and Young, W. R. 1982. Homogenization of potential vorticity in planetary gyres. *J. Fluid Mech.*, **122**, 347–367.

Richards, K. J., Xie, S.-P., and Miyama, T. 2009. Vertical mixing in the ocean and its impact on the coupled ocean-atmosphere system in the eastern tropical Pacific. *J. Climate*, **22**, 3703–3719.

Richardson, P. L. 1980a. Benjamin Franklin and Timothy Folger's first printed chart of the Gulf Stream. *Science*, **207**, 643–645.

Richardson, P. L. 1980b. Gulf Stream ring trajectories. *J. Phys. Oceanogr.*, **10**, 90–104.

Richardson, P. L. 1981. Gulf Stream trajectories measured with free-drifting buoys. *J. Phys. Oceanogr.*, **11**, 999–1010.

Richardson, P. L. 1985. Average velocity and transport of the Gulf Stream near 55W. *J. Mar. Res.*, **43**, 83–111.

Ridgway, K. R. and Dunn, J. R. 2007. Observational evidence for a Southern Hemisphere oceanic supergyre. *Geophys. Res. Lett.*, **34**.

Ridgway, K. R. and Godfrey, J. S. 1994. Mass and heat budgets in the East Australian Current: A direct approach. *J. Geophys. Res.*, **99**, 3231–3248.

Ridgway, K. R., Dunn, J. R., and Wilkin, J. L. 2002. Ocean interpolation by four-dimensional weighted least squares—Application to the waters around Australasia. *J. Atmos. Oc. Tech.*, **19**(9), 1357–1375.

Risien, C. M. and Chelton, D. B. 2008. A global climatology of surface wind and wind stress fields from eight years of QuikSCAT scatterometer data. *J. Phys. Oceanogr.*, **38**, 2379–2413.

Roe, G. H. and Baker, M. B. 2007. Why is climate sensitivity so unpredictable? *Science*, **318**, 629–632.

Roemmich, D. and Gilson, J. 2001. Eddy transport of heat and thermocline waters in the North Pacific: A key to interannual/decadal climate variability? *J. Phys. Oceanogr.*, **31**, 675–687.

Roemmich, D., Hautala, S., and Rudnick, D. 1996. Northward abyssal transport through the Samoan passage and adjacent regions. *J. Geophys. Res.*, **101**, 14039–14055.

Ronski, S. and Budéus, G. 2010. Physical oceanography during POLARSTERN cruise ARK-XVIII/1a. Supplement to: Ronski, S; Budéus, G (2005): Time series of winter convection in the Greenland Sea. *Journal of Geophysical Research-Oceans*, **110**, C04015, doi:10.1029/2004JC002318.

Rooth, C. 1982. Hydrology and ocean circulation. *Prog. Oceanogr.*, **11**, 131–149.

Rossby, H. T. 1965. On thermal convection driven by nonuniform heating from below: an experimental study. *Deep Sea Res.*, **12**, 9–16.

Rossby, H. T. and Miller, P. 2008. Ocean eddies in the 1539 Carta Marina by Olaus Magnus. *Oceanography*, **16**, 77–88.

Rothman, T. 2017. The forgotten mystery of inertia. *American Scientist*, **105**(6), 344.

Rowe, G. D., Firing, E., and Johnson, G. C. 2000. Pacific equatorial subsurface countercurrent velocity, transport, and potential vorticity. *J. Phys. Oceanogr.*, **30**, 1172–1187.

Rudels, B. 2010. Constraints on exchanges in the Arctic Mediterranean—do they exist and can they be of use? *Tellus*, **62A**, 109–122.

Rudels, B. 2015. Arctic Ocean circulation, processes and water masses: A description of observations and ideas with focus on the period prior to the International Polar Year 2007–2009. *Prog. Oceanogr.*, **139**, 22–67.

Rudels, B. 2016. Arctic Ocean stability: The effects of local cooling, oceanic heat transport, freshwater input, and sea ice melt with special emphasis on the Nansen basin. *J. Geophys. Res.*, **121**.

Rudels, B., Eriksson, P., Grönvall, H., Hietala, R., and Launiainen, J. 1999a. Hydrographic observations in Denmark Strait in Fall 1997, and their implications for the entrainment into the overflow plume. *Geophys. Res. Lett.*, **26**, 1325–1328.

Rudels, B., Friedrich, H. J., Hainbucher, D., and Lohmann, G. 1999b. On the parameterisation of oceanic sensible heat loss to the atmosphere and to ice in an ice-covered mixed layer in winter. *Deep Sea Res., Part II*, **46**, 1385–1425.

Saenko, O. A., Schmittner, A., and Weaver, A. J. 2004. The Atlantic–Pacific seasaw. *J. Climate*, **17**, 2033–2038.

Salmon, R. 1990. The thermocline as an "internal boundary layer." *J. Mar. Res.*, **48**, 437-469.

Samelson, R. M. and Vallis, G. K. 1997. Large-scale circulation with small diapycnal diffusion: The two-thermocline limit. *J. Mar. Res.*, **55**, 223–275.

Sarmiento, J. L. and Bryan, K. 1982. An ocean transport model for the North Atlantic. *J. Geophys. Res.*, **87**, 394–408.

Sarmiento, J. L. and Gruber, N. 2006. *Ocean Biogeochemical Dynamics*. Princeton, NJ and Woodstock: Princeton University Press.

Saunders, P. M., Coward, A. C., and de Cuevas, B. A. 1999. Circulation of the Pacific Ocean seen in a global ocean model: Ocean Circulation and Climate Advanced Modelling project (OCCAM). *J. Geophys. Res.*, **104**, 18281–18299.

Schanze, J. J., Schmitt, R. W., and Yu, L. L. 2010. The global oceanic freshwater cycle: a state-of-the-art quantification. *J. Mar. Res.*, **68**, 569–595.

Schmitt, R. W. 1994. Double diffusion in oceanography. *Ann. Rev. Fluid Mech.*, **26**, 255–285.

Schmitz, W. J. 1995. On the interbasin-scale thermohaline circulation. *Rev. Geophys.*, **33**, 151–173.

Schopf, P. S. 1983. On equatorial waves and El Niño. II: Effects of air–sea thermal coupling. *J. Phys. Oceanogr.*, **13**, 1878–1893.

Schott, F. 1983. Monsoon response of the Somali Current and associated upwelling. *Prog. Oceanogr.*, **12**, 357–381.

Schott F. and Leaman K. D. 1991. Observations with moored acoustic doppler current profilers in the convection regime in the Golfe du Lion. *J. Phys. Oceanogr.*, **21**, 558–574.

Schott, F. and McCreary Jr., J. P. 2001. The monsoon circulation of the Indian Ocean. *Prog. Oceanogr.*, **51**, 1–123.

Schott, F., Visbeck, M., and Fischer, J. 1993. Observations of vertical currents and convection in the central Greenland Sea during the winter. *J. Geophys. Res.*, **98**, 14401–14421.

Schott, F. A., Zantopp, R., Stramma, L., Dengler, M., Fischer, J., and Wibaux, M. 2004. Circulation and deep-water export at the western exit of the subpolar North Atlantic. *J. Phys. Oceanogr.*, **34**, 817–843.

Schott, F. A., Dengler, M., Zantopp, R., Stramma, L., Fischer, J., and Brandt, P. 2005. The shallow and deep western boundary circulation of the South Atlantic, at 5°–11°S. *J. Phys. Oceanogr.*, **35**, 2031–2053.

Seidov, D., Antonov, J. I., Arzayus, K. M., Baranova, O. K., Biddle, M., Boyer, T. P., Johnson, D. R., Mishonov, A. V., Paver, C., and Zweng, M. M. 2014. Oceanography North of 60°N from World Ocean Database. *Prog. Oceanogr.*, **132**, 153–173.

Sen Gupta, A. and England, M. H. 2004. Evaluation of interior circulation in a high-resolution global ocean model. Part I: deep and bottom waters. *J. Phys. Oceanogr.*, **34**, 2592–2614.

Sloyan, B. M. and Rintoul, S. R. 2001. The Southern Ocean limb of the global deep overturning circulation. *J. Phys. Oceanogr.*, **31**, 143–173.

Smethie, W. M., Fine, R. A., Putzka, A., and Jones, E. P. 2000. Tracing the flow of North Atlantic Deep Water using chlorofluorocarbons. *J. Geophys. Res.*, **105**, 14297–14323.

Smith, R. D., Maltrud, M. E., Bryan, F. O., and Hecht, M. W. 2000. Numerical simulation of the North Atlantic at 1/10°. *J. Phys. Oceanogr.*, **30**, 1532–1561.

Sokolov, S. and Rintoul, S. R. 2009. Circumpolar structure and distribution of the Antarctic Circumpolar Current fronts: 1. Mean circumpolar paths. *J. Geophys. Res.*, **114**(C11). C11018.

Spall, M. 2003. On the thermohaline circulation in flat bottom marginal seas. *J. Mar. Res.*, **61**, 1–25.

Spall, M. A. 2004. Boundary currents and watermass transformation in marginal seas. *J. Phys. Oceanogr.*, **34**, 1197–1213.

Spall, M. A. and Schneider, N. 2016. Coupled ocean–atmosphere offshore decay scale of cold SST signals along upwelling eastern Boundaries. *J. Climate*, **29**(23), 8317–8331.

St. Laurent, L., Naveira Garabato, A. C., Ledwell, J. R., Thurnherr, A. M., Toole, J. M., and Watson, A. J. 2012. Turbulence and diapycnal mixing in Drake Passage. *J. Phys. Oceanogr.*, **42**(12), 2143–2152.

Stammer, D. 1998. On eddy characteristics, eddy transports, and mean flow properties,. *J. Phys. Oceanogr.*, **28**, 727–739.

Stewart, K. D. and Haine, T. W. N. 2016. Thermobaricity in the transition zones between alpha and beta oceans. *J. Phys. Oceanogr.*, **46**(6), 1805–1821.

Stocker, T. F. 1998. The Seesaw Effect. *Science*, **282**(5386), 61–62.

Stommel, H. 1948. The westward intensification of wind-driven currents. *Trans. Am. Geophys. Union*, **29**, 202–206.

Stommel, H. 1957. A survey of ocean current theory. *Deep Sea Res.*, **4**, 149–184.

Stommel, H. 1960. Wind drift near the equator. *Deep Sea Res.* **6**, 298–302.

Stommel, H. 1961. Thermohaline convection with two stable regimes. *Tellus*, **13**, 224–230.

Stommel, H. 1979. Determination of water mass properties of water pumped down from the Ekman layer to the geostrophic flow below. *Proc. Nat. Acad. Sci.*, **76**, 3051–3055.

Stommel, H. 1989. The Slocum Mission. *Oceanography*, **2**, 22–25.

Stommel, H., Arons, A. B., and Faller, A. J. 1958. Some examples of stationary planetary flow patterns in bounded basins. *Tellus*, **10**, 179–187.

Stone, P. H. 1972. A simplified radiative-dynamical model for the static stability of rotating atmospheres. *J. Atmos. Sci.*, **29**(3), 405–418.

Stone, P. H. 1974. The meridional variation of the eddy heat fluxes by baroclinic waves and their parameterization. *J. Atmos. Sci.*, **31**(2), 444–456.

Straub, D. N. 1993. On the transport and angular momentum balance of channel models of the Antarctic Circumpolar Current. *J. Phys. Oceanogr.*, **23**, 776–782.

Strub, P. T., Mesias, J. M., Montecino-B., V., Rutllant-C., J., and Salinas, S. 1998. Coastal ocean circulation off western South America. In Robinson, A. R. and Brink, K. H. (eds.), *The Sea*, vol. 11. New York, NY: John Wiley & Sons.

Sverdrup, H. U. 1947. Wind-driven currents in a baroclinic ocean; with application to the equatorial currents of the eastern Pacific. *Proc. Nat. Acad. Sci.*, **33**(11), 318–326.

Sverdrup, H. U., Johnson, M. W., and Fleming, R. H. 1942. *The Oceans: Their Physics, Chemistry and General Biology*. Englewood Cliffs, NJ: Prentice-Hall, Inc.

Swallow, J. C. and Worthington, L. V. 1957. Measurements of deep currents in the western North Atlantic. *Nat.*, **179**, 1183–1184.

Talley, L. D. 2003. Shallow, intermediate, and deep overturning components of the global heat budget. *J. Phys. Oceanogr.*, **33**, 530–560.

Talley, L. D. 1985. Ventilation of the subtropical North Pacific: the shallow salinity minimum. J. Phys. Oceanogr., **15**, 633–649.

Talley, L. D. 2008. Freshwater transport estimates and the global overturning circulation: shallow, deep, and throughflow components. *Prog. Oceanogr.*, **78**, 257–303.

Talley, L. D. 2013. Closure of the global overturning circulation through the Indian, Pacific, and Southern Oceans: Schematics and transports. *Oceanography*, **26**(1), 80–97.

Talley, L. D., Reid, J. L., and Robbins, P. E. 2003. Data-based meridional overturning streamfunctions for the global ocean. *J. Phys. Oceanogr.*, **16**, 3213–3226.

Talley, L. D., Pickard, G. L., Emery, W. J., and Swift, J. H. 2011. *Descriptive Physical Oceanography: An Introduction*. 6th edn. San Diego, CA: Academic Press.

Tansley, C. E. and Marshall, D. P. 2001. On the dynamics of wind-driven circumpolar currents. *J. Phys. Oceanogr.*, **31**, 3258–3273.

The MODE Group. 1978. The mid-ocean dynamics experiment. *Deep Sea Res.*, **25**, 859–910.

Thompson, S. R. 1993. Estimation of the transport of heat in the Southern Ocean using a fine-resolution numerical model. *J. Phys. Oceanogr.*, **23**, 2493–2497.

Thorpe, S. A. 2005. *The Turbulent Ocean*. Cambridge and New York, NY: Cambridge University Press.

Thual, O. and McWilliams, J. C. 1992. The catastrophe structure of thermohaline convection in a two-dimensional fluid model and a comparison with low-order box models. *Geophys. Astrophys. Fluid Dyn.*, **64**(1–4), 67–95.

Toggweiler, J. R. and Samuels, B. 1995. Effect of Drake Passage on the global thermohaline circulation. *Deep Sea Res., Part I*, **42**, 477–500.

Tomczak, M. and Godfrey, J. S. 2003. *Regional Oceanography: An Introduction*. 2nd edn. Delhi: Daya Publishing House.

Treguier, A. M., Theetten, S., Chassignet, E., Penduff, T., Smith, R. et al. 2005. The North Atlantic subpolar gyre in four high resolution models. *J. Phys. Oceanogr.*, **35**, 757–774.

Trenberth, K. E. and Caron, J. M. 2001. Estimates of meridional atmosphere and ocean heat transports. *J. Climate*, **14**, 3433–3443.

Tschudi, M., Fowler, C., Maslanik, J., Stewart, J. S., and Meier, W. 2016. *Polar Pathfinder Daily 25 km EASE-Grid Sea Ice Motion Vectors. Version 3*. Tech. rept. National Snow and Ice Data Center, Boulder, Colorado USA.

Tsubouchi, T., Bacon, S., Naveira Garabato, A. C., Aksenov, Y., Laxon, S. W. et al. 2012. The Arctic Ocean in summer: A quasi-synoptic inverse estimate of boundary fluxes and water mass transformation. *J. Geophys. Res.*, **117**.

Tsuchiya, M. 1975. Subsurface countercurrents in the eastern equatorial Pacific Ocean. *J. Mar. Res.*, **33** (Suppl.), S145–S175.

Tsujino, H. and Suginohara, N. 1999. Thermohaline circulation enhanced by wind forcing. *J. Phys. Oceanogr.*, **29**, 1506–1516.

Turner, J. S. 1973. *Buoyancy Effects in Fluids*. Cambridge and New York, NY: Cambridge University Press.

Uehara, H., Suga, T., Hanawa, K., and Shikama, N. 2003. A role of eddies in formation and transport of North Pacific subtropical mode water. *Geophys. Res. Lett.*, **30**.

Valdivieso, M., Haines, K., Balmaseda, M., Chang, Y. S., Drevillon, M. et al. 2017. An assessment of air–sea heat fluxes from ocean and coupled reanalyses. *Clim. Dyn.*, **49**, 983–1008.

Vallis, G. K. 2000. Large-scale circulation and production of stratification: Effects of wind, geometry, and diffusion. *J. Phys. Oceanogr.*, **30**, 933–954.

Vallis, G. K. 2006. *Atmospheric and Oceanic Fluid Dynamics: Fundamentals and Large-Scale Circulation*. 1st edn. Cambridge and New York, NY: Cambridge University Press.

Vaughan, D. G., Comiso, J. C., Allison, I., Carrasco, J., Kaser, G. et al. 2013. Observations: Cryosphere. In Stocker, T. F., Qin, D., Plattner, G.-K., Tignor, M., Allen, S. K. et al. (eds.), *Climate Change 2013: The Physical Science Basis. Contribution of Working Group I to the Fifth Assessment Report of the Intergovernmental Panel on Climate Change*. Cambridge and New York, NY: Cambridge University Press.

Veronis, G. 1966. Wind-driven ocean circulation–Part 2. Numerical solutions of the non-linear problem. *Deep Sea Res.*, **13**, 30–55.

Veronis, G. 1975. *The Role of Models in Tracer Studies*. National Academy of Science.

Veronis, G. 1981. *A theoretical Model of Henry Stommel*. The MIT Press.

Visbeck, M., Marshall, J., Haine, T., and Spall, M. 1997. Specification of eddy transfer coefficients in coarse-resolution ocean circulation models. *J. Phys. Oceanogr.*, **27**, 381–402.

Volkov, D. L., Lee, T., and Fu, L.-L. 2008. Eddy-induced meridional heat transport in the ocean. *Geophys. Res. Lett.*, **35**.

Waite, A. M., Thompson, P. A., Pesant, S., Feng, M., Beckley, L. E. et al. 2007. The Leeuwin Current and its eddies: an introductory overview. *Deep Sea Res., Part II*, **54**, 789–796.

Wajsowicz, R. C. 1993. The circulation of the depth-integrated flow around an island with application to the Indonesian Throughflow. *J. Phys. Oceanogr.*, **23**, 1470–1484.

Wajsowicz, R. C. and Gill, A. E. 1986. Adjustment of the ocean under buoyancy forces. Part I: The role of Kelvin waves. *J. Phys. Oceanogr.*, **16**, 2097–2114.

Ward, M. L. and Hogg, A. McC. 2011. Establishment of momentum balance by form stress in a wind-driven channel. *Ocean Modelling*, **40**, 133–146.

Warren, B. A. 1973. Transpacific hydrographic sections at Lats. 43° S and 28° S: the SCORPIO Expedition–II Deep Water. *Deep Sea Res.*, **20**, 9–38.

Warren, B. A. 1981. Deep circulation of the world ocean. In Warren, B. A. and Wunsch, C. (eds.), *Evolution of Physical Oceanography*. MIT Press.

Warren, B. A. 1983. Why is no deep water formed in the North Pacific? *J. Mar. Res.*, **41**, 327–347.

Weaver, A. J., Marotzke, J., Cummins, P. F., and Sarachik, E. S. 1993. Stability and variability of the thermohaline circulation. *J. Phys. Oceanogr.*, **23**, 39–60.

Weaver, A. J., Saenko, O., Clark, P., and Mitrovica, J. 2003. Meltwater Pulse 1A from Antarctica as a Trigger of the Bolling-Allerod Warm Interval. *Science*, **299**, 1709–1733.

Welander, P. 1971. The thermocline problem. *Phil. Trans. R. Soc. Lond. A*, **270**, 415–421.

Whitehead, J. A. 1998. Topographic control of oceanic flows in deep passages and straits. *Rev. Geophys.*, **36**, 423–440.

Whitworth III, T., Nowlin Jr., W. D., and Worley, S. J. 1982. The net transport of the Antarctic Circumpolar Current through Drake Passage. *J. Phys. Oceanogr.*, **12**(9), 960–971.

Whitworth III, T., Warren, B. A., Nowlin Jr., W. D., Rutz, S. B., Pillsbury, R. D., and Moore, M. I. 1999. On the deep western-boundary current in the Southwest Pacific Basin. *Prog. Oceanogr.*, **43**, 1–54.

Williams, R. 1989. The influence of air–sea interaction on the ventilated thermocline. *J. Phys. Oceanogr.*, **19**, 1255–1267.

Williams, R. G. 1991. The role of the mixed layer in setting the potential vorticity of the main thermocline. *J. Phys. Oceanogr.*, **21**(12), 1803–1814.

Williams, R. G., Spall, M. A. and Marshall, J. C. 1995. Does Stommel's mixed layer "demon" work? *J. Phys. Oceanogr.*, **22**, 3089–3102.

Winton, M. 1996. The role of horizontal boundaries in parameter sensitivity and decadal-scale variability of coarse-resolution ocean general circulation models. *J. Phys. Oceanogr.*, **26**, 289–304.

Wolfe, C.L. and Cessi, P. 2009. Overturning circulation in an eddy-resolving model: the effect of the pole-to-pole temperature gradient. *J. Phys. Oceanogr.*, **39**, 125–142.

Woodgate, R. A., Weingartner, T. J., and Lindsay, R. 2012. Observed increases in Bering Strait oceanic fluxes from the Pacific to the Arctic from 2001 to 2011 and their impacts on the Arctic Ocean water column. *Geophys. Res. Lett.*, **39**.

Woodruff, S. D., Slutz, R. J., Jenne, R. L., and Steurer, P. M. 1987. A Comprehensive Ocean–Atmosphere Data Set. *Bull. Amer. Meteor. Soc.*, **68**, 1239–1250.

Woods, J. D. 1968. Wave-induced shear instability in the summer thermocline. *J. Fluid Mech.*, **32**(4), 791–800.

Woods, J. D. and Barkmann, W. 1986. A Lagrangian mixed layer model of Atlantic 18°C water formation. *Nature*, **319**, 574–576.

Wootton, D. 2015. *The Invention of Science, a New History of the Scientific Revolution.* HarperCollins.

Wunsch, C. 2002. What is the thermohaline circulation? *Science*, **298**, 1179–1180.

Wunsch, C. 2005. The total meridional heat flux and its oceanic and atmospheric partition. *J. Climate*, **18**, 4374–4380.

Wunsch, C. 2006. *Discrete Inverse and State Estimation Problems.* 1st edn. Cambridge and New York, NY: Cambridge University Press.

Wunsch, C. 2011. The decadal mean ocean circulation and Sverdrup balance. *J. Mar. Res.*, **69**, 417–434.

Wunsch, C. 2015. *Modern Observational Physical Oceanography: Understanding the Global Ocean.* 1 edn. Princeton, NJ and Woodstock: Princeton University Press.

Wunsch, C. and Ferrari, R. 2004. Vertical mixing, energy, and the general circulation of the oceans. *Ann. Rev. Fluid Mech.*, **36**, 281–314.

Wunsch, C. and Heimbach, P. 2007. Practical global oceanic state estimation. *Physica D*, **230**, 197–208.

Wunsch, C. and Roemmich, D. 1985. Is the North Atlantic in Sverdrup balance? *J. Phys. Oceanogr.*, **15**, 1876–1880.

Yang, J., Proshutinsky, A., and Lin, X. 2015. Dynamics of an idealized Beaufort Gyre: 1. The effect of a small beta and lack of western boundaries. *J. Geophys. Res.*, **121**, 1249–1261.

Yashayaev, I. 2007. Hydrographic changes in the Labrador Sea, 1960–2005. *Prog. Oceanogr.*, **73**, 242–276.

Young, W. R., and Rhines, P. B. 1982. A theory of the wind-driven circulation II: Gyres with western boundary layers. *J. Mar. Res.*, **40**, 849–872.

Yu, L. and Weller, R. A. 2007. Objectively analyzed air–sea heat fluxes for the global ice-free oceans (1981-2005). *Bull. Amer. Meteor. Soc.*, **88**, 527–539.

Zangenberg, N. and Siedler, G. 1998. Path of the North Atlantic Deep Water in the Brazil Basin. *J. Geophys. Res.*, **103**, 5419–5428.

Zhao, M., Timmermans, M.-L., Cole, S., Krishfield, R., Proshutinsky, A., and Toole, J. 2014. Characterizing the eddy field in the Arctic Ocean halocline. *J. Geophys. Res.*, **119**.

Ziegenbein, J. 1970. Spatial observations of short internal waves in the Strait of Gibraltar. *Deep Sea Res.*, **17**, 867–875.

Zika, J. D., Sommer, J. Le, Dufour, C. O., Molines, J.-M., Barnier, B. et al. 2013. Vertical eddy fluxes in the Southern Ocean. *J. Phys. Oceanogr.*, **43**(5), 941–955.

Zweng, M. M., Reagan, J. R., Antonov, J. I., Locarnini, R. A., Mishonov, A. V. et al. 2013. *World Ocean Atlas 2013, Volume 2: Salinity.* Tech. rept. 74. NESDIS.

Zwillinger, D. 1989. *Handbook of Differential Equations.* 2nd edn. San Diego, CA: Academic Press.

Index

Bold font indicates the primary reference for each term.

β-effect, 67, **68**, 179, 301, 302, 317, 348, 352, 363
  topographic, *see* topographic, β-effect
β-plane, **68**, 111, 113, 123, 124, 219, 220, 312, 313
f-plane, **68**, 96, 123, 303, 312, 364
in-situ sensor, 4, 8, 10

absolute
  salinity, **6**, 6, **31**, 31, 40, 153
  temperature, 145
  vorticity, **55**, 74, 104, **431**
abyss, **xv**, 8, **14**, 14, **17**, 22, 23, 159, 165, 167, 168,
  170, 181, 205, 226, 246, 247, 276, 282, 289, 299,
  302, 355, 356, 361
abyssal plain, 17, 18, 20
acceleration potential, 161
acoustic-doppler current profiler (ADCP), **7**, 8, 10, 11,
  193, 281, 285
acoustics, 4, 6–8, 10, 11, 24, 34
active tracer, *see* dynamically, active
adiabatic, **5**, 29, 31, **37**, 37, 74, **167**, 167, 168, 189,
  265, 266, 273, 311, 340, 374, 399
advection, **32**, 33, 37, 58, 71, 105, 115, 117, 118, 177,
  178, 204, 205, 207, 221, 252, 255, 256, 259, 266,
  267, 308–310, 412
advective velocity, **257**, 267
advective-diffusive balance, **297**, 311, 319, 344
Agulhas
  current, **85**, 87, 102, 241, 323, 388
  leakage, 323
  retroflection, **87**, 102, 103, 240, 323
Alaska stream, 87
Alaskan coastal current, **354**
albedo, 145, 383, 388, 401, 402
alpha ocean, **15**, 155, 187, 188, 356
altimeter, 4, 9, 11, 273, 421
angular
  momentum, 55, 56
  velocity, *see* velocity, angular
Antarctic
  bottom water (AABW), **278**, 281–284, 302,
  322–324, 326, 332, 395, 396

circumpolar current (ACC), **xvi**, **83**, 84, 159, 160,
  240–242, 244, 260, 274, **281**, 320, 323–326,
  336–338, 340–346, 349, 350, 363–365, 373,
  381
  intermediate water (AAIW), **278**, 328
  peninsula, **83**, 132, 281
anticyclonic, **51**, **82**, 82, 83, 92, 158, 165, 240, 242,
  273, 336, 352, 353, 403
Arctic
  circumpolar boundary current, **355**
  mediterranean sea, **359**
  ocean, 75, 132, 187, 243, 244, 287, 348, 352,
  354–361, 363, 380
argo, **11**, 13, 88, 98, 143, 153, 316, 349
aspect ratio, **18**, 42, 207, **432**
Atlantic
  meridional overturning circulation (AMOC), 388,
  392, 396, 398, 399, 402, 407, 414, 416
  ocean, *see* north, Atlantic, ocean and south,
  Atlantic ocean
  water, 75, **355**, 355–357, 361–363, 365–370, 372,
  381
atmospheric GCM, 28
autonomous vehicle, 10
available potential energy, **249**, 250, 251, 254, 255,
  273, 275, 296, 307

Baffin Island current, **354**
Barents
  sea, **287**, 352, 353, 355, 356, 358, 363
  sea opening, 354, 357, 359
baroclinic, **47**, 249, 257, 373
  instability, **249**, 250, 254, 257, 258, 267, 274, 364,
  400
  zone, **268**, 268, 347, 379
barotropic, **47**, 53, 168
  instability, **250**
  streamfunction, 47, **53**
  vorticity equation, 104, 107, 109
basin mode, **23**
bathymetry, 19, 20, 78, 91, 98, 268, 282, 286, 288,
  289, 305, 321, 353, 355, 373, 420, 422